Nuclear Magnetic Resonance Spectroscopy

Volume II

Author

Professor Pál Sohár, Ph.D., D.Sc.

Head
Spectroscopic Department
EGYT Pharmacochemical Works
Professor
Department of Organic Chemistry
Eötvös Lóránd University of Sciences
Budapest, Hungary

CRC Press, Inc.
Boca Raton, Florida

CHEMISTRY

Library of Congress Cataloging in Publication Data

Sohár, Pál.
  Nuclear magnetic resonance spectroscopy.

  Bibliography: p.
  Includes index.
  1. Nuclear magnetic resonance spectroscopy.

I. Title.
QC762.S575   1983   543'.0877   82-9524
ISBN 0-8493-5632-6 (v. 1)   AACR2
ISBN 0-8493-5633-4 (v. 2)
ISBN 0-8493-5634-2 (v. 3)

Direct all inquiries to CRC Press, Inc., 2000 Corporate Blvd., N.W., Boca Raton, Florida, 33431.

© 1983 by CRC Press, Inc.

International Standard Book Number 0-8493-5632-6 (v. 1)
International Standard Book Number 0-8493-5633-4 (v. 2)
International Standard Book Number 0-8493-5634-2 (v. 3)

Library of Congress Card Number 82-9524
Printed in the United States

PREFACE

These books are an updated version of a Hungarian handbook the author published in 1976 with the same title. Since it was the first original NMR handbook in the Hungarian language, I endeavored to embrace, even if briefly, all topics of NMR spectroscopy. When the manuscript was finished, at the beginning of the 1970s, FT spectroscopy was only beginning to be studied, and thus the investigation of nuclei other than protons had much less significance than now. Therefore, many parts of the books discuss NMR phenomena primarily from the aspect of proton resonance.

With the very rapid spread of Fourier transform technique and FT instruments, the investigation of other nuclei, first of all carbon, has become feasible. In addition to chemical shifts, coupling constants and intensities are now sources of NMR information and relaxation times have become one of the starting points of structure elucidation. This progress made it necessary to deal more closely with certain theoretical and technical principles as well as with some relationships between spectrum and chemical structure less important from the aspect of proton resonance.

Therefore, some of the problems are touched in more than one chapter. The resulting, somewhat looser structure of the books has, however, the advantage that the special principles and applications pertaining to the same nucleus may be discussed in a more concentrated and unique manner.

The encyclopedic character of the book has claimed a relatively great volume and a necessarily brief, schematic discussion of some details. But perhaps this distinguishes my book from other handbooks and justifies its publication. The books are meant primarily for chemists, undergraduates, and young spectroscopists who wish to get broad but not too detailed information on all important topics of NMR spectroscopy. I hope that scientists working in related fields, e.g., analysts, biochemists, or physicists will also find useful information in the books. At any rate, the books were written keeping in mind the demands of organic chemists who wish to use NMR spectroscopy as a tool in structure elucidation, but who also want to know, at least schematically, the theoretical foundations of the method applied.

The first chapter is a concise but comprehensive discussion, primarily from the aspect of proton resonance, of NMR theory, which assumes only minimum quantum chemical knowledge, and the mathematical apparatus is limited to the level accessible for average chemists. The main purpose of this chapter is to point out how the magnetic nuclei of molecules may be identified with the individual spin systems, and how can one determine the spectrum parameters given the key to structure elucidation, i.e., chemical shifts and coupling constants, from the measured spectrum. As the quantum theory of NMR spectroscopy is a settled, classical branch of science, and as I endeavored to review the most important principles only, the references in this chapter extend only to original publications of fundamental importance, completely neglecting more recent literature pertaining to special details.

Chapter 2 deals with the operation principles and technical problems of traditional (CW) and modern (FT) spectrometers, and also discusses special techniques of measurement. Of the latter, the measurement and use of various double resonance spectra, temperature-dependent measurements important in the investigation of dynamic phenomena, and the measurement of integrated intensities permitting quantitative analysis have the greatest weight. The theoretical foundations of Fourier transform spectroscopy, the experimental determination of relaxation times, the theory of relaxation, and relaxation mechanisms are also discussed here. In connection with relaxation times, a separate section is devoted to the cancer diagnostic application of NMR and tomography.

Chapter 3 reviews the proton resonance parameters of the most important groups of organic compounds. The huge and ever-increasing amount of data collected in this field prohibits any attempt toward completeness, and thus again I restricted myself to classical results, selecting the illustration material from the most characteristic or the first examples. Therefore, more recent literature is represented in the chapter only by some reviews or extremely interesting papers. An exception is the last section of this chapter, which deals with the medium effects, predominantly with the shift reagent technique. This part discusses the exchange phenomena and hydrogen bonds, as well as the NMR investigation of optical purity and radical reactions (CIDNP).

Chapter 4 discusses, on the basis of similar principles, the most characteristic properties of the NMR spectra of carbon nuclei, in much smaller volume those of nitrogen, and very schematically only the properties of ^{17}O, ^{19}F, and ^{31}P nuclei, with hints to the special aspects, pertaining to the given nuclei, of some theoretical problems, principles, and measurement technique.

Chapter 5 is a collection of problems. All problems are concerned with elucidation of chemical structures taken from the practice of the author. Most of the problems were real, occurring in practice. The 66 problems connected with ^1H NMR spectra and the 12 problems connected with carbon NMR, together with the roughly 200 spectra provide, hopefully, a good training for the chemist in his own spectrum evaluation practice.

I feel it my duty to express my gratitude to all who contributed to the publication of this book.

My first thanks should go to my students, since the subject of the book was collected first for my lectures. Inspired by their interest and questions, many ideas arose for working out certain topics and selecting the material for illustration.

To my chemist colleagues, whose names are listed in the name index, thanks are due not only for the samples synthesized partly by them and given for recording the spectra of the Problems section, but also for the preparation of model derivatives, or purification of compounds in course of structure elucidation, and also for the assistance provided by chemical information about the compounds studied.

For constructive discussions and suggestions on the completion or modification of certain parts, I am indebted to Professors G. Snatzke (Bochum), and H. Wamhoff (Bonn) and Doctors G. Jalsovszky (Budapest), I. Kövesdi (Budapest), and T. Széll (New York).

For the promotion of the English edition my thanks are due to Professor B. Csakvári and I. Kovács, as well as to Publishing House of the Hungarian Academy of Sciences; I wish to express my thanks to Dr. G. Jalsovszky for the tedious, competent, and unrewarding work of translation.

I am indebted to publishers Academic Press, Elsevier, Heyden and Son Ltd., Pergamon Press and Wiley-Interscience, personally to Professors E. D. Becker and T. C. Farrar (Washington), A. F. Casy (Alberta) and H. Wamhoff (Bonn), Doctors H. Schneiders (Bruker, Karlsruhe), D. Shaw (Oxford), C. S. Springer (New York), F. W. Wehrli (Bruker, USA), T. Wirthlin (Varian, Zug) and to firms Bruker Physik AG, JEOL, Sadtler Research Laboratories Inc., and Varian Associates for placing at my disposal illustration material or, respectively, for permitting my use of this material.

My thanks are due to the managers of my firm, EGYT Pharmacochemical Works, personally to Professor L. Pallos, Scientific Director, for ensuring the conditions for compiling this book.

It gives me great pleasure to acknowledge all my previous and present collaborators their enthusiasm, careful and competent work in all phases of the preparation of this book. I want to mention them by name: late Mrs. Dr. Zs. Méhesfalvy, Mrs. A. Csokán, Mrs. J. Csákvári, Mr. A. Fürjes, Mrs. Á. Kiss-Tamás, Mrs. M. Leszták, Mr. Gy. Mányai, Mrs. É. Mogyorósy, Mrs. Dr. K. Ósapay, Mr. I. Pelczer, Miss V. Windbrechtinger, and Mr. J. Zimonyi.

I release these books in the hope that it will benefit all those colleagues who wish to be acquainted with this singularly many faceted, still rapidly developing, in all fields of chemistry extremely efficient and again and again reviving branch of science: NMR spectroscopy.

Pál Sohár
Budapest, April 1981

THE AUTHOR

Pál Sohár, Ph.D., D.Sc. is the Head of the Spectroscopic Department of EGYT Pharmacochemical Works and Professor of the Eötvös Loránd University of Sciences, Budapest, Hungary. He graduated in 1959 at the Technical University of Budapest and obtained his Ph.D. degree in 1962 in physical chemistry with "Summa cum Laude" qualification.

Professor Sohár received the "Candidate of Sciences" degree in physical chemistry from the Hungarian Academy of Sciences in 1967 on the basis of his Thesis "Investigation of Association Structures by Infrared Spectroscopy" and the D.Sc. degree in 1973 for his research work in the application of IR and NMR spectroscopy in the structure elucidation of organic molecules. He has served as Professor at the Eötvös Loránd University since 1975. He was the Head of the Spectroscopy Department of the Institute of Drug Research. He began his career in this Institute in 1959 and received promotions to scientific assistant in 1962, to senior assistant in 1969, and to scientific counselor from 1974 to 1980. He assumed his present position in 1980.

Professor Sohár is a member of the Committees of Physical and Inorganic Chemistry, Spectroscopy and Theoretical Organic Chemistry of the Hungarian Academy of Sciences and the member of ISMAR (International Society of Magnetic Resonance, Chicago, Illinois). He is the Secretary of the Committee on Molecular and Material Structure of the Hungarian Academy of Sciences.

Professor Sohár is the author of more than 180 scientific papers and has been the author or coauthor of six books, among them the first monographs in Hungarian on infrared spectroscopy and nuclear magnetic resonance. He is the coeditor of the series *Absorption Spectra in the Infrared Region* published by Akadémiai Kiadó, Budapest (Publishing House of the Hungarian Academy of Sciences). He has given more than 150 scientific presentations or invited lectures, and was several times invited lecturer of postgraduate courses at the Technical Universtiy of Budapest.

His current major research interests include structure elucidation of organic compounds by nuclear magnetic resonance and infrared spectroscopy.

TABLE OF CONTENTS

Volume I

Chapter 1
Theory of Nuclear Magnetic Resonance Spectroscopy

Introduction 1

1.1.	The Magnetic Resonance Spectrum 3	
1.1.1.	Magnetic Properties of Atoms and Molecules. Magnetic Moment and Susceptibility 3	
1.1.2.	The Quantized Nature of the Angular and Magnetic Moments of Atomic Nuclei 5	
1.1.3.	The Motion of Nuclei in an External Magnetic Field. The Larmor Precession 7	
1.1.4.	Magnetic Levels and the Energy of Transitions Between Them 8	
1.1.5.	Interpretation of Transitions Between Magnetic Energy Levels............. 10	
1.1.6.	Population of the Magnetic Energy Levels 11	
1.1.7.	Macroscopic Magnetization of Spin Systems and the Spin Temperature..... 13	
1.1.8.	Spin-Lattice Relaxation... 14	
1.1.9.	Spin-Spin Relaxation 15	
1.1.10.	Saturation 17	
1.1.11.	Natural Line Width of Nuclear Magnetic Resonance Lines................. 18	
1.1.12.	The Bloch Equations 19	

1.2.	Chemical Shift 24	
1.2.1.	The Phenomenon of Chemical Shift....................................... 24	
1.2.2.	Diamagnetic Shift 26	
1.2.3.	Paramagnetic Shift 30	
1.2.4.	The Magnetic Anisotropy Effect of Neighboring Groups 31	
1.2.5.	Effect of Intramolecular Ring Currents 35	
1.2.6.	Solvent Effects 38	
1.2.7.	Additivity and Determination of Chemical Shifts; Correlation Tables; Reference Substances... 41	

1.3.	Spin-Spin Coupling: Qualitative Interpretation 45	
1.3.1.	The Phenomenon of Spin-Spin Coupling 45	
1.3.2.	The Multiplicity of Spin-Spin Splitting and the Relative Intensity of Lines 45	
1.3.3.	First-Order and Higher-Order Spin-Spin Coupling........................ 49	
1.3.4.	The Coupling Constant 51	
1.3.5.	Geminal Couplings 55	
1.3.6.	Vicinal Couplings 60	
1.3.7.	Long-Range Couplings 62	
1.3.7.1.	Long-Range Couplings in Unsaturated Systems: Allylic and Homoallylic Coupling 62	
1.3.7.2.	Long-Range Couplings in Saturated and Fluorine-Containing Compounds... 67	
1.3.8.	Classification of Spin Systems. Chemical and Magnetic Equivalence 68	

1.4.	The Quantum-Mechanical Treatment of Magnetic Resonance Spectra	73
1.4.1.	The Schrödinger Equation and Hamilton Operator of Spin Systems	73
1.4.2.	Spin Angular Momentum Operators and Their Properties	77
1.4.3.	Eigenfunctions of Spin Systems	79
1.4.4.	Multispin Systems	80
1.4.5.	Transition Probabilities and the Relative Intensities of Spectral Lines. Selection Rules	86
1.5.	The Quantitative Study of Spin System	88
1.5.1.	The A_2 Spin System	88
1.5.2.	The AX Spin System	92
1.5.3.	The AB Spin System	95
1.5.4.	Further Spin Systems Characterized by a Single Coupling Constant	100
1.5.4.1.	The AX_2 Spin System	100
1.5.4.2.	The AB_2 Spin System	104
1.5.4.3.	The AX_3 and AB_3 Spin Systems	107
1.5.4.4.	Spin Systems A_2X_2 and A_2B_2	111
1.5.4.5.	Spin Systems A_2X_3 and A_2B_3	115
1.5.5.	Spin Systems Characterized by Three Coupling Constants	115
1.5.5.1.	The AMX Spin System	115
1.5.5.2.	The ABX Spin System	119
1.5.5.3.	The ABC Spin System	126
1.5.6.	Spin Systems Characterized by Four Coupling Constants	126
1.5.6.1.	The $AA'XX'$ Spin System	126
1.5.6.2.	The $AA'BB'$ Spin System	131

Chapter 2
NMR Spectrometers, Recording Techniques, Measuring Methods

Introduction		135
2.1.	CW Spectrometers and Their Main Parts	135
2.1.1.	Magnetic Field Stability and Field Homogeneity	135
2.1.2.	The RF Transmitter. The Detector and the Phase Correction	138
2.1.3.	Sensitivity, Spectrum Accumulation	140
2.1.4.	Spectrum Integration Signal Intensity Measurement, Quantitative Analysis	142
2.1.5.	Variable Temperature Measurements	144
2.1.6.	Double Resonance in CW Mode	147
2.1.6.1.	Spin Decoupling	148
2.1.6.2.	Selective Decoupling	149
2.1.6.3.	Tickling	152
2.1.6.4.	INDOR	156
2.1.7.	NMR Solvents and Reference Substances. Calibration	158
2.2.	Fourier Transform NMR Spectroscopy	162
2.2.1.	The Fourier Transformation	165
2.2.2.	Pulsed Excitation and the Free Induction Decay (FID): Pulsed Fourier Transform (PFT) Spectroscopy	175
2.2.3.	FT NMR Spectrometers	180
2.2.4.	The PFT Spectrum	183

2.2.4.1.	Accuracy, Dynamic Range, Filtering, Apodization, and Scaling	183
2.2.4.2.	The FT of FID and the Resolution	184
2.2.4.3.	Tailored Excitation	186
2.2.4.4.	Pulse Width	188
2.2.5.	Double Resonance Methods in PFT Spectroscopy	188
2.2.5.1.	Off-Resonance	188
2.2.5.2.	Double Resonance Difference Spectroscopy (DRDS) and Selective Nuclear Population Inversion (SPI)	191
2.2.5.3.	The Overhauser Effect	192
2.2.5.4.	Gated Decoupling. Determination of NOE, and the Absolute Intensities of ^{13}C NMR Signals	194
2.2.5.5.	Noise Modulation Proton Decoupling. Stochastic Excitation	197
2.2.6.	Relaxation Mechanisms	198
2.2.6.1.	The Properties of Molecular Motions Causing Relaxation	198
2.2.6.2.	Dipole-Dipole Relaxation Mechanism	200
2.2.6.3.	Spin Rotational Relaxation Mechanism	201
2.2.6.4.	Relaxation by Chemical Shift Anisotropy	202
2.2.6.5.	The Scalar Relaxation Mechanism	203
2.2.6.6.	Quadrupole Relaxation	204
2.2.6.7.	Nuclear-Electron Relaxations	204
2.2.7.	Measurement of Relaxation Times	205
2.2.7.1.	Measurement of T_1 by Inversion Method	205
2.2.7.2.	Measurement of T_1 by Progressive Saturation	207
2.2.7.3.	Measurement of T_1 by Saturation Method	208
2.2.7.4.	Measurement of NOE	208
2.2.7.5.	Measurement of T_2 by the Spin-Echo Method	208
2.2.7.6.	Measurement of T_2 by the Carr-Purcell Method	209
2.2.7.7.	Measurement of T_2 by the Meiboom-Gill Method	210
2.2.7.8.	Measurement of $T_{1\rho}$ (Rotating-Frame) Relaxation Time	210
2.2.7.9.	Spin Locking	212
2.2.7.10.	J-Echo, J-Spectrum	212
2.2.8.	Relaxation Times in Cancer Diagnosis; FONAR	214
2.2.9.	Further Use of Pulse Sequences. Solvent Peak Elimination. Magic Angle Spinning. Two-Dimensional FT-NMR Spectroscopy	218
2.2.9.1.	Improvement of Sensitivity: DEFT, SEFT	218
2.2.9.2.	Recording the High-Resolution Spectra of Solid Samples by Means of Pulse Sequences. The ''Magic Angle''	219
2.2.9.3.	Elimination of Strong Solvent Signals: WEFT	220
2.2.9.4.	Two-Dimensional NMR Spectroscopy (2DFTS)	221
Abbreviations		229
Notations		231
References		237
Index		269

Volume II

Chapter 3
Proton Resonance Spectroscopy

Introduction 1

3.1.   Saturated Acyclic Alkanes.. 1
3.1.1.   Methyl Groups 1
3.1.2.   Ethyl Groups 3
3.1.3.   Methylene Groups.. 5
3.1.4.   Methine Groups 5
3.1.5.   Compounds with Long Paraffin Chains and Polymers.......................... 9
3.2.   Saturated Cyclic Compounds 11
3.2.1.   Cyclopropane Derivatives 12
3.2.2.   Oxiranes.. 14
3.2.3.   Thiiranes 15
3.2.4.   Aziridines 16
3.2.5.   Cyclobutanes and Four-Membered, Saturated Heterocycles 19
3.2.6.   Cyclopentane Derivatives and Their Heteroanalogues 21
3.2.7.   Cyclohexane Derivatives and Their Heteroanalogues....................... 23
3.2.7.1.   Conformational Analysis by ^{1}H NMR Spectroscopy...................... 23
3.2.7.2.   Cyclohexane Derivatives 27
3.2.7.3.   Steroids 30
3.2.7.4.   Carbohydrates 38
3.2.7.5.   Saturated, Six-Membered Heterocyclic Compounds 43

3.3.   Unsaturated Compounds.. 46
3.3.1.   Acetylene and Allene Derivatives 46
3.3.2.   Olefines 49
3.3.3.   Conjugated Dienes and Polyenes 54
3.3.4.   Cycloolefins 55
3.3.5.   Continuously Conjugated Cyclic Polyenes.................................. 59

3.4.   Aldehydes, Nitriles, Azides, Amides, Oximes, Imines, Azo, Nitro and
           Nitroso Compounds, N-Oxides, etc... 61

3.5.   Aromatic Compounds 67

3.6.   Heteroaromatic Compounds 75
3.6.1.   Five-Membered Aromatic Heterocycles and Their Derivatives 76
3.6.2.   Pyridine and Its Derivatives 82
3.6.3.   Fused and Quasiaromatic Heterocyclic Systems; Analogues Containing
           More Nitrogens... 87

3.7.   Medium Effects; Exchange Processes 92
3.7.1.   Hydrogens Attached to Other Atoms Than Carbon........................... 92
3.7.1.1.   Exchange Processes 92
3.7.1.2.   Hydrogen Bonds 97
3.7.1.3.   Hydroxy Groups... 99
3.7.1.4.   NH and SH Groups 105
3.7.1.5.   Protons Attached to Atoms Other Than C, O, N, or S and Adjacent to
           Magnetic Nuclei... 113
3.7.2.   The Determination of Optical Purity 114
3.7.3.   Shift Reagents.. 115
3.7.3.1.   Paramagnetic Metal Ion Induced Shift; Structure of the Complexes and the

Correlation Between the Complex Concentration and the Induced Shifts ... 115

3.7.3.2.　The Components of Shift-Reagent Induced Shifts 120

3.7.3.3.　The Shift Reagents: The Rare Earth Metal Atoms and the Organic
　　　　　 Complexing Agents 126

3.7.3.4.　The Effect of Shift Reagent on the Chemical Shifts and Coupling Constants
　　　　　 of the Substrate.. 127

3.7.3.5.　Application of Shift Reagent Technique in Structure Elucidation.......... 130

3.7.4.　 Chemically Induced Dynamic Nuclear Polarization: CIDNP............... 134

3.7.4.1.　Dynamic (DNP) and Adiabatic Nuclear Polarization; Theoretical Interpretation
　　　　　 of CIDNP Effect 136

3.7.4.2.　The Sign of CIDNP and the Multiplet Effect in CIDNP: Relative Intensities
　　　　　 in CIDNP Spectra 138

3.7.4.3.　Application of CIDNP in Organic Chemistry............................. 141

Chapter 4
The Resonance Spectra of Nuclei Other Than Hydrogen

Introduction 145

4.1.　　 Carbon Resonance Spectroscopy 146
4.1.1.　 Theoretical Interpretation of Carbon Chemical Shifts..................... 148
4.1.1.1.　Chemical Shifts in Carbon Resonance Spectra 148
4.1.1.2.　The Effect of the Physical Properties of Carbon Nuclei on Shielding 148
4.1.1.3.　The Effect of Structural Parameters on the Shielding of Carbon Nuclei 151
4.1.2.　 The Chemical Shifts of Various Compounds in Carbon Resonance
　　　　　Spectra 159
4.1.2.1.　Alkanes 159
4.1.2.2.　Cycloalkanes 161
4.1.2.3.　Halogen Derivatives.. 167
4.1.2.4.　Alcohols, Ethers, and Acyloxy Derivatives............................... 167
4.1.2.5.　Carbohydrates 171
4.1.2.6.　Amines and Their Salts... 171
4.1.2.7.　Saturated Heterocycles 173
4.1.2.8.　Functional Groups with Unsaturated Nitrogen 173
4.1.2.9.　Alkenes 175
4.1.2.10.　Allene and Acetylene Derivatives 179
4.1.2.11.　Carbonyl Compounds 180
4.1.2.12.　Aromatic Compounds 185
4.1.2.13.　Heteroaromatic Systems 188
4.1.2.14.　Steroids 194
4.1.2.15.　Alkaloids 201
4.1.2.16.　Nucleosides, Nucleotides... 202
4.1.2.17.　Amino Acids and Peptides 203
4.1.2.18.　Polymers 207
4.1.2.19.　Carbonium Cations, Metal-Organic Compounds......................... 207
4.1.3.　 Spin-Spin Interaction in ^{13}C NMR Spectroscopy; $^{13}C-^1H$ and $^{13}C-^{13}C$
　　　　　Coupling Constants.. 208
4.1.3.1.　Multiplicity of Carbon Resonance Signals: The Measurement of $^{13}C-^1H$ and
　　　　　$^{13}C-^{13}C$ Coupling Constants 209
4.1.3.2.　Satellite Spectroscopy 209
4.1.3.3.　On the Theory of Spin-Spin Interactions 210

4.1.3.4. The $^1J(C,H)$ Coupling Constant 213
4.1.3.5. The $^2J(C,H)$ Coupling constant 216
4.1.3.6. $^3J(C,H)$ Couplings...218
4.1.3.7. $^{13}C-^{13}C$ and $^{13}C-X$ (X = C, H) Couplings...........................219
4.1.4. Spin-Lattice Relaxation in Carbon Resonance 220
4.1.4.1. Carbon Resonance Intensities...220
4.1.4.2. Spin-Lattice Relaxation of Carbon Nuclei 221
4.1.4.3. Dipole-Dipole Relaxation..222
4.1.4.4. Other Carbon Relaxation Mechanisms......................................224
4.1.4.5. Nuclear-Electron Relaxation...226
4.1.4.6. Application of Dipolar Spin-Lattice Relaxation Times in Structure
Analysis..227
4.1.4.7. Investigation of Molecular and Segmental Motions.......................230
4.1.5. Carbon Resonance of ^{13}C Labeled Compounds; Investigation of Reaction
Mechanisms and Biosyntheses...232
4.1.6. Shift Reagents in Carbon Resonance 234
4.1.7. Dynamic Carbon Resonance Spectroscopy 234

4.2. Nitrogen Resonance 238
4.2.1. The Basic Problem of Nitrogen Resonance Spectroscopy 238
4.2.2. Chemical Shifts in Nitrogen Resonance 239
4.2.3. Calibration 239
4.2.4. Relationships Between Chemical Shifts and Molecular Structure...........240
4.2.5. The Characteristic Chemical Shift Ranges of Various Compound Types in
Nitrogen Resonance 243
4.2.5.1. Amines, Hydrozines, and Related Derivatives.............................243
4.2.5.2. Carboxylic Amides, Thioamides, Peptides 243
4.2.5.3. Azides, Compounds with sp Nitrogen 244
4.2.5.4. Azoles 245
4.2.5.5. Azines..247
4.2.5.6. Oximes and Nitro Derivatives 249
4.2.5.7. Azo- and Diazo-Derivatives 250
4.2.5.8. Nitroso Derivatives...250
4.2.6. Spin-Spin Interactions and Coupling Constants of Nitrogen Atoms........251
4.2.6.1. $^1F(N,H)$ Couplings 252
4.2.6.2. $^nJ(N,H)$ Couplings ($n > 1$)...254
4.2.6.3. Nitrogen Couplings with Atoms Other Than Protons 256
4.2.6.4. The Quadrupole Relaxation of ^{14}N Isotope; Technical Problems of the
Measurement of ^{14}N Coupling Constants and of Nitrogen Resonance
Spectroscopy 256

4.3. Oxygen Resonance 259

4.4. Fluorine Resonance..261
4.4.1. ^{19}F NMR Chemical Shifts 261
4.4.2. ^{19}F NMR Parameters of Saturated Open Chain Compounds 262
4.4.3. ^{19}F NMR Parameters of Saturated Cyclic Compounds...................265
4.4.4. ^{19}F NMR Parameters of Unsaturated Compounds 266
4.4.5. ^{19}F NMR Parameters of Aromatic and Heteroaromatic Compounds 267

4.5. Phosphorus Resonance 268

Abbreviations..271
Notations..273

References...279

Index..311

TABLE OF CONTENTS

Volume III

Chapter 5
Structure Determination Problems 1

Abbreviations..333
Notations..335

Index..341

Chapter 3

PROTON RESONANCE SPECTROSCOPY

INTRODUCTION

Proton magnetic resonance (^{1}H NMR) spectroscopy is particularly important field of NMR spectroscopy. Most studies of the first two decades of NMR spectroscopy were devoted to proton resonance. The reasons are theoretical and practical. Organic compounds contain, practically without exception, hydrogen atoms. Hydrogen is the most frequent nucleus of molecules, and it occurs in the most varying chemical environments. The large natural abundance of the ^{1}H isotope (99.8%, compare Table 1) and consequently its high sensitivity in nuclear resonance also facilitate the measurements. To record ^{1}H NMR spectra, much simpler conditions are required than for other types of nuclei. In this section we shall discuss the most common types of protons and their spectroscopic features (the characteristic values of chemical shifts and coupling constants) derived theoretically and obtained experimentally.

3.1. SATURATED ACYCLIC ALKANES

An aliphatic compound may contain three different kinds of protons, namely hydrogens in methyl (RCH_3), methylene ($RR'CH_2$), or methine ($RR'R''CH$) groups. The overall ranges of chemical shift for these types of protons are 0 to 4.5, 0.5 to 5.5, and 1.0 to 7.5 ppm, respectively. Owing to the mutual overlaps, great care must be taken in drawing conclusions on the structure from the chemical shifts. Moreover, the chemical shift ranges for various types of protons are always based on the data of relatively few (at most some hundred) compounds, and anomalous chemical shifts may well occur. In these cases, it is also necessary to compare the spectra with those of model compounds similar in structure. Studies on model compounds are also very useful because they provide much narrower chemical shift ranges for particular types of protons in smaller groups of compound (e.g., compare Tables 22 and 23). Furthermore, certain quantitative and semiquantitative correlations may also be drawn from the data obtained by comparison (e.g., between the electron affinities or bulkiness of neighboring groups and the chemical shifts). This permits one to predict certain chemical shifts by calculation.

3.1.1. Methyl Groups

The chemical shifts of methyl groups are shown in Table 22. The signal attributed to a methyl group is a singlet, doublet, or triplet according to the number of hydrogens (0, 1, 2) on the neighboring atom. Exceptions are the compounds in which the methyl groups are attached to unsaturated carbon atoms, where long-range couplings may cause splitting (compare Problems **22, 25, 46**, and **60**).

58

In saturated systems, long-range coupling is a rare phenomenon. An example is 1,2-dibromo-2-phenyl-propane (CH_2–$CBrPh$–CH_2Br)[1212] in which a long-range coupling occurs

Table 22
THE CHEMICAL SHIFT RANGES (PPM) OF METHYL SIGNALS FOR VARIOUS FUNCTIONAL GROUPS

Compound	Chemical shift range	Compound	Chemical shift range	Compound	Chemical shift range
Me–metal	− 1.3—1.8	Me–C(sp)	2.0—2.6	Me–X	2.1—4.3
Me–Si(sp³)	0—0.6	Me–N(sp³)	1.8—3.8	Me–SO₂–	2.5—4.3
CH₄(gas)	0.13	Me–S–	1.8—2.2	Me–SO–	3.2—4.3
Me–C(sp³)	0.7—1.9	Me–CO–	2.0—2.9	Me–O–	3.2—4.6
Me–C(sp²)	1.5—2.8	Me–Ar	2.1—2.8	Me–N(sp²)	

between the methyl group and one of the chemically nonequivalent methylene protons (^4J = 0.65 Hz). By replacing the methylene protons, one by one, with deuterium, it can be shown that, as expected,* in the preferred conformation **58** H-2 being in *trans* position to the methyl group is coupled (''W'' pattern).

In addition, conformers or rotational isomers, etc. with sufficiently long lifetimes may also cause multiplets to occur. Of course, chemically nonequivalent methyl groups in the same molecule produce, depending on whether the groups are isochronous, coinciding or separated signals (compare Problem **3**). The chemical shift of methyl protons usually differs significantly from that of neighboring protons, as compared to the *vicinal* coupling constant (7 Hz). For this reason, the coupling is almost always first order (i.e., A_2X_3 or AX_3 spectra) or very close to it (roof structure).

Systematic investigations of CH_3X-type compounds have shown that the chemical shift of the methyl signal is directly proportional to the electronegativity of substituent X. Thus, for example, the methyl signals of CH_4, CH_3I, CH_3Br, CH_3Cl, and CH_3F molecules have decreasing shieldings (increasing δ-values) in the above sequence (compare Table 2). The same holds for other substituents, only the proportionality constants are different. The chemical shifts of H–Me, C–Me, S–Me, N–Me, and O–Me groups, for example, are increasing in this order (compare Table 22) and are proportional to the electronegativity of the heteroatoms (except for the methylmercapto group, for which the chemical shift is larger than expected).

In MeCOR derivatives, the signal of the methyl protons adjacent to the carbonyl groups is shifted paramagnetically by 1 to 2 ppm, due to the $-I$ effect of the carbonyl, as compared to the signal of methyl groups attached to saturated carbon atoms. For the same type of substituent R, the range of chemical shifts is narrower (see Table 23). Problems are caused, however, by the overlapping ranges. The methyl protons of acetone and acetic aldehyde both absorb at 2.17 ppm in $CDCl_3$ solution. In Table 23 the chemical shifts of some other simple acetic acid derivatives are listed, too. As can be seen, the full range is only 0.64 ppm. Consequently, functional groups do not essentially affect the chemical shift of acetyl protons, explaining why they are generally indistinguishable from one other by ^1H NMR data. For this purpose, the IR or ^{13}C NMR spectra are more suitable. Table 23 also indicates that a carbonyl group affects the chemical shift of the adjacent protons, not only by its $-I$ effect, but its magnetic anisotropy also has a role in determining the resultant shielding. Since the sign and magnitude of the anisotropic effect varies with the mutual steric position, the chemical shift also depends on the steric structure of the carbonyl compounds. This explains, for example, why the acetyl protons of acetic anhydride are more shielded than that of dimethylacetamide.

The singlet of the *t*-butyl groups corresponds to nine equivalent protons and is not split

---

* See Volume I, p. 67.

Table 23
THE CHEMICAL SHIFTS AND SHIFT RANGES (PPM) OF THE ACETYL GROUPS OF ACETIC ACID DERIVATIVES[434]

Compound	$\delta COCH_3$	Compound	Chemical shift range
MeCOOEt	2.03	MeCON(sp^3)	2.0—2.3
MeCOOH	2.10	MeCOOR	2.0—2.5
MeCHO, MeCOMe	2.17		
MeCOOCOMe	2.20	MeCOOAr	2.3—2.7
MeCONMe$_2$	2.28	MeCO–C(sp^2)	2.4—2.7
MeCOPh	2.59		
MeCOCl	2.67	MeCOAr	2.5—2.9

Table 24
THE CHEMICAL SHIFT RANGES (PPM) OF THE METHYL SIGNALS OF ETHYL GROUPS FOR R–CH$_2$–CH$_3$ TYPE COMPOUNDS

R	Chemical shift range	R	Chemical shift range	R	Chemical shift range
Metal	0.60—1.50	CO–	0.95—1.25	O–CO–	1.20—1.45
C(sp^3)	0.70—1.05	O–	1.10—1.25	S–	1.20—1.60
C(sp^2)	0.80—1.25	CON(sp^3)	1.10—1.75	X	1.25—1.90
N(sp^3)	0.90—1.70	Ar	1.20—1.40	OAr	1.30—1.50

(there is no *vicinal* hydrogen); thus its relative intensity, as compared with any other signal appearing in an ¹H NMR spectrum, is extremely high. For this reason, it is very characteristic and easy to recognize. Its position is insensitive to the inductive effects of substituents of the quaternary carbon atom, $\Delta\delta CMe_3$: 0.65 to 1.95 ppm.

The methyl groups attached to heteroatoms, especially the methoxy protons, also give rise to easily recognizable signals. The strong $-I$ effect of oxygen causes a considerable paramagnetic shift (in the order of 2 to 3 ppm), hence the methoxy signal appears in the region of 2.8 to 4.2 ppm. Since there are neither *vicinal* hydrogens nor neighboring magnetic nuclei, a singlet with an intensity corresponding to three protons is one of the most characteristic types of signals.

S-methyl and *N*-methyl groups also have characteristic lines, although apart from some exceptions, their chemical shifts are smaller, and thus they may overlap with other signals more frequently. The *N*-methyl signals, when a proton is also attached to the nitrogen atom, may be split (e.g., see Problem 27). Owing to their high intensities, the singlets of methyl groups attached to quaternary carbon atoms are also characteristic in most cases.

3.1.2. Ethyl Groups

The chemical shifts of the methyl signals of ethyl derivatives still reflect the inductive effect of the substituents, though the magnitude of the effect is decreased considerably by the inserted methylene group, and only the protons of the latter group are affected remarkably (see Table 24). This is evident since the inductive effect is almost fully isolated even by a single methylene group, and the effect on a γ-carbon atom can practically never be observed.

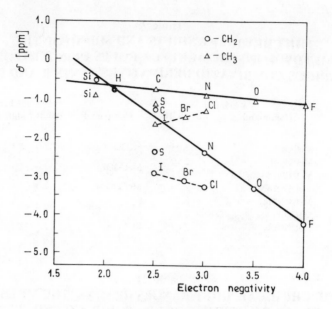

FIGURE 117.   The chemical shifts of the methyl and methylene signals of RCH_2CH_3-type compound plotted against the electronegativity of substituent R. Broken lines refer to halogens.[1354]

By plotting the chemical shifts of methyl and methylene protons against the electron-withdrawing effect of the substituents, a straight line can be obtained with a smaller slope for the former and a larger slope for the latter type of protons (see Figure 117). Consequently, not only δCH_3 and δCH_2, but also their differences are proportional to the electronegativities of the substituents. With decreasing electronegativity, the two signals, more precisely the two groups of lines, gradually approach each other, then overlap, and finally their positions reverse, i.e., the methylene protons are more shielded than the methyl hydrogens. Examples for this situation are $SiEt_4$[649] and $ZnEt_2$.[1024]

The chemical shifts of ethyl halides, although showing a similar proportionality, line up along a line with different slope (see Figure 117, broken lines), and the sulfur compounds also show irregularities. The halides behave abnormally in other respects, too. Whereas the analogous signals of methyl and ethyl halides have increasing shifts in the sequence of iodine-chlorine-bromine, the signal of the methine protons, e.g., of the isopropyl and cyclohexyl halides in *geminal* position to the substituent, shows opposite shift order, and the shift differences are substantially smaller.[1379]

This behavior can be explained as follows. The $-I$ effect of halogens deshields the neighboring protons considerably (even by 3 ppm for α-protons), and this effect increases, of course, in the sequence of iodine, bromide, chlorine, and fluorine. As anisotropic neighboring groups, however, the halogens cause opposite shifts, and the magnitude of the effect varies in the opposite sequence, i.e., the anisotropic effect is the strongest for iodine. The two opposite effects have different contributions depending on the chemical structure and geometry of the molecule, too, and the resulting shift is therefore quite varying, not even its sense being always the same.*

When the ethyl protons are not coupled with other hydrogens (e.g., as in ethoxy, carbethoxy, etc. derivatives, and in all ethyl groups attached to quaternary carbons, except for the unsaturated and aromatic compounds), their presence is always easy to detect from the simultaneous appearance of a symmetric triplet and a quartet, characterized by a coupling

* See Volume I, p. 34-35.

Table 25

THE CHEMICAL SHIFT RANGES (PPM) OF THE
METHYLENE SIGNALS OF COMPOUNDS RCH₂R' FOR
VARIOUS COMBINATIONS OF SUBSTITUENTS R AND R'[720]

R	R'	Chemical shift range	R	R'	Chemical shift range
	Metals	−1.00—+2.00	$C(sp^2)$	$C(sp^2)$	2.70—3.95
	$C(sp^3)$	0.95—2.05	$C(sp^2)$	$N(sp^3)$	2.90—3.80
	$C(sp^2)$	1.85—2.45	Ar	$N(sp^3)$	3.30—4.45
	$C(sp)$	2.10—2.80	$C(sp^3)$	O—	3.35—4.50
$C(sp^3)$					
	$N(sp^3)$	2.25—3.60			
	X	2.25—3.70	$C(sp^2,sp)$, Ar,X	X	3.65—4.60
	S—	2.35—3.00	$C(sp^2,sp)$	O—	4.00—5.10
	Ar	2.60—3.35	Ar		4.30—5.35

From *JEOL NMR Applications*, JEOL Co., Ltd., Tokyo, Japan. With permission.

constant of 6 to 8 Hz (e.g., see Figure 22). It is particularly easy to recognize the quartet of ethoxy groups, since the chemical shift of the methylene protons increases substantially in the neighborhood of the oxygen atom (the methylene quartets of diethyl ether, ethanol, and ethyl acetate appear at 3.35, 3.60, and 4.05 ppm, respectively).[434]

3.1.3. Methylene Groups

The methylene groups are attached to two substituents, resulting in a broader range of shifts and more complicated and varied possibilities of splitting as compared to the case of methyl groups. The multiple splitting and occasional overlaps with other signals may even lead to spectra in which the methylene signal cannot be identified at all (compare Problems **18** and **19**). When the molecule contains several types of methylene groups, their signals or signal groups may merge into an uncharacteristic absorption. In contrast, when the molecule contains only one methylene group, isolated from other spins, its signal is usually singlet with a chemical shift varying in a relatively narrow range (see Table 25).

The signal of isolated methylene protons may be split by long-range coupling or when the two hydrogens are diastereotopic.* The shift difference of diastereotopic methylene protons depends strongly on the solvent. This phenomenon is illustrated by the spectrum of 2-phenoxy-acetaldehyde diethylacetal [PhOCH₂CH(OEt)₂], in which the methylene quartet of the ethoxy groups shows a further *AB* splitting (see Figure 118). Similar examples can also be found in Problems **10, 30,** and **51**.

Not only in the ethyl derivatives but also generally, the change in chemical shift of the methylene signals is proportional to the electronegativities of the substituents, and the effects of the two substituents are additive. This is expressed by the modified Schoolery rule:[1291,1305]

$$\delta CH_2 = 1.25 + \Sigma \rho_i \text{ ppm} \qquad (270)$$

The values of constant ρ_s for some substituents can be found in Table 26. For example, the methylene signal of benzyl bromide (PhCH₂Br) appears at 4.43 ppm,[434] and from Equation 270, $\delta CH_2 = 1.25 + 1.3 + 1.9 = 4.45$ ppm.

3.1.4. Methine Groups

The methine signals can be identified only in the spectra of simpler compounds because

* See Volume I, p. 69.

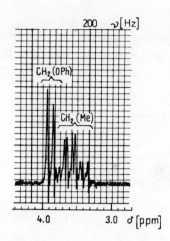

FIGURE 118. Part of the spectrum of 2-phenoxy-acetaldehyde diacetal, PhOCH$_2$CH(OEt)$_2$, with the methylene multiplets. (From *High Resolution NMR Spectra*, SADTLER Research Laboratories, Philadelphia; Heyden and Son, London, 1967-1975. With permission.)

Table 26

SUBSTITUENT CONSTANTS ρ_i (PPM) OF SOME SUBSTITUENTS[1291] FOR ESTIMATION OF CHEMICAL SHIFTS OF ALIPHATIC METHYLENE AND METHINE PROTONS BY EQUATIONS 270 AND 271, RESPECTIVELY

Substituent	ρ_i	Substituent	ρ_i	Substituent	ρ_i
CF$_3$	0.6	SR, CONRR'	1.1	Cl	2.0
COOH	0.7	COR, CN	1.2	OH	2.1
CR=CR'R"	0.8	Ph	1.3	OPh	2.4
C≡CR	0.9	I	1.4	OCOR	2.7
COOR	1.0	OR	1.8	NO$_2$	3.0
		Br	1.9		

it is generally difficult to estimate the combined effect of the three substituents on the chemical shift. Therefore, the chemical shifts determined from the Schoolery rule are in poorer agreement with the experimental results than in the case of methylene groups, and thus Equation 271 can be used to obtain a rough estimate, only:

$$\delta CH = 1.50 + \Sigma\rho_i \ ppm \qquad (271)$$

The values of constant ρ_i are given in Table 26.

Of course, the possibilities for spin - spin coupling are even broader. Further difficulties are caused by the low relative intensity of the methine signal in comparison with the methyl or methylene signals, since the former arises from only one hydrogen. Accordingly, it frequently occurs that the methine signal cannot be identified in the spectrum.

Nevertheless, the signal of certain methine protons may become characteristic if there is no coupling, i.e., the signal is a singlet, and its chemical shift is larger than usual, i.e, it

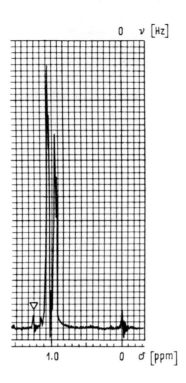

0   ν [Hz]

1.0   0   δ [ppm]

FIGURE 119. Part of the ¹H NMR
spectrum of 3-hydroxy-4-methylpentine,
HC ≡ C–CH(OH)–CHMe₂, with the
methyl signal. (From *NMR Spectra Catalog*, VARIAN Associates, Palo Alto,
Calif., 1962-1963. With permission.)

is not overlapping with other signals. These conditions are met, e.g., by the protons attached
to tertiary carbons where the substituents of the carbon atom cause sufficiently large paramagnetic shift.

Like in the methoxy and methylene groups, the adjacent oxygen atom also causes a strong
paramagnetic shift in the methine signal. For example, in the spectrum of di-isopropyl ether,
it can be found at 3.56 ppm.[259]

59

The effects of more substituents similar in nature are additive. The singlet methine signal
of Compound **59** can be found at 4.65 ppm.[250] Due to the effect of the three *para*-chlorophenoxy groups, the methine signal of Compound **6/I** in Problem **6** is shifted to 6.4 ppm.
The chemical shift of the proton of trinitromethane (7.5 ppm) is even larger.[665]

For those isopropyl derivatives in which the coupling involves only the methyl protons
the methine signal is a symmetric septet, and the chemical shift can be predicted fairly well
by Equation 271. Of course, the methyl signal is a doublet, and the splitting is 6 to 8 Hz,
as usual for the *vicinal* coupling of saturated compounds. If the two methyl groups are
diastereotopic, this doublet splits into a double doublet. This is illustrated by the spectrum
(see Figure 119) of 3-hydroxy-4-methylpentine [HC≡C–CH(OH)–CHMe₂]. The chemical

Table 27
THE CHEMICAL SHIFT
RANGES (PPM) OF
METHYL AND METHINE
PROTONS FOR RCHMe$_2$
TYPE COMPOUNDS

X	Chemical shift range	
	δCH_3	δCH
C(sp^3)	0.80—1.10	1.20—2.40
C(sp^2)	0.85—1.10	2.20—2.80
Ar	1.00—1.30	2.55—3.20
CO—	1.10—1.35	2.60—3.05
O—	1.10—1.50	3.65—4.25

shift ranges of the methyl and methine signals of isopropyl groups are shown for various substituents in Table 27. The data show an analogous tendency to the one already discussed with the ethyl derivatives, i.e., the shift of the methine proton is influenced considerably more by the adjacent substituent than the shift of the more distant methyl protons.

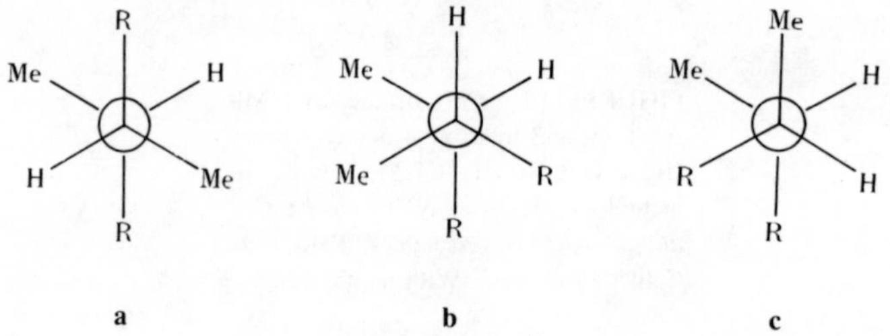

a   b   c

60

The conformational study of 2,3-symmetrically disubstituted butane derivatives (MeCHR–CHRMe) is a classical example for the application of NMR spectroscopy in stereochemistry. These compounds have $AA'X_3X_3'$ spectrum; the A protons and X_3 groups are, namely, magnetically nonequivalent. The coupling constant $J_{AA'}$ permits one to draw conclusions on the relative stability of the rotamers.[39,174] Both the *meso* **60** and the *racemic* **61** compounds have three stable *trans* conformations (**a** to **c**), of which, if substituent R is bulky, the most probable are **60a** and **61a**, respectively. Since the dihedral angles between the C–H bonds of the methine groups are 180 and 60° for conformers **60a** and **61a**, respectively, the correlation between the dihedral angles and the coupling constants* involves that $J_{AA'}$ is larger for the *meso* compound than for the *racemic* derivative. Consequently, the higher the relative weight of conformers **60a** and **61a** (the bulkier substituent R), the larger the difference between the measured values of $J_{AA'}$ of the *meso* and *racemic* derivatives. The above is supported by the experimental data.[174] The difference in $J_{AA'}$ between the *meso* and *racemic* compounds is smaller for R = Cl, than for R = Br, in accordance with the smaller steric requirement of this substituent: for R = Cl, $J_{AA'}$ = 7.9 (*meso*) and 3.0 Hz (*racemic*); for R = Br, $J_{AA'}$ = 6.3 (*meso*) and 3.3 Hz (*racemic*); hence ΔJ(Cl) = 4.9 > ΔJ(Br) = 3.0 Hz.

---

* See Volume I, Equations 79, 80a, and 80b.

61

3.1.5. Compounds with Long Paraffin Chains and Polymers

In the ^{1}H NMR spectra of compounds with longer paraffin chains, the methyl, methylene, and methine signals coalesce into uncharacteristic absorption maxima. It is at most the signals of Me$_3$C , Me$_2$C , and MeC groups, and sometimes of Me$_2$CH , and MeCH groups, that show up against the background and can be identified.

For example, in the spectrum of the relatively simple compound, 3-methyl-pentane (Et$_2$CHMe), the methyl doublet and the two methyl triplets merge into an uncharacteristic maximum (see Figure 120a), and so do the methine and methylene multiplets. In the spectra of compounds containing long paraffin chains, and, of course, in those of normal paraffin as well, the two methyl groups give rise to an asymmetric triplet (B part of an A_2B_3 multiplet), and the methylene protons produce a more or less broad, singlet-like maximum, as illustrated by the spectrum (see Figure 120b) of n-pentadecane [Me(CH$_2$)$_{13}$Me]. The spectrum (see Figure 120c) of 2,2,4-trimethylhexane (Me$_3$C–CH$_2$–CHMeEt) shows clearly how the singlet of the t-butyl group dominates the spectrum, completely concealing the doublet and triplet of the two other methyl groups. The common absorption of the two methylenes and the methine group appears only as a hardly observable maximum.

These examples illustrate that the ^{1}H NMR spectra of paraffins are generally not informative. However, e.g., for normal paraffins, the length of the chain can be determined from the intensity ratio of the methyl and methylene signals.

NMR spectroscopy has an important role in the structure elucidation of polymers, too.[52,322,1042,1043,1278] Of course, high resolution routine measurements can be performed only in cases when a solution of appropriate concentration can be prepared in one of the usual NMR solvents. Then, the NMR spectrum can be used to advantage, e.g., in determining the configurations of certain parts in the polymer chain, the size and recurrence of conformatively homogeneous chain sections, or in finding correlations between the configurations and conformation of the chain. In the polymerization of dienes and olefines, one can study the correlation between the structure of the polymer product or the way of addition and the catalyst or other reaction conditions. With copolymers the composition, with chains composed of dissimilar units, and the regularity or irregularity of structure can also be studied.

Owing to the hindrance of conformational motions, the individual types of protons are not equivalent in the polymer chain, and thus their signals are broadened in comparison with those of the monomers. If the polymer is investigated in sufficiently diluted solutions and at sufficiently high temperatures, the chemically identical protons become equivalent again due to the free motion of the polymer chain (fast interconversion of the conformers), to make the signals sharper. If associations are possible, the spectra vary with the temperature, concentration, and the solvent, too.

62

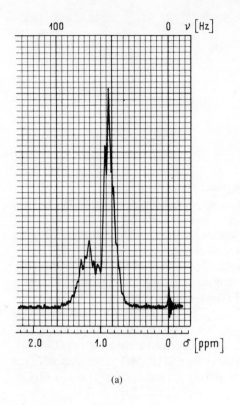

(a)

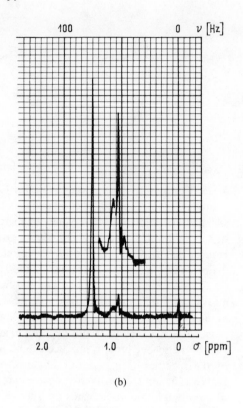

(b)

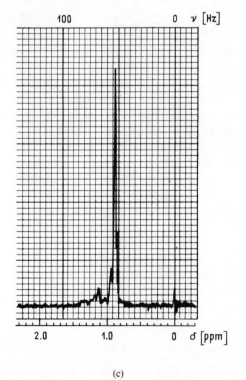

(c)

FIGURE 120.   The spectra of (a) 3-methylpentane, (b) *n*-pentadecane, and (c) 2,2,4-trimethylhexane. (From *High Resolution NMR Spectra,* SADTLER Research Laboratories, Philadelphia; Heyden and Son, London, 1967-1975. With permission.)

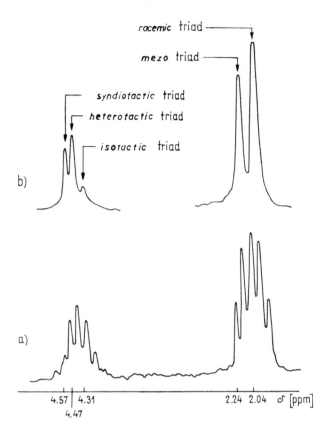

FIGURE 121. (a) The spectrum of polyvinylchloride (**62**) with (b) the decoupled methine and methylene signals.[1290]

In the spectra of polymers, splitting can also be observed in addition to line broadening. Figure 121 shows the spectrum of polyvinylchloride (**62**), the usual and the DR spectrum. It can be seen that in the normal spectrum the methine and methylene signals are more complex than expected (quintet and triplet), and the two signals are still split in the spin-decoupled spectrum (triplet and doublet). The reason is that the methylene and methine groups are, in themselves, nonequivalent due to the different configurations of the carbon atoms. The former groups may be present in two and the latter in three forms, depending on which three adjacent carbon atoms are chosen from the chain. The structure of the spectrum is determined by these *triads*.

From the aspect of methylene hydrogens, two types of *triads* exist, in which, if represented in a planar projection, the two chlorine atoms are located on the same side (*meso triad*) or the opposite sides (*racemic triad*) of the chain. The two cases can be represented, e.g., by the triads 1-2-3 and 5-6-7. The methine hydrogens may be present in three types of triads which are different in the position of the β-chlorine atoms: both chlorines are located on the same side as the methine hydrogen (*isotactic*), the chlorines are on the side opposite to the methine proton (*syndiotactic*), or one chlorine atom is on one side and the other on the opposite (*heterotactic triad*). Examples for the three types are the triads 6-7-8, 2-3-4, and 4-5-6. The intensities of lines corresponding to the various triads permit one to determine the relative weights of the configurations represented by the various triads.

3.2. SATURATED CYCLIC COMPOUNDS

Cyclic compounds are very suitable for configurational and conformational studies, and

NMR spectroscopy is an excellent tool for this purpose, as will be illustrated by a few examples.

3.2.1. Cyclopropane Derivatives

Cyclopropane (C_3H_6) is a symmetric (regular triangular), rigid, planar molecule. The dihedral angles of *vicinal* hydrogens are $\varphi(cis) = 0°$ and $\varphi(trans) = 144°$. The signal of the six equivalent methylene protons is a singlet at 0.22 ppm.[1519] This small value reflects an abnormally strong shielding of the ring protons. As comparison, the methylene signal of propane appears at 1.45 ppm.[1289] The cyclopropyl ring shields not only the ring protons, but other hydrogens of cyclopropyl derivatives, too. In the spectrum of Compound **63** the H-4,5 signal appears at 6.91 ppm, at a chemical shift 0.5 ppm smaller than in model Compound **64**, where the shift is 7.42 ppm.[501]

63   64

The substituents of cyclopropane derivatives affect the chemical shifts of the ring protons strongly, primarily that of the *geminal* proton. Electron-withdrawing substituents deshield the *geminal* ring protons and electron-repelling substituents have opposite effects. Certain substituents (such as Cl, COOR, COOH, and Ar) cause paramagnetic shift in the signal of the *cis vicinal* hydrogens (on the same side of the rings as the substituent) with respect to the signal of the *trans* protons, whereas for other substituents (such as Me, OR, OAr, Br, COAr, and $SiMe_3$) the reversed relationship $\delta H^c < \delta H^t$ is characteristic (see Table 28).[1051]

These relationships can be used to solve configuration problems. Thus, e.g., on the basis of the relative chemical shifts of the ring protons, it is easy to distinguish the *cis-trans* isomers **65** and **66**. Since the *vicinal* ring protons of the carbalkoxy derivatives obey the relationship $\delta H^t < \delta H^c$, consequently one can expect for Compound **65** the chemical shifts δH_A and δH_X to be larger than for Compound **66**. With proton H_B the situation is opposite. In **65** the carbomethoxy group on C-1 is in *cis* position to H_A and H_X, whereas in isomer **66** it is *cis* to H_B. The other group on C-2 is irrelevant in this respect because its position related to the ring protons is the same in the two isomers. The observed values are as follows: $\delta H_A = 1.15$, $\delta H_B = 1.65$, and $\delta H_X = 1.92$ ppm in one of the isomers, and the corresponding shifts in the other isomer are 1.31, 1.20, and 2.22 ppm, respectively.[1088] Consequently, the first set of data corresponds to Compound **66**, and the second corresponds to isomer **65**.

65   66

One may utilize in configuration analysis not only the signals of the ring protons, but also those of certain substituents. Thus, e.g., in the pair **67-68**, the *vicinal* methyl signals are influenced by the phenyl group in a way that the signal of the methyl groups located at

Table 28
THE CHEMICAL SHIFTS (PPM) OF RING PROTONS OF CYCLOPROPANES[1051,1379]

```
 1      3
  \   /
   \5/
  2/ \4
   6
```

Substitution			1	3	5	2	4	6
Mono			Br		0.88	2.84	1.00	
			COOH		1.06	1.58	0.87	
			NH₂		0.35	2.30	0.35	
			Ph		0.74	1.80	0.68	
			COOMe		0.93	1.63	0.93	
Di	gem	symm.	Me		0.20	Me	0.20	
			OAc		1.24	OAc	1.24	
			Cl		1.47	Cl	1.47	
		asymm.	Ph		1.24	COOH	1.65	
			Ph		0.88	Br	1.17	
	vic	trans	COOMe	2.03	1.33	2.03	COOMe	1.33
			CN	2.06	1.58	2.06	CN	1.58
			COPh	3.32	1.68	3.32	COPh	1.68
		cis		COPh	1.50		3.15	2.10
Tri	1,3,4		SiMe₃	Cl	1.12	0.58	Cl	1.43
			OMe	Cl	1.52	3.62	Cl	1.66
			Br	Cl	1.58	3.45	Cl	2.08
			COOH	COOH	2.43	3.17	COOH	2.30
			COOH	Me	0.99	1.39	Me	0.79
	1,2,3		Me	OPh	0.53	Me	3.33	0.64
	1,4,6		COPh		3.75	4.23	COPh	
Tetra				CN	3.47		CN	3.47
Penta			Ph		CN	4.93	CN	

the same side as the substituent is shifted diamagnetically relatively to that of the *trans* methyl group. Since the methyl groups on the same side of the ring have coincident signals, for Compound **67** the weaker of the two methyl signals (at 0.77 ppm) and in the spectrum of **68** the stronger one (at 0.90 ppm, with double intensity) must have smaller shift. The methyl signal of **67** arising from six hydrogens appears at 1.17 ppm, and that of the *trans* methyl group of **68**, with a half intensity, can be found at 1.22 ppm.[287] Similarly, in isomers **69** and **70**, the diamagnetic effect of the phenyl group shields the *cis* carbomethoxy group, causing the methoxy signal to appear at 3.36 ppm in the spectrum of **69** and at 3.68 ppm in the spectrum of isomer **70**.[788] It can be seen from Table 28 that the range of the ring proton signals of cyclopropyl derivatives is very wide, 0.1 to 5.0 ppm.

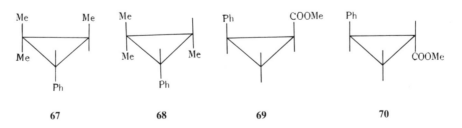

| 67 | 68 | 69 | 70 |

The methylene protons of mono- and 1,2-disubstituted cyclopropanes and, if the $-I$ effects of the substituents are not too strong, the other ring protons as well, have small chemical shifts, like in cyclopropane. In such cases it is easy to recognize the signals of the ring protons in the spectra, since other types of protons do generally not absorb in this region (0.1 to 0.7 ppm). Even the methylene signal of cyclopropene (C_3H_4) appears only at 0.92 ppm. The methine signal is, of course, in the region characteristic of olefinic protons,* around 7.0 ppm.[1519]

Depending on the type of substitution and the nature of substituents, cyclopropane derivatives give rise to various types of spectra. The ring protons of symmetric and asymmetric 1,1-, and *cis* 1,2-, or *trans* 1,2-disubstituted cyclopropanes give A_4 and $AA'BB'$, ABC_2 and $ABCD$, or $AA'BB'$ and $ABCD$ multiplets, eventually corresponding to first order spectra ($AA'XX'$, ABX_2, $ABMX$, etc.). According to the concordant results of systematic studies and theoretical calculations, the value of J^c (7 to 11 Hz) is always greater than J^t (4 to 8 Hz), whereas J^{gem} is negative with a magnitude between 0.5 and 9 Hz in cyclopropanes. These relative magnitudes of *vicinal* coupling constants are in agreement with the *Karplus* relation (Equation 79) and the dihedral angles ($\varphi^c = 0°$ and $\varphi^t = 144°$). Electron-withdrawing substituents generally decrease the coupling constants, and the value of J^{gem} is the most sensitive to substitution.[1526]

3.2.2. Oxiranes

Ethylene oxide (C_2H_4O) is planar and has a geometry very close to a regular triangle. The singlet of its four protons appears at 1.58 ppm,[1187] at a shift 0.80 ppm smaller than the methylene signal of acyclic diethyl ether.[434,649] Hence, the diamagnetic shift of the ring protons, characteristic of cyclopropanes can also be observed here. The shielding effect of the three-membered ring is reflected, e.g., by Compound **71**, for which $\delta H_A = 0.70$ ppm, whereas analogue **72** has its corresponding signal at approximately 1.20 ppm. On the other hand, the signal of H_B is shifted downfield in the spectrum of **71** with respect to δH_B of **72** (from 1.20 ppm to 1.35 ppm),[1412] since the shielding effect of the ring is overcompensated by a deshielding of the C–O bond (compare Figure 16). In substituted derivatives the chemical shift may increase to almost 5.4 ppm (see Table 29). The shielding decreases in the monosubstituted derivatives, and the *geminal* H-2 is deshielded most strongly. For most of the substituents (C_{sp^2}, Ar, CHO, Ac, COOR, and OAc) deshielding is stronger on the *trans* proton H-4 than on the *cis* proton (H-3). Substituents Cl, CH_2Cl, and CN exert opposite relative shielding effects on the *vicinal* protons, i.e., in this case $\delta H^{cis} > \delta H^{trans}$. Alkyl substituents have the same effect on both *vicinal* ring hydrogens.

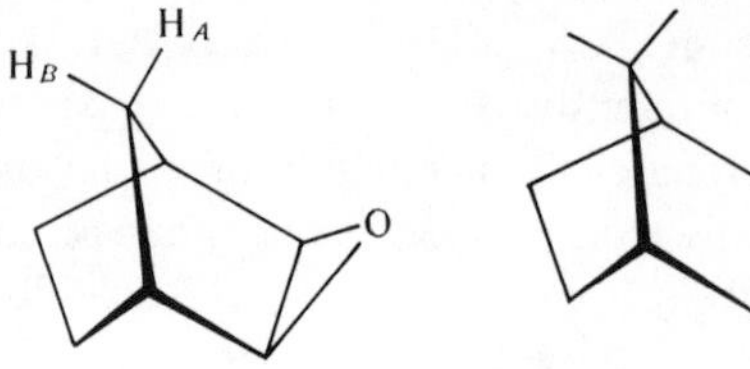

| 71 | 72 |

The monosubstituted oxirane derivatives have *ABC*, *ABX*, or *AMX* spectra. An example for the latter is the spectrum (see Figure 122) of styrene oxide **73a**. The chemical shifts (δH-2, 3.83; δH-3, 2.77; and δH-4, 3.12 ppm)[1044] clearly show the $-I$ effect of the substituent, causing a paramagnetic shift primarily in the signal of the *geminal* ring proton (H-2), and the diamagnetic shielding effect of the ring currents, accounting for the smaller shift in the signal of the *cis vicinal* proton (H-3) in comparison with that of the *trans* proton (H-4).

In Compounds **73b**, the α-protons in the side chain are diastereotopic, and thus their

---

* See p. 49 and Table 36.

Table 29
THE CHEMICAL SHIFTS (PPM) OF THE RING PROTONS OF OXIRANES[1051]

Substitution		Chemical shift or substituent			
		1	2	3	4
Mono		Me	2.66	2.29	
		CMe$_3$	2.60	2.48	
		CN	3.50	3.11	3.02
		Ac	3.37	2.94	3.00
		CH$_2$Cl	3.20	2.84	2.65
		Ph	3.62	2.52	2.82
		COOH	3.47	2.93	2.99
		CHO	3.36	3.10	3.17
		Cl	4.90	2.83	2.75
		OAc	5.33	2.58	2.77
Di	1,2 (*gem*)	Me		2.42	
		Ph	Me	2.55	2.78
	1,3 (*cis vic*)	Me	2.70	Me	2.70
		Ph	3.89	Me	3.12
		Ph	4.17	Ph	4.17
	1,4 (*trans vic*)	Me	2.57		Me
		Ph	3.39	2.79	Me
		Ph	3.67		Ph

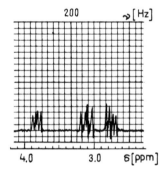

FIGURE 122.   The multiplets of the oxirane protons of stirene oxide (**73**). (From *NMR Spectra Catalog*, VARIAN Associates, Palo Alto, Calif., 1962-1963. With permission.)

signal is split. In several cases, long-range coupling was also observed with the *trans* ring proton (a "W" pattern). The coupling constant ^4J of methyloxirane (**73c**) is 0.5 Hz.[429] The coupling constant of *geminal* hydrogens is probably positive in oxiranes[428] and the most sensitive towards substitution. The range of J^{gem} is 4.0 to 6.5 Hz. In accordance with the *Karplus* relation, $J^c > J^t$ (the former being generally in the range of 2.0 to 5.5 Hz, and the latter being in the range of 1.3 to 3.0 Hz (compare Problem **36**). The coupling constants decrease linearly with the $-I$ effect of the substituents.

3.2.3. Thiiranes
The chemical shift of the protons of ethylene sulfide **74a** with a geometry of planar

R 73a: R = Ph
   b: R = CH₂Cl
   c: R = Me

R R′ 74a: R = R′ = H
       b: R = R′ = Ph
       c: R = H, R′ = Ph

O

S

equilateral (not regular) triangle is smaller (2.27 ppm)[1005] than that of ethylene oxide, corresponding to the lower electronegativity of sulfur. In accordance the coupling constants of the substituted derivatives are at the same time higher. The shift of the ring protons is, similar to the other three-membered cyclic systems, also smaller than that of the linear analogue (the methylene signal of diethyl sulfide appears at 2.49 ppm).[259] From the sparse number of literature data, the following ranges of coupling constants can be derived (e.g., see the data of Problems **63** and **66**): $J^c = 5.0$ to 7.5, $J^t = 5.0$ to 6.6, and $J^{gem} = 0.4$ to 1.5 Hz. Substitution influence the chemical shifts of the remaining ring protons in a manner similar to that of the oxyrane derivatives. The protons of *cis* stilbene sulfide for example, absorb at 3.96 ppm,[1044] whereas the signals of the *geminal*, *cis* and *trans vicinal* protons of phenyl thiirane (**74c**) appear at 3.56, 2.47, and 2.30 ppm, respectively.[915]

3.2.4. Aziridines

In the spectrum of aziridine (**75**), the signals of the amino and methylene protons appear at 0.03 and 1.62 ppm,[1044] at about 0.4- and 0.9-ppm lower shifts than the corresponding signals of the acyclic diethylamine.[1377] The strong shielding around the three-membered ring is, therefore, also present in this case.

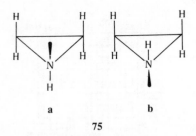

a

b

75

The conclusion on the substituent effect on the chemical shifts and the relative magnitudes of the coupling constants of oxirane and thiirane derivatives also pertains to aziridines. A new aspect is the possibility of the inversion of nitrogen (**75a** ↔ **75b**). The arrangement around the nitrogen atom is not planar (the N–H bond forms an angle of 117° with the plane of the three-membered ring),[1434] and thus the protons "below" and "above" the plane of the ring are nonequivalent. Since, however, the potential barrier of the inversion is very low and thus the inversion is very fast at room temperature, the environments of the four methylene protons are averaged and the hydrogens become equivalent. The signal of the four methylene protons of ethyleneimine is, therefore, a singlet. The rate of inversion depends, of course, on the temperature, and the process can be occasionally freezed-in at low temperature. With certain substituted derivatives of aziridine, this may produce inverses analogous to Structures **75a,b**, and in such cases complicated multiplets may occur instead of the simple spectrum observed at room temperature.

By methods discussed later,* the activation free enthalpy, $\Delta G^\ddagger$, of the inversion can be calculated from the temperature dependence of the spectrum. The usual magnitude of this potential barrier is 40 to 80 kJ/mol, but higher values have also been obtained. The inversion barrier of 1-chloro-2,2-dimethylaziridine, for example, is $\Delta G^\ddagger \approx 98.5$ kJ/mol.[835] In such

* See Section 3.7.11.

cases the inverses* can also be separated preparatively.[210] The temperature dependence of the spectra of 1,2,3-trisubstituted derivatives[176] is very instructive. For instance, the *trans* isomer of 1(*N*)-ethyl-2,3-dimethylaziridine (**76**) shows the expected temperature dependence: at low temperatures the two ring protons and the two methyl substituents are nonequivalent due to the slow interconversion of inverses **a** and **b**. At higher temperatures they become equivalent as a result of a faster inversion and their signals coincide. In contrast, the spectrum of the *cis* isomer (**77**) is temperature invariant. (Although the ring protons of the *cis* isomer are equivalent in both inverses, the chemical shifts of the corresponding singlets are different. Thus, the temperature dependence would manifest itself in a signal shift.) The invariance is due obviously to the fact that the inverse form **77b** is not preferred for energetical reasons (the three bulky substituents are located on the same side of the ring), and thus Structure **77b** may not be present in measurable quantities. In aziridine salts the chemical shifts of the ring protons increase substantially, corresponding to the presence of a positively charged nitrogen atom (e.g., see Problem **42**).

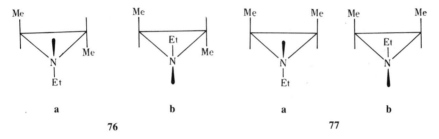

The salts and quaternary derivatives of aziridine are unable to form inverses; in this case the mixture of isomers is easy to distinguish. If, for example, in Compound **78** group R″ is bulkier than R′, isomer **79** is less preferred, and of the signal pairs the weaker must belong to the latter isomer.

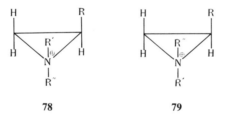

In the case of *N*-substitution, the shielding of ring protons is increased by electron-donating groups (e.g., in the spectrum of 1-ethylaziridine, the *cis* and *trans* protons absorb at 1.1 and 1.7 ppm, respectively; the separation is due to the slow inversion).[1051] In contrast, electron-withdrawing substituents deshield the ring protons (the singlet of ring protons of 1-*para*-tolyl-aziridine appears at 2.40 ppm).[46]

The substituents of the carbon atoms influence the chemical shifts of the remaining ring protons in a manner similar to the cyclopropane or oxirane derivatives. In 2-aryl-aziridines (**80**), for example, the chemical shifts of the ring protons increase in the sequence $\delta A < \delta B < \delta X$.[209]

In 1,2-diphenylaziridine (**81**, R=R′=H), the δH-2 chemical shift (2.83 ppm) is equally increased by 3-chloro or 3-methyl substitution. In the substituted derivatives (**81**, R=Me or

---

* The inverse of aziridine and of its *N*-substituted and symmetrically C-substituted derivatives are identical. In contrast, the inverses of asymmetrically substituted aziridines are optical isomers, and thus they can also be separated by preparative methods.

Cl, R′=H) δH-2: 3.10 and 3.20 ppm. If a further (4-*trans*) methyl or chlorine substituent is introduced (**81**, R=Cl, R′=Me, or Cl), the methyl substitution decreases, and the chlorine substitution increases δH-2 (to 3.10 and 3.53 ppm, respectively).[357] This behavior is similar to that found for cyclopropanes and oxiranes.

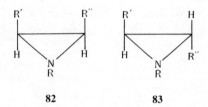

In the case of 1-alkyl-2,3-disubstituted aziridines the paramagnetic shift caused by electron-withdrawing substituents is usually larger in the *trans* isomer.[1120,1433] In the *cis* and *trans* isomers of 1 (*N*)-benzyl-2,3-dibenzoylaziridine (**82** and **83**, R=Bzl, R′=R″=Bz), the chemical shifts of the singlet signals of ring protons are 3.4 and 4.1 ppm. It is interesting that even in the spectra of many asymmetrically substituted derivatives, a singlet or an *AB* quartet very close to the A_2 case can be observed for at least one of the isomers. This is the case, for example, with Compound **82**, R=cyclohexyl, R′=Ph, R″=Bz. In the spectrum of the *cis* isomer, the ring protons give an *AB* multiplet, whereas in the case of the *trans* isomer (**83**, R=cyclohexyl, R′=Ph, R″=Bz), the same signal is a singlet at 3.6 ppm. For the *N*-benzyl analogues (R=Bzl), the singlet of the *trans* isomer appears at 3.65 ppm, and from the *AB* quartet of the *cis* isomer 3.25 and 3.40 ppm is obtained for the chemical shifts of the ring protons.

Of course, the relative chemical shifts of the substituents can also be utilized in structure determination. In the case of the 2-methyl-3-phenylaziridine isomers (**82** and **83**, R=H, R′=Me, and R″=Ph), the chemical shift of the methyl signal is evidently larger for the *trans* isomer, since in the *cis* derivative the phenyl group on the same side of the three-membered ring shields the methyl protons. Indeed, it has been found[209] that Me(*cis*) = 0.53 ppm and Me(*trans*) = 0.83 ppm.

According to the Karplus relation, $J^c > J^t$; J^c is in the range of 6 to 8.5 Hz while J^t = 2 to 4 Hz for the majority of the compounds investigated. *N*-halogenation increases the coupling constants, e.g., for 1-chloro-2-phenylaziridine J^c = 8.5 and J^t = 6.3 Hz,[211] while in position 2 or 3 — similar to other compounds — the electron-withdrawing substituents have an opposite effect (in 2-phenylaziridine, for example, J^c = 6.1 and J^t = 3.3 Hz, whereas in aziridine the corresponding coupling constants are 6.3 and 3.8 Hz).[915] The *geminal* coupling constant is usually small and most probably positive.

The broadening of the signals of certain aziridine derivatives is attributed to the coupling between the nitrogen atom and the ring protons.[1057] The aziridine ring exerts a shielding effect on the neighboring protons, like the cyclopropane and oxirane rings. An example[1413] is the *syn* - *anti* isomer pair **84-85**, in which diamagnetic shift is shown by H_A atoms in isomer **84** and by H_B atoms in isomer **85** relative to **86**: $\Delta \delta H_A$, −0.23 (**84,86**) and 0 (**85,86**); $\Delta \delta H_B$, −0.21 (**84,86**) and −0.39 ppm (**85,86**), respectively.

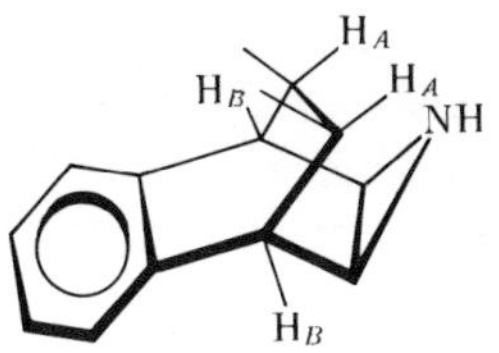

84 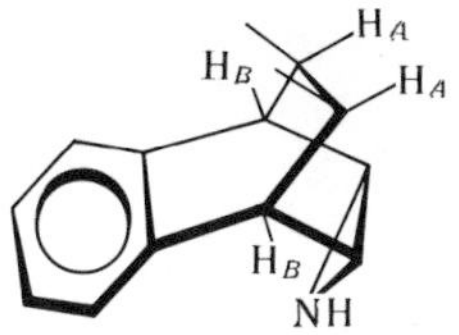　　　　　　　　85 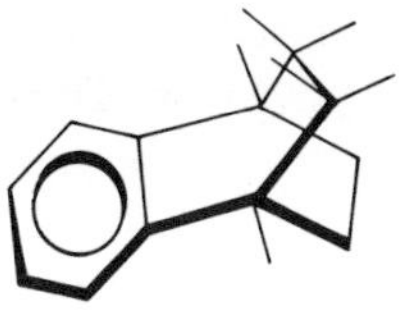　　　　　　　　86

3.2.5. Cyclobutanes and Four-Membered, Saturated Heterocycles

Depending on substituents, the four-membered ring of cyclobutane derivatives is planar or slightly distorted. In the latter case, two types of hydrogens the quasi-*axial* and quasi-*equatorial* ones can be distinguished by ring inversion (**87a** ↔ **87b**).

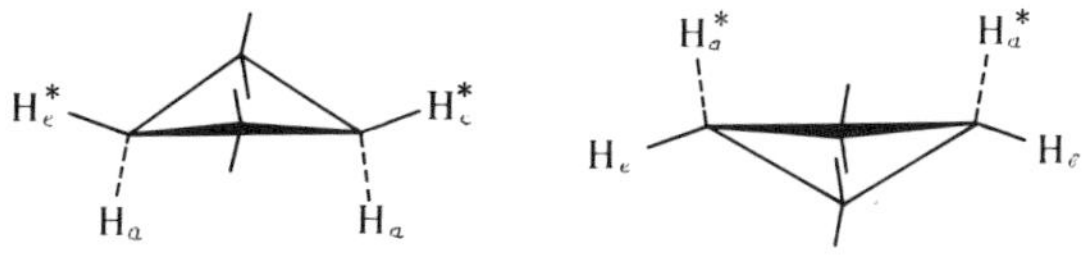

a　　　　　　　　　　b

87

The nonplanar structure of certain cyclobutanes has been proved in addition to IR, X-ray, electron diffraction, and Raman studies, by NMR methods as well. Corresponding to a change in the conformational equilibrium **87a** ↔ **87b**, the coupling constants and the chemical shifts were found to be temperature dependent[812,813] in the nonplanar cyclobutane derivatives, such as in the case of Compounds **88a** and **b** (R=H,Cl, R′=H,D,Br,Me).

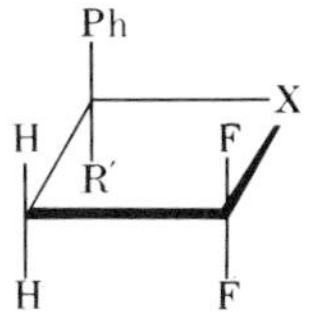

88a: X = CR_2
b: X = C = O

In the case of *cis-* and *trans-*3-isopropyl-cyclobutanol (**89a,b**), the former is conformationally homogeneous (only the conformer bearing the substituents in the position corresponding to the protons H_e^* of formula **87a** is stable), while the *trans* pair is flexible and has a temperature-dependent spectrum.[866] Since the dihedral angles vary with the structure of the ring, the magnitudes of the coupling constants are widely varying ($J^{gem} = -11$ to -15 Hz, $J^c = 4.5$ to 12 Hz, and $J^t = 2$ to 10.5 Hz).[486] For not too strained systems (i.e., where the ring is nonplanar), it can be stated that generally $J^c > J^t$.[1504] By comparing the coupling constants and the dihedral angles determined by independent methods, it has been found that the *Karplus* relation, at least for the relative values, is also valid for cyclobutanes.[812,813] This fact permits one to determine the extent of distortion (angle of torsion) of the four-membered ring from the coupling constants. The results of these calculations are in good agreement with the X-ray diffraction data.

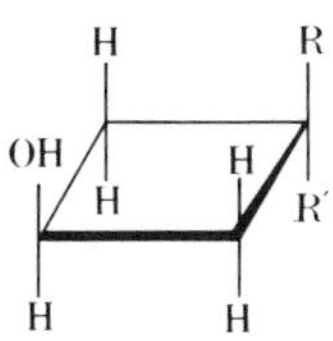

89a: R = $CHMe_2$,
　　　R′ = H
b: R = H,
　　R′ = $CHMe_2$

Cyclobutane gives a singlet at 1.96 ppm[1519] which — assuming a distorted ring — indicates a fast ring inversion. The large diamagnetic shielding characteristic of three-membered rings is, therefore, no more observable, particularly if it is taken into account that the signal of cyclohexane appears at 1.44 ppm.[1519] It has been shown by a systematic investigation of rigid, fused-ring systems that a diamagnetic effect acts in the plane of the ring, whereas "above" and "below" it an opposite effect occurs.[1022]

Depending on the type of substitution, the remaining ring protons may represent various spin systems, giving a tool for structure determination, e.g., the 1,3-R-1′,3′-R′-tetrasubstituted (*cis*) derivatives produce *AA′BB′* or *AA′XX′* multiplets, whereas the corresponding 1,3′-R-1′,3-R′-tetrasubstituted (*trans*) isomers give an A_4 singlet. An interesting illustration is the case[48] of Compound **90**. At the photodimerization of R–CH=CH–COOEt (R = *o*-anisyl) molecule, 11 structures could be assumed differing in the relative (*cis* or *trans*) orientation of the two R and COOEt substituents, respectively. The correct one (**90**) was to be determined by ^{1}H NMR spectroscopy. The nine symmetric structures could be excluded easily on the basis of the four separately appearing methoxy signals, and the two remaining alternatives had to be distinguished from each other by chemical methods.[866]

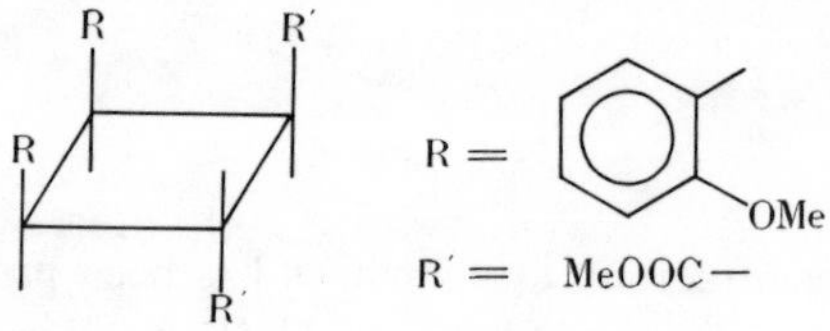

90

Four-membered heterocyclic rings are planar due to the presence of hetero atom. In the spectrum of the oxethane ring (**91**), the triplet-like signal of the α-hydrogens appears at 4.73 ppm and the near quintet of the β-protons appears at 2.72 ppm.[1044] The virtually first-order spectrum is, in fact, a rudimentary $A_2A'_2\,BB'$ multiplet, as indicated by the line broadening and the fine structure observable under higher resolution. The large shift of the α-hydrogens (the methylene signal of diethyl ether is, e.g., at 3.38 ppm[649]) can be attributed to the additive $-I$ effect of the oxygen atom and the anisotropic effect of the C–O bond. From the *AA′BB′* spectrum of 2,2-dideutero-oxethane,[889] $J_{3,3} = -11.15$ and $J_{4,4} = -6.02$ Hz. It can be seen here that the hetero atom with $-I$ effect increases the coupling constant algebraically, i.e., decreases the absolute value of the negative J^{gem}, in agreement with the facts stated earlier.* The magnitude of the positive *vicinal* coupling constants are 8.65 and 6.87 Hz.[889] In contrast to the expectations, only one of these values decreases in comparison with cyclobutane (for which $J^c = 10.4$ and $J^t = 4.9$ Hz.)[973]

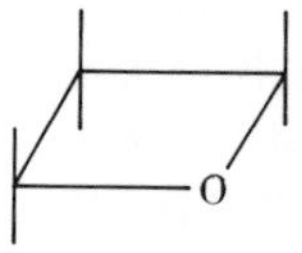

91

In the investigations of azethidines, similar to the aziridines, a central role is played by the study of ring inversion, since the temperature dependence of the NMR spectrum permits the easy determination of the inversion barrier. The investigation of substituted azethidines

* See Volume I, p. 59.

Table 30
THE DIHEDRAL ANGLES BETWEEN THE *CIS* AND *TRANS VICINAL* RING PROTONS OF CYCLOPENTANE IN THE ENVELOPE AND TWIST FORMS[1119]

Position of *vicinal* protons in the ring	Relative position of protons	Dihedral angles			
		Envelope form		Twist form	
1,2≡1,5	cis	46°		15°	
	trans	74°	166°	105°	135°
2,3≡4,5	cis	29°		39°	
	trans	91°	149°	81°	159°
3,4	cis	0°		48°	
	trans	120°		72°	168°

has shown that the extent of ring distortion decreases with the increasing size of the *N*-alkyl substituent, and for bulky substituents planar structure can be expected.[376]

3.2.6. Cyclopentane Derivatives and Their Heteroanalogues

Corresponding to the larger ring size, the conformation motions are less hindered in cyclopentanes. Owing to the nonplanar structure, these compounds may assume various conformations. The conformations can be classified into two types. In the *envelope* form (**92a**), four carbons are in one plane and the fifth is outside this plane. In the other main conformation, the *twist* form (**92b**), only three carbons are in the same plane, and the two remaining are on two different sides of this plane. Ten *envelope* forms are possible, since five carbons may occupy a position "above" or "below" the plane, respectively. They can be converted into one another *via* the *twist* forms by rotation about the C–C bonds. The number of *twist* forms is ten as well. The 20 conformers can be converted into each other continuously in the above manner, and this motion is called *pseudorotation*.[833,1119] The energy difference between the conformers is very small, and thus their simultaneous presence must be assumed. However, the relative positions of the atoms are widely different in the conformers, and so are the dihedral angles of the *vicinal* ring hydrogens (see Table 30).[1119] Therefore, the spectrum parameters are average values, from which the preferred conformation, if any, can be determined only in exceptional, fortunate cases.

a   b

92

Corresponding to the fast conformational motion, the spectrum of cyclopentane consists, at all experimentally available temperatures, of one singlet at 1.51 ppm.[1516] The substituents affect the ring protons in a manner similar to the cyclopropanes and cyclobutanes, but due to the conformational motions the difference in shielding of the *vicinal* hydrogens is usually too low to be of help in the assignment. The signal of the methine hydrogen adjacent to the substituent is shifted paramagnetically to an extent depending on the electron affinity of the substituent. This signal appears, e.g., in the case of iodocyclopentane at 4.32 ppm, whereas the corresponding signal of methylcyclopentane is at 1.81 ppm.[649]

From the relatively limited number of data, the coupling constants have the following values: $J^{gem} = -12$ to -18 Hz, $J^c = 6$ to 9 Hz, and $J^t = 2.5$ to 4.5 Hz. Frequently, "W"-type long-range coupling can be observed, for which $^4J = 0.5$ to 1.0 Hz.

Of the heterocyclic analogues, tetrahydrofuran and its derivatives were studied most extensively, due to their occurrence in carbohydrates of great biological importance. Owing to the adjacent oxygen atom, the α-hydrogens have large chemical shifts, their signal appearing at 3.75 ppm. The relative diamagnetic shift with respect to the oxethane protons is due to the nonplanar structure of the ring. The signal of the β-protons appears at 1.85 ppm.[1044]

The configurational and conformational analysis is generally not easy for tetrahydrofuran derivatives. In the case of 2,5-dimethoxy derivatives, one can hardly find a significant difference between the spectrum parameters of the *cis* and *trans* isomers; the methoxy signals, for example, appear at 3.26 and 3.28 ppm, respectively.[17]

An example for a successful configurational analysis is the structure determination of the three isomers (**93** to **95**) of 2,5-dimethoxy-3,4-dibromotetrahydrofuran.[536] The *trans* position of the bromine atoms and the dominance of the *twist* conformation are evident for steric reasons. Two isomers are symmetric, and the third (**93**) is asymmetric. Consequently, the compound having a spectrum with two methoxy signals (at 3.42 and 3.47 ppm) must correspond to Structure **93**, whereas the two remaining spectra can be paired with Structures **94** and **95** on the basis of the coupling constant $J_{3,4} \equiv J^t$. The ring protons of these compounds produce $AA'XX'$ spectra, from which the coupling constant $J_{AA'}$ can be determined. The dihedral angle between H-3 and H-4 is much smaller (approximately 72°) in **94** than in Structure **95** (approximately 168°, see Table 30), and thus, according to the *Karplus* equation, the spectrum with greater J^t must correspond to Structure **95**. Since from the two spectra $J^t = J_{AA'} = 6.8$ and 1.7 Hz, the pairing of spectra and structures is straightforward. Note that the chemical shift differences are insufficient to distinguish between the isomers.

93   94   95

The α- and β-protons of pyrrolidine absorb at 2.82 and 1.65 ppm, respectively.[1075] To distinguish between the *cis* and *trans* isomers of symmetrically 2,5-disubstituted *N*-benzyl-pyrrolidines (**96**), the benzyl-methylene signals can be utilized. In the *cis* isomers these methylene protons are equivalent, giving a singlet, whereas they are diastereotopic in the *trans* derivatives, to yield and *AB* multiplet.[651]

96

The salts of pyrrolidine may produce a mixture of *cis* and *trans* isomers which can be distinguished in asymmetric molecules. All the signals are doubled, owing to the presence of two isomers, and the relative intensities of the corresponding signals are the same, giving the ratio of the isomers in the mixture. The assignment of the *cis* and *trans* forms to the spectra is usually straightforward from the molecular structure. In the case of isomers **97** and **98**, the former is more stable for steric reasons, and thus the more intense signals in the spectrum of the mixture belong to this isomer. The configurations of several pyrrolidinium salts have been determined on this basis.[96,97]

As a consequence of the conformational flexibility, the *vicinal* coupling constants span a very wide range (0.5 to 12 Hz), and since the conformational equilibrium is also influenced by the pH the ^{1}H NMR spectrum of pyrrolidine derivatives is pH dependent (not only the NH signal, but the chemical shift of the ring protons and the coupling constants as well).[10]

The isolated methylene group of the dioxolane ring (**99**) absorbs at 4.57 ppm, and the other two methylenes absorb at 3.57 ppm.[250] This compound is a good example for the additivity of substituent effects (in this case the paramagnetic effects of the two oxygen atoms). It can be derived from the spectra of the *cis* and *trans* isomers of 2,2,4,5-tetra-methyldioxolane, by analyzing the $AA'X_3X'_3$ multiplets, that (unlike in other five-membered and smaller ring systems) $J^c = 5.85 < J^t = 8.35$ Hz, respectively.[791]

3.2.7. Cyclohexane Derivatives and Their Heteroanalogues

The investigation of cyclohexane derivatives is one of the most extensively cultivated topics in ^{1}H NMR spectroscopy for two reasons. First, a large number of cyclohexane derivatives and their heteroanalogues occur among the organic compounds, and the majority of them are important from biological aspects (terpenes, steroids, carbohydrates with pyranose rings, quinolisidines, etc.). The second reason is the particular value of the ^{1}H NMR method in the configuration and conformation analysis of these compounds. In comparison with the analysis of the ring systems discussed so far, the investigation of cyclohexanes is easier because, on one hand, stable chair conformations may be present, usually with quite high potential barriers between them, and on the other hand, the substituents often force the molecule into a particular conformation, thereby reducing the task to the study of conformationally homogeneous systems. The application of ^{1}H NMR spectroscopy in conformational analysis will therefore be discussed here in some detail.

3.2.7.1. Conformational Analysis by ^{1}H NMR Spectroscopy

In the previous sections, several examples illustrated how the configurations of cyclic compounds with smaller rings can be determined from the chemical shifts and coupling constants. These parameters are very sensitive to the steric structure, and thus they provide two, often independent, sources of information in the conformational analysis. Compounds with different configurations (e.g., the *cis* and *trans* isomers) cannot be converted into one another without breaking the chemical bonds. In contrast, the conformers are atomic arrangements, differing in their steric structures, that are interconvertible by the simple rotation

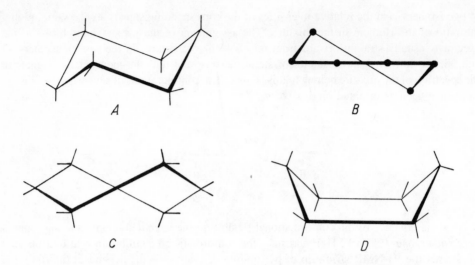

FIGURE 123.   The particular conformations of cyclohexane. (A) The chair, (B) half-chair, (C) twist, and (D) boat conformations.

of atoms about the chemical bonds joining them. For this reason, the determination of conformations is usually difficult, since the rotation about the bonds is generally free at room temperature, and thus the conformers are in equilibrium. Even if this is not the case, the conformers of the same molecule are hard to distinguish due to their very similar structures. Conformational analysis has been revolutionized by NMR spectroscopy. This method enables one to not only determine the steric structures of conformationally homogeneous molecules, but also to simply analyze conformational equilibria. This analysis makes accessible the relative weights (amounts) of the conformers, the activation free enthalpy, $\Delta G^{\ddagger}$, of the conformational motions (i.e., the potential barrier), and all thermodynamical parameters derivable from it. The calculated values are fairly accurate, and the calculations are very simple.

At any rate, it must be known that this method is inferior in accuracy to the X-ray or neutron diffraction methods, but the situation is anyhow more complex in solution than in solid or gas phases, and the technique is much simpler, requiring less time and labor.

From the aspects of conformational studies one may distinguish rigid and flexible molecules. The former may assume only one conformation, while the latter produce and equilibrium between the various conformers, which is temperature dependent.

The most stable conformation of cyclohexane is the chair form (see Figure 123 A). The dihedral angle of the *vicinal* C–H bonds is 180° for two *axial* hydrogens and 60° for the *equatorial* ones or a pair of *axial* and *equatorial* hydrogens. Upon rotating the atoms about the C–C bonds, the half-chair (B) and the twist-boat (C) conformations lead to the boat conformation (D), and through a reversed sequence of conformers a chair conformation (A') is obtained again. This series of motions is called ring inversion, since conformer A' is the inverse of A. The hydrogens exchange positions pairwise during the inversion, i.e., the originally *axial* ones move into *equatorial* position and vice versa. Since there are repulsion forces between the hydrogen atoms, varying with their mutual distances, the true energy minimum is not the A = A' chair form, but a slightly distorted variant of it, A* ≡ A'*,[339] in which the dihedral angles are $\varphi_{a,a} = 174°39'$, $\varphi_{a,e} = 54°39'$ and $\varphi_{e,e} = 65°21'$ (in this conformation the 1,3-*diaxial* interactions are weaker). For the same reason, molecules with unstable chair or distorted chair conformations assume, instead of the boat conformation D, a distorted, twistboat conformation (C). This is illustrated by the schematic curve of the ring inversion in Figure 124. The diagram shows that in most cases only the chair conformations must be taken into account.

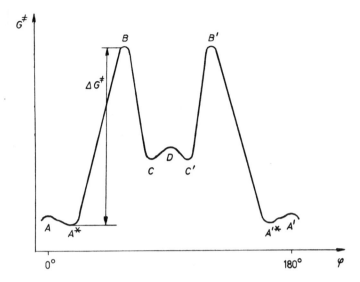

FIGURE 124.   The schematic potential curve of the ring inversion of cyclohexane.

At room temperature, the spectrum of cyclohexane consists of one, sharp singlet at 1.44 ppm, due to the fast ring inversion A $\rightleftharpoons$ A'.[1519] This singlet separates into two broad signals at $-70°C$ with a chemical shift difference of $\Delta\delta \approx 0.5$ ppm. This investigation has also revealed that $\delta H_e > \delta H_a$ and $J_{a,a} \gg J_{a,e} \approx J_{e,e}$.[42,181]

Both inequalities have proved to be of general validity for cyclohexane derivatives and even for the great majority of analogous heterocyclic compounds, and they have become the fundamental principles of configurational and conformational analysis. The relation $\delta H_e > \delta H_a$ was derived from the low temperature studies of appropriately substituted model compounds. The relation $J_{a,a} \gg J_{a,e} \approx J_{e,e}$ is already obvious from the *Karplus* relation, but in the knowledge of the relative chemical shifts, it also follows from the fact that the multiplet of the *axial* hydrogens with the smaller shift is much broader at $-70°C$ than that of the *equatorial* protons. From the coalescence temperature ($-67°C$), the potential barrier of the ring inversion was determined* and $\Delta G^{\ddagger} \approx 40$ kJ/mol was obtained.[717]

To determine the potential barrier, it is necessary to know the relative amounts of the conformers or the equilibrium constant, K.** If the ring inversion is slow, and the signals corresponding to the various conformers do not overlap with one another or with other signals, the intensity ratio of any corresponding line pair can be used to determine the value of K:

$$K = p/(1 - p) \tag{272}$$

where p and $(1 - p)$ are the mole fractions of the two conformers. The mole fractions can be determined from the intensity ratio of any corresponding line pair, regardless whether the signal belongs to one of the ring protons or to the ones of a substituent.

The hydrogens of *axial* and *equatorial* R substituents have different shielding, e.g., just like the *geminal* ring protons. If, therefore, a signal of substituent R is a line pair with a relative intensity of R_eR_a, the intensity ratio H_aH_e for the *geminal* protons will be the same.

If the signals of the conformers are not separated at the temperature of the measurement

* Compare p. 94.
** See p. 94 and 95.

owing to a fast ring inversion, K can also be determined from the average chemical shift, provided that δ_a and δ_e are known, and the differences $\delta - \delta_a$ and $\delta_e - \delta$ are large enough (i.e., the free enthalpy difference of conformers $\Delta G^{\ddagger}$ is large). The average shift can be expressed as

$$\delta = p\,\delta_a + (1-p)\delta_e \tag{273}$$

The problem here is the determination of δ_a and δ_e. This can be accomplished in two ways. The simpler method is to freeze in the ring inversion by decreasing the temperature to a value where the signals of the two conformers separate. It often occurs, however, that the inversion is too fast even at the attainable lowest temperature, or the signals do not separate for other reasons.

The shifts δ_a and δ_e of cyclohexane derivatives can also be determined from model compounds. The most suitable model compounds are the *cis* and *trans* isomer pairs of 4-*t*-butyl derivatives.

100   **101**

According to the concordant results of quantum chemical calculations, X-ray and NMR measurements, the bulky *t*-butyl group is always *equatorial*, and this substitution produces, therefore, conformationally homogeneous systems. The chair conformation containing the *t*-butyl group in *axial* position can therefore be disregarded as well as, for example, the half-chair or, for monosubstituted derivatives, the twist conformations. Thus, the chemical shift, e.g., of the ring hydrogen *geminal* to substituent R in the *cis* isomer (**100**), can be regarded as δ_e^{gem} and in the *trans* isomer (**101**) as δ_a^{gem}. With these values:

$$K = p\,/(1-p) = (\delta_e - \delta)/(\delta - \delta_a) \tag{274}$$

Already known δ_a and δ_e data of related molecules can also be used, but in this case the approximation may become poorer. The equilibrium constant K can also be determined from the average of the coupling constants, if the different values characteristic of the conformers are known. Taking the example of monosubstituted cyclohexane derivatives, the multiplet arising from the coupling of the *geminal* hydrogen with the two *vicinal* methylene groups can be regarded as the X part of an $AA'BB'X$ multiplet (the long-range couplings with more distant protons are negligible). The spacing of the two outer lines of the X multiplet is $2(J_{AX} + J_{BX})$.* Thus, instead of the average of the sum of coupling constants, the signal widths W can also be used to determine K:

$$K = (W_a - W)/(W - W_e) \tag{275}$$

where W is the average signal width measured in the case of conformational equilibrium, and W_a or W_e are the signal widths in the spectra of the pure conformers. The latter parameters can be determined again from low-temperature measurements or from the spectra of model compounds similar to isomers **100** and **101**.

---

* In first approximation, and assuming that $J_{AX} + J_{BX} \approx J_{A'X} + J_{B'X}$.

3.2.7.2. Cyclohexane Derivatives

Conformational and configurational analysis of cyclohexane derivatives are facilitated primarily by the following facts.

1. As a general rule for cyclohexanes, the chemical shift of the *equatorial* ring protons is larger than that of the otherwise identical *axial* ones. The rule pertains equally to the hydrogens with different steric positions in the two conformers of the same compound, the corresponding protons of *cis-trans* isomer pairs, and to the *geminal* hydrogens of methylene groups which are not affected noticeably by the occasional neighboring substituents. As already mentioned,* the relationship $\delta H_e > \delta H_a$ can be attributed to the anisotropy of C–C bonds, which increases the shielding of the *axial* protons. These cases are illustrated by the data of Table 31, also listing the data of some steroids and carbohydrates in addition to the simple cyclohexane derivatives for which the $\Delta\delta_{a,e}$ chemical shift difference is generally 0.3 to 0.9 ppm. The majority of data originate from low-temperature measurements.

2. The statement concerning the relative magnitudes of coupling constants is of general validity as well, i.e., $J_{a,a} > J_{a,e} \geq J_{e,e}$, also holds for cyclohexane derivatives. The ranges of the coupling constants $J_{a,a}$ and $J_{a,e}$ or $J_{e,e}$ for simple cyclohexane derivatives, steroids, as well as for carbohydrates are 4.5 to 12.5 and 2.0 to 4.5, 6.0 to 10.0 and 2.0 to 4.0, as well as 7.5 to 14.0 and 1.0 to 7.5 Hz, respectively. For molecules related in type $J_{a,e}$ is, in general, about 1 Hz greater than $J_{e,e}$.

3. The energy difference between the conformers increases with the size of substituents, and with appropriate substitution (e.g., butyl), conformationally homogeneous systems can be produced even at room temperature. Such systems have been used to prove the general rules given under (1) and (2) and for empirical data collection.

From the spectra of **100** and **101** (R=OH), $\delta H\text{-}1a$ (**101**) = 3.45 and $\delta H\text{-}1e$ (**100**) = 4.00 ppm. For Compound **101** (R=OH) $J_{1,2}^{trans} \equiv J_{a,a} \approx 11$, $J_{1,2}^{cis} \equiv J_{a,e} \approx 4.2$ Hz, whereas from the spectrum of **100** (R=OH) $J_{1,2}^{trans} \equiv J_{e,e} \approx 2.7$ and $J_{1,2}^{cis} \equiv J_{e,a} \approx 3.0$ Hz. These parameters have been determined from the analysis of the *ABX* and *AA'BB'X* spectra of partially deuterated compounds (3,3,4,5,5,6,6-hepta- and 3,3,4,5,5-penta-deuterated derivatives).[40,1425]

The chemical shift of cyclohexane (1.44 ppm)[1519] is slightly larger than that of the methylene signal of *n*-hexane (1.25 ppm).[649] This paramagnetic shift of 0.2 ppm can be attributed to the anisotropy of C–C bonds.** It is noted that the chemical shift of the ring protons is roughly the same in the homologues with larger rings, and the corresponding signal always appears between 1.3 and 1.6 ppm.

It is particularly useful in conformational analysis that, as already discussed with smaller rings, the chemical shift of the *geminal* ring hydrogen is strongly influenced by the substituents, and thus its signal appears at larger chemical shifts, generally well separated from the signals of the other ring hydrogens. Consequently, its accurate chemical shift, line width, and multiplicity can readily be determined, and the resulting data may be used as a starting point in the conformational analysis. The chemical shifts of the *geminal* ring protons of some monosubstituted cyclohexane derivatives are given in Table 32.

The spectra of monosubstituted cyclohexanes are rather complex, and the ring protons, except for the *geminal* H_X, give a featureless, overlapping absorption. Even if one attempts to interpret the structure of the *X* multiplet as the *X* part of an *AA'BB'X* spectrum (neglecting long-range couplings), the problem is still quite difficult. For this reason it is important that, as discussed above, the configuration and conformation, or the conformational equilibrium

---

* See Volume I, p. 34.
** See Volume I, p. 34.

Table 31
THE CHEMICAL SHIFT DIFFERENCES (PPM) BETWEEN THE *EQUATORIAL* AND *AXIAL* RING PROTONS FOR SOME CYCLOHEXANE, ACETYLATED CARBOHYDRATES, AND STEROIDS[1147,1379]

Compound	$\Delta\delta_{a,e}$
Cyclohexane	0.48
Fluorocyclohexane (δH^{gem})	0.45
Bromo- and chlorocyclohexane (δH^{gem})	0.72
Iodocyclohexane (δH^{gem})	0.75
Cyclohexanol (δH^{gem})	0.60
Acetylcyclohexanol (δH^{gem})	0.31
Cis and *trans* 2-*t*-butylcyclohexanol (δH-1)	0.76
Cis and *trans* 3-*t*-butylcyclohexanol (δH-1)	0.64
Cis and *trans* 4-*t*-butylcyclohexanol (δH-1)	0.40
2-Bromo-4-phenylcyclohexanone (δH-2)	0.49
2-Bromo-4-*t*-butylcyclohexanone (δH-2)	0.38
δ-1,2,3,4,5,6-hexachlorocyclohexane (δH-1 to 6)	0.51
ϵ-1,2,3,4,5,6-hexachlorocyclohexane (δH-1 to 6)	0.20
α- and β-D-glucose pentaacetate (δH-1)	0.45
α-L- and β-D-arabinose tetraacetate (δH-1)	0.69
11α- and 11β-hydroxy-progesterone (δH-11)	0.43
Androsterone and epiandrosterone (δH-3)	0.45
6α- and 6β-chloro-5α-cholestan-7-one (δH-6)	0.42
6α- and 6β-fluoro-5α-cholestan-7-one (δH-6)	0.11
2α- and 2β-bromo-lanost-8-en-3-one (δH-2)	0.04

can also be determined from the chemical shift δX or the signal width alone, provided that the X signal appears separately in the spectrum. Fortunately, the signal of the *geminal* ring hydrogen is shifted substantially by hydroxyl, ether, and acetoxy substituents (see Table 32), and the cyclohexane derivatives of biological importance and their heterocyclic analogues usually contain hydroxyl groups. The acetylation of these groups causes further paramagnetic shifts of 1 to 1.5 ppm in the signal of the *geminal* ring hydrogens, which is under all circumstances sufficient for the corresponding lines to occur separately. For acetoxy derivatives, it is a general rule that $\delta CH_3 (OAc_a) > \delta CH_3(OAc_e)$.[260,859,1014]

Of the disubstituted cyclohexane derivatives, the *trans* isomers are conformationally homogeneous and the substituents are *diequatorial* in both the 1,2- and 1,4-disubstituted derivatives. Consequently, their spectra are independent of temperature, unlike the *cis* isomers which, owing also to the *axial-equatorial* positions of the substituents, exist in a conformational equilibrium. In the case of 1,3-disubstituted cyclohexanes, the situation is reversed: the *cis* isomer containing both substituents in *equatorial* position is conformationally homogeneous, whereas in the *trans* isomer one-one substituent is in *axial* and *equatorial* position, respectively, and the spectrum is temperature dependent due to the conformational equilibrium of the ring inverses.

The investigation of dimethylcyclohexanes has revealed some interesting results that can be utilized to advantage in structure determination.

1. The shift of the methyl signal is larger for the *axial* groups, i.e., $\delta(Me)_a > \delta(Me)_e$.
2. The coupling between the methyl protons and the *geminal* ring hydrogen is significantly different for the *axial* and *equatorial* methyl groups (the difference is approximately 1 Hz), and $J_{Me_aH_e} > J_{Me_eH_a}$.
3. *Equatorial* methyl groups cause a diamagnetic shift for the *vicinal axial* ring protons,

Table 32
THE CHEMICAL SHIFTS (PPM) OF THE
GEMINAL RING PROTONS OF
MONOSUBSTITUTED CYCLOHEXANES[168]

Substituent	δH^{gem}	Substituent	δH^{gem}
Me	1.37	OMe	3.05
COOEt	1.68	OH	3.52
COOH	2.30	NHAc	3.75
Ph	2.50	Br	4.07
NH$_2$	2.65	Ts	4.18
SH	2.72	NO$_2$	4.38
SPh	2.98	OAc	4.58

but do not affect to noticeable extent the *equatorial* pairs of these hydrogens (i.e., the protons *trans* to the methyl group).

4. *Axial* methyl groups cause a paramagnetic shift in the signal of the *trans vicinal*, i.e., *axial* hydrogens.

For example, the above effects explain why the *axial* (A) hydrogens of *cis-cis* 1,3,5-trimethylcyclohexane (**102**), i.e., the 2*a*,4*a*,6*a*-hydrogens, are strongly shielded, (δH_A: 0.47 ppm!)[1271] This chemical shift is 1.17 ppm smaller than that of the *equatorial* (B) protons. The small chemical shift of H-4*a* in the *trans-trans* isomer **103** (δH-4*a* = 0.47 ppm) and the corresponding large chemical shift difference for the ring protons 4($\Delta\delta_{a,e} \equiv \Delta\delta AB$ = 1.05 ppm) have the same reason.

102   **103**

The derivatives of *t*-butylcyclohexane with homogeneous conformation are suitable models for studying the effects of substituents on the chemical shifts of the ring protons. In *cis* 1-nitro-4-*t*-butylcyclohexane the nitro group is *axial* and causes a specific paramagnetic shift in the signal of the *vicinal, equatorial* hydrogens (i.e., *cis* to the nitro group). The signal is shifted to 2.58 ppm; the chemical shift of their *axial* pairs is below 2 ppm. In comparison with the spectrum of the *trans* isomer, $\Delta\delta_{a,e}$ = 0.19 ppm[506] for the proton *geminal* to the nitro group. Further illustrations for the conformational analysis of cyclohexane derivatives can be found in Problems **35** and **45**.

104

The bicyclic and tricyclic fused cyclohexane derivatives are rigid molecules and can therefore be used to study the relationships between the spectrum parameters and the steric structure. If, however, the molecule contains no substituent that would affect the chemical shifts, the spectrum consists of overlapping, featureless signals or only one broad or even sharp signal appears in the spectrum, owing to the hardly different chemical shifts. The spectrum of [2.2.0]-bicyclohexane (**104**) consists of one, very broad line ($\Delta v \approx 75$ Hz) around 2.35 ppm.[319] In the spectrum of adamantane (**105**), the hardly separated maxima of the methylene and methine protons appear at 1.80 and 1.77 ppm.[649] Adamantane consists of three cyclohexane rings in chair conformation, whereas the rings of twistane (**106**) are all in the twist-boat conformation. In the spectrum of the latter, the methine signal appears at 1.63 ppm, the signal of the methylene groups inserted between the methine groups is at 1.33 ppm, and the adjacent methylene groups absorb at 1.55 ppm (**17**).[267,1517]

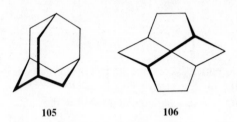

105 **106**

3.2.7.3. Steroids

Due to their biological importance, distinguished attention is paid to this family of compounds. The intensity of interest in this field is indicated by several reviews[1079,1293] and monographs.[131,1032]

Owing to the large number of protons attached to the skeleton, steroids have very complex spectra, in which the methylene and methine signals merge into a featureless, broad absorption between 0.5 and 2.5 ppm. From this broad absorption generally only the signal of the angular methyl groups (positions 18 and 19) emerges. Of course, at higher frequencies, the signals of the protons attached to carbon atoms bearing OH, OR, or O-acyl groups, and the signals of the aromatic and olefinic protons if there are such groups in the molecule occur separately. Moreover, characteristic and easily assignable signals are shown by various substituents (e.g., ethinyl, dimethylamino, t-butyl, methoxy, acetoxy, formyl, etc. groups). The rigid, fused ring system offers several possibilities for studying the relationships between structure and spectrum parameters, and the relationships obtained may give a clue in the solution of several structure determination and spectroscopic problems (configurational and conformational analysis, long-range anisotropic effects and couplings).

In the spectrum of androstan-11-one (**107**), the sharp singlets of the 18- and 19-methyl groups at 0.67 and 1.03 ppm are superimposed on the broad absorption, about 80 Hz in width, of the skeletal hydrogens (see Figure 125). Apart from these signals, only the signals of the 12-methylene protons and the H-1e are separated at 2.27 and 2.55 ppm, respectively.[1044] The paramagnetic shift of these signals is due to the anisotropic effect of the carbonyl group.

107

The methyl signals are in the usual high-field range (compare Table 22), and their positions are strongly influenced by the steric structure of the skeleton. Corresponding to the six

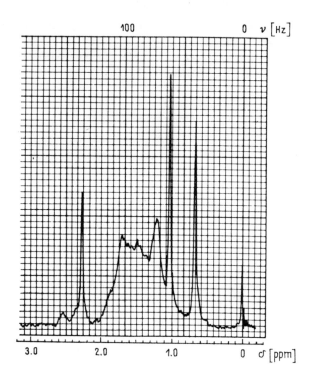

FIGURE 125. The spectrum of androstan-11-one (**107**). (From *NMR Spectra Catalog*, VARIAN Associates, Palo Alto, Calif., 1962-1963. With permission.)

centers of asymmetry of androstane, there are, in principle, 32 possible anomeric pairs (diastereomers) with different steric structures. However, only four of these diastereomers must be taken into account in practice because the other isomers and their derivatives are very rare among the steroids known so far. The chemical shifts of the methyl signals of the four stereoisomers are given in Table 33. It is clear from the data that the shielding of the 18-methyl group depends on the relative configuration of the C and D rings, whereas that of the 19-methyl group is affected by the relative position of the A and B rings. The differences are caused by the anisotropy of the cyclohexane rings, i.e., of the C–C bonds.*

In the normal (5α,17α) steroids δMe(18) ≈ 0.7 < δMe(19) ≈ 0.8 ppm. The chemical shift of the methyl signal is influenced, sometimes to an even greater extent (by 0.7 to 0.8 ppm), also by the substituents. They usually increase the chemical shifts, and the stronger the effect the closer the substituent to the methyl groups, e.g., a 1-oxo group increases the chemical shift of the 19-methyl group by 0.38 ppm, but increases by only 0.02 ppm the shift of Me(18). On the other hand, in 17-oxo derivatives, δMe(19) increases by 0.02 ppm and δMe(18) increases by 0.17 ppm.[131]

On the basis of investigations extended to the spectra of almost 200 steroids, the substituent effects have been studied systematically by Zürcher,[1581] and it has been found that the effects of substituents are additive. Further data can also be found in the literature.[131,1079]

Although the additivity is not exclusively valid and a number of exceptions can be found, it is very useful in the structure elucidation of unknown steroids, e.g., according to Zürcher's data, the 11-oxo group of Compound **107** causes a shift of the Me(18) and Me(19) signals by +0.21 and −0.03 ppm, respectively, whereas the measured chemical shifts of 1.03 and

* Compare Volume I, p. 34.

Table 33

THE CHEMICAL SHIFTS (PPM) OF THE METHYL SIGNALS OF DIASTEREOMERIC ANDROSTANES[1581]

Configuration	Ring anellation		δMe(18)	δMe(19)
	A/B	C/D		
5α,14α	trans	trans	0.69	0.79
5α,14β	trans	cis	0.99	0.77
5β,14α	cis	trans	0.69	0.92
5β,14β	cis	cis	0.99	0.90

0.67 ppm [see Figure 125 (**27**)][1044] differ by +0.24 and −0.02 ppm from the basic values given in Table 33.

The exceptions from the additivity can all be traced back to the deformation of the ring system and can be observed mainly when conjugation, unsaturated bonds, or fused strained rings (primarily oxiran rings) occur or polar, bulky substituents get close to one another (e.g., in 12,20-diones). Thus, the anomal effects can be forecast, and in the possession of appropriate models even their magnitudes are predictable.

The use of the additivity of substituent effects is illustrated by the structure identification of an unknown obtained by microbiological hydroxylation of the steroid **108** (R=H). Since the spectrum contains no signal with a large chemical shift characteristic of methine protons *geminal* to the hydroxyl group, it is clear that a tertiary hydroxyl group has been formed, i.e., only an 8-, 9-, or 14-substitution is compatible with the spectrum. As, however, the relative shifts of δMe(18) and δMe(19) with respect to the starting material are about the same in the spectrum of the product (+0.18 ppm),[1415] the substituent must be located symmetrically between these groups, i.e., in position 8 (**108**, R=OH). The data of Zürcher predict the same magnitude of shift.[131,1581] A 9-hydroxyl group would not affect the signal of Me(18) and would cause a change of +0.14 ppm in δMe(19), whereas a 14-hydroxyl substituent would leave δMe(19) unaffected and cause a paramagnetic shift of 0.12 ppm in the singlet of Me(18). Of course, the substituents also influence the shielding of the neighboring skeletal hydrogens.

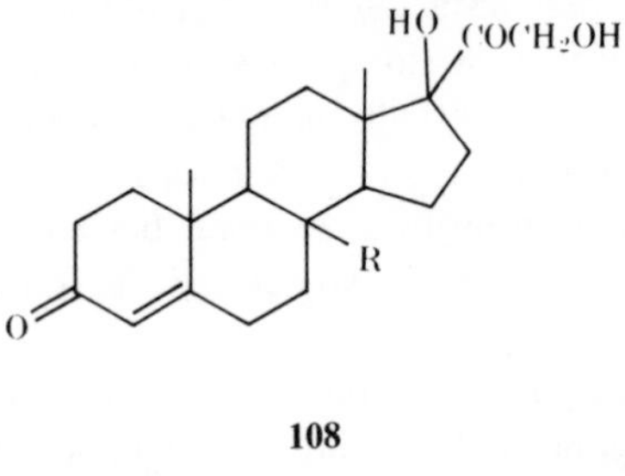

108

A carbonyl group generally deshields the adjacent protons, i.e., a paramagnetic shift can be expected, and the signals appear most probably in the range of 1.9 to 2.6 ppm, as in the case of **107** for the methylene protons in position 12. The effect is usually noticeable for only the α-hydrogens, and thus the signals separated from the broad absorption of the skeletal hydrogens are equal to the number of hydrogens adjacent to the carbonyl group. It must be considered, however, that, depending on the position of the carbonyl group, other hydrogens

Table 34
THE CHEMICAL SHIFTS AND SHIFT DIFFERENCES (PPM) OF H-6 IN THE SPECTRA OF ANOMERS 109[1039]

X	H-6a (X_α)	H-6e (X_β)	$\Delta_{a,e}\delta$H-6
F	4.51	4.40	0.11
Cl	4.43	4.01	0.42
Br	4.63	4.13	0.50
I	4.91	4.37	0.54

may also be affected by it. It has already been mentioned that with Compound **107** the δH-1e shift is even larger than that of the 12-methylene protons in α-position. The same situation occurs in the spectra of androstan-7-one and its derivatives. In these spectra the H-15e signal is also shifted, not only those of the 6-methylene and 8-methine hydrogens in α-position.[128] If in a given steric structure certain protons get into the shielding cone of the carbonyl group, opposite shifts may also occur, e.g., in Compound **107** H-9 is in such a position, and the paramagnetic shift arising from its α-position is overcompensated by the anisotropic effect. The signal appears at 0.78 ppm.[1044]

In the spectra of α-halogenated keto derivatives the signals of the α-hydrogens are well separated in the range of 4.0 to 5.1 ppm. The rule $\delta H_e > \delta H_a$ generally true for cyclohexanes (and also for steroids) is reversed for these compounds. The investigations of the anomers of cholestane derivatives with Structure **109** have led to the results listed in Table 34.[1039] An analogous phenomenon was also observed earlier for α-bromocyclohexanones.[1505]

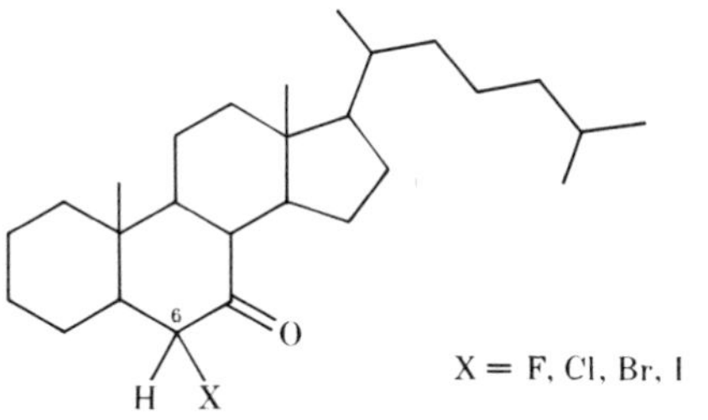

X = F, Cl, Br, I

109

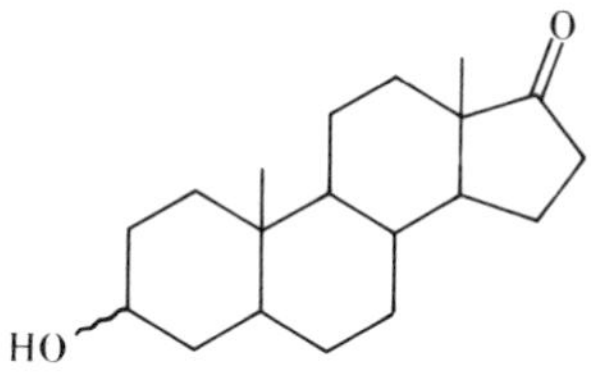

110

Like in halogenated steroids, the signals of the methine protons adjacent to hydroxy groups are also separated from those of the skeletal hydrogens. These protons absorb in the range of 3.5 to 4.5 ppm, and upon acetylation a further paramagnetic shift of about 1 ppm can be observed. Hence, the range of the analogous signals of acetoxy derivatives is 4.6 to 5.6 ppm.[1293] The anomers can be identified on the basis of the rule $\delta H_e > \delta H_a$. For anomers **110**, δH-3a = 3.58 and δH-3e = 4.15 ppm; $\Delta\delta_{a,e}$ = 0.57 ppm.[1293]

In addition to the chemical shifts, the magnitudes of coupling constants may also be useful in the determination of configurations. The relation $J_{a,a} > J_{a,e} \approx J_{e,e}$ is invariably valid, for steroids, too. The range of *diaxial* couplings is approximately 7.5 to 14 Hz, and that of the $J_{a,e}$ and $J_{e,e}$ couplings is 1 to 7.5 Hz. Thus, from the splittings of the signal of the hydrogens adjacent to the functional groups, or if the lines of these signals overlap from

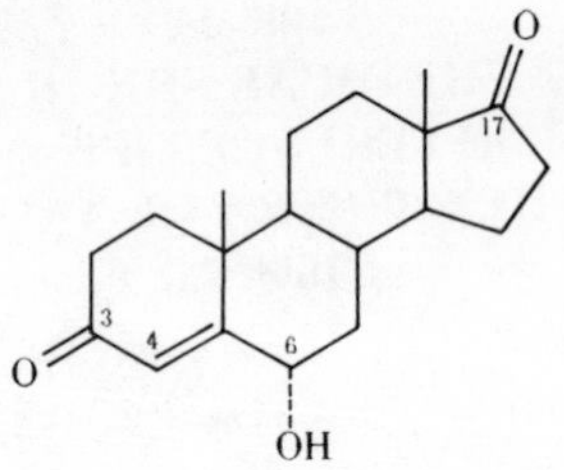

111

the width of the latter, one can determine the steric positions of the substituents. The splitting or signal width of the multiplets may also be of help in the assignment when the molecule contains more identical (e.g., acetoxy) groups.

In the spectrum of **111** the multiplet of H-6, split into four lines, appears at 5.05 ppm (see Figure 126). The line frequencies are 300, 302, 304, and 306 Hz. Accordingly, $J_{AX} + J_{BX} \leqslant 6$ Hz. (The interaction of H-6 and the 7-methylene protons produces an *ABX* multiplet.) Consequently, it can be excluded that either J_{AX} or J_{BX} would correspond to a *diaxial* coupling (in the *AMX* approximation $J_{AX} \approx 2$ Hz and $J_{MX} \approx 4$ Hz). Since, however, one of the hydrogens in position 7 is certainly *axial*, H-6 may not be *axial*, only *equatorial* (β). Accordingly, the compound is the 6α-hydroxy derivative. The further assignment is δMe(18) = 0.98 (from the data of Zürcher, exactly the same shift is obtained), δMe(19) = 1.58 (according to Zürcher 1.45), δOH = 2.77, and δ($=C_4H$) = 5.96 ppm.

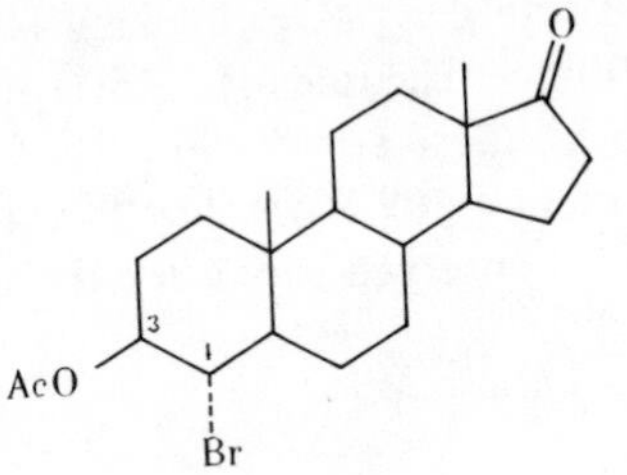

112

Depending on the position of the substituent, the *geminal* hydrogen may be adjacent to two, three, or four protons, and the corresponding multiplet varies in complexity accordingly (the X part of an *ABX*, *ABMX*, or *ABMNX* spectrum). The above example illustrated the simplest (*ABX*) case. If more complicated multiplets occur, their lines usually overlap, and the magnitudes of the coupling constants can be estimated only from the line widths. In the spectrum of **112** e.g., H-3 is coupled with the 2-methylene protons and with H-4. Thus, its multiplet corresponds to the X part of an *ABMX* spectrum (see Figure 127). The complex

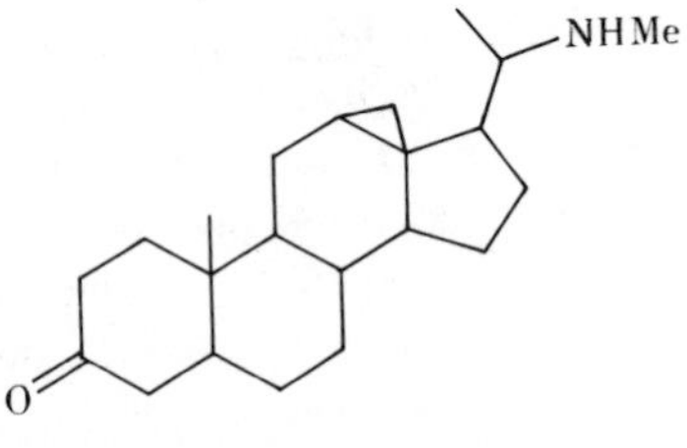

113

multiplet at 4.83 ppm can be distinguished from the noise only at high gains, and its fine structure is quite uncertain. However, the width of the multiplet (approximately 25 Hz)

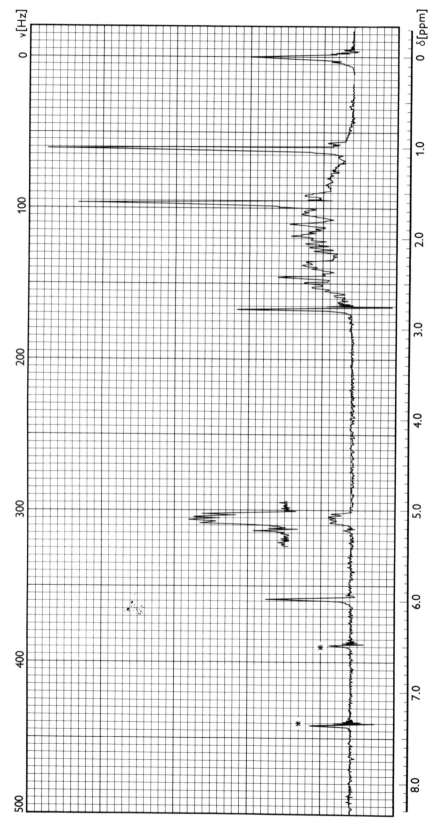

FIGURE 126. The spectrum of 6α-hydroxy-androsta-4-en-3,17-dione (111).

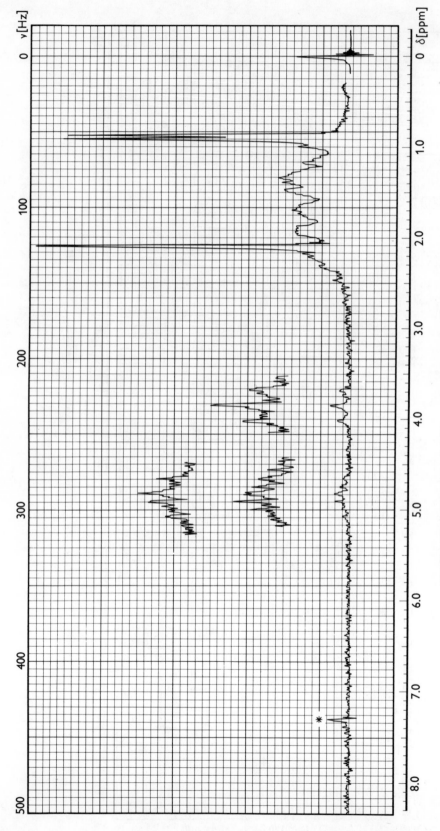

FIGURE 127. The spectrum of 3β-acetoxy-4β-bromoandrostan-17-one (**112**).

indicates that H-3 is *axial*. This is proved unambiguously by the H-4 triplet at 3.85 ppm with a splitting of approximately 10 Hz. This triplet splitting may be conceived only by assuming that $J_{3,4} \approx J_{4,5}$, and the magnitude of splitting also indicates that both couplings are *diaxial*. Accordingly, the 3β-acetoxy-4α-bromo configuration (in which both substituents are *equatorial*) can be derived from the spectrum. Since H-4 has only two neighbors, the fine structure of its signal is simple (the *X* part of an *AMX* spectrum in which $J_{AX} = J_{MX}$), and the line groups are clearly discernible. The signals of Me(18) and Me(19) groups appear at 0.87 and 0.90 ppm, and the methyl singlet of the acetoxy group can be found at 2.08 ppm.[1321]

In the spectrum of steroid **113**, the methylene hydrogens of the fused cyclopropane ring produce two maxima around -0.1 (!) and 0.3 ppm,[549] indicating that the cyclopropyl protons are strongly shielded in condensed systems as well. The methylene protons of the cyclopropane ring of Compound **114** produce an *AB* multiplet, from which $\delta A = 0.36$ and $\delta B = 1.27$ ppm and $J_{AB}^{gem} = 4.2$ Hz.[775] According to the investigations on a number of 5,10-methylenesteroid analogues, $J^{gem} < 6$ Hz. The methyl ether **115** of enone **114** is liable to a valence isomery yielding derivative **116**. The two equally possible alternative Structures **115** and **116** can be distinguished from one another on the basis of the coupling constant J_{AB}. The value of 10 Hz is compatible with Structure **116** only.* This structure is also confirmed by the large chemical shift difference of the methylene protons (1.06 and 3.13 ppm) caused by the shielding of the aromatic π-electron sextet on H_A, which is above the double bonds in **116**.**

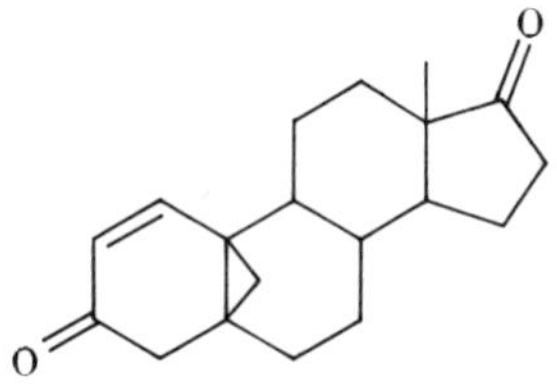

114 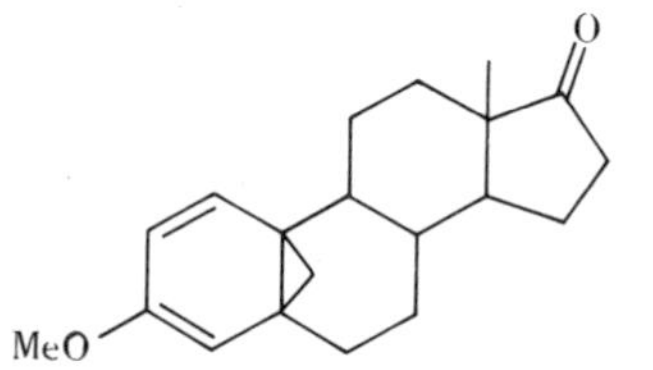   115 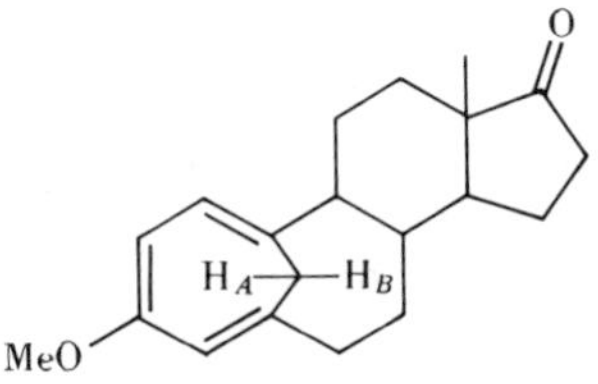   116

A similar shielding effect can also be observed in other bridged structures. In Structure **117** one of the 10-methylene protons (*A*) is shielded by the C=C double bond, explaining the large shift difference between the methylene protons ($\delta A = 3.50$ and $\delta B = 4.20$ ppm, $J_{AB} = 8$ Hz). It is interesting to note that H-6e is coupled with only one of the 7-methylene hydrogens (a doublet at 4.70 ppm, $J = 5.5$ Hz), owing to the neighboring oxygen atom, the $-I$ effect of which decreases the coupling constant.*** The unusually low value of J_{AB}^{gem} can be attributed to the same effect. The assignment of the other separately observable signals is $\delta Me(18) = 1.97$ and $\delta(=C_4H) = 5.78$ ppm.[1321]

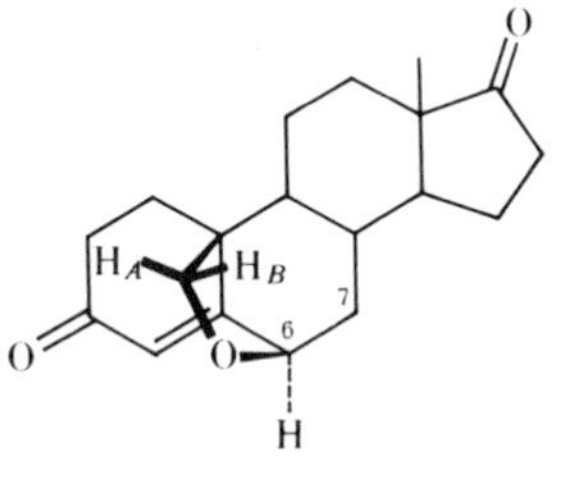

117

---

* Compare p. 14.
** See also p. 75, the case of the homotropylium cation.
***See the very similar case of 2,3-epoxy-tetrahydrofuran, Volume I, p. 61-62.

The NMR investigation of the steroids involves a number of interesting problems.* We must stress the importance of up-to-date NMR techniques in steroid research. Because of the very complex multiplets of skeletal hydrogens, even the most simply recognizable methyl signals may be hard to identify in 60 MHz ^{1}H NMR spectra. For example, the methyl signals of cholestane are already not separated, the doublet of Me(20) cannot be identified. The low quantity of compounds of biological origin and the poor solubility require the application of higher measuring frequencies (with super conducting magnets) and the FT technique. The investigation of the ^{13}C NMR spectra is especially important and useful in steroid chemistry.

3.2.7.4. Carbohydrates

The easily available carbohydrates of great biological importance are excellent models for configurational and conformational studies. This explains why hydrocarbons are among the first subjects of the application of NMR spectroscopy in organic chemistry. The relationship between the dihedral angles and the *vicinal* coupling constant of which numerical representation is the Karplus relation was observed first for carbohydrate acetates.[838]

The NMR spectra of most carbohydrates are rather complex, due to the wide variety of configurations and conformational equilibria and to the presence of many and similar functional groups (e.g., see References 49, 254, 404, 619, 697, and 698 and Problems **43, 48, 52, 58, 59, 63, 65,** and **66**).

The starting point of configurational and conformational analysis of pyranoses is the practically identical geometry of the tetrahydropyrane ring and cyclohexane.[422] The identical molecular structure involves that, for example, the carbohydrates with tetrahydropyrane rings have two stable chair conformations, one of which is usually preferred, and besides these conformations the twist-boat one may be stable only rarely.

The spectrum of tetrahydropyrane contains two groups of signals. The multiplet of the α-hydrogens (H-2,6) has, of course, larger chemical shift (approximately 3.6 ppm) than the broad signal of the remaining H-3,4,5 protons, which appears at about 1.6 ppm.[168] In 2-substituted derivatives, the chemical shift of the *geminal* ring proton increases further, depending on the nature of substitution. Thus, in the spectra of the 2-methoxy, 2-acetoxy, 2-chloro, and 2-bromo derivatives, δH-2 is, in turn, 4.45, 5.88, 6.17, and 6.62 ppm.[35,170] The glucosidic hydrogen (H-1) of acetylated sugar derivatives produces a signal around 6 ppm, which is well separated from most of the other hydrogens and is therefore easy to identify.

A number of empirical rules established by Lemieux et al.[838] — which due to precautions — are well applicable in the structure identification of unknown carbohydrates.

1. $\delta H_e > \delta H_a$ for any such isomeric or conformer pairs in which the position of only one hydrogen, attached of course to the same carbon atom, is changed. $\Delta \delta_{e,a} = \delta H_e - \delta H_a \approx 0.6$ ppm.
2. $J_{a,a} > J_{a,e} \approx J_{e,e}$ for any *vicinal* hydrogen pairs, generally with $J_{a,a}$ about three times larger than $J_{a,e}$ or $J_{e,e}$ ($J_{a,a} \approx 6$ to 10 Hz, whereas $J_{a,e}$ and $J_{e,e} \approx 2$ to 4 Hz).**
3. $\delta(Me)_e < \delta(Me)_a$ for the methyl signals of the acetoxy groups in *equatorial* and *axial* positions.

   Using these rules, it is easy to distinguish between pentaacetyl-α-D-glucopyranose (**118**)** and -β-D-glucopyranose (**119**) on the basis of the magnitude of $J_{1,2}$ (which is ~3 and ~8 Hz, respectively, corresponding to the $J_{1,2} \equiv J_{a,e}$ and $J_{1,2} \equiv J_{a,a}$ types of coupling), provided that the molecules are in the more stable **a** conformation, or at least this conformation is the predominant one in the solution investigated.

---

* Compare p. 87 and Volume I, p. 41, 63, and 68, and Problem **34**.
** Compare p. 27.
***In Formulas **118** to **125**, — and — stand for $-OAc$ and $-CH_2OAc$ groups; in case of **125**, they also stand for OH and CH_2OH, respectively.

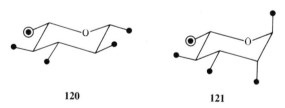

118

It is already more difficult to differentiate the chair conformers of the anomers with the C-1 and C-2 substituents in *cis* position because $J_{a,e}$ and $J_{e,a}$ are not much different, although some systematic differences can be observed:[839] $J_{1,2} \equiv J_{e,a} \approx 2.5$ to 3.5 Hz and $J_{1,2} \equiv J_{a,e} \approx 1.0$ to 1.5 Hz.

119

4. If an *equatorial* acetoxy group in α-(*vicinal*) or β-position to an *axial* hydrogen is replaced by an *axial* substituent, this results in a paramagnetic shift of 0.2 or 0.25 ppm, respectively, in the signal of the skeletal hydrogen. The effect is opposite in direction for *equatorial* skeletal protons, leading to a shift of −0.2 ppm in both α- and β-positions. For Compound **120**, δH-1 = 5.76 and δH-5 = 3.9 ppm, whereas for **121** δH-1 = 6.1 and δH-5 = 4.1 ppm.[254] The corresponding chemical shifts of the latter from the parameters of **120** making use of empirical correlations (1) and (4) are for H-1, 5.76 + 0.6 (*a → e* change in the position of H-1) −0.2 (the acetoxy group *vicinal* to H-1*e* has become *axial*) = 6.16 ppm. For H-5, 3.9 + 0.25 (the acetoxy group in β-position to H-5*a* changes from *equatorial* to *axial*) = 4.15 ppm. The data obtained are in fair agreement with the experimental shifts.

120   **121**

5. For the analogues that contain free hydroxyl groups, paramagnetic shifts of 0.3 and 0.35 ppm, respectively, arise in the signal of the *axial* skeletal hydrogen if an *equatorial* hydroxyl group in *vicinal* or β-position is replaced by an *axial* group. The position of neighboring hydroxyl groups does not affect, however, the chemical shifts of the *equatorial* ring hydrogens.

The investigation[404] of acyloxy-carbohydrates at various temperatures has shown that with the great majority of these compounds, at room temperature, the less stable conformation also takes part in the equilibrium to a measurable extent (>5%). The free energy difference between the conformers is less than 4 kJ/mol, except for the case of two-three models which are practically homogeneous in conformation and in which this difference is in the magnitude of 8 kJ/mol. The conformation equilibrium of the pentaacetyl pyranoses changes for the analogous pentabenzoyl derivatives usually in favor of the conformer containing an *axial* substituent on C-1 which is the minor component in the original equilibrium.[404]

NMR studies have confirmed also in other cases, the fact already established chemically, that in the conformational equilibria of certain carbohydrates (e.g., the 1-halosubstituted triacylpentoses) the conformer that contains more substituents in *axial* positions and believed therefore to be less stable is preferred.[683] It can be derived, e.g., from the spectrum of Compound **122**, that all of the *vicinal* coupling constants are less than 4 Hz,[667] and thus conformer **b** must be dominant. The reason is the *anomeric effect* [413,836] arising from the electrostatic repulsion between the dipoles represented by the cyclic C–O bond and the bond between C-1 and the substituent. Since in conformations of type **b** these dipoles may be more distant, the effect makes such conformations preferred. The magnitude of the anomeric effect can be determined from the free energy difference of the anomeric pairs or by comparing these values of the compounds showing the anomalous equilibrium to those of the analogues containing substituents with no anomeric effect. The anomeric effect is usually less than 2 kJ/mol.

a   b

122

The 5-methylene protons of **122** are nonequivalent, producing an *AB* multiplet. The corresponding chemical shifts do not obey the rule $\delta H_e > \delta H_a$ since the two *axial* acetoxy groups in positions 3 and 4 and the halogen atom increase δH-5a to such an extent that the order of shifts is not only reversed, but a remarkable difference (approximately 0.6 ppm) occurs in the opposite sense. Similarly, in the benzoyl anomers **123** and **124**, the δH-1 shifts are[699] 6.55 (**123**, R=Bz) and 6.40 ppm (**124**, R=Bz), hence $\delta H_a > \delta H_e$. In the *N*-acetyl analogues (R=H), δH-1 are 5.77 (**123**) and 6.20 ppm (**124**), thus, here is no irregularity: the chemical shift is smaller for Structure **123** where H-1 is *axial*.

NRAc   AcRN

123   124

The solutions of unsubstituted aldohexoses contain, in principle, five components, including the open aldehyde form and the pyranose and furanose anomers. The open aldehyde form, however, can generally be neglected. The bulky hydroxymethyl group gives rise to a preferred pyranose conformation in which this substituent is *equatorial*, and thus the systems can be regarded as practically homogeneous in conformation. An exception is α-D-idopyranose (**125**, R=H), since the four *equatorial* secondary hydroxyl groups stabilize conformation **b**.[50] With the corresponding pentaacetyl derivative (R=Ac), however, conformation **a** is preferred, due to the anomeric effect.[130]

In the heavy water solution of D-idose, the pyranose and furanose anomers have been

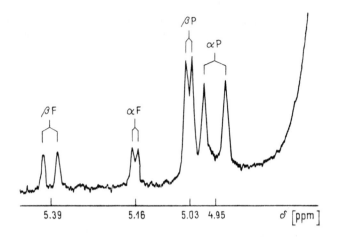

FIGURE 128. The 100-MHz spectrum of D-idose on D$_2$O, with the signals of the anomeric hydrogens. αF, βF, αP, and βP refer to the doublets of the anomeric protons corresponding to α- and β-furanose and to α- and β-pyranose, respectively.[49]

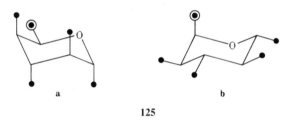

125

detected,[50] and the relative amounts of the four components have also been determined from the relative intensities of the H-1 signals (see Figure 128). In the identification of the H-1 signals corresponding to the four components, it was utilized that, according to model investigations,[49] δH-1 (pyranose) < δH-1 (furanose) for the same carbohydrate. Of course, the signal of pyranoses with the smaller shift (H-1a) but larger splitting ($J_{a,a}$) can be assigned to the α-anomer, and that with the larger shift (H-1e) and smaller splitting ($J_{e,a}$) can be assigned to the β-anomer on the basis of the rules δH$_e$ > δH$_a$ and $J_{a,a}$ > $J_{e,a}$. The H-1 signals of the furanose anomers can also be assigned by the splitting, considering, that J^c > J^t.* The signal of the glycosidic hydrogen (H-1) can generally be utilized in a similar manner to interpret the spectra of other aldoses. The free energies of the components can also be determined from the equilibrium constants,** and the differences measured for the anomeric pairs are always lower than 4 kJ/mol, indicating the safe detectability of the less stable anomer, too. Furanose anomers are always less probable,[49,839] and for many derivatives, the amount of even the preferred anomer hardly reaches the limit of detection (approximately 5%).

It is worth noting the investigations using computer methods, in which the spectra of the components known in pure form are "subtracted" from the spectrum of a multicomponent system to obtain that of the components unknown in pure form.[308] (The principle is somewhat similar to the optical compensation also known in IR spectroscopy.[671]) The spectrum parameters of acyclic carbohydrates or the temperature dependence of parameters permit the determination of the relative stabilities of rotamers (e.g., see Problem **66**).[1329]

In Compound **126** a virtual coupling*** was observed between H-1 and H-3 (W pattern!).

---

* Compare p. 22.
** See p. 94 and 95.
***Compare Volume I, p. 63.

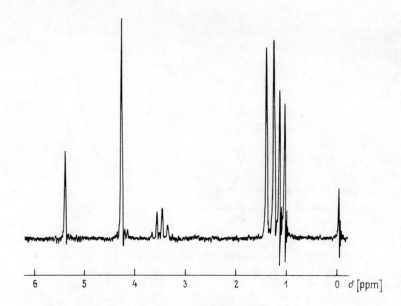

FIGURE 129.   The spectrum of 1,4-anhydro-6-desoxy-2,3-*O*-isopropylidene-β-L-talopyranose (**127**).[208]

Accordingly, the H-1 signal is a triplet at 5.63 ppm, and the H-2 and H-3 signals coincide at 4.48 ppm.[682] H-3, being in α-position to a keto group, can readily be deuterated with NaOD (through the enol form), and in the spectrum of the partially 3-deuterated derivative the H-1 signal is a doublet with a splitting of 3 Hz.

126

   In the spectrum of talo-derivative **127** (see Figure 129), only H-5 and the 6-methyl signals are split (the H-5 quartet appears at 3.45 ppm, and the methyl doublet appears at approximately 1.1 ppm, with J = 7 Hz). The H-2,3,4 signal is a common singlet around 4.3 ppm due to an accidental isochrony, and the signal of H-1 is also a singlet at 5.4 ppm. On the other hand, the two diastereotopic methyl groups are chemically nonequivalent (absorbing at 1.35 and 1.5 ppm).[208] The dihedral angles of proton pairs 1,2, 3,4, and 4,5 are nearly 90°, and thus, according to the Karplus relation, there is no observable coupling between them. Furthermore, since H-2 and H-3 are isochronous, the splitting corresponding to coupling $J_{2,3}$ does not occur, either.

127

The anomers of di- and oligosaccharides can also be differentiated by the signal of the glycosidic hydrogen. The corresponding doublets of α-glucose and β-maltose appear at 6.3 and 5.2, as well as at 6.6 and 4.6 ppm, respectively.[253,1438] The NMR study of nucleosides and nucleotides is a topic of importance. From the anisotropic effect of the heteroaromatic ring on the carbohydrate protons, the preferred conformations of pyrimidine-nucleosides can be determined.[325]

3.2.7.5. Saturated, Six-Membered Heterocyclic Compounds

With pyperidine derivatives, as compared to cyclohexanes, further problems may be caused by the nitrogen inversion and the configurations of protonated and quaternary derivatives. Also taking into account ring inversion, pyperidine can be regarded as the equilibrium mixture of four forms. With unsubstituted piperidines, two of these are identical, and only the nitrogen inverses containing the lone electron pair in *axial* and *equatorial* position, respectively, are distinguishable, from which the former is more stable due to the smaller *1,3-diaxial* interaction between hydrogens.[168] The nitrogen inversion is, however, very fast and therefore the inverses are not detectable. Even the ring inversion can be freezed-in only well below room temperature.

Like in the spectrum of tetrahydropyrane, the α-hydrogens of piperidine also produce a broad signal (δH-2,6 ≈ 2.7 ppm) with a larger shift than the also broad maximum of H-3,4,5 (appearing at approximately 1.5 ppm). The chemical shift of the NH signal depends on the concentration, pH, temperature, and the solvent* and appears in the range of 1.3 to 2.0 ppm.[1503]

For the derivatives, the rule $\delta H_e > \delta H_a$ holds. The chemical shift difference $\Delta\delta_{e,a}$ is 0.85 ppm for the 2,6-protons of N-t-butylpiperidine[169] and 0.57 ppm for the 4-t-butyl derivatives.[1147] The large difference in the former case can be attributed to a *trans-diaxial* interaction between the lone pair of the nitrogen and the *axial* α-hydrogens. Similarly $\Delta\delta_{a,e} = 0.9$ ppm for the α-methylene hydrogens in Compound **128**.[623] The approximate ranges of the coupling constants characteristic of piperidine derivatives are[1147] $J_{2,2}^{gem} = -11$ to -14, $J_{a,a}^{2,3} = 10$ to 11.2, $J_{e,e}^{2,3} = 2.0$ to 3.2, $J_{a,e}^{2,3} = 1.9$ to 3.0, and $J_{e,a}^{2,3} = 4.5$ Hz.

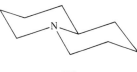

128

Similar to pyrrolidines,** in the case of N-benzyl-2,6-symmetrically disubstituted piperidines, the methylene signal of the benzyl group is a singlet for the *cis* isomer, but it is an AB quartet (with $J_{AB} \approx 14$ Hz) for the corresponding *trans* isomer.[651]

Due to the anisotropic and $-I$ effects of carbonyl and thiocarbonyl groups (compare Problems **2** and **9**), the α-hydrogens of N-acetyl-piperidine absorb at 3.40 ppm, and those of the N-thioacetyl derivative at 3.75 ppm.[1503] In the spectra of N-acyl derivatives the signals of both the *axial* and the *equatorial* α-hydrogens are split, due to the hindered rotation. *** The chemical shifts of the identical 2- and 6-methylene hydrogens are $\delta H_e^{2,6} = 4.57$ and 3.83 ppm, $\delta H_a^{2,6} = 3.03$ and 2.57 ppm at room temperature, whereas at 120°C the signal pairs collapse, to yield $\delta H_e^{2,6} = 4.2$ and $\delta H_a^{2,6} = 2.8$ ppm.[893]

In the case of slow proton exchange, or even a fast process with retention on the nitrogen, protonated asymmetric piperidines form *cis-trans* isomers. Thus, the spectrum indicates the

* See Section 3.7.1.4, p. 105-113.
** Compare p. 22.
***Compare Volume I, p. 71.

presence of two isomers of 1,2-dimethylpiperidine in which the protons are bonded in different steric positions (**129** and **130**), since both the *C*-methyl and the *N*-methyl doublets are doubled.[254] Of course, the stronger line pairs correspond to the *trans* isomer (**129**) containing *equatorial* methyl groups (see Figure 130a,b). Accordingly, the relation $\delta(NMe)_e$ $> \delta(NMe)_a$ observed for the bases also holds for *N*-methyl-piperidinium salts, although in the latter case also exceptions have been found. For the *trans* isomer, the **a** conformation is clearly preferred, but in the *cis* isomer, the **a** and **b** conformers may, in principle, both exist. In heavy water solution, the *N*-methyl signal is a singlet, owing to a fast proton exchange which eliminates the splitting caused by the coupling of the NH and methyl protons. Since, however, the methyl signals are still doubled (see Figure 130b), it must be assumed that the proton exchange proceeds with a retention of the conformation.[175,968]

129

130

The salts of *cis* and *trans* *N*, 2,6-trimethyl piperidines are different. In the spectra of the salts, the *C*- and *N*-methyl signals are doubled for the *cis* isomer, but are not for the *trans* compound. From the intensity ratio of the pair of *N*-methyl signals (doublet in $CDCl_3$ and singlet in D_2O) of the 2,6-*cis* isomer, a ratio of 2:1 can be obtained for the protonated isomers, of course, in favor of the *equatorial* *N*-methyl isomer (*trans* to the also *equatorial* 2,6-methyl groups). In contrast, the 2,6-*trans* isomer, has *no* geometric isomers (*N*-inverses), even assuming the retention of conformation, due to a fast ring inversion. Namely, the ring-inverses of the "two" *N*-inverses are pairwise identical. Accordingly, one can observe two (CMe_e and CMe_a) methyl doublets in $CDCl_3$ and one in D_2O and only one *N*-methyl signal (doublet and singlet, respectively) in both solutions. Similarly, in the spectra of quaternary *N,N '*-dimethyl-substituted analogues, the fast ring inversion produces one *C*-methyl and one *N*-methyl signal for the 2,6-*trans* isomer, whereas the signals of the *N*-methyl groups appear separately in the spectrum of the 2,6-*cis* isomer.[760,761]

A ^3J coupling may often occur between the nitrogen and the protons attached to the β-carbon atoms of quaternary piperidines.* Accordingly, e.g., in the spectrum of 1,1,2-trimethylpiperidinium iodide, both lines of the doublet of the 2-methyl group at 1.35 ppm split further into 1:1:1 triplets with a spacing of 1.7 Hz.[1147]

In the case of sulfur-containing, six-membered heterocycles, the rule $\delta H_e > \delta H_a$ does not hold for the α-hydrogens, its opposite being true.[11,30,240] This pertains equally to 1,4-oxathianes, 1,3-dithianes, and to their fused derivatives. On the other hand, the shifts of the *geminal* α-protons of the corresponding sulfoxides are regular again.[502]

The relation $J_{a,a} > J_{a,e}, J_{e,e}$ also holds for heterocycles containing more than one heter-

* See p. 105 and Section 4.2.6.2, p. 254-256.

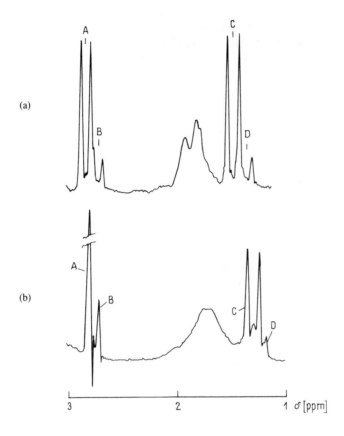

FIGURE 130. The spectrum of 1,2-dimethylpyperidine hydrochloride with the methyl signals. (a) In $CDCl_3$ and (b) in heavy water. (From Casy, A. F., *PMR Spectroscopy in Medicinal and Biological Chemistry*, Academic Press, New York, 1971. With permission.)

oatom. Coupling constants $J_{2,3}$ are substantially different e.g., for Compound **131** and **132**:[406] $J_{2,3} \equiv J_{a,a} = 8.8$ Hz (**131**) and $J_{2,3} \equiv J_{a,e} = 2.7$ Hz (**132**). In the spectrum of 1,3-dioxane (**133**), the singlet of the 2-methylene hydrogens appears at 4.70 ppm, the H-4,6 multiplet is at 3.8, and H-5 at 1.7 ppm. The former data indicate the additivity of the $-I$ effects of the hetero atoms. The coupling $J_{2,2}^{gem} \approx -6$ Hz is smaller than $J_{4,4}^{gem} \equiv J_{6,6}^{gem} \approx -11$ Hz and $J_{5,5}^{gem} \approx -13$ Hz, due to the effect of the electron-withdrawing hetero atoms decreasing the coupling constants in absolute value.* The *vicinal* spin-spin couplings of derivatives give the following ranges for $J_{4,5}$: $J_{a,a} \approx 9$ to 12, $J_{e,a} \approx 4$ to 7, $J_{a,e} \approx 2.5$ to 3.5, and $J_{e,e} \approx 0.5$ to 2 Hz,[1147] also in agreement with the data obtained for other six-membered heterocyclic systems.**

131   **132**   **133**

* Compare p. 20 and Volume I, p. 59.
** See p. 27.

Table 35

**THE CHEMICAL SHIFTS (PPM) OF
ACETYLENIC PROTONS FOR RC≡CH
TYPE COMPOUNDS[434,1379]**

R	δ(≡CH)	R	δ(≡CH)
OMe	1.35	CH_2OH	2.35
Me,H	1.80	CH_2Cl	2.40
CHO	1.90	$pNH_2-C_6H_4$	2.71
CH_2I	2.20	Ph	2.90
CH_2CN	2.25	$pNO_2-C_6H_4$	3.20

3.3. UNSATURATED COMPOUNDS

The ^{1}H NMR spectra of unsaturated compounds are characteristically different from those of aliphatic compounds. The chemical shifts of the hydrogen atoms attached to unsaturated carbons are generally much larger than those of the aliphatic protons. The presence of π-electrons opens wider possibilities of significant long-range couplings.

Three types of protons attached to unsaturated carbons can be distinguished, depending on whether the carbon atom is in sp or sp^2 hybrid state and whether in the latter case the group is terminal or an intermediate member of the chain. The chemical shift ranges characteristic of these proton types are δ(≡CH), 1.0 to 3.3, δ(=CH$_2$), 3.5 to 7.0, and δ(=CH), 4.5 to 10.0 ppm.

There is a quite large overlap between the chemical shift regions of saturated and unsaturated hydrogens, but as this pertains mainly to rarely occurring acetylenic and terminal ethylenic protons, which can usually be recognized easily, e.g., by their splitting pattern or from the IR spectrum, it is generally easy to detect the presence of unsaturated groups in the molecule under study.

3.3.1. Acetylene and Allene Derivatives

The ^{1}H NMR signal of acetylene appears at 1.80 ppm ($J = 9$ Hz).*[1316] The strong shielding can be attributed to the anisotropy of the triple bond** which compensates most of the deshielding caused by the $-I$ effect of the unsaturated bond. The signal of mono-substituted derivatives can be found in the range of 1.0 to 3.3 ppm. Electron-withdrawing substituents increase the chemical shift (see Table 35), like in aliphatic compounds (compare Figure 117). The signals of the α-methylene hydrogens in compounds of type HC≡C–CH$_2$R appear in the range of 2 to 5 ppm.

Although the chemical shift of acetylenic protons overlaps with that of aliphatic hydrogens, the very sharp singlet can be identified easily if there is, as generally, no coupling with other nuclei. The assignment of these compounds is dubious only when the molecule also contains monomeric hydroxyl group, e.g., as in the case of steroid **134**. In this case the

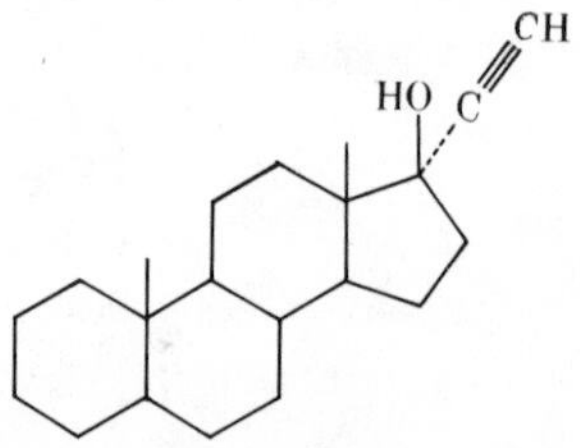

134

---

* For the way of obtaining the coupling constant, see p. 210 and Equation 78.
** Compare Volume I, p. 33.

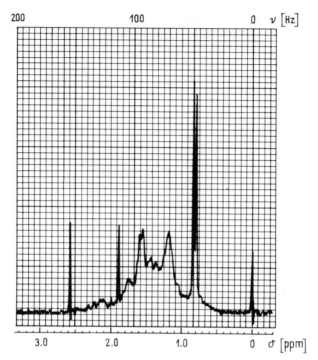

FIGURE 131. The spectrum of 17α-ethinyl-17β-androstanol (**134**). (From *NMR Spectra Catalog*, VARIAN Associates, Palo Alto, Calif., 1962-1963. With permission.)

hydroxy signal is also a sharp singlet with an intensity of one proton (see Figure 131), and upon the addition of few drops of heavy water, both signals disappear.* Thus, it can be determined only by acetylation that the assignment is δOH = 1.88 and $\delta(\equiv CH)$ = 2.55 ppm.[1044] The solvent effect can also be used very efficiently to identify the signals of acetylenic protons. These signals are shifted considerably more in polar solvents with respect to the signals measured in inert solvents than the signal of any other proton — except for the acidic protons.** This phenomenon is attributed to collisional complex formation.*** With respect to the shift found in CCl_4, a diamagnetic shift of 0.2 ppm can be observed in benzene, and paramagnetic shifts of about 1 ppm can be measured in pyridine and dioxane. The ¹H NMR spectrum of methylacetylene in $CDCl_3$ consists of one singlet at 1.80 ppm[1044] due to an accidental isochrony of the methyl and acetylenic hydrogens. If the spectrum is measured in a 1:1 mixture of $CDCl_3$ and benzene-d_6, an AB_3 multiplet appears.[281]

Transferred by the π-electrons, long-range couplings may often occur in compounds of type HC≡CCHRR', for which ^4J = 2.5 to 3 Hz. In the spectrum of propargyl alcohol (HC≡C–CH₂OH), e.g., from the triplet of acetylenic proton at 2.50 ppm or from the methylene doublet at 4.28 ppm, the ^4J coupling constant is 2.5 Hz.[1044] Polyacetylenes show couplings between even more distant protons.†

The role of hyperconjugation in long-range coupling is indicated by the fact that for methyl-acetylene ^4J is much larger (3.5 Hz) than for vinyl-acetylene (CH₂=CH–C≡CH), where ^4J = 2.5, $^5J^c$ = 0.5, and $^5J^t$ = 1.0 Hz.[1315] In the spectrum of Z–HC≡C–CH=CHOMe molecule (see Figure 132), the acetylenic and olefinic protons represent an AMX spin system (the signal of the methoxy protons is, of course, a singlet at 3.80 ppm), and J_{AX} = 1, J_{AM}

* Compare p. 97 and 101.
** See Volume I, p. 40 (point 2 and accompanying footnote), and Problem **34.**
***Compare Volume I, p. 38.
† Compare Volume I, p. 65.

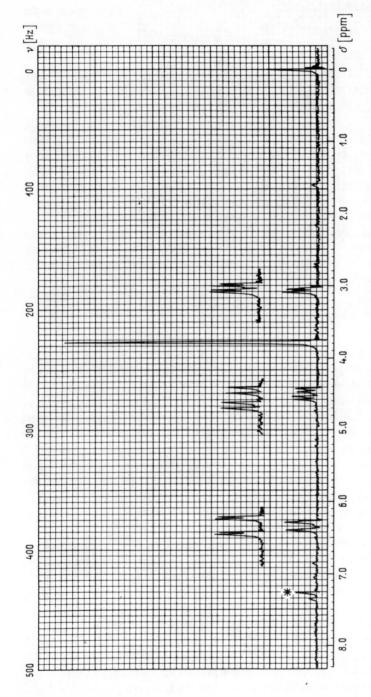

FIGURE 132.   The spectrum of Z-1-methoxybuten-3-ine (HC≡C–CH=CHOMe). (From *NMR Spectra Catalog*, VARIAN Associates, Palo Alto, Calif., 1962-1963. With permission.)

$= 8$, and $J_{MX} = 3$ Hz, with J_{AX} corresponding to a spin-spin interaction between the acetylenic and the more distant (β-) olefinic proton.[1044]. The assignment follows from the magnitudes of coupling constants. A coupling of 8 Hz may exist only between *vicinal* olefinic hydrogens, consequently those two quartets arise from the olefinic protons in which the distance of the outer doublets is 8 Hz. Of the doublets, the one with the larger splitting must be due to the α-olefinic proton (^4J), since the β-hydrogen is more distant from the acetylenic protons (^5J). Thus, $\delta A = \delta H_\beta = 6.35$ ppm $> \delta M = \delta H_\alpha = 4.52$ ppm $> \delta X = \delta(\equiv CH) = 3.08$ ppm, as expected.*[1044]

The singlet of allene[1513] appears at 4.67 ppm. Electron-withdrawing substituents cause deshielding, primarily in the vicinity of the *geminal* proton, e.g., the methylene and methine signals of bromoallene ($H_2C=C=CHBr$) appear at 4.95 and 5.97 ppm, respectively.[1512]

The magnitude of coupling constants ^4J is about 6 Hz[1513] (in allene 7 Hz). In monosubstituted and 1,1-disubstituted derivatives, $J^{gem} < 2$ Hz, similar to the case of *geminal* olefinic protons.** The coupling involving the α-hydrogens of substituents can be characterized by $^3J = 6$ to 7 and $^5J = 2$ to 3 Hz.*** The ^{1}H NMR spectra of allenes are not really characteristic. The detection of this functional group can be attempted rather by ^{13}C NMR† or IR spectroscopy.

3.3.2. Olefins

The singlet of the protons of ethylene appears at 5.29 ppm.[1174] The large chemical shift in comparison with the aliphatic and even with the acetylenic protons is explained by the $-I$ effect of the unsaturated bond. The signal of substituted derivatives appears most frequently in the range of 4.5 to 6.5 ppm, well separated from the signals of other proton types. The fixed steric position of the substituents attached to the double bond enables the investigation of the characteristic differences between the *geminal, cis,* and *trans vicinal* coupling constants offer an excellent tool for differentiation of geometric isomers.

The three hydrogen atoms of vinyl derivatives form *ABC* spin system if the possible interactions with the protons of the substituents are disregarded. The *ABC* system may be simplified into *ABX*, or, less frequently, *AMX* systems. Vinylchlorosilane ($CH_2=CHSiCl_3$) produces, owing to an accidental isochrony, a nearly singlet[1044] spectrum, vinylbromide[1250] produces an *ABX*, whereas vinylacetate ($CH_2=CHOAc$) produces an *AMX* multiplet.

If the eight lines of the *AMX* spectrum are close to one another, e.g., like in the spectrum (see Figure 133) of vinylacetate, it is possible to assign the related lines (the two double doublets) by simply diluting the solution. With a change in concentration, the multiplets of the different protons undergo different shifts, but the distances of the lines of the same proton, of course, do not change.

The effect of substituents on the chemical shifts of the vinyl protons is illustrated in Table 36. It can be seen that (taking also into account some other data) the chemical shift range of vinyl protons is 3.5 to 7.8 ppm. Within this range, $\delta H^{gem} = 5.4$ to 7.8, $\delta H^c = 3.9$ to 6.6, and $\delta H^t = 3.5$ to 6.2 ppm. The chemical shifts of olefin protons of di- and trisubstituted ethylenes can be predicted fairly well (± 0.15 ppm) by the following empirical formula, expressing the additivity of substituent effects:[281,952,1087]

$$\delta(=CH) = 5.25 + \Sigma \, \rho^i \tag{276}$$

The ρ^i constants are listed in Table 37.

A linear correlation has been found between the Taft constant ρ_M and expression $\Delta = \delta H^{gem} - (\delta H^c + \delta H^t)/2$ formed from the chemical shifts of the vinyl protons.[80] However,

---

* Compare Tables 36 and 37, p. 51 and p. 46.
** See p. 52.
***See p. 53 and Volume I, p. 67.
† See p. 179.

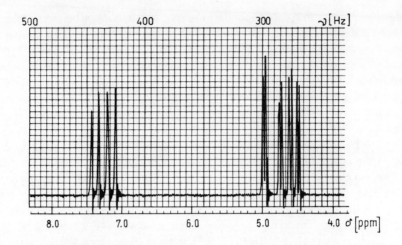

FIGURE 133.   Part of the spectrum of vinylacetate (CH₂=CHOAc) with the multiplets of olefinic protons. (From *NMR Spectra Catalog*, VARIAN Associates, Palo Alto, Calif., 1962-1963. With permission.)

the anisotropic neighbor effect must be taken into correction, e.g., in the spectrum (see Figure 29) of crotonaldehyde (**15**), the signal of the proton *geminal* to the carbonyl group appears at 6.13 ppm and that of the *vicinal* proton in *cis* position to the former appears at 6.87 ppm. The corresponding shifts calculated by Equation 276 are 6.09 and 6.65 ppm.

The estimation of the chemical shifts of olefinic protons by Equation 276 using the substituent constants of Table 37 is less reliable for trisubstituted ethylenes than for mono- or disubstitution. In such cases one can utilize in the assignment, the solvent effect.* ASIS ($\Delta\delta_{CCl_4} - \delta_{C_6D_6}$) is -0.17 and -0.53 ppm, respectively, on the shifts of the *vicinal* methyl groups of isomers **135** and **136**.[1219] ASIS is greater on the positively polarized part of the molecule, i.e., on the protons more distant from the carboxyl group, causing therefore a larger diamagnetic shift in the methyl signal of isomer **136**. The ASIS is well observable on the olefinic protons too. Whereas for isomer **135** there is a difference of -0.43 ppm between the shifts measured in CCl_4 and in C_6D_6, the difference is only -0.02 ppm for isomer **136**. The signal of the methyl group *geminal* to the carboxyl group shows weaker ASIS, which differs hardly for the isomers (-0.13 and -0.15 ppm). A linear correlation has also been observed between the dipole moments of the substituents and the chemical shifts of the signals of *vicinal* hydrogens, increasing proportionally with the former.[1177]

$$
\begin{array}{cc}
\underset{Me}{\overset{H}{\diagup}}C=C\underset{COOH}{\overset{Me}{\diagdown}} & \underset{H}{\overset{Me}{\diagup}}C=C\underset{COOH}{\overset{Me}{\diagdown}} \\
\textbf{135} & \textbf{136}
\end{array}
$$

As indicated by the data of Tables 36 and 37, the signal of *geminal* hydrogens is always shifted paramagnetically with respect to ethylene. If substituent R is attached to the vinyl group by a carbon, this shift is related to the anisotropy of C–C bonds, and if it is attached by a hetero atom, the shift is due to its $-I$ effect. Of course, the shift is much larger in the latter case. If the anisotropic effect of the substituent causes diamagnetic shift, e.g., like with nitrile groups, this compensates in part the $-I$ effect, leading to only a smaller paramagnetic shift (0.17 ppm for acrylnitrile).

* ASIS, see Volume I, p. 38 and 40.

Table 36
THE CHEMICAL SHIFTS (PPM) OF VINYL PROTONS FOR CH$_2$=CHR TYPE COMPOUNDS[179,254,1379]

R	δH^{gem}	δH^{cis}	δH^{trans}	R	δH^{gem}	δH^{cis}	δH^{trans}
H	5.29	5.29	5.29	Ph	6.64	5.65	5.15
CN	5.46	5.94	5.81	Cl	6.26	5.48	5.39
Me	5.73	4.96	4.88	Br	6.36	5.75	5.83
F	5.84	3.92	3.62	Ac	6.63	6.52	6.11
COOH	5.89	6.20	5.71	NO$_2$	6.81	6.16	5.61
OMe	6.21	3.93	3.74	OAc	7.07	4.63	4.35

Table 37
SUBSTITUENT CONSTANTS δ^a (PPM) FOR THE ESTIMATION OF CHEMICAL SHIFTS OF VINYL PROTONS BY EQUATION 276[952]

Substituent	δ^{gem}	δ^{cis}	δ^{trans}	Substituent	δ^{gem}	δ^{cis}	δ^{trans}
Alkyl	0.45	−0.22	−0.28	CO −[b]	1.06	0.91	0.74
CH$_2$N(sp^3)	0.58	−0.10	−0.08	CO −	1.10	1.12	0.87
CH$_2$O −	0.64	−0.01	−0.02	NR −[b]	1.17	−0.53	−0.99
Cycloalkyl	0.69	−0.25	−0.28	OR[b]	1.21	−0.60	−1.00
CH$_2$CO −	0.69	−0.08	−0.06	OR	1.22	−1.07	−1.21
COOR[b]	0.78	1.01	0.46	C(sp^2)[b]	1.24	0.02	−0.05
NR	0.80	−1.26	−1.21	CON(sp^3)	1.37	0.98	0.46
COOR	0.80	1.18	0.55	Ar	1.38	0.36	−0.07
C(sp^2)	1.00	−0.09	−0.23	NCO −	2.08	−0.57	−0.72
CH$_2$Ar	1.05	−0.29	−0.32	OCOR	2.11	−0.35	−0.64

[a] The terms *gem*, *cis*, and *trans* refer to the relative position of the substituent and the olefinic hydrogen.
[b] Conjugated.

If only the inductive effect is to be taken into account, the shifts of *vicinal* and *geminal* hydrogens are affected by the substituent to a similar extent, but mostly in opposite sense (see Tables 36 and 37). The electron density around the *vicinal* protons is increased by electron-repelling substituents (such as alkyl), and thus diamagnetic shift occurs. Due to a hyperconjugation, this effect is particularly strong in the case of propene (−0.33 and −0.41 ppm).

Shielding is caused by substituents with +M effects (−NR−, −OR, −NCO, −OCO), and the diamagnetic shift is larger for the *trans* hydrogen. The delocalization of electrons in vinyl-methyl ether (CH$_2$=CHOMe) is also proved, besides the strong shielding of *vicinal* hydrogens, by the long-range coupling of olefinic protons with the methoxy protons ($^4J^{gem}$ = 0.35, $^5J^{cis}$ = 0.3, and $^5J^{trans}$ = 0.2 Hz).[472] For methoxy hydrogens there is usually no measurable splitting.

On the *cis vicinal* proton of aromatic vinyl derivatives, the weak +M effect is overcompensated by the deshielding effect of the ring, thus $\delta H^{cis} > \delta H^{trans}$ and $\rho_{cis} > 0$, $\rho_{trans} < 0$. If a carbonyl substituent is attached to the vinyl group, the main factor is the −I effect, and deshielding of the *cis vicinal* hydrogen is increased further by the anisotropic effect ($\delta H^{cis} \gg \delta H^{trans}$, $\rho_{cis} \gg 0$, and $\rho_{trans} > 0$).

The chemical shift range of the protons of 1,1-disubstituted olefins is 4.5 to 6.3 ppm.*

* Compare p. 46 and Problem **30**.

Table 38

THE COUPLING CONSTANTS (Hz) OF VINYL PROTONS FOR CH₂=CHR TYPE COMPOUNDS[1379]

R	J^{trans}	J^{cis}	J^{gem}	R	J^{trans}	J^{cis}	J^{gem}
F	12.7	4.7	− 3.2	OMe	14.0	6.7	− 2.0
Cl	14.3	7.4	− 1.5	NO₂	15.0	7.6	2.0
Br	15.5	7.1	− 2.0	COOH	15.4	9.7	0.8
Me	16.8	10.0	2.1	Ph	17.6	10.6	1.1
Li	23.9	19.3	7.1	Ac	17.8	10.8	1.1
OAc	13.6	6.4	− 1.6	CN	18.3	12.1	0.8

The shielding of olefinic protons in Z - and E -1,2-disubstituted derivatives (even in the case of identical substituents) are different, and the stronger the anisotropy of the substituent, the larger the difference of chemical shifts. The reason is that the hydrogens are closer to the *vicinal* substituent in the E -isomer, and thus the anisotropic effect of the latter is more predominant. This makes clear why the chemical shift of the olefinic protons is 0.55 ppm smaller (6.28 ppm) in the spectrum of diethyl maleate (Z-EtOOC–CH=CH–COOEt) than in that of the E-isomer, diethyl fumarate.[1044]

The interval of chemical shifts is narrower for the Z isomers of 1,2-disubstituted olefins (5.2 to 6.3 ppm) than for their E counterparts (5.2 to 8.0 ppm). The olefinic signal of trisubstituted ethylenes appears between 5.0 and 8.1 ppm.

In structure analysis of ethylenes, the characteristic difference of the coupling constants of olefinic protons can be utilized: $J^{trans} > J^{cis} > J^{gem}$. They decrease, both separately and in their sum, proportionally with the electronegativity of the substituent.[635] This behavior can be attributed to the $-I$ effect of the substituents, which reduces the electron densities transferring the spin-spin interactions, and thereby decreases the magnitude of interaction.*

Table 38 shows the coupling constants of some vinyl derivatives. It can be seen from a comparison of the first five rows that the coupling constants decrease with the increasing $-I$ effect of the substituents. When R contains π-electrons, they increase the coupling constants;** this is illustrated by the second part of the table. The coupling constants of vinyl-methyl ether do not fit in either scheme. This is related to the $+M$ effect of lone pairs contributing to the spin-spin interaction.

It is generally true for α-carbonyl ethylenes (–CH=CH–CO–) that the chemical shift of the β-olefinic proton is larger than that of the α-one, due to the $-M$ effect of the carbonyl group, which decreases the electron density around the β-hydrogen, and increases it for the α-one: –CH=CH–C=O ↔ –C⊕H–CH=C–O⊖.

The coupling constants J^{trans}, J^{cis}, and J^{gem} usually fall in the ranges of 12 to 18, 5 to 14, and 0 to $+3.5$ Hz, but irregular values can also be found (e.g., see the data of CH₂=CHLi in Table 38).

On the basis of the coupling constants, it is very simple to distinguish between Z and E 1,2-disubstituted olefins, and, of course, the small coupling constants immediately reveal the case of 1,1-disubstitution (compare Problem **30**). The magnitudes of ^3J *vicinal* coupling constants of the interaction between the α-hydrogens of the side chain and the olefinic protons (type⟩CH–CH=) are generally in the range of 6 to 8 Hz, and they depend on the mutual steric position*** of the hydrogens. The greatest coupling constants (11 to 12 Hz)

---

* See Volume I, p. 61.
** Compare Volume I, p. 56 and 62.
***Dihedral angle, see Volume I, p. 60 and Equations 79, 80a, and 80b.

can be expected for the *trans* (**137a**) form, and the smallest ones can be expected for the *gauche* (**137b**) rotational isomer. If the rotation of the substituents is free, an average value can be measured in the spectrum. Bulky substituents prefer the rotational isomers of type

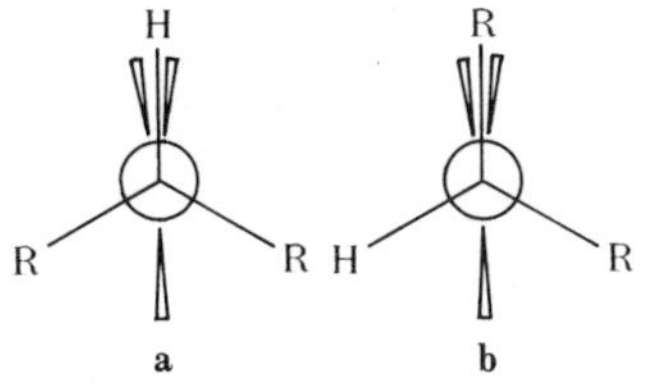

137

137a, in which the hydrogen atoms are in *transoid* position, with higher coupling constant. If the rotation of the substituent is hindered, the ^4J (allylic) coupling constants of ethylidene hydrogens and the chemical shift differences of these protons are temperature dependent.[635] It has been shown, e.g., in this way, that the s-*trans* conformer of vinylcyclopropane (**138**) is more stable than the s-*gauche* form by 4 kJ/mol.[350]

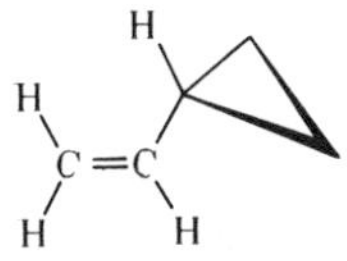

138

The allylic and homoallylic long-range couplings of 0.5 to 2.0 Hz, occurring very frequently with olefins, have already been discussed.*

In vinyl formate (**139**), $^5J_{1,4} = 1.7$, $^5J_{2,4} = 0.8$, and $^4J_{3,4} = 0.6$ Hz.[1248] The relative magnitudes of coupling constants are in agreement with the principle already discussed** that the strongest long-range couplings occur in a ''W'' arrangement and that the inclusion of hetero atoms increases these splittings. In the spectrum (see Figure 134) of 2,3-dibromopropene ($CH_2=CBr-CH_2Br$), the doublet of the proton in *cis* position to the bromomethyl group shows a further triplet splitting, producing by the overlap of the two outer lines of

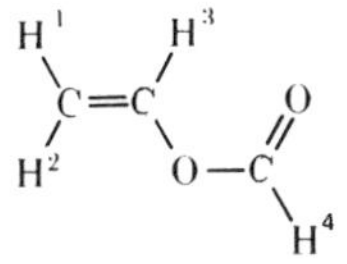

139

the latter an 1:2:2:2:1 quintet. This is in agreement with the rule*** that the coupling constants of *cisoid* allylic interactions are larger. The *transoid* coupling causes no significant splitting, and thus the signal of the corresponding proton is a doublet at 5.62 ppm (AMX_2 system, in which $J_{AX} \approx 0$). It can be seen from the data of Table 36 that the chemical shift is larger

* See Volume I, Section 1.3.7.1, p. 62-67.
** See Volume I, Section 1.3.7.1, p. 62-67.
***See Volume I, p. 62.

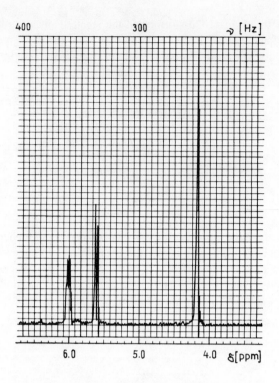

FIGURE 134.   The spectrum of 2,3-dibromopropene-1 (CH₂=CHBr–CH₂Br). (From *NMR Spectra Catalog,* VARIAN Associates, Palo Alto, Calif., 1962-1963. With permission.)

for the olefinic hydrogen *trans* to the bromine attached to the sp² carbon (the quintet appears at 6.02 ppm), and the same conclusion can also be drawn from literature data.[281,1087]

The olefin signal of ethylidene derivatives (CH₂=CRR′) is often a broad singlet, like in the spectrum (see Figure 135) of 2-propenyl acetate (CH₂=CMe–OAc), because the two ethylenic hydrogens are accidentally isochronic and the splitting caused by the *cisoid* and *transoid* couplings with the methyl protons is insignificant. The weak allylic coupling is shown not only by the broadened olefinic signal, but also by the different heights of the two methyl signals. The signal of the methyl substituent attached to the olefinic group is broadened by the allylic coupling, and thus the sharper acetyl methyl signal is much "higher".*

3.3.3. Conjugated Dienes and Polyenes

The chemical shifts and coupling constants of the terminal protons of conjugated olefins are similar to those of the derivatives with one double bond, but the protons located inside the chain have about 1 ppm larger chemical shifts than in ethylene derivatives (compare Problem **62**). The H-3 signal of isoprene (CH₂=CMe–CH=CH₂) appears at 6.35 ppm,[649] and the singlet of 1,1,4,4-tetraphenylbutadiene can be observed at 6.80 ppm.[1044]

The ⁵J long-range couplings of the terminal protons of butadiene derivatives illustrate well that, also in this case, the interaction is stronger via "W"-pathway.** In compounds of type **140** $J_{1,4}$ coupling constants are in the range of 1.3 to 1.8 Hz, whereas those of type $J_{1,3} \equiv J_{2,4}$ and $J_{2,3}$ are 0 to 0.8 Hz.[1147]

---

*   Compare Volume I, p. 64-65.
** Compare Volume I, p. 67.

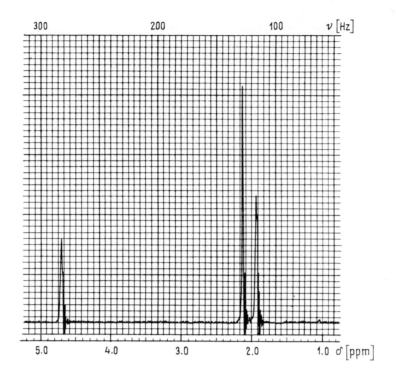

FIGURE 135. The spectrum of 2-propenyl acetate (CH$_2$=CMe–OAc). (From *NMR Spectra Catalog*, VARIAN Associates, Palo Alto, Calif., 1962-1963. With permission.)

140

In certain cases the structure of large molecules can also be elucidated, e.g., it was possible to determine the configuration of a polyene with Structure **141** on the basis[338] that the signals of the *vicinal* methyl groups in *trans* position to the olefinic hydrogens appear at lower frequencies (1.57 to 1.61 ppm) than those of the *cis* substituents (1.65 to 1.69 ppm). Since in the spectrum of Compound **141** the 1.58 ppm signal has a threefold intensity as the signal at 1.65 ppm, all –CH=CMe– groups must be in *trans* configuration (there must be six *trans* and two *cis* methyl groups).

141

Arising from the various spin-spin interactions of olefinic protons, the spectra of conjugated polyenes are usually rather complicated. The analysis of such spectra can be simplified by means of DR technique, partial deuteration, addition of shift reagent, or with the use of model compounds.

3.3.4. Cycloolefins
The signals of the cycloolefinic protons appear between 5.25 and 7.0 ppm and of the

methylene groups adjacent to double bonds in the region of 1.5 to 4.0 ppm. In the spectrum of cycloheptatriene (**142**), the multiplets of the olefinic protons in α-, β-, and γ-positions to the methylene group can be found at approximately 5.28, 6.12, and 6.55 ppm, respectively, indicating that also in this case the chemical shifts are larger inside the polyene chain. The lines of the triplet of the methylene group (2.18 ppm) merge into a broad signal at lower temperatures, and upon further cooling it splits into an *AB* pattern (δ*A* = 1.57 and δ*B* = 2.80 ppm).[41,718] This phenomenon is due to a fast ring inversion (**a** ⇌ **b**) at room temperature, which freezes upon cooling. The inversion barrier is higher in 1,2-substituted derivatives, since the steric hindrance between the substituents in the planar transition state is greater. In **142** $\Delta G^{\ddagger} \approx 25$ kJ/mol,[41,718] but 65 kJ/mol for derivative **143**.[648] The potential barrier is also higher in the fused derivatives, e.g., $\Delta G^{\ddagger} \approx 38$ kJ/mol[1046] for dibenzocycloheptatriene (**144**).

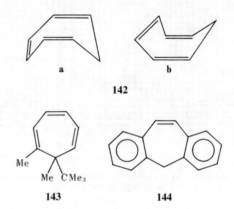

a   b

142

Me

Me CMe₃

143   144

With larger rings, the conformational equilibria prevent, even in simpler cases, the formation of multiplets with sharp lines, and compounds of this type have more or less broad maxima. In the spectrum of cycloocta-1,5-diene (**145**), the half-band widths of the singlet-like δCH₂ and δ(=CH) signals at 2.4 and 5.6 ppm, respectively, are about 7 and 10 Hz,[1044] instead of the usual ~1 Hz. The spectra of **145** and related compounds are, of course, temperature dependent owing to the conformational equilibria.

145

In the spectrum of cyclopropene, the signal of the olefinic hydrogens appear at about 7.0 ppm, the one of methylene protons is at 0.92 ppm,[1519] due to the stronger shielding generally occurring with three-membered rings. $J^{cis} = 1.8$ Hz.[1312] With cyclopropene-3-one (**146**) δH = 9.0 ppm and $J^{c} = 3$ Hz.[198]

O

146

Cyclobutene gives an *AA'BB'B''B'''* spectrum, from which δ(=CH) = 6.03 and δCH₃ = 2.57 ppm, and $J^{c} = 2.9$ Hz.[650] It follows from these data that, the coupling constants of *cis* olefinic protons increase with the size of the ring.* Analogous changes have been observed in the homologous series of cycloalkenones.[644]

The ^3J coupling constants of type RR′CCH–CH= are conformation dependent and usually small, e.g., in cyclobutene a value of 1 Hz was measured.[650] Cyclic dienes often show long-range couplings transferred by π-electrons. The range of the ^5J couplings of type –CH=C–C=CH– in conjugated systems is 0.6 to 0.9 Hz.[173] The coupling constants between the methylene hydrogens on C-3 and C-6 of cyclohexa-1,4-diene (**147**) are $^5J^t = 8$ and $^5J^c = 9.6$ Hz.[548] The surprisingly large values can be attributed in part to the planar structure, in which the dihedral angles are approximately 180°, in part to the presence of four π-electrons,** and finally to the two possible pathways (3-2-1-6 and 3-4-5-6) of the interaction.***

147

The spectra of bi- and polycyclic olefins are often simpler than expected. Thus, the isochronous methylene, methine, and olefinic hydrogens of bicycloheptadiene (**148**) give rise to an $A_2M_2X_4$ spectrum (see Figure 136) in which a symmetric triplet, septet, and triplet, respectively, correspond to the three groups at 2.00, 3.58, and 5.75 ppm.[1044] The spectrum can be interpreted, e.g., by assuming that the methylene, methine, and olefinic hydrogens are, practically, magnetically equivalent ($J_{1,6} \approx J_{2,6} \approx J_{6,7}$) and the interactions 1,2 ≡ 4,5 do not cause observable splitting.

148

Valence isomery is frequently observable in polycyclic olefins. Thus, cyclooctatrienone (**149**) is in equilibrium with Compound **150** through valence isomerization, and the relative weight of the two isomers is 19:1.[541] The valence isomerization of 1,2-disubstituted cyclooctatetraenes takes place already at accessible temperatures. The ratio of the isomers (**151** and **152**) can be determined (e.g., it is 17:1 at room temperature for R=Me and R′=COOMe), and from the temperature dependence of the spectrum for the above substituents ΔG‡ of isomerization is 80 kJ/mol.[43]

149   **150**

Bullvalene (**153**) undergoes a fast, reversible valence isomerization.[1262] This compound has more than 1.2 million equivalent structure isomers! At 100°C, the ¹H NMR spectrum

---

* Compare Volume I, p. 60.
** Compare Volume I, p. 62.
***Compare Volume I, p. 67.

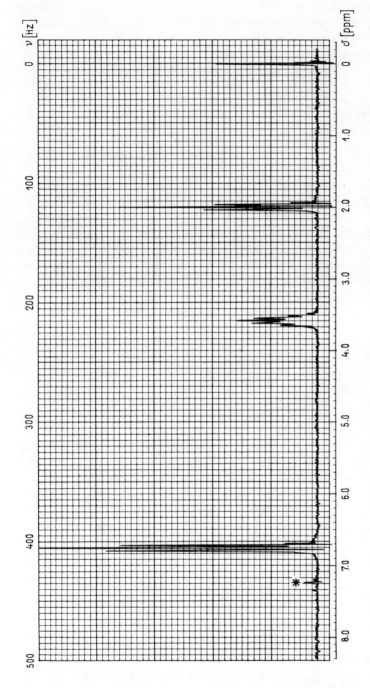

FIGURE 136. The spectrum of bicycloheptadiene (**148**). (From *NMR Spectra Catalog*, VARIAN Associates, Palo Alto, Calif., 1962-1963. With permission.)

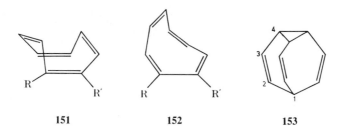

151 152 153

of the compound consists of one sharp singlet (4.2 ppm), indicating that isomerization is so fast that the chemical environments of all protons (aliphatic and olefinic) are averaged, and thus the protons are isochronous. The signals of the aliphatic (H-1,4) and olefinic (H-2,3) protons separate at −25°C, shifting to 2.1 and 5.7 ppm (intensity ratio, 2. The monosubstituted derivatives may have, in principle, four different structures bearing the substituent on the C-1 atom or on one of the three C-2, C-3, or C-4 atoms. The singlet spectrum (e.g., at 4.22 ppm in the case of the bromo derivative) indicates that the four structural isomers interconvert so rapidly that all protons and substituent positions become equivalent.[1072] If the spectrum of bromobullvalene is recorded at −45°C, the complicated multiplets of the aliphatic and olefinic protons appear separately, and the intensity ratio is 7:11, expressing that Br replaces neither an olefinic nor an aliphatic hydrogen (since in these cases the intensity ratio would be 3:6 or 4:5), that is, the site of substitution cannot be localized.

Among the cycloolefins the symmetric (C_2) *trans* isomers of the compounds with larger rings (**154**) can occur in two enantiomeric forms (**a** and **b**), which may interconvert (racemize) by rotation about the C–C single bonds adjacent to the double bond.[136] The process can be monitored, e.g., by NMR method in optically active solvents.* The half-times of racemization are $6·10^5$ years, 70 s, and $8·10^{-4}$ s for $n = 6$, 7, and 8, respectively, i.e., they decrease abruptly with the ring size.

H H

H H

(CH$_2$)$_n$ (CH$_2$)$_n$

a b

154

The 9-11 membered N-substituted lactams are heteroanalogues of **154** with a substituent R and carbonyl oxygen instead of the two olefinic protons, respectively, and the replacement of one sp^2 carbon by the amide nitrogen. The enantiomers are, however, not observable here due to a fast racemization via the E isomer. This was confirmed by the fact that the methyl signals of N-methyl lactams of this type are split due to the presence of geometric isomers (the methyl groups of the Z and E isomers absorb at about 2.9 ppm and between 3.0 and 3.15 ppm, respectively).[1000]

The Z-E isomers of compounds containing exocyclic double bonds, e.g., of type **155** and **156** can be differentiated on the basis of the effect of the oxo group on the chemical shifts,[507] since in isomers **156** the olefinic proton is deshielded (7.7 to 8.2 ppm) relatively to its counterpart in compounds **155** (6.4 to 7.0 ppm).

3.3.5. Continuously Conjugated Cyclic Polyenes

In continuously conjugated ring systems with planar structures, internal ring currents may

---

* Compare Volume I, p. 41, point 5.

155   156

occur, and thus anomalous shieldings may be possible.* The protons of azulene (**157**) have characteristically large chemical shifts, similar to the aromatic hydrogens. The multiplet appears in the range of 400 to 510 Hz at 60 MHz.**

157

Of the annulenes C_nH_n, the aromatic ones with $(4n + 2)\pi$-electrons and the antiaromatic homologues with $4n$ π-electrons have characteristically different spectra. The spectrum of the latter type is temperature dependent, and at room temperature is a singlet with a chemical shift similar to those of olefinic protons. The signal of compound $C_{24}H_{24}$, e.g., appears at 6.84 ppm at room temperature.[238] In the spectra of aromatic homologues, separate signals arise from the outer and inner protons. In the spectrum of analogue $C_{18}H_{18}$ (**11**) they appear at 8.9 and -1.8 ppm,*** with an intensity ratio of 2:1.[706] It is thus clear that the $C_{24}H_{24}$ homologue may not be aromatic, and there is a rapid interconversion of valence isomers followed by the interchange of the "inner" and "outer" protons. The same complex motion proceeds in aromatic homologues at higher temperatures. This was confirmed by the temperature dependence of the spectrum of the mononitro $C_{18}H_{18}$ annulene.[545] Whereas the signals of **11** separated at room temperature give a common signal at about 5.3 ppm at higher temperatures, the nitro derivative gives a spectrum in which the signals of inner and outer protons at about -3 and 10 ppm with intensity ratio 6:11 are separated at $-70°C$. However, at higher temperatures, also two signals appear at about 8.5 and 3.5 ppm of intensity ratio 5:12. This can be explained so that the interchange of inner and outer protons proceeds in such a way that the former alternate their position with six of the latter. Two of the three groups of six hydrogens in **11** interchange in one step. In the case of nitro-derivative, the five protons belonging to the same group as the nitro substituent cannot alter their outer position, since the bulky nitro group has no room inside the ring. Consequently, only the other 12 hydrogens take part in isomerization and give the signal at 3.5 ppm corresponding to the averaged environment. The excluded 5 protons have a signal at 8.5 ppm, characteristic of outer hydrogens. Hence, the aromatic or antiaromatic character of annulenes can be determined very simply from the ^{1}H NMR spectrum.

At low temperatures (around $-80°C$), the spectrum of Compound **11** does not change

---

*   Compare p. 91 and Volume I, p. 36-38.

** The multiplets from which the chemical shifts have not been determined are characterized by the frequency range in hertz in which the lines occur. In such cases, of course, the measuring frequency must be given. The chemical shift measured in parts per million has always a definite value for a given nucleus. However, the positions of lines in a multiplet can be given only in hertz, since one nucleus may produce several lines, and only from the position, number, and intensity of these lines can one determine the single value of chemical shift belonging to the nucleus.

***Compare Volume I, p. 36.

significantly, but in the spectra of compound $C_{24}H_{24}$ the signal is split into two maxima, appearing at 4.7 and 12.0 ppm, and corresponding to the outer and inner protons, respectively.[238] The reversed order of chemical shifts proves unambiguously the antiaromatic nature of the compound, which, according to quantum mechanical calculations,* may give rise to paramagnetic ring currents in the system.[1133] The aromatic, i.e., planar[1059] or in another case[1436] the antiaromatic structures of dehydroannulenes, were determined similarly.

In the spectrum of dehydroannulene **158**, the signal of the *trans* olefinic protons appears at 10.9 ppm, due probably to an isomerization **a** $\rightleftharpoons$ **b** fast at room temperature.[1133] The six other protons give rise to signals between 4.2 and 5.0 ppm. If a similar chemical shift (4.5 ppm) is assumed for the outer *trans* proton, the shift of the inner hydrogen is obtained to be 17.3 ppm, indicating a strong antiaromatic character.

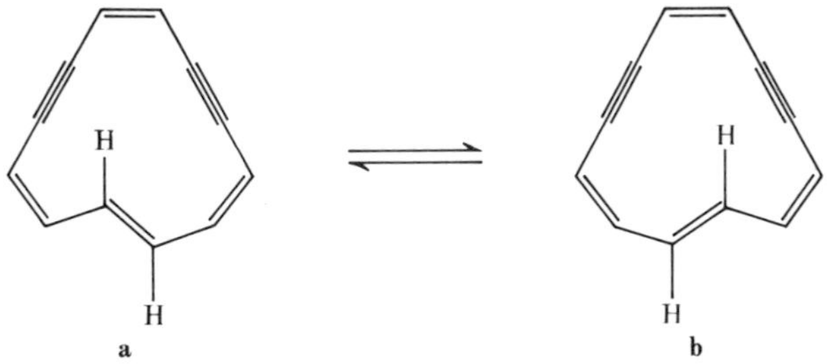

158

3.4. ALDEHYDES, NITRILES, AZIDES, AMIDES, OXIMES, IMINES, AZO, NITRO AND NITROSO COMPOUNDS, *N*-OXIDES, ETC.

Protons attached to unsaturated atoms or situated in α-position to such atoms in widely different functional groups, but with characteristic spectrum parameters, will be discussed in this section, e.g., of the carbonyl compounds, only the *aldehydes* will be mentioned, since** — the various carbonyl compounds have no characteristic ^{1}H NMR spectra.

In the spectra of aldehydes, the signal of the hydrogen attached directly to the carbonyl group has a large chemical shift, because the anisotropic effect of the carbonyl group, which is here very strong on the coplanar hydrogen (compare Figure 16), is added to its $-I$ effect. Of the protons attached to carbon atoms, the aldehydic protons are the most deshielded ones, making aldehydes easy to recognize from the ^{1}H NMR spectrum, unless they are present in hydrate, dimeric, or polymeric forms.

The signals of aliphatic aldehydes appear in the range of 9.5 to 9.8 ppm. Formyl derivatives have, of course, similar large chemical shifts (around 8 ppm). The corresponding signals of methyl formate (HCOOMe) and DMFA appear at 8.08 and 8.02 ppm, respectively.[1044] The ^3J coupling constants of type $-CH-CH(=O)$ are in the range of 1 to 3 Hz.

The anisotropic effect of aromatic rings*** causes an even stronger deshielding. The signal of aromatic aldehydes appears, between 9.6 and 10.5 ppm, and the signal of 9-antraldehyde (**159**) appears at 11.5 ppm,[750] which can be attributed to the accumulated anisotropy of the fused system† and to the hydrogen bond involving H-1 or H-8.

The aldehyde protons may often take part in long-range couplings.‡ In aromatic aldehydes there is no significant coupling with the protons in *ortho* and *para* positions (^4J, $^6J < 0.2$

* Compare Volume I, p. 38.
** See p. 2.
***Compare Section 1.2.5, Volume I, p. 35-38.
† Compare p. 72.
‡ See p. 53 and 82 and Volume I, p. 50.

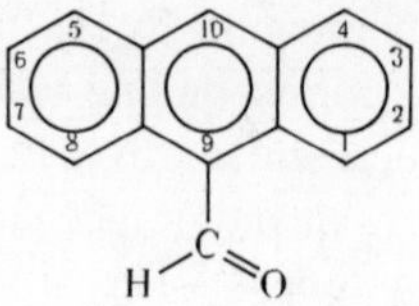

159

Hz), whereas the ^5J coupling with the *meta* ring proton produces a splitting of ~0.7 Hz[345,750] via "W" pathway (**160a**). This can be proved by investigating appropriately substituted derivatives. 2,5-Disubstituted benzaldehydes, if the substituent is bulky enough, may exist only in form **b**, and in this case $^5J \approx 0.35$ Hz, similar to the 2,6-disubstituted derivatives, in which the aldehyde group is perpendicular to the plane of the ring and the signal of the aldehyde hydrogen is a triplet. Of course, the largest coupling constants (0.7 to 0.9 Hz) can be observed in the spectra of 2,3-disubstituted benzaldehydes, in which structure **a** is fixed.

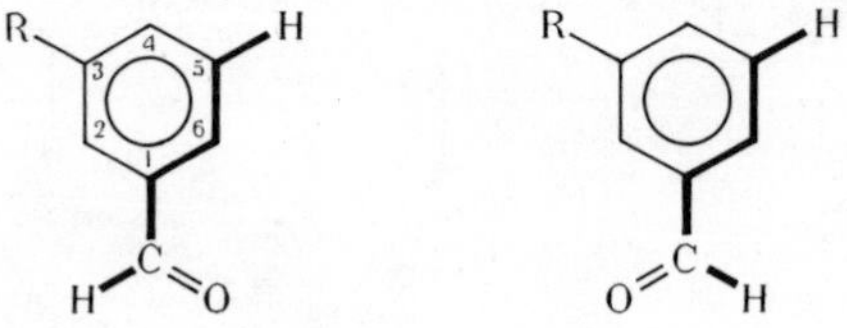

160

The signal of aldehyde protons may be shifted substantially (by some tenth of parts per million) due to the solvent effect.[772] In TFA or other strong acids at low temperatures, the aldehydes (and the ketones as well) may be protonated, and in this case Z-E isomery may occur (**161a,b**). With aldehydes, it is easy to distinguish between the isomers on the basis of the magnitude of *vicinal* coupling constants, due to the relation* $J^{trans} > J^{cis}$, which also holds for aldehydes. For protonated acetaldehyde,[212] $^3J = 8.5$ (Z) and 18.5 Hz (E).

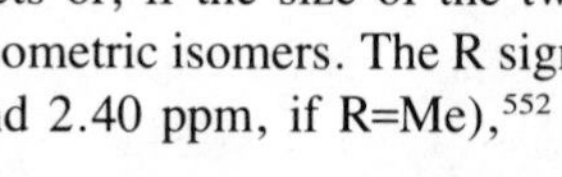

161

Like in the case of olefins, the isomeric forms of protonated *ketones* can be distinguished by utilizing the substituent effects or, if the size of the two substituents are different, the different concentrations of the geometric isomers. The R signals of $R_2C=SO$-type compounds appear separately (e.g., 2.38 and 2.40 ppm, if R=Me),[552] proving the nonlinear geometry of the bond system C=S→O.

The *nitrile* group, as electron-withdrawing substituent, causes deshielding of the adjacent hydrogens. However, as anisotropic neighbor, it has an opposite effect, similar to that of acetylenes,** along the axis of the triple bond (see Figure 16). For this reason, the methylene protons of compounds $R-CH_2-CN$ are relatively strong shielded (their chemical shift is 2.0 to 2.7 ppm). The methylene signals of allyl cyanide derivatives ($RR'C=CR''-CH_2CN$) appear

---

*   See p. 52 and 72.
**   See p. 46 and Volume I, p. 32.

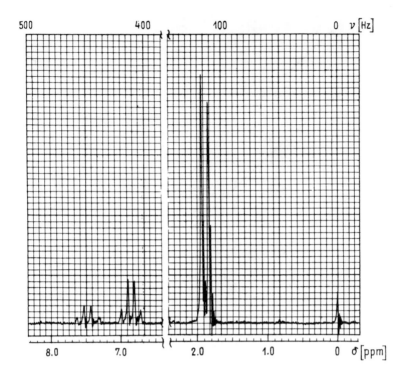

FIGURE 137.   The spectrum of acetaldoxime (**162**). (From *NMR Spectra Catalog*, VARIAN Associates, Palo Alto, Calif., 1962-1963. With permission.)

around 3.2 ppm, and the *geminal* hydrogens of nitrile-substituted cyclohexanes absorb at approximately 3.6 ppm.

Among the simplest aliphatic nitro compounds, nitro methane, ethane and *i*-propane give methyl signal at 4.38, 1.48, and 1.55 ppm, respectively, while the methylene and methine protons of the two latter absorb at 4.29 and 4.67 ppm.[259,1044] The two signals of ethyl nitrite are at 1.39 and 4.78 ppm, respectively.[1044]

The signal of *diazo* derivatives (R–CHN$_2$) appears in the range of 4.5 to 5.0 ppm, and that of *diazoketones* (R–CO–CHN$_2$) appears in the range of 6.0 to 6.5 ppm (compare Problem **12**). The signal is sharp and its position solvent dependent.

The methyl signal of methylazo derivatives appears between 3 and 4 ppm, e.g., the signal of dimethyl azide (MeN=NMe) is at 3.68 ppm.[513] This very large shift is characteristic of methylazo compounds.*

Azomethine groups (–CH=N–) absorb, similar to aldehydes, at large δ-values. The signal of phenylglioxal aldoxime (Ph–COCH=NOH) is at 8.13 ppm,[1044] and that of benzaldazine (Ph–CH=N–N=CH–Ph) is at 8.55 ppm.[434]

Compounds containing C=N and N=N bonds very frequently show geometric isomery, e.g., thus, the methyl doublet and the methine quartet of acetaldoxime are doubled at 1.86 and 1.89 as well as at 6.84 and 7.44 ppm, respectively (see Figure 137),[1044] corresponding to the *syn* and *anti* forms (**162a,b**). The related doublet and quartet pairs can be selected by their relative intensities. It follows from the chemical shifts that the more intense signals correspond to isomer **a**. The methine quartet can be expected downfield in the spectrum of isomer **b**, whereas the opposite holds for the methyl signal, owning to the anisotropy of the hydroxyl group which causes paramagnetic shift.** The resulting assignment is in agreement

---

* Compare with the *N*-methyl signals of *N*-methyl triazoles and tetrazoles with similar large shifts (p. 90).
** Compare Volume I, p. 33 and structures **4a** to **4c**.

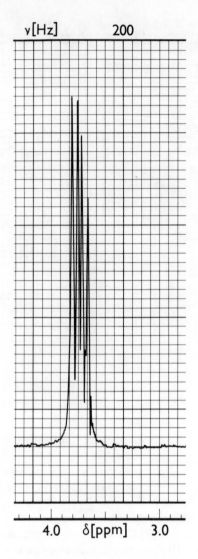

FIGURE 138. Part of spectrum of 2,2',4,4'-tetramethoxy-benzo-
phenon-oxime (**163**) with the methoxy signals.

with the experience that the lone pair of sp² nitrogens must be regarded as a bulkier ''substituent'' than a hydrogen attached to the nitrogen.[809]

Of course, the above considerations also pertain to ketoximes, e.g., explaining the presence
of four methoxy signals in the spectrum (Figure 138) of oxime **163**. In the spectra of *syn*
and *anti* 2-acetobenzofuran oxime (**164a,b**), the singlets of the 3-methine groups appear at
7.96 ppm and 7.20 ppm, respectively (see Figure 139), wherefrom the isomeric structures
of the samples can be derived easily.[1339] The anisotropic effect of the hydroxy group causes

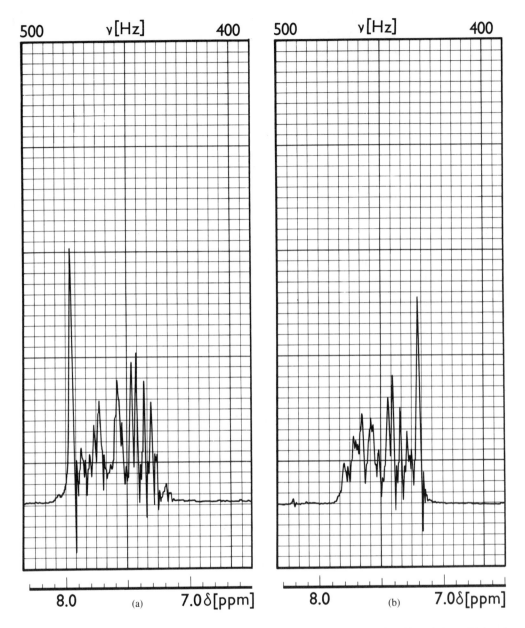

FIGURE 139.   Parts of the spectra of (a) *syn* and (b) *anti* 2-acetobenzofuran-oxime (**164**) with the multiplet of the ring protons.[1339]

much larger paramagnetic shift to H-3 in the *syn* isomer than in the *anti* compound, due to their smaller distance.

163

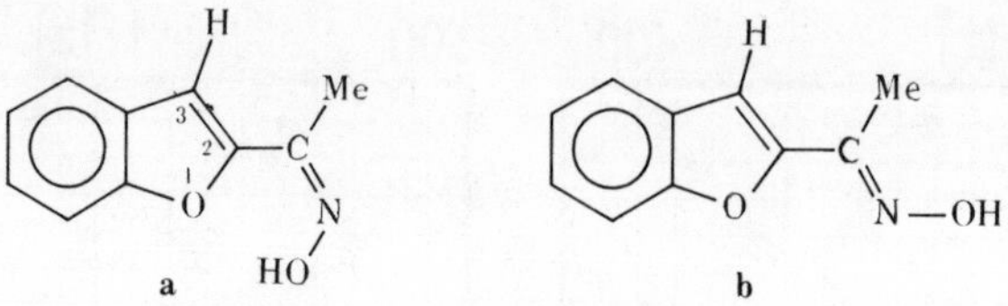

164

The presence of isomeric mixtures or the phenomenon of hindered rotation has been detected by NMR methods for several *C*-nitroso, *N*-nitroso, azoxy, amide, nitrite, and other similar derivatives. In some instances these investigations required low-temperature measurements, of course. The two methyl signals of compounds $(MeNO)_2$, Me_2N-NO and $[MeN=NO]^{\ominus}$, are at 4.00 and 4.20, 3.09 and 3.82, and at 3.20 and 3.70 ppm, respectively.[649,1044]

From the NMR studies of nitrogen-containing unsaturated compounds, it is possible to discuss here only some arbitrarily chosen further results. The chemical shift difference between the *equatorial* and *axial* α-hydrogens of *N*-nitroso piperidine (**165**) is approximately 3 ppm, one of the largest differences observed between a pair of methylene protons.[274]

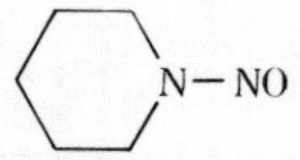

165

Z-E isomers may also occur among imines, due to hindered rotation and inversion, respectively. The assignment of the signals to the isomers is straightforward when the relative stabilities of the two isomers are different, e.g., if substituent R of imine **166** is bulky enough, and R′ is larger than R″, isomer **a** is more stable, and the stronger signals belong to it.

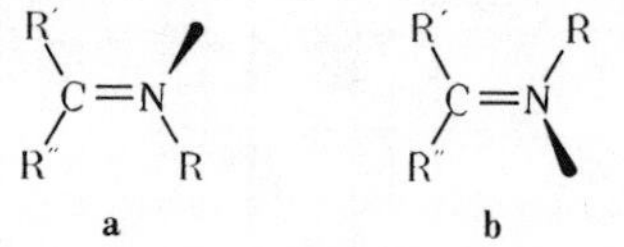

166

When R′=R″, geometric isomery may not occur, but the signals of R′ and R″ occur separately due to a slow **a** ⇌ **b** change. They can be assigned, for example, on the basis of the solvent effect (ASIS). In benzene solutions the signal of the group *trans* to the lone pair is shifted more with respect to the signal measured in indifferent solvent.[1574]

If the effects of substituent R are substantially different on the *vicinal* groups, the spectrum can also be interpreted on this basis. When, for example, R′=R″=Me and R=Ph, R″ is more deshielded in isomer **a,** in a coplanar structure.

The signals of a protonated mixture can be assigned in a similar manner, e.g., the *O*-methyl and *S*-methyl signals of Compound **167** are doubled in TFA, appearing at 4.50 and

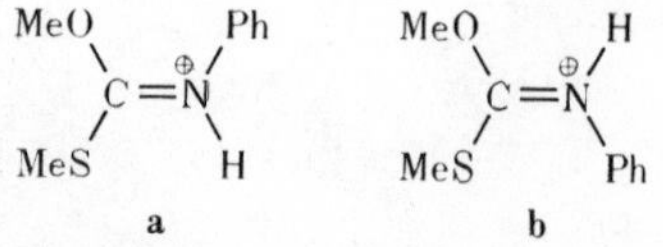

167

4.15 ppm and at 2.50 and 2.90 ppm, respectively.[1463] The signals at 4.50 and 2.50 ppm are stated to belong to isomer **a**, and those at 4.15 and 2.90 ppm are stated to belong to isomer **b**.

Protonation and quaternerization deshield the α-hydrogens of the substituents attached to the nitrogen and to the carbon, and the effect on the latter is stronger (!). The methyl signals of imine **166** (R=R'=R"=Me) appear, for example, at 1.75, 1.88, and 2.95 ppm in the spectrum of a CCl_4 solution, whereas in acidic solutions the corresponding signals of the protonated salt appear at 2.73, 2.82, and 3.66 ppm.[1065]

In the case of amides one can determine whether the nitrogen or the oxygen atom is protonated, e.g., if the nitrogen atom of DMFA is protonated, the double bond character of the C–N bond is eliminated, and the anisochrony of the methyl groups is removed, whereas it is retained upon *O*-protonation. As the methyl signals occur separately in acid solutions,[504] the latter structure must be present. This has been found not only for DMFA but also for a number of *N,N*-disubstituted amides.*

3.5. AROMATIC COMPOUNDS

Owing to the coplanarity of delocalized π-electron system and the attached substituents capable of conjugation, the spectrum parameters of aromatic systems are substantially different from those of other compounds. The spectrum of pure benzene contains one sharp line at 7.37 ppm (in a 2% CCl_4 solution at 7.27 ppm), i.e., the chemical shift is more than 1.5 ppm larger than, for example, that of nonaromatic cyclooctatetraene (5.69 ppm).[434] The reason for this difference is the anisotropy of the aromatic ring, which deshields** the coplanar protons.

The signal of the aromatic protons is seldom a singlet in the spectra of monosubstituted derivatives (e.g., toluene). They have mostly *ABB'CC'* (e.g., phenol) or sometimes *ABB'XX'* spectra (e.g., nitro benzene). These complex spectra can be characterized by three chemical shifts (δH^o, δH^m, and δH^p) and six coupling constants ($J^{o,o'}$, $J^{o,m}$, $J^{o,m'}$, $J^{m,m'}$, $J^{o,p}$, and $J^{m,p}$). The analysis of these systems is rather complicated, and the positions and relative intensities of certain lines cannot be given by explicit formulas. These cases must be solved by trial and error techniques.

By means of high-resolution CW spectrometers it becomes possible to separate the individual lines, enabling a complete analysis of the spectra. Thus, for example, the first complete analysis of the aromatic multiplet of toluene was performed in 1968, on the basis of a 100-MHz spectrum.[1529]

The electron affinity ($-I$ and $-M$ effect) of the substituents is expressed by the linear relationships[366,509,818,1355] between the chemical shifts and the *Hammett* and *Taft* constants and other parameters characteristic of the electron affinity of the substituents (dipole moment, electron density, etc.). Certain substituents, primarily the carbonyl, nitro, and nitroso groups as well as the halogens also influence the shielding of the *ortho* protons by anisotropy. The above relationship is, therefore, very accurate for *para* protons, but valid only approximately and with exceptions for the hydrogens in *ortho* positions.

The situation is more complicated with the *meta* hydrogens because both the *I* and *M* effects and the anisotropic effect are weaker and comparable. Accordingly, no similar relationship could be found for δH^m, although in reaction kinetics the substituent constants can be used equally well to interpret the behavior in the *para* and *meta* position.

The above are illustrated in Figure 140, in which the chemical shifts, related to that of benzene, of some monosubstituted benzene derivatives are shown in the sequence of the increasing electron-withdrawing effect of the substituents.[1355] It can be seen from the diagram

---

* Compare p. 108.
** See Volume I, Section 1.2.5 (p. 35-38).

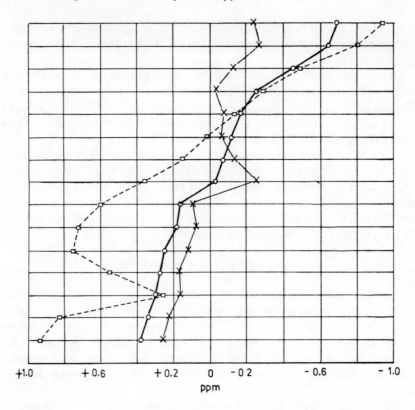

FIGURE 140.   The relative chemical shifts of the *ortho, meta* and *para* protons of monosubstituted benzene derivatives, with respect to benzene, in the sequence of the increasing
electronegativity of the substituents. □ : *ortho*, x: *meta*, o: *para* ring protons.[1355]

that the substituent effect is the strongest on the *ortho* protons and the smallest on the *meta*
ones. Furthermore, the shifts of *para* and partly also of *ortho* protons (except for strong
electronegative substituents) are in linear relationship with the electronegativity of the substituent, while the *meta* protons show quite unsystematic behavior.

According to a spectrum analysis of benzene performed by the ^{13}C side-band technique,*
$J^o = 7.54$, $J^m = 1.37$, and $J^p = 0.69$ Hz.[1173] Model investigations have shown that these
three kinds of coupling constants are without exception characteristically different: $J^o > J^m$
$> J^p > 0$, and in substituted derivatives they fall into the ranges of 7.0 to 9.2, 1.0 to 3.1,
and 0.2 to 0.7 Hz.[604,1205] Hence the substituents have no major effect on the coupling constants
(the ranges are narrow), and J^o is similar to that of *vicinal* aliphatic or *cis* olefinic protons.
This coincidence already indicates the fact, also derived theoretically, that the *ortho* couplings
are transferred mainly by the σ-electrons. This is the reason for the low value of J^m (similar
to the ^4J couplings of saturated compounds). In the *para* coupling the π-electrons also play
an important part, as generally in unsaturated compounds.[1379] Within the given range, the
value of J^o increases proportionally with the sum of *I* and *M* effects of the substituents.[306]

The interactions of the aromatic protons with α-hydrogens of the substituents are weak,
and they cause only line broadening at most. These coupling constants are in the range of
0 to 1 Hz, exceptionally 1 to 2 Hz (e.g., see Figure 37). The same holds for the interactions
between protons belonging to different rings of fused aromatic compounds.[1379]

Investigations on simple disubstituted benzenes have shown that the effects of the substituents are additive.[366] The chemical shifts of the *ortho, meta,* and *para* hydrogens of

* See Section 4.1.3.2 (p. 209-210).

Table 39
SUBSTITUENT CONSTANTS ρ_i (PPM) FOR ESTIMATION OF CHEMICAL SHIFTS OF AROMATIC RING PROTONS BY EQUATION 277[302,460,938]

Substituent	ρ_i^o	ρ_i^m	ρ_i^p	Substituent	ρ_i^o	ρ_i^m	ρ_i^p
NHMe	−0.9	−0.2	−0.7	Br	0.2	−0.1	−0.05
NH$_2$	−0.8	−0.25	−0.65	C(sp)	0.2	−0.05	−0.05
NMe$_2$	−0.65	−0.2	−0.65	Ph	0.2	0.05	−0.05
OH	−0.5	−0.1	−0.5	CN	0.25	0.2	0.3
OMe	−0.5	−0.1	−0.45	I	0.4	−0.25	0
OR	−0.35	−0.05	−0.3	NHCOR	0.4	−0.2	−0.3
F	−0.3	0	−0.25	N$^+$H$_3$	0.4	0.2	0.2
Me	−0.15	−0.1	−0.15	COAr	0.45	0.1	0.2
CMe$_3$	−0.1	0	−0.25	CHO	0.55	0.2	0.3
OCOPh	−0.1	0.05	−0.1	COR	0.6	0.1	0.2
SH	−0.05	−0.1	−0.2	COOR	0.7	0.1	0.2
Cl	0	−0.05	−0.1	CONH$_2$	0.7	0.2	0.25
SR	0.1	−0.1	−0.2	COCl	0.8	0.2	0.35
C(sp^2)	0.15	0	−0.15	COOAr	0.9	0.15	0.25
OCOR	0.2	−0.1	−0.2	NO$_2$	0.95	0.25	0.4

bromobenzene, related to that of benzene, are $+0.22$, -0.09, and -0.03 ppm. The corresponding relative shifts in nitrobenzene are $+0.95$, $+0.21$, and $+0.33$ ppm. These values were used to calculate the shifts of H-2-6 of *meta*-nitrobromobenzene. The calculated (1.17, 0.92, 0.12, and 0.55 ppm) and the experimental shift differences (1.11, 0.89, 0.10, and 0.53) agree within 0.06 ppm.[938] The same differences for *para*-fluoro-chlorobenzene are ≤ 0.08 ppm.[366,1355]

Using Equation 277 and the substituent constants ρ_i given in Table 39,[302,460,938] it is possible to predict the approximate chemical shifts of the ring protons of any benzene derivative:

$$\delta ArH = 7.27 + \Sigma \rho_i \tag{277}$$

This additivity rule can be utilized to decide between substitution isomers (e.g., see Problems **28** and **49**).

As an illustration, the spectrum of a mononitro derivative, obtained by nitration from *ortho*- aminophenol, is shown (see Figure 141). Two of the theoretically possible four isomers (1,2,4-trisubstituted derivatives) can be discarded immediately on the basis of the multiplicities of the lines because their spectrum would consist of two doublets and one double doublet, owing to the insignificant splitting due to $J^p < 1$ Hz.*

One cannot distinguish, however, between alternatives **168** and **169** by the splitting. From Equation 277, δH-4 = 7.47, δH-5 = 6.77, and δH-6 = 6.92 ppm for isomer **168**, and δH-4 = 7.47 and δH-5 = δH-6 = 6.77 ppm for isomer **169**. As the measured shifts are 6.58, 7.08, and 7.61 ppm, the product is isomer **168**.

168   **169**

---

*   Compare Volume I, p. 116-118.

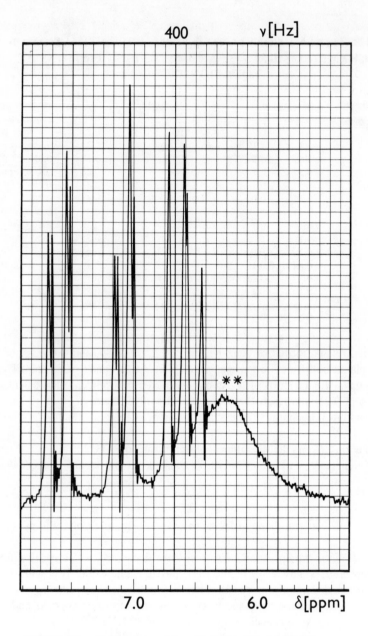

FIGURE 141.   The spectrum of 2-amino-3-nitrophenol (**168**). The overlapped, common signal of acidic protons arising from the solvent and solute,
respectively, are marked here and throughout the text with two asterisks (**).

A difference exceeding 0.1 ppm can generally be observed between the experimental and
calculated values in cases when there are identical substituents in *ortho* position or one of
the substituents of the *ortho* pair is a bulky group (e.g., nitro or dimethylamino). In the
former case the substituent effects are perturbed mutually, whereas in the latter the bulky
group assumes a forced conformation different from that of the monosubstituted benzenes,
thereby changing the conjugation and anisotropic effect.[366] In general, the additivity rule
does not hold for benzene derivatives having great dipole moments.[1379]

When applying the additivity rule, one must take into account that the data of Table 39 pertain to data measured in indifferent solvents. The solvent effect* may influence the chemical shifts and their differences, and thus noticeable deviations may occur.[366]

From the data of Table 39 it follows that the largest shifts occur in nitrobenzene derivatives, and the smallest ones occur in phenols, anilines, and anisols. The signals of *para*-dinitro-benzene, 2,4,6-trinitrotoluene, and picric acid (**170**) appear at 8.45,[434] 8.84,[1044] and 9.13,[649] respectively (calculated values: 8.47, 9.47, and 9.47 ppm, and the two latter compounds represent the most irregular cases, since they are strongly polar and contain three adjacent bulky substituents).

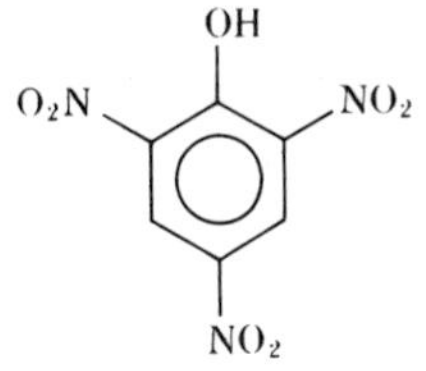

170

The aromatic signals of 3,4,5-trimethoxyaniline and 3,4,5-trimethylaniline appear at 5.89 and 6.18 ppm, respectively, and the aromatic protons of 2,6-dimethylanisole (**171**) absorb at 6.80 ppm[649] and those of *para*-diethoxybenzene absorb at 6.67 ppm.[434] When the sub-stituents affect the chemical shifts of the ring protons similarly or not at all or the effects of the two substituents compensate each other, *ortho*-disubstituted benzene derivatives may give rise to singlets, e.g., by *ortho*-xylene (**172**) at 7.10 ppm or by *ortho*-anisidine (**173**) at 6.80 ppm.[1044] In the case of *meta*-disubstitution singlet occurs less frequently, but *meta*-xylene gives a singlet-like signal at approximately 7.0 ppm.

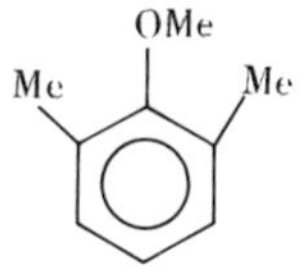

171 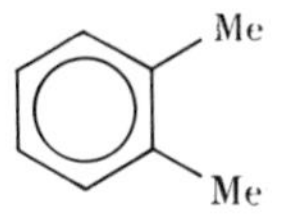   **172** 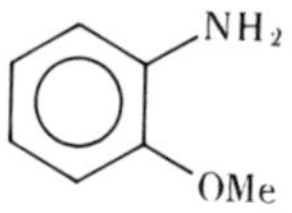   **173**

For symmetric *ortho*-disubstitution the *AA'BB'* multiplet is characteristic as in case of *ortho*-dichlorobenzene (see Figure 69). Complex (*ABCD*) spectra are produced, e.g., by *ortho*-nitrobenzaldehyde and *meta*-chloroaniline and first-order (*AMPX*) spectra (see Figure 142) by N-acetyl anthranilic acid (**174**) and *meta*-nitrobenzenesulfonic amide (**175**).

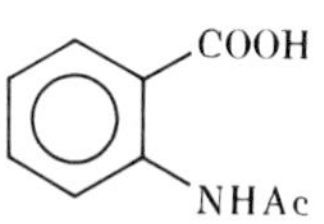

174 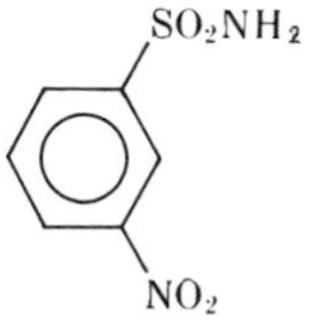   **175**

The spectra of *para*-disubstituted derivatives in the case of symmetric substitution are singlets. A singlet may also arise from accidental isochrony, in the case of asymmetric substitution, such as in the spectrum of *para*-chloro-mercaptobenzene at 7.23 ppm.[1044] In the spectra of asymmetrically substituted derivatives *AA'BB'* or *AA'XX'* multiplets occur

---

* See Volume I, Section 1.2.6, p.38-40.

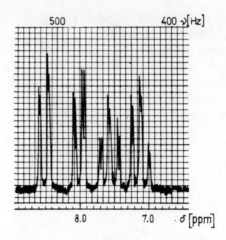

FIGURE 142.   Part of the spectrum of *N*-acetyl-anthranilic acid (**174**) with the multiplets of the aromatic protons. (From *High Resolution NMR Spectra*, SADTLER Research Laboratories, Philadelphia; Heyden and Son, London, 1967-1975. With permission.)

from which the chemical shifts and the coupling constant J^o can be given by *AB* or *AX* approximation.*

In fused aromatic systems, as already mentioned**, there is no noticeable spin-spin interaction between the protons attached to different rings. Thus, naphthalene (**176**), anthracene (**177**), and triphenylene (**178**) give rise to *AA'BB'* spectra, pyrene (**179**) produces an AB_2, perilene (**180**) produces an *ABX*, and coronene (**181**) produces a singlet spectrum. The spectrum of phenanthrene (**182**) is more complex (an *ABCD* multiplet). Atoms H-9,10 of anthracene and phenanthrene and H-3,4,8,9 of pyrene do not interact with the other protons, and thus their signals are singlets.

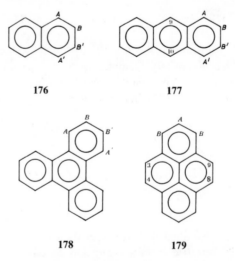

The chemical shifts are influenced by all rings, causing stronger deshielding of ring protons

*   See Volume I, p. 131 and 133 and Figures 64 and 65.
**   See p. 69.

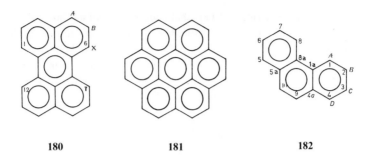

180   **181**   **182**

than in simple benzenes. The effect of adjacent rings is inversely proportional to the third power of distance.[124]

Shifts larger than generally can be observed in the following cases:[732]

1. For the "inner" protons of linearly fused aromatic systems (e.g., the δH-9,10 signal of **177** appears at 8.3 ppm)
2. For the sterically hindered "peri" protons (e.g., H-1,6,7,12 of **180** absorb at 8.1 ppm)
3. For systems in which further delocalization may occur besides the "periferic" one (e.g., the shift of the singlet of **181** is 8.85 ppm.

The examples cited so far to demonstrate the diamagnetic shift caused by aromatic ring anisotropy* will be completed here with a few further cases to illustrate the possibilities for applications in structure elucidation. In the spectrum of Z stilbene (**183**), a singlet corresponds to the aromatic protons[1044] [δArH = 7.18 and δH(olefin) = 6.55 ppm], whereas the spectrum of E stilbene (**184**) contains an $AA'BB'C$ multiplet[1044] in the region of 430 to 460 Hz [δH(olefin) = 7.10 ppm] at 60 MHz.

$$\begin{array}{cc} Ph & Ph \\ \diagdown \quad \diagup \\ C=C \\ \diagup \quad \diagdown \\ H & H \end{array} \qquad \begin{array}{cc} H & Ph \\ \diagdown \quad \diagup \\ C=C \\ \diagup \quad \diagdown \\ Ph & H \end{array}$$

183   **184**

¹H NMR investigation of conformations is very powerful also in the case of macrocyclic aromatic compounds. Compound **185** (R=i Pr, R'=Me) has two stable conformations: the symmetric propeller (**a**) and the helical one (**b**). The conformational equilibrium was proved by temperature-dependent NMR measurements. In conformation **a** the aryl-methyl groups are equivalent, giving one common signal, unlike in conformation **b**, where three singlets correspond to them.[388,389,1070]

a   **b**

185

* See Volume I, Section 1.2.5, p. 35-38.

The protons of the terminal rings of heptahelicene (**186**), which are "above" the other ring, are more shielded than those of the homologues comprising less rings. The signal of H-1 (peri) in phenanthrene appears at 8.7 ppm, and in benzophenanthrene the shift is increased in a normal way to approximately 9.2 ppm. In the five-ring analogue, a strong shielding can already be observed, the signal is around 8.5 ppm, and the chemical shift decreases to 7.2 ppm (!) in the case of heptahelicene.[940]

186

A specific ASIS can also be observed with helicenes. Upon changing an indifferent solvent (CCl_4 or $CDCl_3$) with benzene, the lower homologues show the usual diamagnetic shift (approximately 0.2 ppm). Starting with heptahelicene, however, this effect decreases, first on the protons of the terminal rings, and then, gradually in the higher homologues, on the "inner" protons as well, and finally the ASIS ceases completely. The obvious explanation is that in the higher homologues the solvent molecules have access only to one side of the fused rings, and thus their diamagnetic effect prevails to only a lower extent, if at all.[940]

In the spectrum of Compound **187a**, the acetyl-methyl signal appears at 1.55 ppm, at a chemical shift value 0.53 ppm smaller than in the spectrum of acetamido-malonester [$AcNHCH(COOEt)_2$], where this signal is in the region characteristic of acetamides, at 2.08 ppm.[1328] The ethyl signals are similarly stronger shielded, the triplet being shifted from 1.29 to 1.01 ppm, and the methylene multiplet (the methyl hydrogens and the diastereotopic methylene protons form an ABX_3 spin system) being shifted from 4.29 to 3.95 ppm. The NH signal of acetamido-malonester appears at $\sim$7.4 ppm, and that of **187a** appears at $\sim$6.5 ppm. The reason for the shifts is the shielding of the aromatic rings, since the hydrogens of the acetamido-malonester group in a *quasi-axial* position are located "above" the plane of the ring (due to the bulkiness of this group this arrangement is preferred). The formamide analogues show a similar phenomenon, i.e., the formyl signal appears at 8.3 ppm in formamide-malonester, whereas in the spectrum of **187b** it appears at 7.61 ppm, shifted diamagnetically by 0.7 ppm.[1328]

187a: R = Ac
b: R = HCO

Shielding anisotropy may also occur in quasi-aromatic systems. This is the reason for the unprecedented large chemical shift difference (approximately 6 ppm) between the methylene hydrogens of the homotropylium cation (**188**), where $\delta H_A = -0.6$ and $\delta H_B = 5.2$ ppm,[762] and also for the large chemical shifts (9.3 ppm) of the protons of aromatic character attached to unsaturated carbons.

Table 40
THE CHEMICAL SHIFT (PPM) OF THE RING PROTONS OF HETEROAROMATIC AND ANALOGOUS NONAROMATIC COMPOUNDS FOR ILLUSTRATION OF THE EFFECT OF AROMATICITY ON SHIELDING[251,779,1044]

Compound	190a	191a	190b	192	190c	191b
δH_α	7.42	6.22	6.68	5.73	7.30	6.06
δH_β	6.37	4.82	6.22	4.72	7.10	5.48
$\Delta\delta H_\alpha H_\beta$	1.05	1.40	0.46	1.01	0.20	0.58
$\Delta\delta H_\alpha(1,2)^a$		1.20		0.95		1.24
$\Delta\delta H_\beta(1,2)^a$		1.55		1.50		1.62

[a] Shift difference of H_α and H_β, respectively, for compound pairs **190a-191a**, **190b-192**, and **190c-191b**.

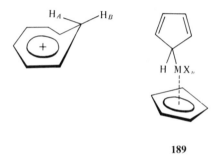

189

The mixed σ-π complexes of cyclopentadiene, e.g., compounds of type **189**, give two singlets at room temperature, proving that there is no fast σ ⇌ π isomerization, whereas interconversion the valence isomers of the σ-complex is fast,[1515] and thus the five-ring protons give rise to a singlet (the metal-carbon σ-bond is not fixed, but it is in a fast oscillation between the five cyclopentadienyl carbons). A similar valence isomerization also takes place in methyl-trimethyl-silyl cyclopentadiene, in the spectrum of which, a singlet can be assigned to the ring protons.[340] At low temperatures the isomerization process slows down, and the ring protons give an *ABCX* multiplet.

3.6. HETEROAROMATIC COMPOUNDS

Similar to benzenes, the ring protons of heteroaromatic compounds are deshielded relatively to olefinic protons, owing to the electron delocalization. The presence of heteroatoms has two further consequences:

1. The chemical shifts of the ring protons depend on their positions relative to the hetero atom: α-protons are strongly deshielded by the heteroatom.
2. The electron distribution is asymmetric in the ring: the electron density is higher in the neighborhood of the heteroatom, thereby lowering the aromaticity of the hetero ring with the electronegativity of the heteroatom. This is illustrated by the data of Table 40, in which the chemical shifts of the α- and β-protons of furan, pyrrole, and thiophene (**190a to c**) are listed, in comparison with those on nonaromatic molecules (**191a, b, 192**) with similar structures.

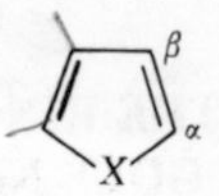

190a: X = O
   b: X = NH
   c: X = S

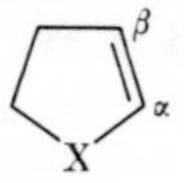

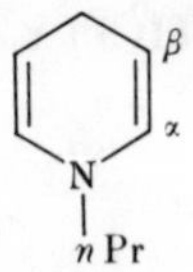

191a: X = O
   b: X = S

192

The deshielding effect of the hetero atom on the ring protons is indicated by the large $\Delta\delta H_\alpha H_\beta$ differences. This difference decreases with an increasing aromatic character, in the expected sequence of furan, pyrrole, thiophene. The deshielding (anisotropic) effect of the aromatic ring on the ring protons is reflected by the large values of $\Delta\delta H$ (**190a, 191a**), $\Delta\delta H$ (**190b, 192**), and $\Delta\delta H$ (**190c, 191b**).

3.6.1. Five-Membered Aromatic Heterocycles and Their Derivatives

Of the five-membered aromatic heterocycles systems, furan (**191a**) can be expected to have the simplest spectrum (compare Table 40) because of the large chemical shift difference between the α- and β-protons ($\Delta\delta\alpha,\beta > 1$ ppm), resulting in a first-order coupling to produce an *AA'XX'* multiplet. The spectrum of furan in CDCl$_3$ (see Figure 143a) consists of only two triplets, but in acetone, already two quartets can be observed. In accord with the small chemical shift difference between the ring protons, a higher-order coupling occurs in thiophene (**191c**), giving rise to an *AA'BB'* spectrum (Figure 143c).

The spectrum (Figure 143b) of pyrrole (**191b**) is substantially more complex than that of furan and thiophene. Further complications arise from the spin-spin coupling between the NH proton and the ring protons and from the heteronuclear coupling with the nitrogen atom, which causes line broadening. Owing to these interactions, the spectrum of liquid pyrrole is a complex multiplet of broad lines. The structure of the spectrum depends on the solvent and on the concentration. In benzene on dilution the original multiplet is replaced by a slightly structured broad maximum.[1251] This is a consequence of ASIS, which is stronger on the α-protons placed closer to the positive pole (in pyrrole the positive center is the nitrogen) of the molecules.* Hence, the chemical shift difference $\Delta\delta H_\alpha H_\beta$ can be expected to decrease, corresponding to the original shift relation $\delta H_\alpha > \delta H_\beta$. This is proved, e.g., by the *N*-methyl signal of 1,2,5-trimethylpyrrole, (**193**, R=Me), which is shifted upfield by 0.72 ppm in benzene solution. The large and opposite shift of the ring proton signals ($\Delta\delta H_\beta = 0.47$ ppm!) is worth noting.[1218] On the other hand, the ASIS on the 2,5-methyl groups is insignificant ($\Delta\delta CH_3 < 0.05$ ppm).

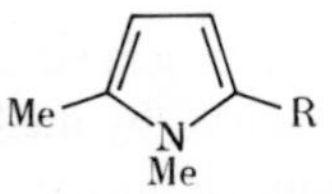

Me ⎯ N ⎯ R
     |
     Me

193

In acetone or CDCl$_3$,[1044] two well-separated 1:3:3:1 quartets can be observed (see Figure

---

*   Compare Volume I, p. 40, point 2.

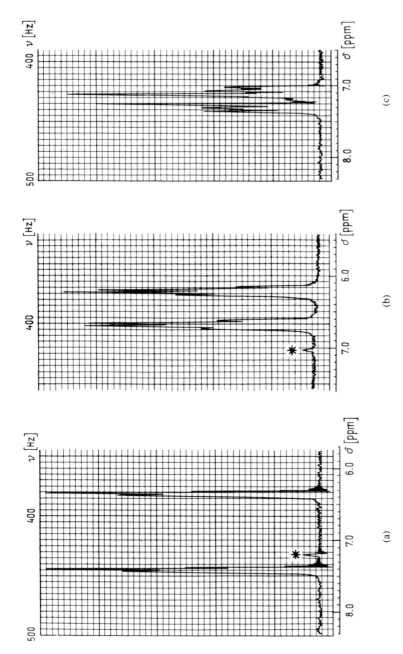

FIGURE 143. The spectra of (a) furan (**190a**), (b) pyrrole (**190b**) and (c) thiophene (**190c**). (From *NMR Spectra Catalog*, VARIAN Associates, Palo Alto, Calif., 1962-1963. With permission.)

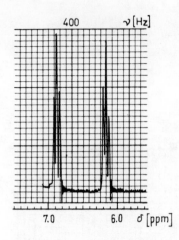

FIGURE 144.   Part of the spectrum of 1-*para*-tolylpyrrole (**195**) with the multiplet of the heteroaromatic protons. (From *High Resolution NMR Spectra*, SADTLER Research Laboratories, Philadelphia; Heyden and Son, London, 1967-1975. With permission.)

143b). The NH signal cannot be identified.* The presence of heteronuclear N–CH coupling can be eliminated by heteronuclear DR,** whereupon the lines become sharper.[1379] The splitting due to the NH–CH interaction is absent in the spectra of *N*-substituted derivatives comprising two triplets (e.g., see Figure 144). Also a pair of triplets occurs[592] in solvents promoting fast proton exchange.*** The ring protons of substituted derivatives usually give very simple spectra, the monosubstituted ones having *ABC, ABX,* or *AMX* multiplets, whereas in case of disubstitution *AB, AX,* or singlet spectra occur.

   Like with benzenes the substituents cause shifts of the signals of ring protons in the same direction. Electron-withdrawing substituents cause downfield shifts, whereas upfield shifts are characteristic for substituents with $+I$ effects. The magnitudes of shifts are, however, widely different. A 2-substituent affects δH-5 about half as much as δH-3, while the H-4 signal remains practically unchanged. A 3-substituent has the strongest effect on δH-2 and the weakest one on δH-5. The latter shift is not influenced to observable extent by electron-withdrawing groups in position 3, and the effect of electron-repelling groups in this position is about four times stronger on δH-2 than on δH-5.

   For the case of 2-substitution, a correlation has been found between δH-4 and the H-3 and H-5 shifts, respectively. Accordingly, for furan derivatives:[434]

$$\Delta\delta\text{H--3} = 3.1\cdot(\Delta\delta\text{H--4}) - 0.4 \text{ ppm} \tag{278}$$

$$\Delta\delta\text{H--5} = 1.2\cdot(\Delta\delta\text{H--4}) + 0.1 \text{ pmm} \tag{279}$$

The shift differences are related to the chemical shifts of furan. The values $\Delta\delta$H-3 and $\Delta\delta$H-5 of 2-substituted furan derivatives can be used to calculate the corresponding shifts of thiophene analogues:[593]

---

*   Compare p. 106.
** See Volume I, p. 49.
***Compare p. 95, 105, and 106.

Table 41
THE CHEMICAL SHIFT RANGES (PPM) OF THE RING PROTONS OF FURAN, PYRROLE, AND THIOPHENE DERIVATIVES[179,592,663,1044,1379]

Hetero ring	Ring proton	Chemical shift range		
		Generally	2-Substitution	3-Substitution
190a	δH-2	6.8—8.6	—	6.9—8.5
	δH-3	4.8—7.8	4.9—7.7	—
	δH-4	5.8—7.0	5.8—7.0	6.0—7.0
	δH-5	6.5—8.5	6.7—8.1	7.0—8.0
190b	δH-2	6.3—7.7	—	6.3—7.7
	δH-3	5.5—7.2	5.7—7.1	—
	δH-4	5.5—6.8	5.9—6.4	5.8—6.8
	δH-5	6.3—7.2	6.3—7.2	6.4—6.8
190c	δH-2	6.0—8.5	—	6.0—8.5
	δH-3	6.0—8.0	6.0—8.0	—
	δH-4	6.5—7.8	6.6—7.2	6.5—7.8
	δH-5	5.8—8.0	6.2—7.8	6.9—7.6

$$(\Delta\delta H\text{-}3)_{furan} = 1.3(\Delta\delta H\text{-}3)_{thiophene} \qquad (280)$$

$$(\Delta\delta H\text{-}5)_{furan} = 0.6\,(\Delta\delta H\text{-}5)_{thiophene} \qquad (281)$$

The above equations permit the shifts to be calculated quite accurately for compounds not investigated so far. The chemical shift ranges for ring protons of furan, thiophene, and pyrrole are summarized in Table 41. The chemical shift ranges for pyrroles are much narrower than the corresponding intervals of furans and thiophenes, which can be attributed to the electron-buffering behavior of nitrogen atom.

Since the substituent effects are additive also in this case, the chemical shifts of di- and trisubstituted furan, thiophene, and pyrrole derivatives can be calculated approximately from the shifts ρ_i of monosubstituted derivatives. In $CDCl_3$ solution

$$\delta H = \delta H_{\alpha,\beta} + \Sigma\rho_i^{\alpha,\beta}\ ppm \qquad (282)$$

where δH_α and δH_β are the shifts of monosubstituted compounds (see Table 40); i is 3, 4, or 5 in the subscript of ρ^α and 2, 4, or 5 in the subscript of ρ^β. Some of the constants (for $CDCl_3$ solutions) are given in Table 42. The substituent effects for pyrrole derivatives are, within a range of 0.1 ppm, the same as for thiophenes,[592] and thus common ρ_i constants can be used to estimate the chemical shifts of the ring protons of thiophene or pyrrole derivatives.

It is stressed again, however, that Equation 282 and other similar formulas give, even if the measurements are performed in the same solvent, only a rough estimate of the chemical shift, and their practical use lies not in the calculation of actual chemical shifts, but rather in the determination of their relative magnitudes. The measured chemical shifts δH-4 and δH-3 of 2-acetyl-5-methylfuran (194) are 6.17 and 7.12 ppm,[1044] whereas the shifts calculated

194

Table 42
SUBSTITUENT CONSTANTS ρ_i (PPM) FOR THE ESTIMATION OF THE CHEMICAL SHIFTS OF SUBSTITUTED FURAN, PYRROLE, AND THIOPHENE DERIVATIVES BY EQUATION 282[281,649,663,1379]

Hetero ring	Substituent	ρ_α			ρ_β		
		$i = 3$	$i = 4$	$i = 5$	$i = 2$	$i = 4$	$i = 5$
190a	OMe	−1.35	−0.25	−0.70	−0.45	−0.30	−0.40
	Me	−0.45	−0.15	−0.20	−0.30	−0.20	−0.15
	Br	−0.05	0.05	−0.05			
	SMe	−0.10	−0.05	−0.10	−0.20	−0.05	−0.15
	I	0.10	−0.15	−0.05	−0.15	0.05	−0.20
	SCN	0.40	0.05	0.10	0.20	0.20	0.05
	Ac	0.80	0.25	0.20	0.45	0.35	−0.15
	COOR	0.85	0.20	0.25	0.45	0.30	−0.15
	CN	0.85	0.30	0.30	0.45	0.20	−0.05
	CHO	0.95	0.30	0.35	0.45	0.35	−0.10
	COOH	0.95	0.35	0.40	0.90	0.55	0.35
	NO₂	1.20	0.55	0.50			
	OMe	−0.95	−0.45	−0.80	−1.10	−0.40	−0.20
	Me	−0.35	−0.20	−0.30	−0.45	−0.20	−0.20
	Br	−0.05	−0.25	−0.10	−0.10	−0.05	−0.10
	SH	0.00*	−0.20*	−0.05*	−0.20	−0.20	−0.10
190b[a]	I	0.15	−0.35	0.00	0.05	0.00	−0.20
and	CN	0.50*	0.00*	0.30*	0.65	0.20	0.15
190c	Ac	0.60*	0.00	0.30	0.70	0.50	0.25
	CHO	0.65*	0.10*	0.45*	0.80	0.45	0.05
	COOR	0.75	0.00	0.20	0.80	0.50*	0.00*
	COOH	0.80	0.10	0.40	0.70	0.35	0.00
	NO₂	0.85*	0.00*	0.30*	0.95	0.60	0.05
	SO₂Me	1.05	0.20	0.80	0.95	0.50	0.45

[a]   For pyrroles ρ_i values increased by 0.1 ppm, in case designated by asterisks (*) by 0.2 ppm are to be used .

by Equation 282 from the data of Table 42 are 6.17 and 7.02 ppm, i.e., the deviations are 0.00 and 0.10 ppm. Hence, the observed and calculated differences of chemical shifts $\Delta\delta_{3,4}$ are 0.95 ppm and 0.85 ppm, respectively.

From the spectrum of 2,3-dibromothiophene, the chemical shifts are[649] 6.91 and 7.23 ppm. The corresponding estimates obtained from the data of Table 42 are 6.80 and 7.15 ppm. The measured and calculated differences are 0.11 and 0.08 ppm, again in quite good agreement. Figure 75 shows the spectrum of 2-formyl-3-bromothiophene (**52**). The measured chemical shifts are δH-4 = 7.18, δH-5 = 7.75, and δCHO = 10.03 ppm. The calculated values of the former two shifts, 7.15 and 7.55 ppm, show discrepancies of −0.03 and −0.2 ppm. The theoretical and experimental differences in shift are 0.57 and 0.40 ppm, still in satisfactory agreement.

The solvent effect must be taken into account, since it may substantially affect both absolute and relative values of the chemical shifts, not only for pyrrole, but also in case of furans and thiophenes. In the spectrum of thiophene the difference in chemical shift of 0.2 ppm measured in CDCl₃ increases to 0.28 ppm in acetone-d₆.[663] It was shown (see Figure 44) how the AB spectrum (more particularly the chemical shift difference $\Delta\delta AB$) of 5-nitro-2-furancarboxylic acid (**46**) varies with the solvent.

N-substitution of pyrroles causes an opposite shift than C-substitution, e.g., in the spectrum (Figure 144) of 1-*para*-tolylpyrrole (**195**), the triplets of the α- and β-hydrogens appear at

Table 43
THE COUPLING CONSTANTS OF RING PROTONS IN FURANES,[281,434,649,1044,1379] IN PYRROLES,[3,273,592] AND IN THIOPHENES[663,1044,1379]

	Coupling	Coupling constant [Hz]		
		190a	190b	190c
	$J_{2,3}$	1.8—2.0	2.4—3.1	4.8—6.0
	$J_{2,4}$	0.7—0.9	1.3—1.5	1.2—2.0
CH–CH	$J_{2,5}$	1.3—1.5	1.9—2.2	2.8—3.2
	$J_{3,4}$	3.2—3.8	3.4—3.8	3.4—4.6
	$J_{1,2}$	—	2.3—3.0	—
CH–NH	$J_{1,3}$	—	2.0—2.5	—

7.16 and 6.18 ppm, i.e., $\Delta\delta H_\alpha = 0.48$ and $\Delta\delta H_\beta = -0.04$ ppm with respect to pyrrole. For 1-methylindole (**196**) $\Delta\delta H_\alpha = 0.28$ and $\Delta\delta H_\beta = 0.14$ ppm, whereas for 2- and 3-methylindoles these differences are -0.4 and -0.19 ppm, respectively. This phenomenon is due to the substantial effect that a substitution may have on the electron affinity and conjugation ability of the pyrrole nitrogen.

195   196

The spectrum of pyrroles is affected basically by the pH, due to its effect on the rates of the NH $\rightleftharpoons$ NH exchange processes, on the association,* and on C-protonation. The latter occurs on the α-carbon atom,[4] and the methylene protons of the dihydro derivative, formed reversibly, are obviously more shielded than the other ring protons. Thiophenes show a similar behavior.[666] In sulfuric acid solution, for example, the signals of pyrrole are shifted with respect to those measured in CDCl$_3$ as follows:[273] $\Delta\delta H$-2 $= 1.52$, $\Delta\delta H$-3 $= 1.22$, $\Delta\delta H$-4 $= 2.18$, and $\Delta\delta H$-5[$\delta CH_2 - (\delta(=CH)] = -1.45$ ppm!

For furan $J_{2,3} = 1.4$ and $J_{2,4} = 1.2$ Hz[1175] were found. The coupling constants of thiophene are $J_{2,3} = 4.7$, $J_{2,4} = 1.0$, $J_{2,5} = 2.8$, and $J_{3,4} = 3.3$ Hz.[580,663] For pyrrole, measured values are follows: $J_{2,3} = 2.6$, $J_{3,4} = 3.4$, $J_{2,4} = 1.4$, and $J_{2,5} \approx 2$ Hz. The NH–CH interactions are characterized by $J_{1,2} = J_{1,3} = 2.4$ Hz.[3] The similar values of coupling constants explain the simplicity of the spectra of pyrrole and its derivatives.

The coupling constants of furan, thiophene, and pyrrole derivatives are summarized in Table 43. The derivatives do not show remarkable differences (compare Table 43). The coupling constants decrease, as usual, with the increasing electron affinities of the substituents.[228] The interactions with the protons of the substituents are usually insignificant, and the corresponding values of J are always <1 Hz.

In the spectrum (Figure 145) of furfurol (**197a**), however, a well-observable splitting occurs due to long-range interaction between the formyl hydrogen and ring hydrogen 5 (transmitted by the π-electrons via "W"-pattern),** and the splitting due to a coupling with H-3 is also discernible. The sequence of chemical shifts is $\delta H_5 > \delta H_3 > \delta H_4$. From the coalescence temperature of the formyl signal, $\Delta G^\ddagger = 45.6$ kJ/mol.[326]

---

* Compare p. 106.
** Compare Volume I, p. 67.

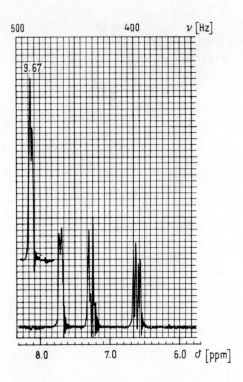

FIGURE 145.   The spectrum of furfurol (**197a**). (From *NMR Spectra Catalog*, VARIAN Associates, Palo Alto, Calif., 1962-1963. With permission.)

197a: X = O
b: X = NH

Whereas 2-formyl-pyrrole (**197b**) gives complicated multiplets due to interactions between the aldehyde and the ring protons (similar to **197a**), 1,5-dimethyl-2-formylpyrrole (**193**, R=CHO) produces an *AX* spectrum, indicating that *N*-substitution weakens the interaction with the formyl hydrogen. From the positions of the two doublets, δH-3 = 6.81 and δH-4 = 6.02 ppm.[1044]

3.6.2. Pyridine and Its Derivatives

Pyridine has, to a first approximation, an *AA'BXX'* spectrum, in which the lines are ordered into three groups of 2:1:2 intensity, where $\delta X > \delta B > \delta A$. It is clear that the α-protons are strongly deshielded and thus the *X* part of the multiplet corresponds to them. The chemical shifts of α-, β-, and γ-protons are 8.61, 7.12, and 7.50 ppm,[251] hence the δH_α and δH_γ shifts are substantially larger than those of benzene (7.27 ppm).

The larger chemical shift of the γ-proton can be attributed to the relatively lower electron density around it caused by the well-known asymmetric electron distribution in pyridine. From the spectrum of pyridine *N*-oxide chlorohydrate, $\delta H_\alpha = 8.68$, $\delta H_\beta = 8.10$, and $\delta H_\gamma = 8.41$ ppm. The shift differences with respect to pyridine are characteristically different (see Problem **17** too): at H_α it is minimum ($\Delta\delta H_\alpha = +0.07$ ppm), and at H_β and H_γ it is very large ($\Delta\delta H_\beta = +0.98$ and $\Delta\delta H_\gamma = +0.91$ ppm). Thus, the uneven electron density

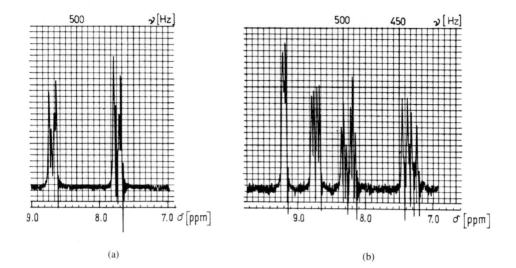

FIGURE 146. Part of the spectrum of (a) ethyl isonicotinate (**198**) and (b) ethyl nicotinate (**199**), with the multiplets of the ring protons. The asymmetric structure of the multiplet on Figure 146a is due to a heteronuclear spin — spin coupling between the α-ring protons and the nitrogen. (From *High Resolution NMR Spectra*, SADTLER Research Laboratories, Philadelphia; Heyden and Sons, London, 1967-1975. With permission.)

of pyridine becomes more symmetric and the shift differences of the proton pairs α-β and α-γ decrease substantially (instead of 1.49 and 1.11 ppm for pyridine, they are 0.58 and 0.57 ppm). The proton-nitrogen coupling is responsible for the broad lines of the H_α multiplet.

For pyridine, the following coupling constants were determined from the analysis of the $AA'BXX'$ multiplet obtained by heteronuclear DR:[251] $J^o_{\alpha,\beta} = 4.88$; $J^o_{\beta,\gamma} = 7.67$; $J^m_{\alpha,\gamma} = 1.84$; $J^m_{\beta,\beta}' = 1.37$; $J^m_{\alpha,\alpha}' = -0.13$, and $J^p_{\alpha,\beta}' = 0.96$ Hz. These values indicate that pyridine derivatives do not obey the relationship $J^o > J^m > J^p \approx 0$, valid for benzene derivatives, and that the coupling constants are not so insensitive towards substitution.

Among monosubstituted derivatives, the simplest spectra ($AA'BB'$ or $AA'XX'$) arise from γ-substituted pyridines. The spectrum (Figure 146a) of ethyl isonicotinate (**198**) comprises an $AA'XX'$ multiplet simplified into four lines, allowing only the calculation of δA, δX, and J_{AX} in the AX approximation. The signals of the ring protons are often separated in the spectra of α- and β-substituted pyridines as well, and in such cases the spectrum parameters can be calculated easily by the $AMPX$ approximation. As illustrations, the spectrum of ethyl nicotinate (**199**) is shown in Figure 146b.

COOEt

198   **199**

In β-substituted pyridines, H-2 is coupled only weakly, if at all, with the other protons, thereby giving rise to ABX or AMX spectra. The simple (often first-order) spectra of di- and trisubstituted pyridines are always easy to interpret.

2-Ethoxy-3-bromopyridine gives rise to an AMX spectrum (Figure 147a), similar to the 2-chloro-5-nitro derivative, but for the latter $J_{3,6} \approx 0$, and thus the two double doublets are simplified into doublets (Figure 147b). The AMX spectrum (Figure 147c) of 2-methoxy-6-

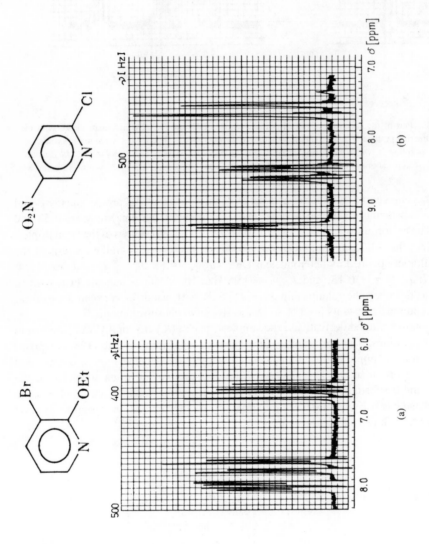

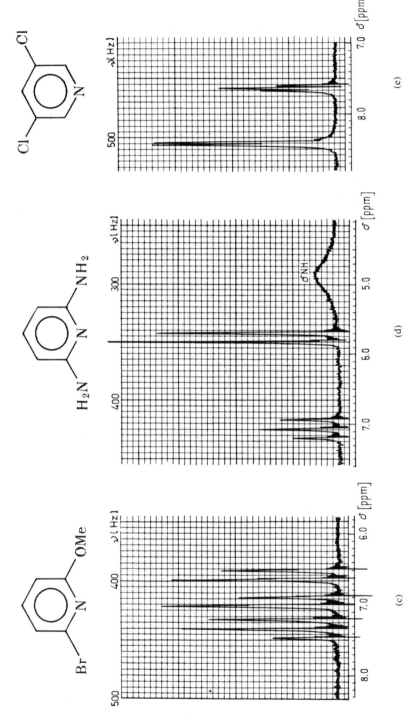

FIGURE 147. Part of the spectra with the multiplets of the ring protons of (a) 2-ethoxy-3-bromopyridine, (b) 2-chloro-6-methoxypyridine, (c) 2-bromo-6-methoxypyridine, (d) 2,6-diaminopyridine (the maximum around 4.9 ppm arises from the amino group), and (e) 3,5-dichloro pyridine. (From *High Resolution NMR Spectra*, SADTLER Research Laboratories, Philadelphia; Heyden and Sons, London, 1967-1975. With permission.)

Table 44
SUBSTITUENT CONSTANTS ρ_i (ppm) FOR THE ESTIMATION OF THE CHEMICAL SHIFTS OF SUBSTITUTED PYRIDINES BY EQUATION 283[281]

Substituent	ρ_α				ρ_β				ρ_γ	
	$i = 3$	$i = 4$	$i = 5$	$i = 6$	$i = 2$	$i = 4$	$i = 5$	$i = 6$	$i = 2,6$	$i = 3,5$
NH_2	-0.70	-0.30	-0.80	-0.50	-0.05	-0.50	0	-0.35	-0.35	-0.75
OH	-0.70	0	-1.00	-0.90	0	-0.35	0.15	-0.25		
OR	-0.55	0	-0.50	-0.30					0	-0.30
Me	-0.10	0	-0.15	0.10	0	-0.05	-0.10	0	0	-0.10
$C(sp^2)$	0.10	-0.15	-0.10	0.05					0.10	0.15
CH_2NH_2	0.20	0.05	-0.10	0.05	0.15	0.15	0.05	0	0	0.05
Cl	0.30	0.30	0.30	0.20	0.20	0.25	0.20	0.10	0	0.05
CH_2OH	0.35	0.30	0	0.05	0.10	0.15	0.05	-0.05	-0.05	0.15
Br	0.40	0.15	0.20	0	0.20	0.45	0.35	0.20	0.10	0.35
CH=NOH	0.40	0.30	0	0.15	0.40	0.45	0.20	0.15	0.25	0.35
COPh	0.60	0.55	0.30	0.30	0.45	0.55	0.35	0.35	0.35	0.40
Ac	0.80	0.35	0.40	0.30	0.70	0.70	0.3	0.35	0.40	0.60
COOR	0.85	0.40	0.50	0.45	0.70	0.65	0.25	0.25	0.35	0.55
CN	0.90	0.40	0.55	0.40	0.65	0.70	0.45	0.50	0.45	0.60
NHAc	0.95	0.15	-0.20	-0.10	0.35	0.50	0.05	-0.15	-0.05	0.30
CHO	0.95	0.40	0.50	0.45	0.45	0.40	0.10	0.20	0.45	0.60
$CONH_2$	1.05	0.45	0.45	0.30						
NO_2	1.10	0.65	0.75	0.25						

bromopyridine contains two doublets and one triplet, since $J_{3,5} \approx 0$, and $J_{3,4} \approx J_{4,5} \cdot AX_2$ spectra are produced by 2,6-diaminopyridine (Figure 147d) and 3,5-dichloropyridine (Figure 147e), but the coupling constant J_{AX} is much larger in the former case (*ortho* coupling), and $\delta A > \delta X$, i.e., $\delta H_4 > \delta H_{3,5}$, whereas in the latter one the *meta* splitting is smaller and the sequence of shifts is reversed ($\delta A < \delta X$). The large shift of α-protons may help in the assignment. One must always take into account, however, the presumable effect of substituents as well! Thus, H-4 and H-6 of 2-ethoxy-3,5-dinitropyridine absorb at 9.20 and 9.42 ppm, respectively.[649]

The substituent effect on the signals of ring protons is about the same as in benzene derivatives. The ranges of relative shifts are as follows: δH_α, 8.0 to 9.5; δH_β, 5.9 to 8.5; and δH_γ, 7.2 to 8.6 ppm.[1379]

The substituents influence primarily the shifts of the adjacent ring protons and influence to only a smaller extent those of the *para* protons. Of course, electron-withdrawing substituents decrease and electron-repelling ones increase the shielding. In the case of 4-substitution, however, the δH_α shifts change to a greater extent than δH_β in *ortho* position, with no regard to the $-I$ or $+I$ effect of the substituent.[1379]

The substituent effects are additive. The use of the shifts of monosubstituted pyridines as substituent constants (see Table 44) enables estimation of the chemical shifts of polysubstituted derivatives:

$$\delta PyH_i = \delta H_{\alpha,\beta,\gamma} + \Sigma \rho_{i\alpha,\beta,\gamma} \tag{283}$$

where α,β, and γ refer to the position of the substituent and i refers to the position of the ring hydrogen in question (i.e., $i = 2,3,4,5$, or 6), and $\delta H_\alpha = 8.59$, $\delta H_\beta = 7.38$, and $\delta H_\gamma = 7.75$ ppm (in 30% DMSO-d_6). The error is <0.3 ppm in most cases, if the data are measured in the same solvent in similar concentrations. As an example, the shifts of H-3,4,6 of 2-amino-5-bromopyridine are 7.03, 7.90, and 8.29 ppm from the data of Table 44, whereas the observed values[649] are 7.10, 7.98, and 8.08 ppm, respectively.

Table 45
THE CHEMICAL SHIFTS OF
THE RING PROTONS (PPM) OF
4-METHYL-PYRIDINE IN
VARIOUS SOLVENTS[1251,1358,1379]

Solvent	δH_α	δH_β	$\Delta \delta_{\alpha,\beta}$
CCl_4	8.35	6.98	1.37
$CDCl_3$	8.45	7.09	1.36
$(CD_3)_2SO$	8.60	7.28	1.32
$(CD_3)_2CO$	8.41	7.15	1.26
D_2O	8.42	7.31	1.11
C_6D_6	7.95	6.09	1.86

The nitrogen atom of pyridine, being very basic, associates with the solvent molecules or is protonated in acidic solvents. Table 45 shows the chemical shifts of the ring protons of 4-methylpyridine in various solvents. As can be seen from the data, there are quite substantial differences (disregarding the ASIS by benzene, the shifts of α-hydrogens may change by 0.25 ppm, and the variation of δH_β with the solvent is 0.33 ppm). In polar solvents, which decrease the asymmetry of the electron distribution of the ring, the difference $\Delta \delta_{\alpha,\beta}$ also decreases.[1358]

Pyridine itself, similar to benzene, is able to induce substantial solvent effect. In the spectrum of steroid **200**, e.g., recorded in $CDCl_3$, the 4β,17,19 and 4α,18 methyl signals, respectively, merge into overlapping bands at 1.75 and 1.40 ppm. In pyridine solution all the five signals appear separately (at 0.53, 0.77, 0.95, 1.12, and 1.75 ppm).[1310] This was the first application of ASIS to separate overlapped signals.

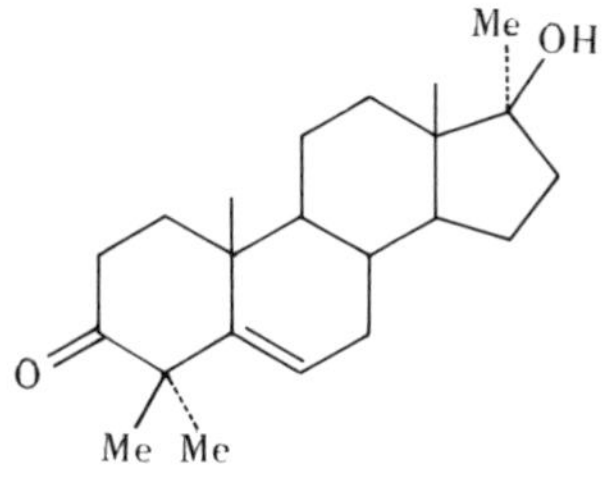

200

The magnitudes of coupling constants of pyridine derivatives[1379] are as follows: $J^o_{2,3}$, 4.0 to 6.0; $J^m_{2,4}$, < 2.5; $J^p_{2,5}$, 0.3 to 1.5; $J^m_{2,6}$, <0.6; $J^o_{3,4}$, 6.5 to 9.2; $J^m_{3,5}$, <2.0; and $J^p_{3,6}$, 0.4 to 2.5 Hz. The data explain that, the spectra of 2-amino-4,6-dimethylpyridine and 3,4-dicarbethoxy-5-methoxypyridine consist equally of two singlets at 6.02 and 6.23 ppm, ($J^m_{3,5} \approx$ 0) and 8.51 and 8.85 ppm ($J^m_{2,6} \approx$ 0), respectively.[649]

3.6.3. Fused and Quasiaromatic Heterocyclic Systems; Analogues Containing More Nitrogens

In heterocycles fused with aromatic rings, the neighboring ("*ortho*") hydrogens, in both the aromatic and the heteroaromatic rings, are deshielded due to the mutual anisotropic effects of the rings. In contrast, the signals of the protons more distant from the fused aromatic ring will be shifted diamagnetically, because the aromatic character decreases in the fused system, e.g., thus, in indole δH-3 is 0.12 ppm larger and δH-2 is 0.14 ppm smaller than the H_α and H_β shifts of pyrrole.[713] If there are more heteroatoms, their accumulated

Table 46
THE CHEMICAL SHIFTS (PPM) OF RING PROTONS IN FIVE- AND SIX-MEMBERED HETEROCYCLES AND IN THEIR DERIVATIVES[51,140,141,237,267,281,613,649,666,759,922,949,1044,1092,1093,1095,1096,1176,1305]

−*I* effects decrease further the electron density around the remaining protons, thereby resulting in deshielding. Nevertheless, the observable increase in chemical shift is not substantial in comparison with the corresponding heterocycles with one nitrogen, since this effect is mostly compensated by a decrease in aromatic character with the incorporation of further heteroatoms. The chemical shift of some representatives of this type are given in Table 46. The spectrum parameters of these compounds, too, are sensitive to the conditions of recording.

The efficiency of the NMR method in the investigation of heterocyclic compounds is illustrated by some examples. In the spectrum of 1,2,4-triazole (**201**) taken in dry DMSO-

201

d_6, there are only two singlets at 13.9 and 7.85 ppm, corresponding to the NH and CH groups and proving the presence of tautomeric structure **a**.[315] The spectrum recorded in HMFA at $-34°C$ shows two CH signals (δH-3 = 8.85, δH-5 = 7.92, and δNH = 15.25 ppm), indicating that the equilibrium is shifted in favor of form **b**.

A CH $\rightleftharpoons$ NH tautomery was observed with quinoxalines **202** (**a** $\rightleftharpoons$ **b**) in which one of the methylene proton in the side chain migrates to the cyclic nitrogen atom.[988] The equilibrium measured in CDCl$_3$ is shifted in favor of form **a** in DMSO-d_6, and in TFA the relative amount of tautomer **a** increases further, e.g., in the derivative where R=CN, the olefinic signal appears at 5.03 ppm and the methylene singlet appears at 4.25 ppm. Their relative intensity gives the relative amounts of the tautomeric forms.

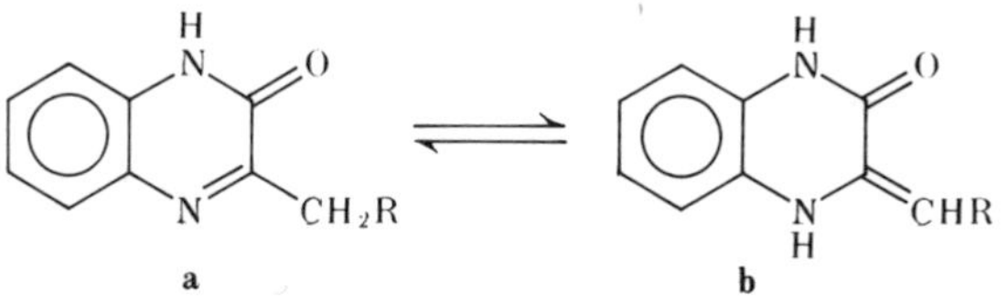

202

Azidopurine **203a** is capable of chain $\rightleftharpoons$ ring tautomery giving the **203b** azide.[1398] In DMSO-d_6 the tetrazole **b** is present (δH-8 = 8.62 and δH-5 = 10.2 ppm), whereas in TFA the azide **a** is predominant (in which δH-2 = δH-8 = 9.13 ppm). In a 1:1 mixture of the solvents, isomers **a** and **b** are in equilibrium, and the chemical shifts are δH-2 (**a**) = 9.25, δH-5(**b**) = 9.88, δH-8(**a**) = 8.93, and δH-8(**b**) = 9.08 ppm. The anisotropic effect of the condensed aromatic tetrazole ring on the chemical shifts is well observable, primarily, of course, on the adjacent hydrogen: $\Delta\delta$H-8(**a**),(**b**) = 0.63 ppm, and $\Delta\delta$H-2(**a**),(**b**) = 0.15 ppm. This chemical shift difference is the basis of pairing the samples to the two structures.

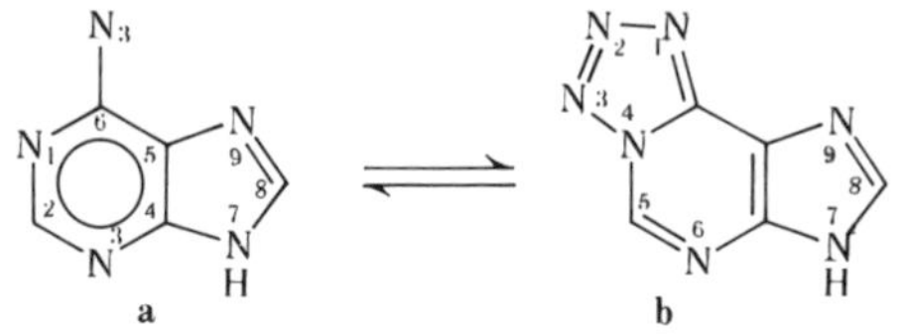

203

In fused systems, the anisotropic effect of lone pairs on nitrogen causes very large paramagnetic shifts in the signals of the protons sterically close to the nitrogen atom of the fused hetero ring, e.g., thus the ring protons of Compound **204** (R=Me) give rise to the following

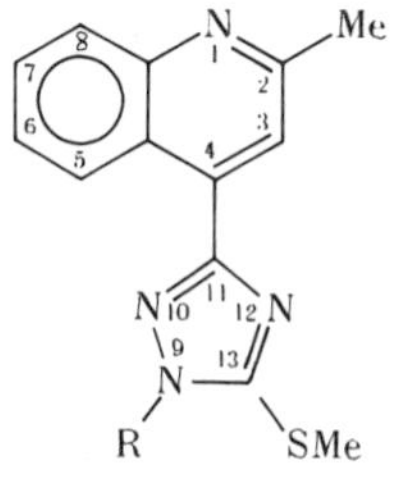

204

signals.[375] The multiplet of H-6 and H-7 appears between 440 and 470 Hz at 60 MHz ($\sim$7.6 ppm), that of H-8 is shifted by the adjacent ring to the region of 475-495 Hz ($\sim$8.1 ppm), and the singlet of H-3 appears at 8.00 ppm (also deshielded due to the proximity of the triazole ring). The multiplet of H-5 can be found at very high frequencies, between 540 and 555 Hz (~9.1 ppm), owing to the interaction with the lone electron pairs of the triazole nitrogens. Whereas the *C*-methyl and *S* -methyl signals appear very close to one another at 2.75 and 2.80 ppm, the *N*-methyl signal has a large chemical shift (3.80 ppm), which is characteristic of *N*-methylazoles.

Large δ-values have also the chemical shifts of the *N*-methyl signals of 1-methylimidazole (3.67 ppm) and 3,5-dimethyltetrazole (4.30 ppm). A number of further examples indicate that the methyl signals of *N*-methylazoles may appear in the same region as the methoxy signals or occasionally at even larger chemical shifts. As further examples, Compounds **205** and **206** are mentioned for which δNMe is 3.95 ppm and 4.20 ppm, respectively.[1362]

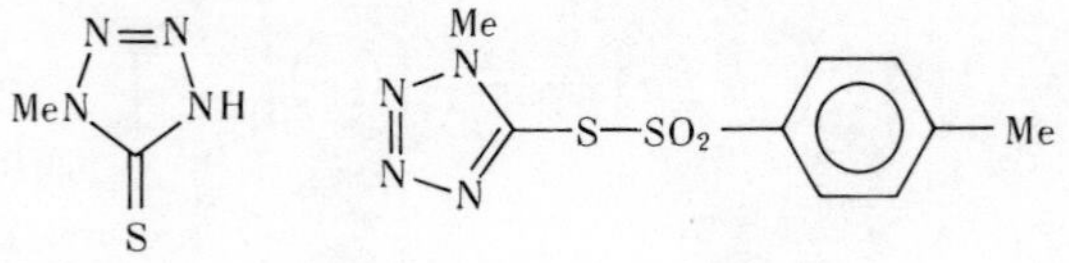

205   **206**

The chemical shifts of the β-hydrogens (H-3,3′ and H−5,5′) of α,α′-bipyridil (**207**) are much different (by 1.28 ppm). The reason is that H-3 or H-3′ get close to the lone pair of the adjacent ring, which causes strong deshielding.[1358]

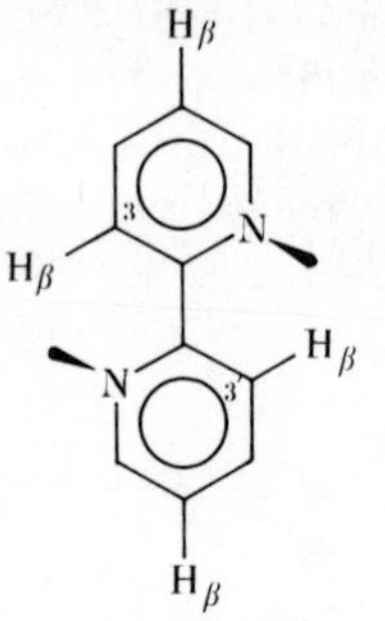

207

In acid solutions, indolisine (**208**) is protonated in position 3 (**b**, R=H), whereas its 3-substituted derivatives (R ≠ H) appears as a mixture of forms **b** and **c**, and the equilibrium depends on the pH.[55,508]

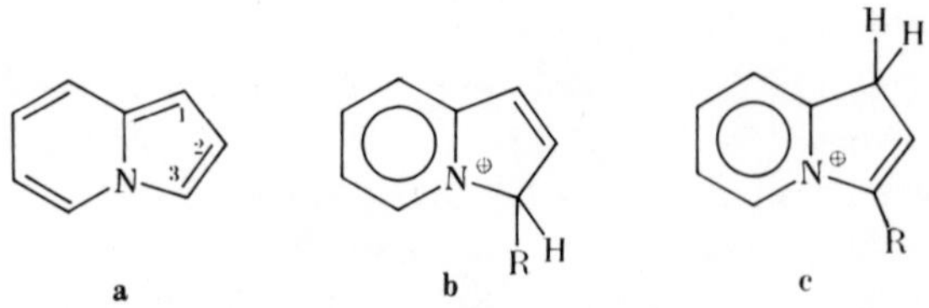

a   b   c

208

This type of compound frequently involves interesting conformational problems, too. The rotation barrier of *N*-acylpyrroles (**209**) is low (ΔG‡ ≈ 50 to 58 kJ/mol),[948] indicating in accord with the IR investigations that the double bond character of the CN bond is weaker than in simple amides.[671] The aromatic system, namely, also uses the lone pair of the nitrogen, decreasing the conjugation of the latter towards the oxygen.

209

The conformational properties of adenosine and cytidine derivatives may have physiological importance. For purine derivative **210**, $\Delta G^{\ddagger} = 64$ kJ/mol was determined for the dimethylamino group. (At room temperature, the signal of the dimethylamino group is a singlet at 3.72 ppm, and it is split at $-12°$C, after a gradual broadening, into two sharp lines with chemical shifts of 3.50 and 3.94 ppm at $-50°$C.[1029])

210

The conformational equilibrium of cyclophanes **211** involves two forms, a chair and a boat type (**a** and **b**). Investigating various aromatic systems, it has been found that the inversion barrier changes with the type of the aromatic rings, e.g., furanophane (**211**, X=O) is flexible ($\Delta G^{\ddagger} = 70$ kJ/mol for **a** $\rightleftharpoons$ **b**),[548] unlike thienophane (**211**, X=S), which occurs in the **a** conformation ($\Delta G^{\ddagger} \approx 113$ kJ/mol).[487]

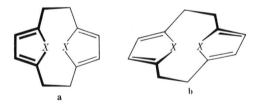

211

Quasiaromatic porphyrins, with skeleton **12**, can be regarded as pyrrole derivatives in which — as mentioned* — anomalous chemical shifts are caused by the ring currents induced in the continuously conjugated system similar to the case of annulenes.** In the spectra of the various derivatives, the methine signals fall between 8.5 and 11.0 ppm, and the NH signals fall between -3.6 and -4.1 ppm.[101,432]

The methine and ring proton signals, as well as the signals of the substituents replacing them, are all singlets, proving the equivalence of the rings, and thereby indicating a fast valence isomerization.[103] In N-substituted porphyrins, as could be expected, the N-alkyl groups are strongly shielded. For the N,N'-dimethyl and N,N'-diethyl porphyrin derivatives, $\delta NCH_3 = -4.89$ and -2.37 and $\delta NCH_2 = -5.16$ ppm, respectively.[256] In these compounds, the four olefinic protons and the four pairs of ring protons or the substituents replacing them are again nonequivalent, producing, therefore, split signals.

The spin-spin coupling between the hydrogens of different rings does not cause observable

---

* See Volume I, p. 36.
** See Section 3.3.5 (p. 59-61).

splitting, similar to benzenes. In exceptional cases, however, such a splitting can be observed, e.g., like with Compound **212**, where $J_{4,8} \approx 0.9$ Hz,[1094] due to the ''W''-arrangement, the extended conjugation, and the lone pairs of the nitrogens transferring the interaction.*

212

The lone pairs on the hetero atoms may give rise to substantial differences in the relative magnitudes of coupling constants as compared to the values obtained from the dihedral angles, e.g., of the partially saturated hetero compounds by the Karplus relation. Thus, $J^c = J^t = 8.4$ Hz for dihydroindole (**213**). If, however, the effect of the lone pair is absent, the usual relation $J^c > J^t$ is reobtained: the salt of **213** has a spectrum from which $J^c = 9.0$ and $J^t = 6.6$ Hz. The same arguments can be applied to the case of dihydrothiophene (**214**), in which the coupling constants are much closer to one another ($J^c = 10.0$ and $J^t = 7.5$ Hz) than for its sulfone derivatives **215**, where $J^c = 9$ and $J^t = 4.2$ Hz.[1402]

213   214   215

3.7. MEDIUM EFFECTS; EXCHANGE PROCESSES

3.7.1. Hydrogens Attached to Atoms Other than Carbon

The mobile hydrogens (protons of OH, NH, and SH groups) have signals with properties characteristically different from those of the CH signals, due to their ability to participate in exchange processes and to form hydrogen bonds. These phenomena are time, temperature, and concentration dependent, and they also depend strongly on the nature of the solvent (pH, polarity).

3.7.1.1. Exchange Processes

The proton types discussed so far remain generally unchanged during the NMR measurement, unless the sample is a reaction mixture or a compound that changes, decomposes, or reacts with the solvent in a fast process. In contrast, the acidic protons are in a continuous dynamic exchange with one another; there is a fast transition between the alternate states. The rate of the exchange processes has a basic influence on the spectra of these compounds, although they are chemically homogeneous and remain ultimately unchanged during the measurement. As, however, the spectrum depends on the rate of these exchange processes, the rate constant can be determined from the spectrum. (Of course, this must be distinguished from kinetic investigations, not discussed here, in which the relative concentrations of the various species participating in the reaction are determined from the relative intensities of the corresponding NMR signals.[134,1182])

The exchange process may involve the same hydrogen of two (identical) molecules or

* See Volume I, p. 62, 67, and 68.

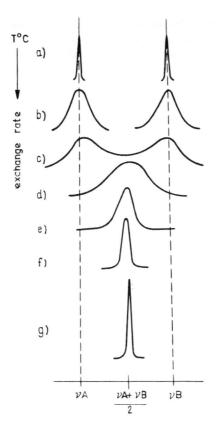

FIGURE 148. Schematic change in the signal shape, having equal width and intensity for two interacting nuclei, as a function of the rate of the exchange.

two or more chemically different protons. These latter may be different protons of the same molecule, but they may also arise from the solvent or from another substance added to the solution. If the two protons involved are labeled with H_A and H_B, the exchange is process $R_A H_A + R_B H_B \rightleftharpoons R_A H_B + R_B H_A$, where R_A and R_B denote the chemical environments of H_A and H_B.

If there is no exchange between the species $R_A H_A$ and $R_B H_B$, protons H_A and H_B each give rise to one sharp signal at frequencies vA and vB. When these protons do not interact with other nuclei, two sharp singlets occur. When the two species have the same concentration and the line widths are the same, the heights of the two signals are also identical (see Figure 148a).

With gradually increasing rates of exchange, the resonance absorption takes place at different transition states of the exchange process, e.g., when H_A has not been removed completely from its original environment R_A, but it has already entered more or less the environment of R_B. In this case line vA is broadened, and the same happens, of course, to line vB (see Figure 148b and c). At a certain rate of exchange, the two lines merge into one, very broad absorption maximum with a total width equal to the original distance of lines vA and vB, i.e., the chemical shift difference ΔvAB (see Figure 148d).

With a further increase in the rate of exchange, the broad maximum becomes gradually narrower (see Figure 148e and f), and when the exchange becomes very fast on the NMR time scale, i.e., the average lifetimes of the two states are shorter than the speed of the resonance process, the two types of protons can be observed as being identical, giving coincident signals. In this case one sharp line is obtained at a frequency of $(\nu A + \nu B)/2$, i.e., in the mid-point of the original two lines (see Figure 148g). Such change in the experimental spectrum upon raising the temperatures was observed first for dimethylnitrosamine (Me_2N-NO_2) between 25 and 200°C.[884]

The theoretical treatment of this problem[603,607,608,618,959,1132] based on the Bloch equations (see Equation 53a to c) yields a complicated expression for the shape of the signal, and it can be seen that the shape depends on the product $\tau (\nu A - \nu B) = \tau \Delta \nu$,[603] where τ is the mean lifetime of the exchanging states A and B. The rate constant of exchange is $\kappa = 1/\tau$.

Assuming that the mean lifetimes of the two states are identical ($\tau_A = \tau_B = \tau$) and that their probabilities of occurrence (the relative amounts of forms $R_A H_A$ and $R_B H_B$) are also the the same ($p_A = p_B = 1/2$) and finally that for large mean lifetimes νA and νB are sharp singlets, since the relaxation times T_2 are sufficiently large,* i.e., $1/T_2(A) \approx 1/T_2(B) \approx 0$, the signal shape varies with κ or τ (for constant $\Delta \nu$) in a manner shown in Figure 148. It can be shown that the lines merge into one another (see Figure 148d) at a mean lifetime of

$$\tau = \sqrt{2}/(\pi \Delta \nu) \qquad (284)$$

corresponding to an exchange rate of

$$\kappa = \pi \Delta \nu / \sqrt{2} = 2.22 \, \Delta \nu \qquad (285)$$

for a given value of $\Delta \nu$. Of course, τ varies with the temperature. When a system gives two signals at room temperature, τ of the states can be decreased by raising the temperature to reach the coalescence point, and when the system gives a coincident signal at a temperature, τ can be increased to separate the signals by decreasing the temperature.

Reaching from either direction the coalescence point belonging to the values of τ defined by Equation 284, one can determine the activation energy E_a of the exchange process $A \rightleftharpoons B$ (potential barrier) or the free enthalpy of activation $\Delta G_c^{\ddagger}$ from the coalescence temperature T_c, with the use of Arrhenius Equation:

$$\kappa_c = Ae^{-E_a/RT_c} \qquad (286)$$

or Eyring Equation

$$\kappa_c = \frac{kT}{h} e^{-\Delta G_c^{\ddagger}/RT_c} \qquad (287)$$

where A is a constant, k is the Boltzmann constant, n is the Planck constant, R is the universal gas constant, T_c is coalescence temperature, and κ_c is the rate constant. Expressing $\Delta G_c^{\ddagger}$ from Equation 287 and substituting κ_c from Equation 285, we obtain;[764]

$$\Delta G_c^{\ddagger} = RT_c \ln(kT_c\sqrt{2})/(h\pi\Delta\nu) = 19.13 \, T_c[9.97 + \log (T_c/\Delta\nu)] \text{ [kJ/mol]} \qquad (288)$$

In the case of spin systems with higher-order coupling, the determination of κ_c is more complicated, and the analogues of Equation 287, which express the relationship between κ_c and the spectrum parameters, can be derived only by quantum mechanical methods. For AB spin systems, Equation 289 can be used:[796,1060]

---

* See Volume I, Section 1.1.9 (p. 15-17) and compare p. 19 and 23.

$$\kappa_c = \pi\sqrt{[(\Delta\nu AB)^2 + 6J^2_{AB}]/2} \qquad\qquad (289)$$

wherefrom

$$\Delta G^{\ddagger}_c = RT_c\ln[kT_c\sqrt{2/(\Delta\nu^2 + 6J^2)}/(h\ \pi)] \qquad\qquad (290)$$

On the basis of the known laws of thermodynamics, $\Delta G^{\ddagger}$ calculated by Equations 288 or 290 may be used to calculate the other activation parameters (energy, enthalpy, and entropy of activation).

In the range of practically accessible temperatures (between -150 and $+200°C$), the exchange processes of such spin systems can be studied for which the mean lifetimes of the spin states are between 1 and 10^{-3} s, i.e., the rate constants fall between 1 and 10^3 s^{-1}. The free enthalpies of activation of such systems are in the range of approximately 10 to 100 kJ/mol.

In practice, one may seldom encounter the case of the broad signal shown in Figure 148d which makes possible the simple calculation of rate constants. Instead, some signal shapes between the limiting cases a and d or d and g can be observed. Although rough estimates can also be given on the basis of these spectra,[764,921] it is more advantageous to vary the conditions of measurement until coalescence is obtained. This can be done by changing the measuring frequency (and hence $\Delta\nu$). Since, however, this is impossible with the majority of spectrometers, the rate of exchange is varied instead, by changing the temperature, solvent, concentration or pH, or sometimes by the addition of catalysts (acid, alkali, or water). The simplest method is temperature-dependent NMR investigation.*

If τ_A and τ_B (the average lifetimes of protons A and B) are different, or more protons, A_n and B_m, different in number ($n \neq m$) occupy the state A and B, their equilibrium concentrations and hence line intensities will be different. In such cases the intensities are proportional to mole fractions $p_A = \tau_A/(\tau_A + \tau_B)$ and $p_B = \tau_B/(\tau_A + \tau_B)$.

As a result of exchange, the signals of acidic protons may assume widely varying forms and appear in almost the complete spectrum range measured. The shapes are extremely varying, too, from the sharp signals, through the more or less broadened maxima to diffuse absorptions that hardly emerge from the baseline or cannot be identified at all.

Exchange processes may also involve the elimination of splitting due to spin-spin coupling.[607] Therefore, the signals of acidic protons are usually not split by spin-spin coupling, nor are, of course, the signals of the adjacent protons. The reason is that the exchange process averages the spin states of acidic protons, and since the energy differences corresponding to the individual spin orientations are eliminated, there is no spin-spin splitting. The molecules differing in the spin orientation of the hydroxy proton of an R_2CHOH-type alcohol can be regarded as being in different states (A and B), and their exchange is

$$R_2CH_A-OH_A + R_2CH_B-OH_B \rightleftharpoons R_2CH_A-OH_B + R_2CH_B-OH_A$$

The fast exchange averages spin states A and B, causing both the methine and the OH signal to appear as singlets. In such cases the mean lifetime of the spin states is $\tau = \sqrt{2}/(\pi J_{AB})$.

The Figure 149a shows the changes in the hydroxy signal of $R-CH_2OH$ as a function of the mean lifetime of the spin states of the hydroxy proton schematically.[57] In the experimental spectrum of ethanol (see Figure 149b), the hydroxy proton gives a singlet at 2.58 ppm, and the methyl and methylene groups have the usual triplet and quartet ($J = 7$ Hz) of ethyl groups at 1.22 and 3.70 ppm.[1044] There is no spin-spin interaction between the hydroxy and the adjacent methylene protons, due obviously to the fast exchange of protons. In the spectrum

---

* See Volume I, Section 2.1.5, p. 144-147.

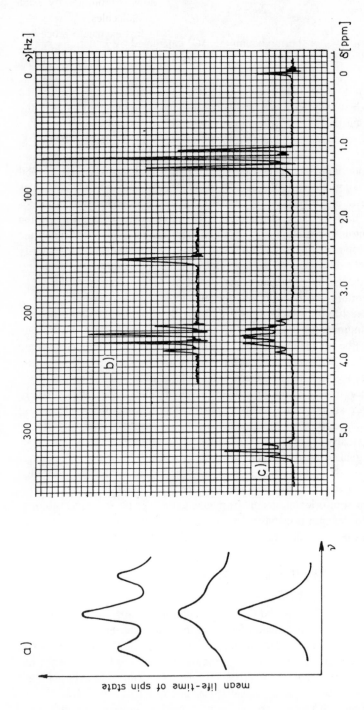

FIGURE 149. (a) Schematic change in line shape of the OH signals of RCH$_2$OH type molecules as a function of the mean lifetime belonging to the spin states of the hydroxyl proton.[57] (b) The spectrum part of 96% ethanol with the methylene and hydroxy signals[104a] and (c) the spectrum of anhydrous ethanol.

of anhydrous ethanol (see Figure 149c), the methyl signal remains a triplet ($J = 7$ Hz) at 1.19 ppm, but the hydroxy proton gives a triplet as well ($J = 5$ Hz), and the methylene signal at 3.65 ppm splits into eight lines: $(n_1 + 1)(n_2 + 1) = 4 \cdot 2 = 8.$* This also indicates the general experience that the exchange of acidic protons is promoted by water through the exchange processes involving the hydrogens of water molecules.

Note that the averaging of spin states may be due not only to fast exchange processes, but also to fast spin-lattice relaxation. In the case of quadrupole nuclei, the spin-lattice relaxation time is shortened to such an extent that the average lifetimes of the spin states become too short for the individual observation of the transitions between them. Chlorine causes, therefore, no splitting in the ^{1}H MNR signals.** With medium relaxation times, line broadening can be observed instead of splitting. This is the case for ^{14}N nucleus, always broadening the NH signals.***

The exchange phenomena permit the simple assignment of the resonance signals arising from acidic protons. As mentioned, exchange may also involve the acidic protons or deuterons of the solvent (TFA-d, D_2O, CD_3OD, etc.) or the protons of the water content of the solvent (a frequent case with hygroscopic solvents, for example, DMSO-d_6). Therefore, adding acid or alkali, or simply heavy water† to the solution (to produce H $\rightleftharpoons$ D exchange), the rate of exchange alters, and the site and shape of the signals originating from acidic protons of the solute will change.

The rate of exchange can also be influenced by changing the temperature, the solvent, or the concentration, and the signals disappearing or changing in shape or position are certainly due to acidic protons while the other signals remain unchanged. This method of detecting acidic protons is very often used in practice,‡ as illustrated by Problems **4, 18, 44,** and **55** (addition of D_2O) or **16, 27, 34, 42, 46,** and **60** (addition of acid).

This coalescence method or line shape analysis based on the investigation at variable temperature of the NMR spectra can, of course, also be applied to any other equilibrium process connected with the interconversion of states in which the change in $\Delta G^{\ddagger}$ is 10 to 100 kJ/mol. Several examples have already been mentioned in connection with the ring inversion, ⧺ the hindered rotation, + valence isomerization, + + and the inversion around, nonplanar primarily nitrogen atoms. + + + Depending on the rate of interconversion between chemically nonequivalent sites, the spectra of the inverses, rotamers conformers, valence isomers, etc. may contain separated signals (slow exchange), overlapping, diffuse maxima (medium lifetime), or sharp signals at the mean chemical shift (fast exchange). The computer analysis[1382] of time-dependent processes (i.e., processes depending on the mean lifetime) and of the corresponding spectra permit the study of systems in which the exchange involves more than two states and the spin systems are complex, with several nuclei and higher order couplings. It is no more necessary to reach the temperature of coalescence or vary the temperature at all because κ and the thermodynamic parameters can be determined from any spectrum. The program is available for a wide variety of spin systems.

3.7.1.2. Hydrogen Bonds

As known, acidic protons readily form hydrogen bonds with suitable atoms of electron

---

* Compare Volume I, Equation 75, p. 48.

** Compare Volume I, p. 49 and 203.

*** Compare Volume I, p. 203.

† The addition of D_2O is also useful when the solvent (CCl_4, $CDCl_3$, etc.) is immiscible with water. In such instances D_2O must be throughly shaken with the solvent, whereupon the signal of the acidic protons in the spectrum of the emulsion broadens, shifts, or disappears.

‡ Compare p. 42.

⧺ See p. 19, 21, 24, 43, 44, 54, 74, 91, etc.

+ See p. 66, 91, and 107, etc.

+ + See p. 59-61, 75, and 91, etc.

+ + + See p. 16-17, 43, and 44, etc.

Table 47
ASSOCIATION SHIFT, $\Delta\delta_{0a}$ FOR
SOME SIMPLE MOLECULES[1260]

Compound	$\Delta\delta_{0a}$ (ppm)	Temperature (°C)
C_2H_6	0.00	− 88
C_2H_4	0.43	− 60
NH_3	1.05	− 77
C_2H_2	1.30	− 82
HCl	2.05	− 86
HI	2.55	− 5
H_2O	4.58	0
HF	6.65	− 60

donor character in solid and liquid phases and in most cases in solutions as well. Molecules which contain acidic protons occur in monomeric (unassociated) state only in dilute, apolar solutions, in the gas state, and at higher temperatures, depending on the strength of their association ability. (Exceptions are, of course, the acidic protons incapable of association for steric reasons, e.g., see the case of compound **1**.) Owing to the hydrogen bonds the signals of the protons participating in them are shifted downfield, as compared to the signals of monomers. The main reason for this shift is that in a hydrogen bond of structure X–H . . . Y, the average bond length $\overline{XH}$ increases, thereby decreasing the electron density and hence the shielding around the proton.[117,1054,1292] The experience that deshielding is about twice larger[1135] in the case of intramolecular associations indicates that the shielding of the proton is affected considerably by the anisotropy of the bonds X–H and H . . . Y as well, although in the nonlinear, intramolecular hydrogen bonds, the electron density around the proton is obviously also changed. A linear bond X–H . . . Y, like in linear molecules (e.g., acetylene, HCN), shields the proton,* thereby compensating, in part, the opposite effect of the decrease in electron density.

It is quite complicated to determine the chemical shift of an XH group in monomeric and associated states because the assumption that all molecules are monomers is justified only in gas phase or in solutions of infinite dilution. The chemical shift of the monomer, δ_o, is usually determined by extrapolation from the chemical shifts measured in solutions of various concentrations. The shift δ_a characteristic of the associated state could be determined, in fact, only from the investigation of solid samples. The shifts obtained in molten state, near the melting point, are generally assumed as characteristic of the associated state.

The shift $\Delta\delta_{ao}$ due to association (the association shift) is approximately proportional to the strength of the hydrogen bond, i.e., to the acidity of the proton involved,[1260] as illustrated by the data of Table 47. Accordingly, the hydroxy signals of acids, alcohols, phenols, chelates, etc. have more or less different chemical shifts, and thus they may be characteristic of the type of compound. Moreover, from the position of the XH signal conclusions can be drawn on the length or strength of the hydrogen bond RX–H . . . YR', or, in the case of the same XH group, the donor strength of the electron donating group YR' can be estimated. Choosing the same donor, one may determine the acidity of the hydrogen atom in the case of different RX substituents. (Accordingly, there is a proportionality between the IR frequencies νXH and the chemical shift δXH.[116,293,577,999,1135])

The inter- or intramolecular character of the hydrogen bonds, similar to IR spectroscopy, may be determined from the solvent, concentration, or temperature dependence of the spectrum. Since the hydrogen bonds are broken at higher temperatures or upon dilution, an

---

*   Compare Volume I, p. 32-33 and Volume II, p. 46 and 62.

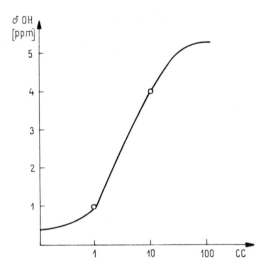

FIGURE 150.   The chemical shift of the hydroxy signal of ethanol in CCl_4 solution, as a function of concentration.[106]

upfield shift of δXH can be observed when the sample is heated or the concentration of the solute is decreased[59] (compare the shift of the hydroxy signal in Figure 149b with the one in Figure 149c). Similarly, since the molecules always associate with polar solvents, the shift δXH also varies with the solvent, increasing with its increasing polarity. The solvent effects are, therefore, due partly to changes in the association conditions. The NMR investigation of hydrogen bonds has extensive literature, including monographs reviewing the theoretical[815,1220] and organic chemical applications.[1013,1113]

3.7.1.3. Hydroxy Groups

Hydroxy protons are capable of association and exchange, and therefore their shifts cannot be characterized by so narrow and thus characteristic ranges, e.g., for acids, phenols, oximes, etc., than as the various types of protons attached to carbon atoms. When ethanol is diluted with CCl_4, e.g., δOH varies over a range of 4 ppm,[106] increasing with the concentration (see Figure 150). If δOH measured in an inert solvent at various concentrations is extrapolated to infinite dilution, the hydroxy signals of most diverse types of alcohols usually fall into the range of 1 to 2 ppm.[1379] In more concentrated solutions, the lower limit of the range increases to approximately 5 ppm.[293,1238] In polar solvents, (e.g., acetone, pyridine, DMSO), where the dissolved alcohol may form strong hydrogen bonds, the chemical shifts may increase very much, even exceeding 4 ppm.

This is shown in Table 48 which gives δOH of phenol in various solvents of the same concentration. It can be seen that certain solvents form weaker hydrogen bonds with phenol than the phenol molecules with one another; in such cases the shift of the signal is smaller than in the spectrum of the pure liquid.

Owing to the stronger ability of phenol to form hydrogen bonds, larger relative shifts (usually 4.5 to 5.5 ppm) can be observed at the same concentration, and higher dilution is required to form monomers than in the case of alcohols. Bulky groups may prevent, or at least hinder, the formation of associates. For this reason, the change in signal shift of trityl alcohol (Ph_3COH) is one order smaller than that of benzyl alcohol[293] in a similar concentration range in the same solvent.

If the association is intramolecular, there is no shift at all upon a change in concentration[1565] because the hydrogen bonds remain intact even in dilute solutions.[1013] For similar reasons,

Table 48
THE δOH SHIFT (PPM) OF PHENOL IN VARIOUS
SOLVENTS, WITH RESPECT TO THE LIQUID
PHASE VALUE IN 10^{-2} MOL/ℓ SOLUTIONS[577]

Solvent	Association shift	Solvent	Association shift
—(Gas)	− 3.5	Dioxane	0.3
CCl$_4$	− 3.0	Et$_2$O	0.6
C$_6$H$_6$	− 2.5	Me$_2$CO	0.7
CS$_2$	− 2.3	PhCOMe	1.2
PhNO$_2$	− 1.3	Me$_2$SO	1.7
MeCN	− 0.8	Pyridine	3.6
MeCOOEt	− 0.1	Et$_3$N	4.0

the signal shifts of hydroxyl group in intramolecular association are also insensitive of the solvent. The value of δOH of Compound **216** is e.g., 11.75 ppm in DMSO-d$_6$ and 11.85 ppm in CDCl$_3$.[252] The association conditions of α- and β-hydroxyketones (e.g., **217**) have been elucidated by studying the variation of the spectrum with the concentration, from which the equilibrium of the free (monomeric) molecules and the intra- and intermolecular associates could be determined.[1551] In this way, by the extrapolation of the chemical shift vs. concentration curve to infinite dilution, the dimerization constants of several phenol derivatives have been determined.[690]

216   **217**

The spectrum of *o*-chlorophenol, since the intramolecular association between the hydroxyl group and the chlorine atom is weak, proved to be concentration dependent.[690] Investigations on intramolecular OH . . . π hydrogen bonds have also been reported.[770] The δOH shift of cyclohexanols depends on the configuration and conformation,[1160] and the investigations on conformationally homogeneous systems, e.g., *cis* and *trans* 4-*t*-butyl cyclohexanol, have shown that δOH$_e$ > δOH$_a$. In DMSO-d$_6$ the δOH$_e$ signal appears in the range of 4.0 to 4.5 ppm, whereas δOH$_a$ appears between 3.8 and 4.2 ppm. The latter signal, corresponding to weaker hydrogen bonds, is less sensitive of concentration.[782,1074] When couplings of type CH—OH are observable, $^3J(H_a,OH_e) \approx$ 4 to 6 > $^3J(H_e,OH_a) \approx$ 3 to 4.5 Hz and their magnitudes also depend on the conformation.*

The relationships between the dihedral angles and the coupling constants $^3J(H,OH)$ are qualitatively similar to those of the protons attached to carbon atoms. This conclusion has been drawn from investigations on compounds with structures fixed by intramolecular hydrogen bonds. The coupling constants corresponding to dihedral angles of 180° are large (around 12 Hz).[93,766,1370]

However, the spin-spin coupling of OH protons is only seldom observable,** owing to

---

* See Reference 1160 and Volume I, p. 62.
** Compare p. 95.

the generally fast exchange processes.[217] Such couplings can usually be detected in anhydrous DMSO-d_6 solutions. The exchange processes are so slow in this solvent that the signals of the different hydroxyl groups are separated, and the effect of the coupling with adjacent methylene or methine groups is also observable. This is important in deciding[268] whether the hydroxyl group is attached to secondary (triplet OH signal), tertiary (doublet), or quaternary (singlet) carbon atom.

The water content may speed up the exchange processes; however, one may also observe CH–OH splittings in the presence of water, but in such cases the signal corresponding to the water content of the solvent is separately observable (the exchange between the protons of water and of the solute is slow; see Problems **34** and **58**).

Owing to the exchange processes, the hydroxy signal of the solute overlaps very frequently with the signal corresponding to the water content (e.g., in DMSO-d_6) or light isotope content (e.g., CD$_3$OH in CD$_3$OD) of the solvent (e.g., see Problems **28** and **31**). When the hydroxy signal is relatively sharp, the wide range of its possible appearance often causes problems in distinguishing it from other signals. In such cases, to help the assignment, acid, alkali, or heavy water* should be added to the solution, or the temperature or concentration should be varied, whereupon the hydroxy signal shifts or disappears (e.g., see Problems **4, 16, 18, 27, 34**, and **44**).

The solvent effect induced by pyridine causes a paramagnetic (!) shift in the hydroxy signals (due to hydrogen bonds). The signals of the neighboring protons are also shifted, downfield, e.g., with phenols, the extent of this downfield shift decreases in the series ortho, meta, para ($\Delta\delta$ = 0.5, 0.15, and 0.0 ppm, respectively).[349]

The locations, shapes, and multiplicities of hydroxy signals are, of course, always temperature dependent.[1051] With increasing temperatures, due to the elimination of hydrogen bonds, the chemical shift decreases and the signals become sharper due to the acceleration of exchange processes.

The hydroxy signals of enols and chelates, like in the IR spectra, have special properties and they appear at extremely large shifts, generally between 11 and 17 ppm. These signals, similar to weaker intramolecular hydrogen bonds, are insensitive to the variation of solvent, concentration, and temperature.[975]

The hydroxy signal of Compound **216**, as we have seen, reflects the properties of chelates, in regard to both the extent of shift and the invariance of the shift with respect to experimental conditions. The signal of completely enolized acetyl-acetone (**218**; R=Me) appears at 15.5 ppm,[1184] and those of di-p-bromobenzoylmethane (**218**; R, p-bromophenyl) appear at 16.6 ppm.[1044] The phenol group of salicylic aldehyde (**219**) absorbs at 10.98 ppm.[649]

218a: R = Me
  b: R = pBr–C$_6$H$_4$

219

Extremely large δOH shift was measured for Compound **220** (18.15 ppm).[398] In contrast, δNH is not larger than expected generally for amides and thioamides** because the sp^3 nitrogen is not involved in strong chelation, as also proved by IR studies.[1345] The large shift

---

* When heavy water is added (compare the fourth footnote on p. 97), the signal of HOD is expected to appear around 5 ppm, which can be recognized easily from its increasing intensity upon the addition of further amounts of heavy water.
** Compare p. 107.

characteristic of enols can be explained on one hand by the decrease in electron density around the proton due to the very strong hydrogen bond and on the other hand by the anisotropic effect of the carbonyl group which is close to the enolic hydrogen.[1047] However, e.g., as shown by the spectrum of Compound **221**, the latter effect is not the decisive one, since in this spectrum the hydroxy signals appears at 6.7 ppm,[1044] although the enolic hydrogen is also close to the carbonyl group, but the hydrogen bond is weak.

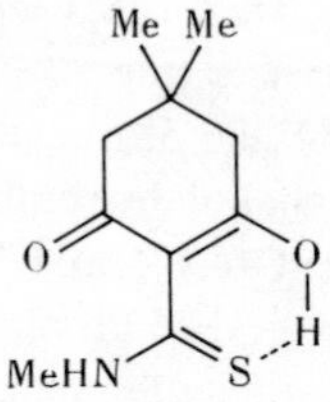

220 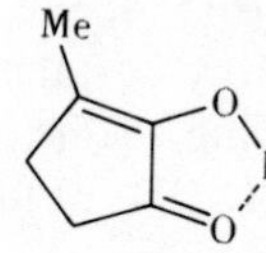   **221**

Enolic hydroxy signal is expected in the spectra of β-dioxo compounds (existing in the enol form) and *o*-carbonyl phenols. In the spectrum of completely enolized 2,5-dicarbethoxy-cyclohexane-1,4-dione (**222b**) and 2,4,5-trihydroxy-butyrophenone (**223**), the δOH signal of the chelated hydroxyl groups appear at 12.2 and 12.6 ppm, respectively.[1320] (Note that the signal of the two phenolic hydroxyl groups of Compound **223** has much smaller shift, 7.30 ppm.)

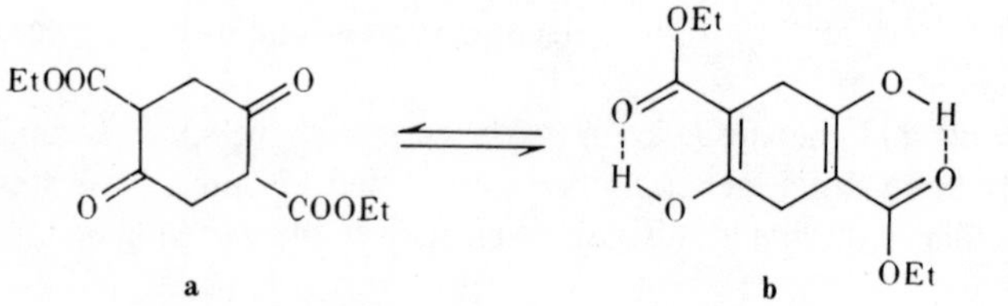

222   **223**

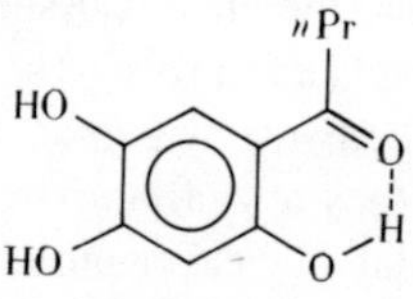

In the case of keto-enol equilibrium, the ratio of the tautomeric forms can readily be determined from the intensities of the NMR signals. As an example, the spectrum of aceto-acetic acid diethylamide (AcCH$_2$CONEt$_2$) is shown (see Figure 151). The signal of the enolic hydroxyl group cannot be identified in the spectrum, and the methylene quartets and methyl triplets corresponding to the keto and enol forms also overlap, but the relative intensities of the singlets (1.95 and 1.65 ppm) of acetyl groups occurring alternately in the tautomers and the intensity of the δ(=CH) signal (4.85 ppm) of the enol tautomer still give two independent data on the ratio of the tautomeric forms, which is approximately 1:1.[1383]

The tautomeric equilibrium of acylmalonates: R–CO–CH(COOEt)$_2$ ⇌ HO–RC=(COOEt)$_2$ was studied similarly, and it has been found that the equilibrium is a function of substituent R, e.g., thus, if R=Me, the keto-enol ratio is approximately 1:2, and if R=*t*-Bu, it is approximately 1:0.[167]

In the spectrum of a deuterochloroform solution of Compound **224**, which contains two chelate rings in the enol form **a**, the signals of the two hydroxy protons appear separately at 13.65 and 14.95 ppm, whereas a third maximum at 13.70 ppm correspond to the keto form **b**.[400] The separate signals are due to the fact that in the case of chelates the intramolecular and intermolecular exchange processes are slow. For similar reasons, chelate signals are generally sharp.[398]

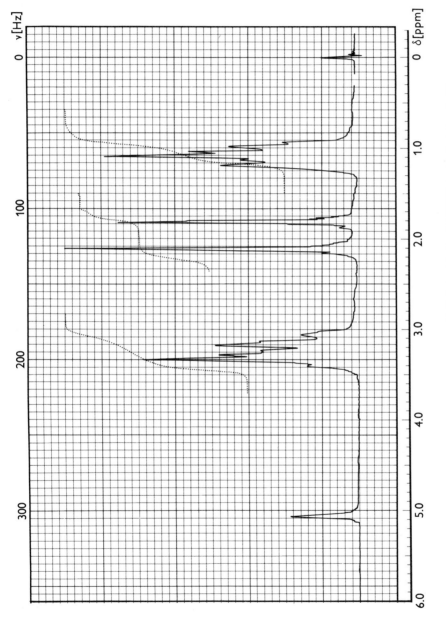

FIGURE 151. The spectrum of acetoacetic acid diethylamide (AcCH$_2$CONEt$_2$) with the integrals.[1383]

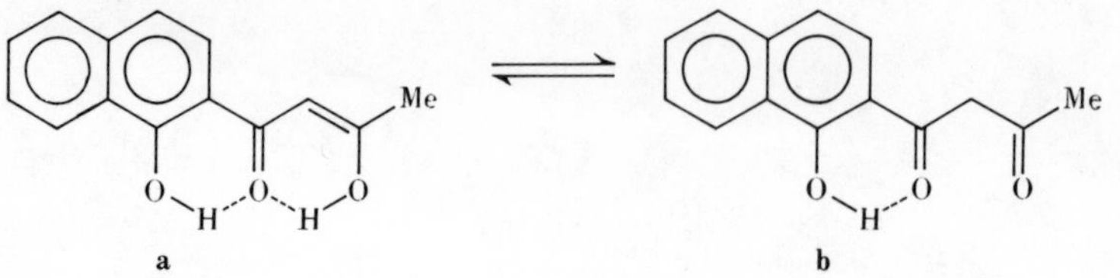

224

In the NMR investigation of oximes, the main problem is the signal duplication* due to *syn-anti* isomery, the mechanism and kinetics of isomerization and the anisotropic effect of the hydroxyl group.** In these investigations, however, the hydroxyl signal itself scarcely plays a role. The usually broad signal appears in the region of 9 to 13 ppm.[401,1044,1379]

The hydroxyl signal of *syn* camphor monoxime (**225a**) appears at 12.0 ppm in $CDCl_3$ solution and at 11.3 ppm in acetone-d_6, whereas the corresponding chemical shifts of the *anti* isomer (**b**) are 7.4 and 10.95 ppm.[270] In the former isomer, the intramolecular hydrogen bond causes a strong downfield shift, which decreases slightly in the very polar acetone solution owing to the breaking of these bonds and the association between the solvent and the solute. The chemical shift of isomer **b** is thus very similar to that measured for isomer **a** in acetone-d, whereas in $CDCl_3$ much smaller shifts correspond to the weak intermolecular hydrogen bonds.

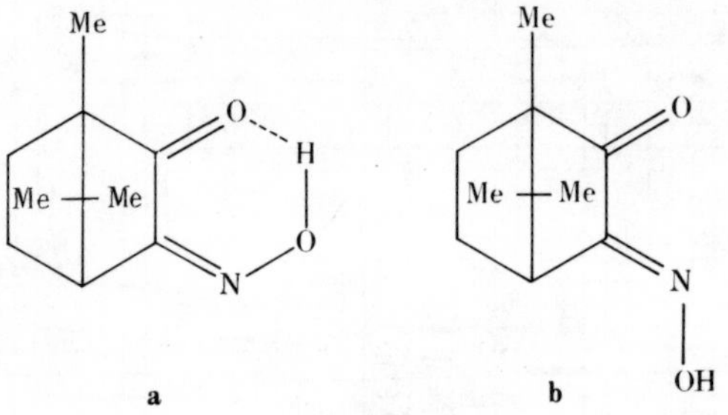

225

The hydroxyl protons of carboxylic acids, following from the acidic character, are strongly deshielded and their signal appears between 9.5 and 13 ppm.[1044,1185] Of course, the large shifts are due in part to the fact that, in nonpolar solutions carboxylic acids are present as dimers, at least up to a concentration of 5%. For this reason, in inert solutions, the position of the hydroxy signal of carboxylic acids does not change with dilution (at concentrations higher than 5 to 10%). Concentration dependence can be observed only in very dilute solutions,[1181] when the dimers dissociate and the chemical shift decreases by at most 1 ppm.[1185] If the chemical shifts are plotted against the concentration (see Figure 152a), the equilibrium constant of the monomer-dimer equilibrium can be determined from the slope of the curve obtained.[1085] The concentration dependence shown by formic acid, acetic acid, and their halogenated derivatives is somewhat different from fatty acids with longer chain, insofar as the chemical shift decreases at higher concentrations as well, since besides the dimers higher associates, containing weaker hydrogen bonds, are also formed.[1181] Consequently, in such cases the curve has a maximum (see Figure 152b). In more polar solvents the dimers disappear already at lower concentrations, and the location of the maximum depends on the solvent.[1185]

---

* Compare p. 64-66.
** Compare p. 64-66 and Volume I, p. 32.

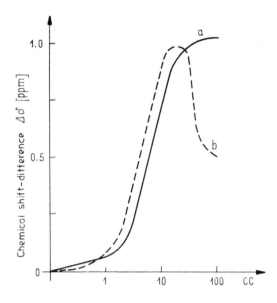

FIGURE 152.   The change in chemical shift of the hydroxy signals of formic acid, acetic acid, and its halogenated derivatives (broken lines) and their higher homologues (solid line) in CCl_4 solution, as a function of the concentration.[1181]

It must be taken into account in the determination of association shifts that the fast exchange between the acid proton and other mobile protons (e.g., the hydroxy protons of hydroxy-carboxylic acids or the acidid protons of the solvent) have a common signal with an average shift, and this latter has completely different behavior (may be shifted, e.g., to much greater extent upon the variation of concentration).[1011] When pyridine or other proton acceptor solvent is used, the acid proton signals are shifted anomally due to salt formation. This phenomenon is, however, useful in recognizing the signals of acid protons.

3.7.1.4. NH and SH Groups

The signals of protons attached to nitrogen atoms are extremely varying both in position, shape, and in multiplicity. In addition to the hydrogen bonds and exchange phenomena, the NH hydrogens are in spin-spin interaction with the nitrogen, the quadrupole moment of which may affect the shape and splitting of the NH signals in various ways, depending among others on the temperature. Moreover, mainly in the spectra of salts and quaternary derivatives and the compounds which contain sp^2 nitrogen atom, further complications may be caused by configurational and conformational problems (inversion and rotation).

The spin-spin interaction of N–H type splits the proton resonance signal into a triplet with a ratio of 1:1:1,[1056] since the spin quantum number of ^{14}N is $I = 1$ (see Table 1), provided that the exchange processes among the NH protons are slow enough. This type of splitting has been observed only in some cases: in the spectra of completely dry ammonia and of some alkylamine salts. The coupling constant $^{14}N–H$ is* between 50 and 60 Hz.[1210] Moisture or addition of base even in trace amounts accelerates the exchange processes to such an extent that sharp NH signals appear, whereas acids slow down the process.[1210] Thus, the line shape and the splitting are pH dependent (see Figure 153).[596] The splitting corresponding to CH–NH couplings is observable mainly for dry DMSO-d_6 solutions (compare Problem **27**). In other cases only line broadening occurs.

---

* Compare p. 252-254 and 258.

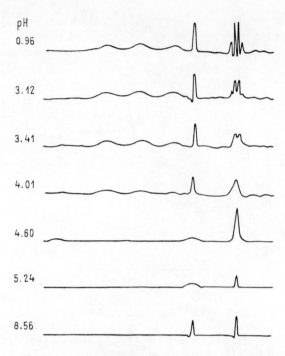

FIGURE 153.   The pH dependence of the spectrum of methylamine in aqueous solution.[596]

The line broadening caused by spin-spin interactions decreases at higher temperatures owing to faster exchange.* It has been observed, however, that the lines of certain NH signals are broader at higher temperatures.[1210] Consequently, in these cases the line broadening is not related to spin-spin interactions. Instead, the quadrupole moment of the nitrogen decreases the relaxation times, causing a line broadening, which increases with the temperature.**

Since amines and amides show strong tendency to form hydrogen bonds, the NH signals vary strongly, independently of the above, with the concentration, temperature, pH, solvent, etc., and sharp signals may occur just as frequently as very broad ones, which disappear completely in the baseline of the spectrum. In dilute solutions of apolar solvents, the NH signals of aliphatic amines occur generally between 0.5 and 2 ppm,[1044] whereas in more concentrated solutions and in liquid phase the self-association may cause an increase of 1 to 2 ppm in the chemical shift.[471] In strong polar solvent, the downfield shift may become several parts per million.[1378]

Due to the additivity of the ring anisotropy and $+M$ effect (partial positive charge) of the nitrogen, the NH signals of aromatic amines appear at approximately 2 ppm larger shifts, and they also depend strongly on the solvent.[1378] In solvents which contain mobile protons (e.g., acids, water, etc.), common absorption can be observed at a shift corresponding to the weighted average of the chemical shifts of the solvent and the solute.

In acid solution, the protons attached to the ionized nitrogen atom are strongly deshielded. These signals are usually sharper and may be split (mainly in the case of primary amines) due to the N–H coupling.*** With unsaturated, aliphatic imines the *Z-E* isomery, occurring

* Compare p. 93-94.
** See p. 97 and 258.
***Compare Volume I, p. 160.

upon protonation, may cause signal doubling. The same phenomenon may be observed if inverses, conformers, and geometric isomers of cyclic amides and their salts* or various association structures are present simultaneously. The resulting separated line pairs (except for those arising from geometric isomers) usually collapse with increasing temperature.

Amide protons give broad maxima at large chemical shifts (the lower limit is 12 to 13 ppm), and the signals are often impossible to identify.[1210] In not too polar solutions, the signals of primary amides occur between 5.0 and 8.5 ppm.[1044] In some instances, the separate NH signals of rotational isomers may also be observed due to hindered rotation.[1111]

The NH signal of *anti* isomer (**a**) of Compound **226** appears at 7.8 ppm in CDCl$_3$. The corresponding signal of the chelated syn isomer (**b**) is shifted to 12.0 ppm.[420]

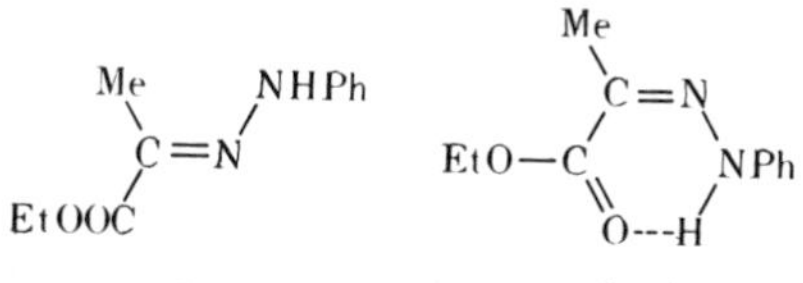

226

The configurations of the vinylogue amides (X=O) and thioamides (X=S) of the general formula **227** depend on the solvent. In apolar solvents (CCl$_4$, CDCl$_3$), the Z form (**a**) is present, and, according to the chelate structure, the NH signal can be found at large chemical shifts (between 9 and 15 ppm). The E isomer (**b**) can be detected only in strongly polar solvents (DMSO-d$_6$, HMPT), and in these spectra the NH signal is shifted upfields (to 8 to 12 ppm). The change in configuration is also indicated by the coupling constant ^3J of olefinic hydrogens, which is, according to the rule $J_Z < J_E$,** much smaller in isomer **a**, e.g., when R=Me, R′=Ph, and X=O, $^3J = 7.6$ and 12.4 Hz in CDCl$_3$ and HMPT.[478] When the electron donor group is weaker, the Z - E isomers are in equilibrium in apolar solutions. Thus, with compounds of type RNH–CMe=CH–SO$_2$R′, two NH signals appear in the spectra of CDCl$_3$ solutions,[1431] in the regions of 7.3 to 7.4 ppm (**a**) and 4.5 to 4.8 ppm (**b**). The upfield shift of the signal is due to the weaker hydrogen bond. The temperature dependence of the NH signals of the **a** forms present as intramolecular associates is weaker than that of the **b** forms.[1445]

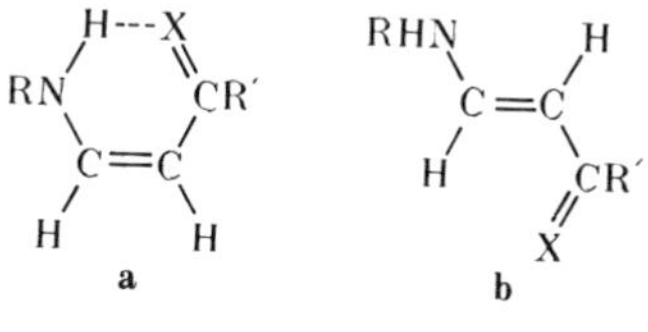

227

Among the numerous examples (compare Problems **33** and **51**), the investigation dealing with the hindered rotation of N-isopropyl-N-benzyl-mesityl carboxylic amide (**228**) is men-

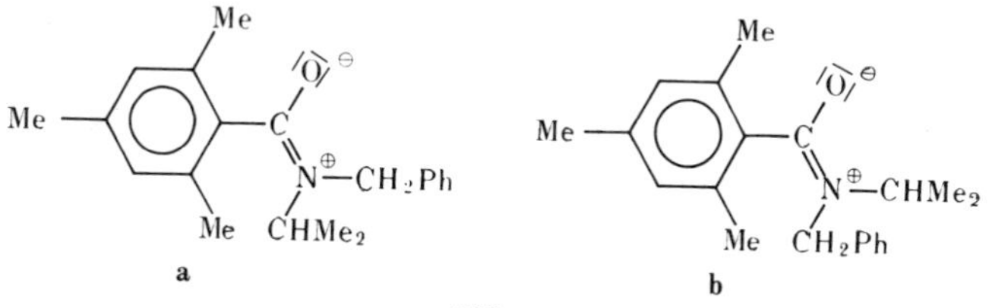

228

---

* See p. 17, 22-23, 44, 66, 66-67, 90, 108, etc.
** See p. 52.

tioned. The rotational isomers (**a** and **b**) can be separated exceptionally at a temperature below 39°C by preparative methods.[921] The real possibility of this separation was predicted from the high potential barrier (95 kJ/mol) determined by NMR methods.

Although in a very narrow range of pK (between $+1$ and -1), acid-base equilibria can also be studied by the NMR investigation of protonated amines. In aqueous solutions a fast equilibrium $R_3N^\oplus H + H_2O \rightleftharpoons R_3N + H_3O^\oplus$ sets in. The chemical shifts of groups R are, of course, different in the salt and in the base (the $-I$ effect of the ionic group in the former causes strong deshielding). Due to the rapid acid $\rightleftharpoons$ base interconversion, the measured shifts are average values. If the parameters of the base and the salt are known, the salt-base ratio, and hence the value of pK_a characteristic of the base, can be determined from the average.[18] The chemical shift varies linearly with the salt-base ratio, whereas the former gives a typical titration curve when plotted against the pH.[597]

When a molecule may be protonated at various sites, also the site of protonation can often be determined, e.g., from the number and relative intensity of the signals. The N-protonated derivative of acetamide (Me–CO–N$^\oplus$H$_3$) must give two signals with equal intensities corresponding to the equivalent protons of the methyl and N$^\oplus$H$_3$ groups. In contrast, the O-protonated form H$_2$N$^\oplus$=C(OH)Me produces three signals of 1:2:3 intensity. As in FSO$_3$H solution, at -92°C, the latter signals appear,[557] the proton must be attached to the oxygen atom.*

C-protonation may take place in the case of several azocyclic and other heterocyclic compounds. The presence of the C-protonated derivative (**b**) of Compound **229a** is proved by the triplet methyl signal ($J = 7.5$ Hz) in CDCl$_3$ of the hydroiodide. When the solution is diluted with D$_2$O, the triplet collapses into a singlet, since the mobile methylene protons in compound **b** are rapidly replaced by deuteria.[108]

229

Some representatives of enamines show an equilibrium between the N- and C-protonated forms. Thus, the spectrum of Compound **230** in HCl contains three methyl doublets. Of these, the splitting of two signals with equal intensities is 2 Hz, and that of the third one is 7 Hz. The former two signals correspond to the two nonequivalent (Z and E) methyl groups of the N-protonated form **a** (^4J allylic coupling), whereas the third doublet arises from the isopropyl group of the C-protonated form **b** (*vicinal* coupling). The intensity ratio of the signals of isomers **a** and **b** indicates that the N-protonated form is the predominant.[419]

230

In the 220-MHz spectrum of the macrocyclic tertiary amines **231** below -95°C, the α-

* Compare p. 67.

methylene protons give two signals ($\Delta\delta NCH_2$ = 0.45 ppm), corresponding to the inverse **a** and **b** (the barrier of the inversion is 32 kJ/mol). The spectrum of the dihydrochloride is time dependent, which is related to the isomerization **231c** $\rightleftharpoons$ **d**. Inverse **d** captures one of the anions, which is located in a cage formed by the three chains.[1304] The NH signal of inverse **c** formed primarily appears at 8.9 ppm, whereas that of inverse **d** at 4.6 ppm. The smaller shift can be attributed to the anisotropy of the methylene chains.

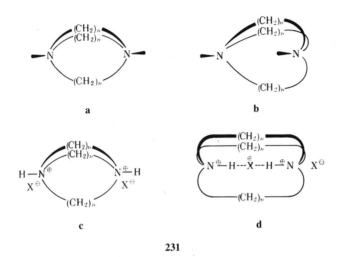

231

In contrast to the quite uniform IR spectra, the ^{1}H NMR spectra of oligopeptides enable the presence and number of certain functional groups or amino acid units to be detected or determined. Owing to the various possibilities of association, the spectra of amino acids and polypeptides vary with the temperature, concentration, and solvent. The signals of $-N^{\oplus}H_3$ groups can be identified most frequently only in strong acid media. The most convenient solvents are D_2O, DMSO-d_6, and TFA. The literature of the NMR investigations on amino acids and the peptides is very extensive,[185,254,781,1226,1403] according to the biological importance of these compounds. The ^{1}H NMR spectra of natural amino acids measured at different frequencies have also been published in collections.[74,178,235,1226]

The identification of amino acids is, of course, easy, and the qualitative analysis of their mixtures is also simple.[1272] Here the strong signals with characteristic shifts give the greatest assistance, e.g., the intense doublet of the isopropyl groups of valine and leucine and analogous, but less intense, signals of alanine, isoleucine, and threonine, the S-methyl singlet of methionine, and the aromatic signals of phenylalanine, tyrosine, tryptophan, and histidine. Figure 154 shows the spectrum of the alanyl-glycine-type dipeptide EtOOC–CH$_2$–NHCO–CHMe–NHCO–OCH$_2$Ph, in which the triplet and doublet methyl signals are at 1.23 and 1.36 ppm; the methylene quartet, doublet, and singlet appear at 3.97, 4.16, and 5.09 ppm, respectively; the methine quintet is at 4.27 ppm; the doublet and triplet of the adjacent and the other NH groups appear at 5.85 and 7.05 ppm; and the signal of the phenyl hydrogens is at 7.32 ppm.[649]

The amino acids can also be identified by indirect methods. The methoxy signals of methyl esters shift substantially (by 0.2 to 1.0 ppm) upfields upon the tritylation of the amino group, and the extent of shift is characteristic of the amino acid.[1485] Thus, the signal of glycine shifts by approximately 0.25 ppm, and those of aromatic amino acids shift by nearly 1 ppm. The phenomenon can be attributed to the ASIS.*

The pH dependence of the spectra can be used to determine terminal groups. The

---

* Aromatic solvent induced shift; Volume I, p. 40.

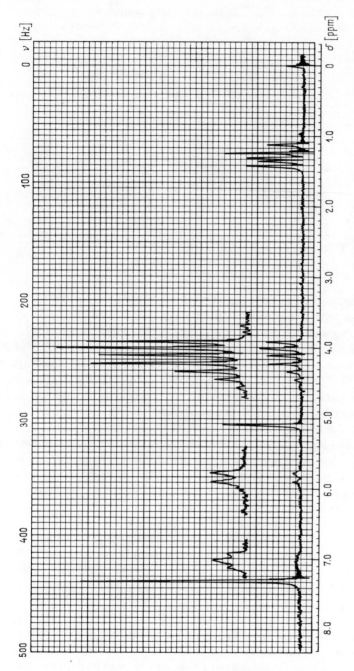

FIGURE 154.   The spectrum of *N*-benzyl-*N*-(*N*-carboxyalanyl)-glycine ethylester) (EtOOC–CH₂–NHCO–CHMe–NHCOOCH₂Ph). (From *High Resolution NMR Spectra*, SADTLER Research Laboratories, Philadelphia; Heyden and Sons, London, 1967-1975. With permission.)

δH_α signals of free carboxyl groups of peptides with zwitterion structure in neutral solutions shift downfields upon acidification, since the carboxylate ion is replaced by a more electron-withdrawing carboxyl group. In alkaline media opposite shifts can be observed, since the also adjacent, strongly electron-withdrawing (positively charged) $-N^{\oplus}H_3$ group is changed for a primary amino group of opposite character. Glycylalanine ($H_3N^{\oplus}-CH_2-CONH-CHMe-COO^{\ominus}$) and its structural isomer alanylglycine ($H_3N^{\oplus}-CHMe-CONH-CH_2COO^{\ominus}$) are easy to differentiate, since the methine and methyl signals of the former and the methylene signal of the latter shift downfield upon acidification, whereas in alkaline media the methylene signal of the former and the methyl and methine signals of the latter undergo upfield shifts.[1287]

Several papers deal with the conformational analysis of dipeptides[235,917,1301] which is performed by the usual methods (extent of anisotropic neighbor-group effect, variation of coupling constants with the dihedral angles, etc.).

The NMR spectrum may often offer valuable information of more complicated polypeptides as well. The spectrum of cyclo-hexapeptide **232** recorded in DMSO-d_6 contains two NH signals at 8.6 and 7.6 ppm,[1267] both with intensities of two protons. This has been interpreted by assuming that two intramolecular hydrogen bonds are formed in the ring between the NH protons attached to the proline carbonyl groups and the two opposite carbonyl oxygens of glycines, whereby the two other NH groups of glycines, forming stronger hydrogen bonds with solvent molecules, have larger chemical shifts. This is supported by the spectrum of the dipeptide cycloglycylproline (**233**), in which the NH signal can be found at 8.5 ppm in DMSO-d_6.

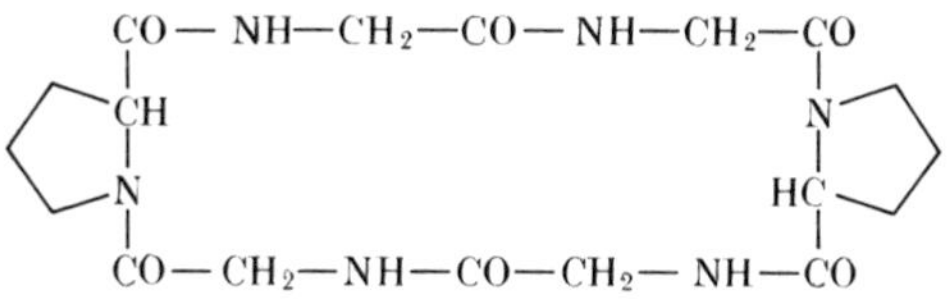

232 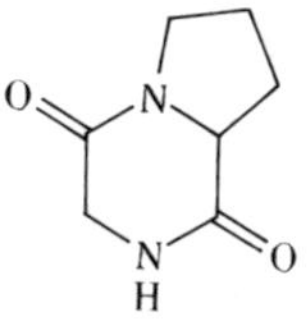   **233**

The 220-MHz NMR spectrum of oxytocine (**234**) comprising eight different amino acids can be seen in Figure 155.[1468] Most of the signals are separated in the 220-MHz spectrum and their assignment is possible using also DR experiments. The lines shown in the figure join the signals that proved to be related by DR measurements.*

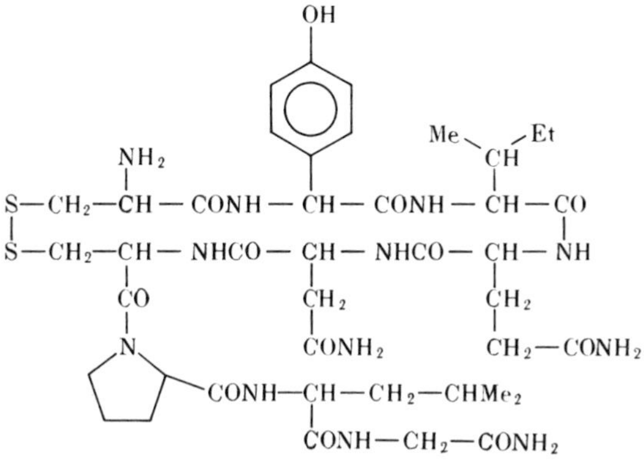

234

---

* Compare Volume I, p. 40.

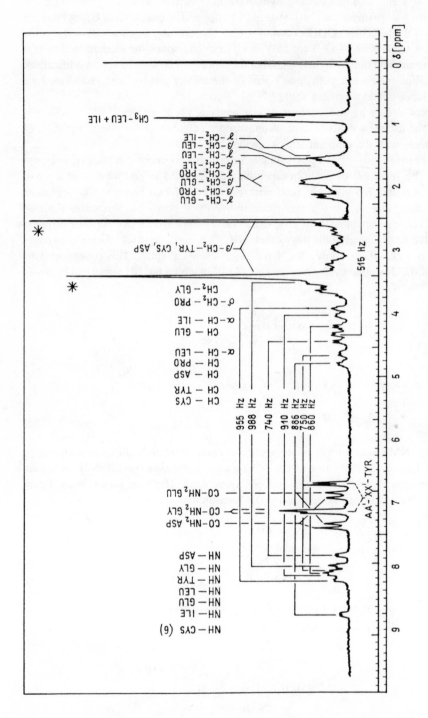

FIGURE 155. The 220-MHz spectrum of oxytocine (**234**) in DMSO-d$_6$ solution.[1467]

In the study of polypeptides, the 60- to 100-MHz spectra give only very restricted possibilities. Besides the overlapping signals of several protons similar in type, the anisotropic dipole interactions, * causing considerable line broadening due for more viscous solutions of less mobile, large molecules, and the poor solubility are additional difficulties in the investigations. For these reasons, the FT technique and higher measuring frequency have an outstanding importance in this field.

The SH signals of mercaptans can be expected to appear between 1 and 2 ppm in dilute solutions.[1044] The signal width is extremely varying and may increase to such an extent that the signal disappears. The association shift is generally less than 0.5 ppm, and even with thioacids and thioenols the chemical shift increases only by approximately 3 to 6 ppm, since the SH groups form only weaker hydrogen bonds.[1379] The SH signal of thioacetic acid (MeCOSH) is at 4.73 ppm in $CDCl_3$.[1044] Studies on the keto-enol equilibria of β-thioketo-thioesters of type **235** (R=alkyl) have shown that the signal of the chelated sulfhydryl groups in the enol form **b** can be found in the region of 7.10 to 7.40 ppm.[405] In spectra recorded in CCl_4, the ^4J coupling between the methyl and SH groups could be observed for all the compounds studied ($^4J \approx 1$ Hz). Of course, the SH signals, too, vary with the concentration, temperature, solvent, and pH.[500,925,1171,1225] Owing to the slower exchange processes, the SH–CH couplings are observable generally more often (besides the DMSO-d_6 solutions, e.g., in CCl_4, as well) than the CH–OH couplings (compare Problem **8**).

235

3.7.1.5. Protons Attached to Atoms Other Than C, O, N, or S and Adjacent to Magnetic Nuclei

Hydrogens attached to silicon atoms give usually multiplets in the region of 3.0 to 6.5 ppm.[411,434,1379] The chemical shift increases with the electronegativities of the substituents attached to the silicon atom, except for vinyl and phenyl groups. This is attributed to hyperconjugation of type p-dπ.[1486] The derivatives of germanium have roughly the same chemical shifts than the silicon analogues.[1254] The signals of lead, tin, phosphorus (compare Problem **50**), and boron hydrides fall into the spectrum regions also usual for organic compounds,[434,1379] whereas the protons of metal hydrides are often extremely shielded, having signals between -5 and -18 ppm.[434] These signals are split by the adjacent magnetic nucleus. The splitting is often very large (several hundred or several thousand hertz!), and thus certain lines may appear beyond the usual spectrum range (see Problem **50**). $^1J(X,H)$ in $(BH_2)_2$, SiH_4, SnH_4, and $PbMe_3H$ are 125, 202, 1842, or 1933 (for isotope ^{117}Sn and ^{119}Sn) and 238 Hz, respectively.[434,1379]

In metal-organic compounds, the signals of the adjacent hydrogens are split by the H-metal spin-spin interaction.** The spectrum of tetramethyltin ($SnMe_4$) contains, on both sides of the very strong central signal corresponding to the isotope ^{118}Sn, two weak doublets with about equal intensities. The outer maxima of these doublets correspond to the methyl groups coupled with the ^{119}Sn isotope and the inner ones correspond to those spin-spin interacting with the ^{117}Sn nuclei.[1044] The intensity ratios follow from the relative natural abundances of the three isotopes, which are, in the sequence of mass numbers about 8, 83, and 9%.[434]

The mid-point of the spectrum of $PbEt_4$[1025] corresponds to the broad absorption of the

---

* Compare Volume I, p. 46 and 219, and Volume II, p. 150.
** Compare p. 210 and Problems 43 and 50.

ethyl groups attached to the inactive [208]Pb isotope because of incidental isochrony of the methyl and methylene protons. On the two sides of this broad signal, one can find the A_2B_3 multiplet of the ethyl groups which are attached to the magnetically active ($I = 1/2$) [207]Pb isotope, and due to its natural abundance of 23%, the lines are relatively intense. From the splittings, $^3J(Pb,H)$ for the methyl hydrogens is 125 Hz, whereas $^2J(Pb,H)$ for the methylene protons is 41 Hz. This anomal relationship between the coupling constants $^2J < ^3J$ can be attributed to the transferring role of the d orbitals of the metal atom: the hyperconjugation with the electrons of the methyl groups increases the magnitude of ^3J interactions. The same phenomenon has been found with other, similar compounds (of type MEt_n). For M = [117]Sn, [119]Sn, Pb, Hg, and Tl, ^2J and ^3J are 31 and 68, 32 and 71, 41 and 125, 88 and 115, and 198 and 396 Hz, respectively.[434]

3.7.2. The NMR Determination of Optical Purity

In the investigation of amino acids and peptides, the determination of their optical purity, and the racemization upon various reactions is a very frequent problem which, therefore, is discussed at this point. Enantiomers can be differentiated by the NMR technique, and thus the optical purity of a given sample and the ratio of the antipodes in a mixture of enantiomers can be determined quantitatively.[979,1154]

The NMR spectra of enantiomers in achiral solutions are identical.* If however, they are dissolved in optically active solvents,[1115] the spectra become different.**

The chemical nonequivalence of analogous nuclei of enantiomers in optically active solvent was first observed in the ^{19}F NMR spectrum of α-trifluoromethyl-benzyl alcohol[PhCH(OH)CF$_3$], and this example called attention to the possibility of determining optical purity by NMR spectroscopy.[1115] The trifluoromethyl signal, which is a doublet in CCl$_4$ solution and in *racemic* 1-phenyl-ethylamine, is split into two doublets of identical intensity in optically active 1-phenyl-ethylamine [PhCH(NH$_2$)CH$_3$]. The reason for this is that the associates formed between the dissolved enantiomers and the solvent molecules are in diastereomeric relationship (e.g., with a solvent of s-configuration RS and SS diastereomers are formed). In these solutions the mutual steric positions of analogous groups, and thus their shieldings, are different. In the *racemic* solvent, the exchange between the collision complexes is fast which leads to an averaging of the signals of the diastereomers. The ratio of the enantiomers of partially resolved substances can be obtained from the intensity ratio of these signals also in *racemic* solvent.

The extent of shift differences between the enantiomers depends on the anisotropic effects of the substituents of the chiral molecule and on the differences between the relative occurrences of the conformers in the RS and SS forms. Nevertheless, these differences are usually small, and this makes it easy to understand why this phenomenon was observed first in a ^{19}F NMR spectrum, where the shifts and their differences*** are a magnitude larger. However, in several cases the differences are large enough in ^{1}H NMR as well to allow the quantitative determination of optical purity. In optically active α-trifluoromethyl benzlyalcohol [Ph–CH(CF$_3$)OH] as solvent, the *C*-methyl doublet, the methine quartet, and the methoxy singlet of partially resolved s-alanine methyl ester hydrochloride (H$_3$N$^\oplus$CHMe–COOMe + Cl$^\ominus$) are all doubled.[1117] Of the *C*-methyl signals, the downfield one is stronger, and in the other two pairs the upfield members (see Figure 156), reveal not only the ratio of enantiomers (approximately 3:2), but also the absolute configurations of enantiomers, because, according to systematic investigations, within a family of compounds the direction of shift in the analogous signals is characteristic of the s or R configuration.[1116]

---

*   See Volume I, p. 69.
** See Volume I, p. 69 and p. 41, point 5.
*** Compare Volume I, p. 26.

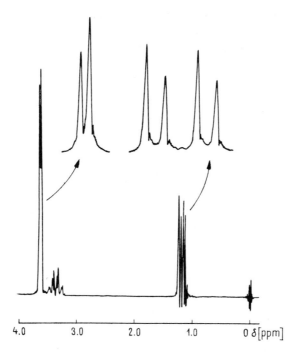

FIGURE 156. Part of the 100-MHz spectrum of partially resolved (s)-alanine-methylester hydrochloride
(MeCHN⊖H₃COOMeCl⊖) in (−)-1-trifluoromethyl-benzyl alcohol solution.[1117]

In the determination of optical purity, NMR measurements can, of course, also be combined with preparative methods. The molecule under study can be reacted with appropriate
optically active compounds to yield diastereomers. These compounds have, of course, different spectra already in achiral solvents. The results of investigations combined with chemical reactions may, however, be falsified by the racemization which may occasionally occur
in the reaction. This possibility must, therefore, always be taken into account!

According to the above, the possible racemization taking place upon the binding of amino
acids can also be investigated by NMR methods. This is illustrated by a partial spectrum of
the dipeptide alanyl-phenylalanine. The methyl doublets of L-alanyl-L-phenylalanine and L-
alanyl-D-phenylalanine are separated well in aqueous solutions at pH ~3,[622] and from the
intensities a diastereomeric ratio of 1:3 can be obtained (see Figure 157). Optically active
shift reagents are also used to enhance shift differences between enantiomers.*

3.7.3. Shift Reagents[493-498,673,1196,1302]

*3.7.3.1. Paramagnetic Metal Ion Induced Shift; Structure of Complexes and the Correlation
Between the Complex Concentration and the Induced Shifts*

Shift reagent technique achieved one of the fastest and most remarkable careers among
the NMR methods. It is extremely simple and cheap and thus easily available for everyone.
By this technique similar or often better results may be obtained in the separation of overlapping and complicated multiplets or in the simplification of high-order spectra to cases
enabling first-order analysis than by using high frequency (≥200 MHz) instruments which
were expensive and hardly accessible at the spreading time of the shift reagent technique
(early 1970s). The technique offers the possibility of clearing up conformations (geometry)

* Compare p. 132-133.

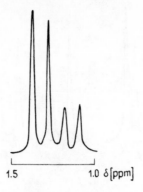

FIGURE 157.   Part of the spectrum of L-alanyl-L-phenylalanine and L-alanyl-D-phenylalanine in D_2O (pH ≈ 3) with the methyl signal.[622]

of flexible molecules in solution, determining absolute configurations and optical purity, solving complicated structural and assignment problems, etc.

It was recognized already in 1960 that in the presence of paramagnetic metal ions the resonance frequencies, i.e., chemical shifts of magnetic nuclei, undergo significant and specific changes. The change may be very large, even 10 to 20 ppm in proton resonance. Specificity means here that the signals of the same nuclear type, e.g., protons in ¹H NMR spectroscopy, within the same molecule are shifted in extremely varying extents, thereby also changing the relative positions of the signals.[707] The relative differences in the shifts may be interpreted and thus predicted theoretically. The term "shift reagent" was first applied on paramagnetic metal ions and other additives that, on mixing them into the sample solution, change the relative chemical shifts of the magnetic nuclei of the solute molecules.

It has been found that lanthanides induce larger shifts than the metal ions used in the early times as shift reagent, and the simultaneously occurring line broadening is also smaller.[858] Since the salts of rare earth metal ions are insoluble in commonly used organic solvents, they were converted into metallo-organic complexes using appropriate organic compounds (of chelate type), thereby producing well soluble shift reagents.[410] These lanthanide complexes are bonded to certain groups (Lewis bases in character) of the solute molecules with simultaneous coordination expansion, and in the resulting anisotropic local magnetic field the chemical shifts of the magnetic nuclei of the substrate undergo specific changes. The phenomenon is completely analogous, e.g., to ASIS,* and can be regarded, similarly, as a special case of medium effect.

It was observed in 1969 by Hinckley[652] that the shift reagents may be employed in the separation of overlapping NMR signals. In the unresolved ¹H NMR spectrum of cholesterol almost all signals became well separated upon the addition of the dipyridine adduct of europium-tris-dipivaloyl-methane. This recognition started the explosive spread of shift reagent technique. It had rapidly turned out that pyridine is unnecessary, europium-tris-dipivaloyl-methane is soluble just as well and induces even higher shifts than the pyridine adduct.[1234] The "dispersion" of signals also simplified the structures of complicated overlapping signals corresponding originally to higher order spin-spin interactions, since owing to the lower $J/\Delta\nu$ ratio** simplify into symmetrical multiplets corresponding to first-order interaction, yielding the individual coupling constants at a glance. Such simplifications

---

*   Aromatic solvent induced shift; see Volume I, p. 40.
** See Volume I, p. 51 and compare Equation 109.

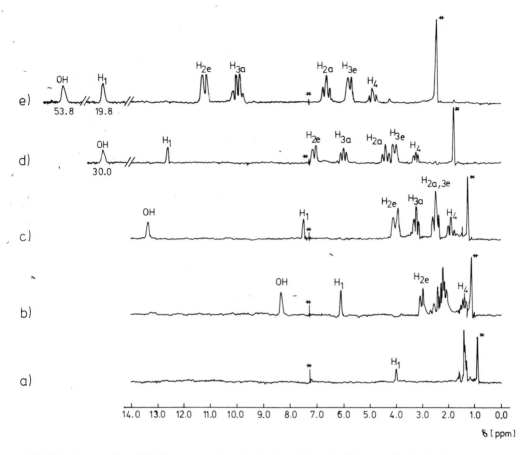

FIGURE 158. (a) The 100-MHz spectrum of *cis* 4-*t*-butyl-cyclohexanol (**100**, R = OH) in CDCl$_3$ and (b) to (e) in the presence of 10.3, 16.0, 33.1, and 60.2 mg Eu(DPM)$_3$.[348]

formerly required high-frequency or complicated, tiresome, and not always successful DR experiments.

It was found that some of the lanthanides induce a downfield shift in almost all NMR signals (paramagnetic shift reagents), whereas others induce opposite shifts (diamagnetic shift reagents).[202] This is in correlation with the periodically changing electronic structure of rare earth metals and the related, and thus also periodically varying magnetic properties.

By now the term "shift reagent" (SR) has been restricted almost exclusively to lanthanide complexes of type **236**. The most widespread representatives are europium-tris-dipivaloyl-methane [Eu(DPM)$_3$] or europium-tris-2,2,6,6-tetramethylheptane-3,5-dione [Eu(TMHD)$_3$], (**236**, Ln = Eu, R = R′ = CMe$_3$) and europium-tris-1,1,1,2,2,3,3-heptafluoro-7,7-di-methyloctane-4,6-dione [Eu(FOD)$_3$], (**236**, Ln = Eu, R = CMe$_3$, R′=*n*C$_3$F$_7$).[1222]

$$\begin{array}{c} R \\ \diagdown \\ C-O \\ HC \diagup \quad \diagup \\ \quad [Ln]_{1/3} \\ C-O \\ \diagup \\ R' \end{array}$$

236

Figure 158 shows the spectrum of *cis* 4-*t*-butyl-cyclohexanol (**100, R=OH**) in the presence

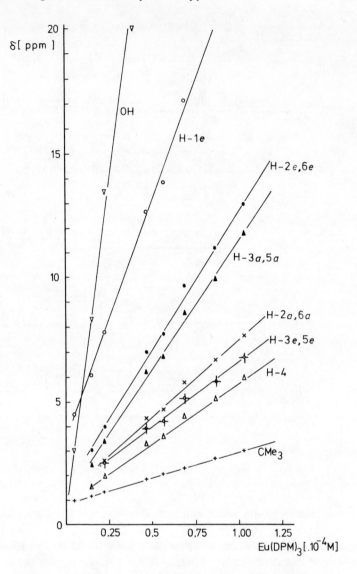

FIGURE 159.   The change in chemical shift of the various protons of *cis* 4-
t-butyl-cyclohexanol (**100**, R = OH) as a function of Eu(DPM)₃ concentration.[348]

of various amounts of Eu(DPM)₃.[348] The figure illustrates well that the shift caused by the
reagent (Δ_i) increases with the amount of SR; it is specific (different in magnitude) for the
various nuclei and does not affect couplings.

   If the chemical shifts of the various protons are plotted against the amount of SR added
(see Figure 159), straight lines of different slopes are obtained.[348] This, quite generally valid
proportionality between Δ_i and the amount of shift reagent (L) indicates that exchange
processes take place between the free substrate (S) and its complex formed with the shift
reagent (LS) and that the process is fast on the NMR time scale,* i.e., $1/\tau \gg |2\pi\Delta_{Ti}|$,
(where τ is the mean lifetime of the complex and Δ_{Ti} is the change in chemical shift of
nucleus i in the complex with respect to the free substrate: $\Delta_{Ti} = \delta LS_i - \delta S_i$, where δS_i
and δLS_i are the chemical shifts of nucleus i in the molecule studied and in its complex
formed with the SR, respectively).

*   See p. 93.

Assuming that L coordinates with the substrate only and that only one type of complex is formed (e.g., 1:1 molar ratio), one may express the equilibrium constant of the fast S + L $\rightleftharpoons$ SL equilibrium as $K = [S][L]/[SL]$ and Δ_i measured experimentally for nucleus i, which is the average of δLS and δS weighted by the concentrations can be given as

$$\Delta_i = \Delta_{Ti}[SL]/S_T \tag{291}$$

where S_T is the total concentration of substrate, $[SL]$ is the concentration of complex, and Δ_{Ti} is the maximum shift induced by the SR (the shift difference δLS $-$ δS measured for nucleus i in the pure complex). If the concentration of the complex is proportional with the amount of SR, i.e., $\sigma = [SL]/L_T = $ constant, for any ratio of $\rho = L_T/S_T$, Δ_i is proportional with the SR-substrate ratio:[655]

$$\Delta_i = \sigma\Delta_{Ti}\rho \tag{292}$$

In this case the measured values of Δ_i, plotted against ρ, yield straight lines with slopes proportional to Δ_{Ti}, characteristic of nucleus i. In practice, low amounts of SR are added to the solution of substrate ($S_T = $ constant), and Δ_{Ti} values are obtained by extrapolation from the Δ_i vs. ρ plots. The resulting Δ_{Ti} data are regarded as basic information concerning structure. One may often meet Δ_i vs. δSL or Δ_i vs. νSL plots or the corresponding semilogarithmic or log-log diagrams. A common shortcoming of all these methods is that they do not take into account the further equilibria possible in addition to LS $\rightleftharpoons$ L + S.

The shift Δ_{Ti} measured in practice is a weighted average of the Δ_i values corresponding to these equilibria. The most important further equilibria* are the formation of 2:1 substrate-SR complexes (S + LS $\rightleftharpoons$ LS$_2$),[34,459,652,730,1194,1282] the di- or occasionally trimerization of shift reagents [L + $(n - 1)$L] $\rightleftharpoons$ L$_n$,[354,1136] and the complexation of eventual impurities I, most often the water content** of the solution (L + I $\rightleftharpoons$ LI).[1193] Taking into account the above equilibria, the observed shift may be given as

$$\Delta_{Ti} = f_S\Delta S + f_{LS}\Delta LS + f_{LS_2}\Delta LS_2 \tag{293}$$

where f_S, f_{LS}, and f_{LS_2} are the substrate mole fractions for the various species (free molecules, 1:1 and 2:1 complexes), and the Δ-s are the Δ_i values of nucleus i in the species concerned.

The consideration of dimerization, complexation of impurities, and solvation is important in the determination of the appropriate mole fractions.[56] The ρ vs. Δ function is linear in the $\rho < 0.7$ range, even if 2:1 complexes are formed in addition to the 1:1 species. In this case the slope of line is higher than for the pure 1:1 complex. In contrast, coordination with the solvent or with impurities decreases the slope.[1196]

Since the shift reagents can bind the substrate in two, equivalent and independent coordination positions, it is easy to show that $K_{2:1}$, the equilibrium constant of reaction LS + S $\rightleftharpoons$ LS$_2$, is four times the $K_{1:1}$ constant. If there are n equivalent and independent coordination possibilities,

$$K = [S](n\,L_T - [SL])/[SL] \tag{294}$$

and

---

* If there is a strong interaction between the solvent and the solute, too, with finite dissociation constant, it should also be taken into account.

** Therefore, in the SR technique, the careful drying of samples (SR and substrate) and the solvent is of utmost importance!

$$\Delta_i = \Delta_{Ti}[LS]/S_T = np\Delta_{Ti}/(1 + K/[S]) = np\Delta_{Ti}/[1 + K/(S_T - [SL])] \tag{295}$$

If the concentration of SR, and therefore that of complexed species, is low (i.e., $S_T \gg [SL]$), Δ_i varies linearly with p. Thus, the Δ_{Ti} values obtained by extrapolation from the Δ_i vs. p plots or from considering Δ_i corresponding to $p = 1$ contain an error factor of $n/(1 + K/S_T)$. This error may lead to even a twofold overestimation of Δ_{Ti}.

The exclusive formation of 1:1 complexes was assumed only, whereas the presence of 2:1 complexes was proved experimentally, too.[459,730,1194,1282] Although a part of the experiments proved the occurrence of 2:1 complexes only at low temperatures or in solid phase and only for the case of $Eu(FOD)_3$, and no 1:2 $Eu(DPM)_3$-substrate stoichiometry has been found so far, the problem of stoichiometry may by no means be disregarded in precise studies. When, however, configuration questions are relevant *only*, this problem may be neglected, since the relative values of Δ_{Ti} and Δ_i are not influenced significantly by the stoichiometry of the complexes, the complexation of the shift reagent.

The above problems may often be bypassed, since:

1.   If only high or only low values of p are taken into account, the 1:1 or 2:1 complexes, respectively, dominate in the equilibrium.
2.   Dimerization should be considered only for $Eu(FOD)_3$; $Ed(DPM)_3$ is not self-associating.
3.   Errors due to impurities may be reduced by the careful drying and purification of components.

Formerly, in SR experiments almost exclusively constant substrate concentration (S_T is constant) was used, and the concentration of SR was varied (adding the SR to the sample in increasing amounts). The disadvantage of this method is that at the beginning of experiment the measured values of Δ_i are smaller than the true values owing to the shift reagent consumption of impurities (water): the Δ_i vs. p line shifts to the right (toward higher p values), and if the dissociation constants of the complexes with the substrate and the impurity are similar in magnitude, even the straight line may be distorted. Moreover, the slope also depends on the absolute value of S_T:[56] if S_T is higher, Δ_i is also higher. It is even more important that the relative Δ_i values are also a function of S_T, which may affect the conclusions drawn on molecular geometry.

By performing the SR experiments under the conditions $[L]_o \ll [S]_o$ and $[L]_o =$ constant, it may be ensured that complex formation takes place in a unique, single-step process, characterized by a single dissociation constant: $L + nS \rightleftharpoons LS_n$, where n is 1 or 2, but only one of them. Since $[L]_o$ is small, there is no significant dissociation, either. In this case,

$$[S]_o = [L]_o\Delta_{Ti}/\Delta_i - (1/K + [L]_o) \tag{296}$$

By plotting substrate concentration $[S]_o$ as a function of $1/\Delta_i$, a straight line can be obtained, the slope of which yields Δ_{Ti} and the intercept K. This method is also insensitive of impurities (i.e., the absolute value of $[L]_o$).

3.7.3.2. The Components of Shift-Reagent Induced Shifts

For the successful application of SR technique in structure elucidation, to avoid the risk of misinterpretation, the theoretical reasons for chemical shift changes should be known. The magnetic nuclei of the complex interact with the unpaired electron spin of paramagnetic metal ion and the resulting change in chemical shift has at least two, occasionally three, main components:

$$\Delta_T = \Delta_c + \Delta_p + \Delta_d \tag{297}$$

Δ_c is the Fermi-contact shift, which occurs when the unpaired electron is on s orbital and the electron-nucleus interaction takes place *through chemical bonds*.

In contrast, the pseudo-contact shift, Δ_p, is a result of an interaction *through space*, thus being a function of the distance and mutual position of the various nuclei of the substrate molecule and the SR, and thus it depends on the geometry of the complex. The third term, Δ_d, is the diamagnetic component of the shift. It includes all further effects on Δ_T that do not arise from the paramagneticity of the metal atom.[1421]

It is assumed[1237] that the *diamagnetic contribution* of the shift arises mostly from the steric strains developed in the complex. Its magnitude for a given SR is the same as the shift measured for the lanthanum analogue containing the same organic component: $(\Delta_d)_{Ln} = (\Delta_T)_{La}$.

Since the $La^{3\oplus}$ ion is diamagnetic, the shift $(\Delta_T)_{La}$ induced by the reagent containing lanthanum nucleus may be taken as the Δ_d component of Δ_T measured for the other lanthanides. Such measurements have shown that in certain cases the Δ_d contribution mostly neglected, is very significant, and may even reverse the sign of Δ_T.[1421] In carbonyl compounds as substrates, Δ_d is roughly 1/5 to 1/6 of Δ_T for the case of Eu and Pr chelates.[261]

As a result of the interaction between unpaired electron spins and nuclear spins leading to *Fermi-contact shift* (Δ_c), the nuclear energy levels split, which, in principle, also splits the NMR signals, like in the case of spin-spin interactions. The corresponding scalar hyperfine coupling constant, A (in hertz units), is specific for the individual species of nuclei (its magnitude and sign being different for the different species).

Since, however, the spin-spin relaxation of electrons is fast, i.e., $T_{1e} \ll 1/A$, there is no splitting, only a change in chemical shift in accordance with the distribution of nuclei among the split levels:[858,1197]

$$\Delta_c = (2\pi\beta/3kT)J(J + 1)g(g - 1)(\nu A/\gamma) \tag{298}$$

The second term in Equation 298 depends on the electron; the third depends on the nucleus. Since hyperfine coupling constants A can be determined experimentally for a given nuclear spin distribution, using their known values,[568] the value of Δ_c relative to protons may be determined for the individual nuclei. For ^1H, ^{13}C, ^{14}N, ^{17}O, ^{19}F, and ^{31}P isotopes, the values of Δ_c are 1, 8.8, 15.2, 24.1, 35.9, and 17.8.

It can be seen that the Δ_c contribution is relatively small for protons. Moreover, the hetero atoms are usually closer to the paramagnetic ion in the complex (it is attached mostly to these atoms) than the peripheric hydrogen atoms, and the larger difference in level populations further increases the value of Δ_c. Therefore the Δ_c contribution of protons may usually be neglected (the above relative data involve the minimum of Δ_c values for the different nuclei), but for other nuclei it must always be taken into account.

The Fermi-contact term has numerous experimental evidences. In proton resonance, the *ratio* of shifts of opposite sign for quinoline and isoquinoline substrates caused by Eu(DPM)$_3$ and Pr(FOD)$_3$ SRs is always the same, whereas the ^{13}C NMR shift ratios are different.[657] The same SR may cause opposite shifts in the ^{13}C NMR signals of different carbon atoms in the same substrate.[1196,1511]

The Δ_i values measured in the ^{13}C NMR spectrum of 4-phenoxybutyronitrile (PhO–CH$_2$–CH$_2$–CH$_2$CN) for C-1 to C-4 atoms in the presence of Eu(FOD)$_3$ are[1576] -40, 3.7, 10.4, and 4.7 ppm.

If only the Δ_p term existed, gradually changing shifts of the same sense would occur along the chain (see below). Different directions of shift were observed in ^{31}P resonance[916] with SR containing the same lanthanide, Pr(DPM)$_3$ and Pr(FOD)$_3$, which can be attributed to the high Δ_c term corresponding to the stronger bond present in the Pr(DPM)$_3$ complex. Since this term is opposite in sense to the Δ_p component, the resulting Δ_i has different sign.

Larger shifts, which indicate larger Δ_c contribution, were observed for the nuclei situated in the neighborhood of the metal atom in the complex in many other cases.[203,348,723]

The Δ_c term may be determined from the Δ_i values of $Gd^{3\oplus}$ complexes, since there is no Δ_p contribution for this metal (see below): $(\Delta_c)_{Ln} = (\Delta_T)_{Gd} - (\Delta_T)_{La}$ provided that the organic chelating agent of the Ln, Gd, and La complexes is the same.[143,144] The contact shift may also be determined by applying a pair of complexes causing dipolar shifts of different sense, e.g., $Eu(DPM)_3$ and $Pr(DPM)_3$, and calculating Δ_c from the relative value of $\Delta\Delta_T$ using the relationships $(\Delta_T)_{Eu} = \Delta_c + \Delta_p + \Delta_d$ and $(\Delta_T)_{Pr} = \Delta_c - \Delta_p + \Delta_d$.

In ^{13}C NMR resonance, the Δ_c term was shown to increase in the order Yb < Tm < Dy < Tb < Er < Ho < Nd < Eu,[1195] with the Δ_c/Δ_p ratio[1196] approximately 0.05 (Yb), 0.1 (Dy), 0.15 (Pr), 0.2 (Ho), 0.2 (Tb), 0.25 (Er), 0.7 (Nd), and 0.8 (Eu). The Δ_p/Δ_c ratios have also been determined in ^{17}O resonance[1196] for the $Ln^{3\oplus} + 3ClO_4^\ominus$ SR, by measuring the ^{17}O shifts: 1340 (Yb), 1000 (Pr), 870 (Dy), 503 (Tb), 447 (Ho), 277 (Nd), 238(Er), and 160 (Eu).

It can be seen that not only the ratios but even the order of lanthanides is different for ^{13}C and ^{17}O, and while for the former the contact shift term has a decisive role, for the latter, like in proton resonance, it is negligible.

The most important component of Δ_i is the *pseudo-contact* or *dipolar shift*, Δ_p, which arises from the anisotropy of the magnetic susceptibility of SR bound to the substrate in a specific orientation. In the case of isotropic susceptibility (and eventually isotropic electron configuration,* the dipolar interactions are averaged, whereas in the anisotropic case, in complete similarity to the anisotropic neighboring group effect, they cause specific chemical shift.

The explosive spread of SR technique was due to the assumption that with respect to the dipolar shift (Δ_p) the other contributions of Δ_i are negligible, and that the magnitude of Δ_p can be calculated for the various nuclei of the substrate from the simplified McConnel-Robertson equation[960] pertaining to axial symmetry (compare Equation 72). In this case the Δ_i values may be used to calculate the geometry of the substrate molecule (configuration, for flexible molecules solution phase conformation) very simply. It has turned out later that the validity of this assumption is far from general, but, primarily in the most frequented field, proton resonance, these simplified applications proved to be successful, leading — in favorable cases — to the solution of various structure elucidation, configuration, conformation, and assignment problems (compare Problems **64** and **65**).[314,678,1337]

According to the complete McConnel-Robertson equation,**

$$\Delta_p \approx \Delta_T = -(3\cos^2\Theta - 1)r^{-3}[\chi_z - (\chi_x - \chi_y)/2] + 3\sin^2\Theta\cos2\varphi(\chi_z - \chi_x)/2 \qquad (299)$$

The authors of the equation assumed originally that it should be applied for solid phase only, while in solutions, assuming the axial symmetry of anisotropy ($\chi_x = \chi_y$), the application of the simplified form Equation 72 is sufficient. In the equations, χ_x, χ_y, and χ_z are the components of magnetic susceptibility, r is the distance between the metal atom (in the origin of the coordinate system) and the atom of the substrate attached, θ is the angle between the z axis and the metal-substrate coordination bond, and φ is the angle defining the mutual position of the SR and the molecule in the $[x,y]$ plane (the angle between the projection of the coordination bond in the $[x,y]$ plane and the x or y principal axis of the anisotropy ellipsoid, see Figure 160.

---

* The isotropy of electronic configuration (and the "g -tensor" characteristic of it) does not necessarily involve the isotropy of magnetic susceptibility,[677] as it was erroneously assumed.

** Several variants of the equation are known, but they are fundamentally the same: the mutual position of substrate and shift reagent in the complex are defined by three variables, one distance and two angles.[34,806]

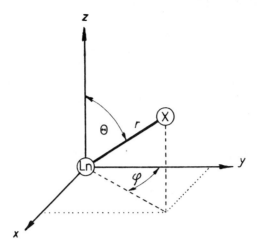

FIGURE 160. The parameters (r, Θ, and φ) determining the mutual position of rare earth metal (Ln) and the atom of substrate molecule attached to it on complexation (X), in the coordinate system of the anisotropy of magnetic susceptibility, where Ln is in the origin.

It follows from Equation 72 pertaining to axial symmetry that the magnitude of dipolar shift Δ_p in a given complex is independent of the nuclei, more particularly of their chemical nature, but depends strongly on the distance from the metal atom and on the position with respect to the coordination bond or the main symmetry axis, i.e., on the value of θ. This is shown in Figure 159, in which the steepest lines corresponding to the largest shifts belong to the OH and H-1 signals closest to the coordination center, and the shifts in the other signals arising from more distant atoms, H-2, H-3, H-4, and methyl groups, are gradually smaller. Since Δ_p is a function of $\cos^2\theta$, not only the magnitude but also the sign of the shift may vary with θ, changing sign at $\theta \approx 55°$. If $\Delta_p > 0$ for the range $0° < \theta < 55°$, then $\Delta_p < 0$ for the $55° < \theta < 125°$ range, and again $\Delta_p > 0$ for $125° < \theta < 180°$ angles (see Figure 161).

The rapid decrease of Δ_i with the distance from the coordination site is illustrated well by the mean shifts measured for normal alcohols in the presence of Eu(DPM)$_3$.[1155] The values of Δ_i for protons attached to C-1 to C-7 are 23.0, 13.4, 8.6, 4.2, 2.4, 1.5, and 0.9 ppm.

The dependence of Δ_i on molecular geometry had been revealed already in an early stage of the spread of SR technique, when deviations were observed from the r^{-3} dependence, e.g., for borneol (**237**), an $r^{-2.2}$ dependence was measured[348] with Pr(DMP)$_3$. When variation

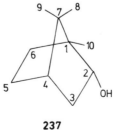

237

with θ was also taken into account, the mean deviation of the measured and calculated Δ_i values decreased to approximately 5%. This was the first practical application of Equation

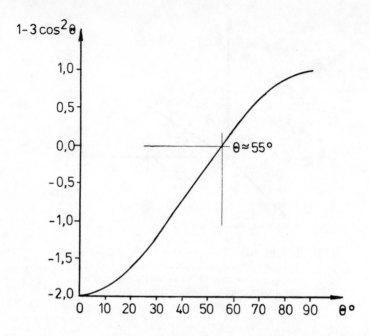

FIGURE 161.   The value of $3\cos^2 \Theta - 1$ as a function of Θ.

72 and of SR technique in the determination of molecular geometry.[723] For the calculations, the Pr–O bond length was taken as 0.3 nm, and the Pr–O–C angle was 126°. On the basis of the iterative calculation, 25° was obtained for the Pr–O–C_2–H dihedral angle corresponding to θ.

The validity of Equations 72 and 299 describing the dependence on θ was soon proved very impressively by measurements in which the SR otherwise always causing paramagnetic shifts produced diamagnetic shifts for certain atoms of certain molecules. The H-3' and H-4' signals, e.g., of *cis* 3-(α-naphthyl)-cyclohexanol derivative **238**, undergo diamagnetic shifts upon the effect of Eu(DPM)$_3$, while the other signals, like *all* signals of the *trans* isomer, show the usual paramagnetic Δ_i (see Figure 162).[1281]

238

Likewise, one of the isopropyl-methyl doublets of 2,6-diisopropyl acetanilide (**239**), which are nonequivalent owing to hindered rotation, is shifted diamagnetically upon the effect of Eu(DPM)$_3$.[1300]

The main advantage of SR technique is the possibility for simple determination of molecular geometry, however, this is also the point where the oversimplified treatment of the problem has caused most of the disappointments, leading to quantitatively or even qualitatively false conclusions. Predictions on the geometry, even if the application of Equation 72 is justified, may be correct only if the magnitude of Δ_p contribution to Δ_T is known,

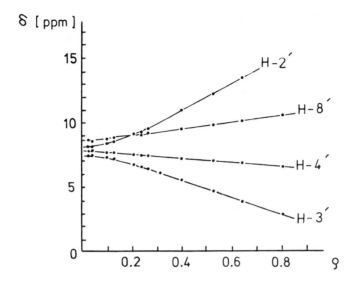

FIGURE 162. The shifts in signals H-2',3',4',8', as a function of ρ in the spectrum of **238** upon the effect of Eu(DPM)₃.[1281]

239

i.e., if the Δ_c and Δ_d terms are also taken into account and the stoichiometry of the complex is considered, too. Moreover, one should know the length of the coordination bond and its position with respect to the substrate (i.e., angle θ). Even with these data in possession, the assumption of axial symmetry is hard to justify.

X-ray diffraction studies have shown on several complexes formed with shift reagents[314,678,1531] that the axis of the molecule (if any) and the coordination bond do not coincide with the principal axis of the anisotropic susceptibility ellipsoid of the shift reagent, i.e., there is an inevitable φ=dependence, requiring the application of Equation 299.[34,806,960] Occasionally, the second term may reach 20 to 80% of the total value of Δ_p.[313]

In fortunate case the geometric changes of the complex, i.e., the transformation of one conformer into another, may already take place in solid phase,[407] and it becomes even more probable in solution.

The free and fast rotation of the substrate molecule about the coordination bond results in, of course, an effective axial symmetry.[56] Here, by "fast" rotation we mean motions fast as compared to Δ_i which is in the order of 1000 Hz for protons (the mean life time of a given conformer, $\tau < 10^{-3}$ s). It was shown, however, that even for not too bulky substrates the rotational barrier is quite high, > 29 kJ/mol.[312] For an effective axial symmetry, however, it is sufficient if the complex conformers exchange rapidly in processes [LS*] + S ⇌ LS + S* or [L*S] + L ⇌ [LS +] + L*.[205,676]

The justification of the assumption of axial symmetry may be investigated most simply by calculating the variation of the difference between theoretical and experimental Δ_i values as a function of φ. If this has a minimum for given φ angles or sets of $r - θ - φ$ values, it is reasonable to assume hindered, partial rotation. Such investigations[56] indicated, e.g., that for aniline as substrate axial symmetry may be assumed, whereas diacetone glucose (**240**) forms a rigid complex of fixed conformation with Eu(DPM)$_3$. For Compound **241** there is no good agreement between the calculated and measured data for any set of parameters, suggesting hindered, partial rotation or slow exchange between the individual conformers.

240   **241**

The connection between the steric hindrance in the complex or the size of the substrate and the presence of effective axial symmetry is obvious. Therefore it is easy to judge whether in the case of a given substrate Equation 72 is applicable instead of Equation 299 without the risk of serious error.

Applying the methods discussed above for the determination of real stoichiometry and the Δ_c and Δ_d contributions and adopting the geometry of the complex from crystallographic investigations (r and θ), the SR technique is successfully applicable for the quantitative investigation of the solution-phase conformations of substrates and the steric structures of complexes. Although the method is not as simple as believed at the beginning, its application with due precautions may be very powerful in structure elucidation.

3.7.3.3. The Shift Reagents: The Rare Earth Metal Atoms and the Organic Complexing Agents

Of the rare earth metals, in the vast majority of cases, europium (atomic mass 152.0), causing paramagnetic shifts, and praseodymium (atomic mass 140.9), causing diamagnetic shifts, are employed, since these two lanthanides induce the largest shifts when related to the simultaneous line broadening: $\Delta\nu/\Delta_T = 0.005$ for Eu and 0.002 for Pr in hertz per hertz units.[15] For the rest of the rare earth metals this ratio is one to two orders higher. A great advantage of Eu is that it induces paramagnetic shifts, and thus at smaller shifts the original sequence of most of the signals is preserved, and a great part of the signal remains in the usual chemical shift range.

The complexes of Pr cause two to three times larger shifts, also depending on the substrate in the opposite direction.[784] Since the shifts are usually larger for signals with originally higher δ (the nuclei adjacent to heteroatoms have larger shifts), the Pr complexes may result in the "compression", occasional overlap of signals, unlike the complexes of Eu causing scattering of signals. Moreover, the signal(s) arising from the organic component of the complex and those of the substrate shift in opposite directions, and thus the "own signals" of the Eu complex, move out of the spectrum region containing most of the signals (diamagnetic shift), whereas those of the Pr complex tend to overlap with the signals of the substrate on increasing concentration. For the same reason, it is a further advantage of Eu complexes that the signal of the organic component (for DPM the *t*-butyl signal) is at higher field than those of other lanthanide complexes, in the neighborhood of the TMS signal [at approximately 0.5 ppm in dilute solutions of Eu(DPM)$_3$], and moves upfields with increasing SR concentration. The joint application of Eu and Pr complexes may split almost all accidentally overlapping signals, making possible therefore the interpretation of extremely complex spectra.

The other lanthanides play a role only in the theoretical investigation of phenomena connected to SRs, save for the application of La and Gd in the determination of Δ_c and Δ_d components of Δ_i. The complexes of Yb are applied more frequently, since they have no tendency of dimerization, have low Δ_c contribution to Δ_i, and form more stable complexes owing to the smaller ionic radius.[407]

In aqueous solutions, the SRs are generally inapplicable, since they coordinate with the water molecules instead of the substrate. If the application of aqueous solutions is unavoidable (e.g., with phosphoroorganic compounds), $Eu(NO_3)_3 \cdot 6H_2O$ is a well applicable SR.[1236]

It was published, erroneously, that phenols and carboxylic acids decompose the SRs,[1234] and therefore with these compounds $BPh_4^{\ominus}$ was used as anion.[1253] Although it has been proved since then that even in the case of phenol or carboxylic acid substrate there is no risk of decomposition of usual SRs, either, $BPh_4^{\ominus}$ anion as SR occasionally still emerges in literature.

In order to make otherwise insoluble rare earth metals soluble in organic solvents, they must be converted into metallo-organic complexes. The organic component must be capable of complex coordination (par excellence representatives of this type are the enols of β-dioxo compounds), ensure a good solubility in organic solvents (contain lipophylic groups large enough for shielding the polar metal atom), and should have none or only one or two of its own signals in the spectrum.

These requirements are best fulfilled by DPM and FOD, and, disregarding the representatives of optically active shift reagents to be mentioned later, they are used nowadays as exclusive chelating agents in the commercial SRs. In proton resonance their fully deuterated analogues may be used to advantage, and with the spread of ^{13}C NMR technique, they also appear in a form enriched in ^{12}C.

FOD complexes have generally an order of magnitude higher solubility and form stronger complexes. Therefore, their apparent Δ_i values are higher, but the true Δ_T values are generally higher for DPM. FOD complexes are more easily dimerized than their DPM analogues, and in the case of polyfunctional ligands, simultaneous coordination has greater importance. Therefore, for conformational studies, it is more advantageous to use DPM complexes, since conclusions on geometry may be well founded only if the coordination structure is unique, when the coordination centers are independent (i.e., possibly far from one another and having widely different dissociation constants; in the latter respect, the application of more loosely bonded DPM complexes is more advantageous). It is to note that the SRs are hygroscopic, and thus the disturbing effects of water, discussed above, must be borne in mind.

3.7.3.4. The Effect of Shift Reagent on the Chemical Shifts and Coupling Constants of the Substrate

The SRs are complex Lewis acids with six ligands, able to bind one or two substrate molecules of Lewis-base via coordination expansion. The relatively weak coordination makes fast exchange phenomena possible corresponding to equilibria $S + L \rightleftharpoons SL$, $SL + S^* \rightleftharpoons S^*L + S$ and $SL + L^* \rightleftharpoons SL^* + L$, which lead to a concentration weighted average of the chemical shifts of the individual nuclei in the free substrate and in the complex. Consequently, the SR technique is applicable to any compound containing atoms with lone electron pair (i.e., hetero atom) or acidic protons (OH, NH, SH, etc.).

The magnitude of Δ_i was investigated in identical concentration, upon the effect of the same SR, on compounds of otherwise analogous structure but containing different functional groups. The resulting sequence of groups, according to increasing values of Δ_T, reflects simultaneously the magnitude of dissociation constant and, to smaller degree, the geometry (sterically preferred or hindered nature of coordination):[1235]

$$-NH_2 > -OH > {>}C{=}O > -CHO > R{-}O{-}R > R{-}S{-}R > R{-}COOR' > -CN$$

Nitrogen-containing compounds are good substrates, such as *N*-oxides, nitroso derivatives, nitriles, azo-derivatives, nitrogen-containing heterocycles, etc. Of compounds with other hetero atoms, sulfoxides and phosphoroorganic derivatives are to be mentioned.

A linear relationship was found[449] between the pK_a values and Δ_i parameters of amines. With *N*-methyl substitution, Δ_i decreases although basicity increases, indicating the role of steric factors.

Although ethers are weaker Lewis bases than alcohols, e.g., in reserpine, Eu(FOD)$_3$ is attached mostly to the methoxy groups[1559] attributed most probably to the two possibilities of coordination.[1560] This is also suggested by the strong shifting of thioethers.[1003]

With oximes of type RR'C=NOH, Δ_i is equally higher for the R and R' groups of the *anti* isomer than for those of the *syn* isomer, making probable the coordination on the nitrogen atom.[118,1429] It is noted that opposite opinions, supporting the coordination on oxygen, may also be found in the literature.[1554]

In *N*-disubstituted amides (RR'N-COOR″) Δ_i is significantly different for the R and R' groups.[94] This is a result of hindered rotation which is enhanced by the coordination of SR through the negative polarization of the carbonyl group and thus the increase in CN bond order. The group *cis* to the coordinated oxygen, being closer to the metal atom, is much more exposed to the chemical shift changing effect of the latter.

With polydentate ligands, the Δ_T vs. ρ function often departs from linear, even if the complexes have unique stoichiometry. When, however, the coordinating groups are "independent", the shift varies linearly with the [L]/[S] concentration ratio.

A distortion of linearity was observed for both isomers of Compound **242** (see Figure 163), where the −COOMe and −OH groups compete for the shift reagent.[485] In contrast, the polyfunctional carbohydrate **243** proved to be a monodentate ligand on the basis of the Δ_i values of protons, which decreased in the sequence of $5 > 4 > 1 > 3 > 2$,[559] indicating a preferred coordination on the hydroxy group.

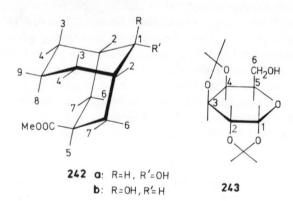

242 **a**: R=H, R'=OH
         b: R=OH, R'=H

243

The complications arising from the competition of coordination sites are often avoidable by a simple chemical transformation of the substrate. Thus, if one of the groups originally having similar dissociation constants is converted into a much weaker Lewis base, the condition of independent functional groups can be realized. The dissociation constants of thiocarbonyl and hydroxy groups may be decreased significantly, e.g., by converting the former into dithioketal [${>}C{=}S \rightarrow {>}C(SR)_2$] and the latter into trifluoroacetoxy (−OH → −OCOCF$_3$) groups.[320]

It is noted that by chemical methods compounds having no Lewis base character may also be converted into "shifting" compounds. Thus, olefins become "shifting" compounds upon the addition of silver salt.[458] Adamantane and some simple hydrocarbons were enclosed into cyclodextrin cages to form compounds capable of coordination.[906]

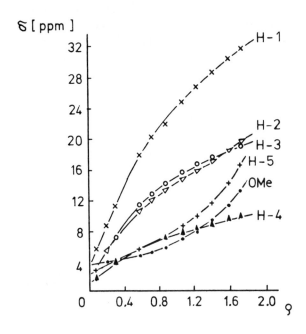

FIGURE 163. The dependence of the proton shifts of Compound **242b** on ρ in the presence of Eu(DPM)₃.[485]

The SR-induced shifts Δ_i may show temperature dependence. This is often connected with changes in conformational equilibria, that may, however, also arise from changes in substrate stoichiometry (i.e., changes due to the effect of temperature on the equilibrium of free substrate and complexes of various stoichiometry) and from the temperature dependence of the paramagneticity of SR. This latter is characteristically different for Eu and the other lanthanides. The paramagneticity of other lanthanides is due to a first-order Zeeman effect which plays no role in the case of Eu owing to the diamagnetic, nondegenerated $^7F_0$ ground state. Instead, a second-order Zeeman effect produces paramagneticity here, which occurs owing to a significant population of the relatively low-energy $^7F_1$ excited state ($\Delta E_{F_1 - F_0} = 350$ cm^{-1}, which is very small, and thus population is approximately 36% at room temperature).[1308] This type of paramagneticity, which is due to the population of an excited state, shows exponential temperature dependence, i.e., the nature of temperature dependence is basically different from that of other lanthanides.[1439,1502]

It is noted that the NMR signals of alcohols and their α-deuterated derivatives split in the presence of shift reagent, and Δ_i is greater by 2 to 3% for the molecules containing the heavy isotope.[655,1311] This isotope effect is attributed to higher basicity and thus lower dissociation constant, stronger complex bond, and probably to changes in Δ_c on deuteration.[654,675]

The coupling constants and, in general, splittings are not affected by SRs. Since, however, the paramagnetic metal ion increases the relaxation rate of nuclei through electron-nucleus interactions, when one (e.g., A) of two spin-spin interacting nuclei is so close to the metal ion that the condition of splitting $J_{AB} > \pi T_{1A} \sqrt{2}/2$ is already not fulfilled because of the low value of relaxation time T_{1A}, the signal of A is suppressed in the spectrum (broadens) and the splitting of signal B disappears.* This is the explanation for some "unexpected" line broadenings and collapsing of multiplets upon the effect of SR, e.g., as observed with quinoline in the presence of Eu(DPM)₃.[1198]

---

* Compare with the similar effect of scalar relaxation on splitting, Volume I, Section 2.2.6.5.

The SR as "electron-withdrawing substituent" may slightly affect coupling constants, too, by decreasing the electron density in the substrate on complex formation, e.g., thus, it increases the absolute value of $^2J^{gem}$-type couplings, like the substituents with $-I$ effect.[1283] The $J^o_{1,2}$ and $J^o_{2,3}$ coupling constants of Compound **244** decrease in the presence of $Eu(DPM)_3$, like upon electrophylic substitution.[1089]

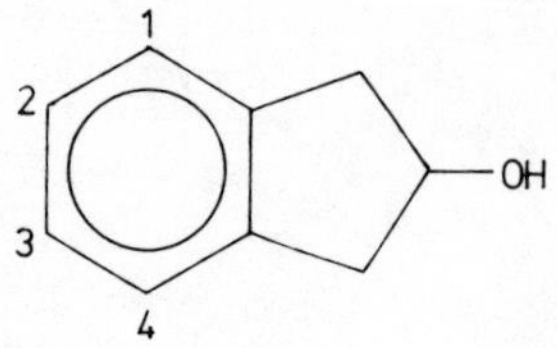

244

For steric reasons, the SRs may change the conformational equilibria of flexible substrate molecules (e.g., the rotamer population of carboxylic amides or the inversion rate of non-planar cyclic compounds) and thereby affect the magnitudes of coupling constants.[172,245,312,741,1444,1558] It is easy to see that the equilibrium of rotamers **a** and **b** of **245** shifts in favor of form **b** with increasing SR concentration.[1395]

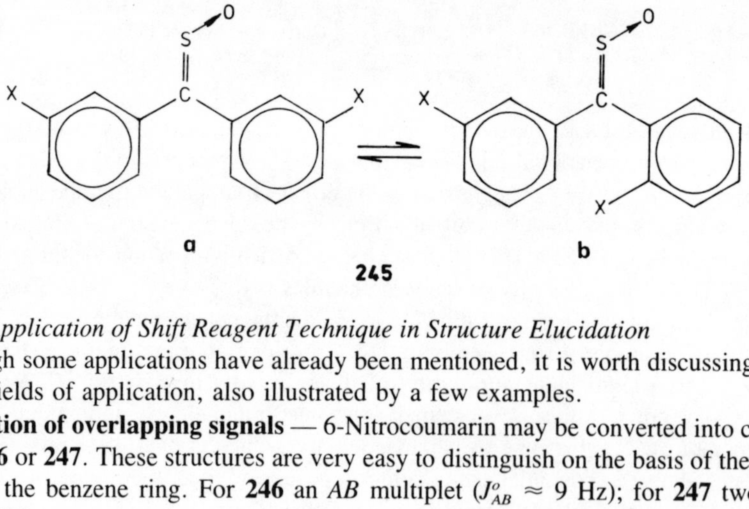

a   **b**

245

3.7.3.5. Application of Shift Reagent Technique in Structure Elucidation

Although some applications have already been mentioned, it is worth discussing the most frequent fields of application, also illustrated by a few examples.

Separation of overlapping signals — 6-Nitrocoumarin may be converted into condensed system **246** or **247**. These structures are very easy to distinguish on the basis of the ^1H NMR signals of the benzene ring. For **246** an AB multiplet ($J^o_{AB} \approx 9$ Hz); for **247** two singlets ($J^p_{AB} < 1$ Hz) are expected to occur. However, in the common proton resonance spectrum the aromatic signals overlap, and thus it turns out only after the addition of SR that Compound **246** was formed.[880]

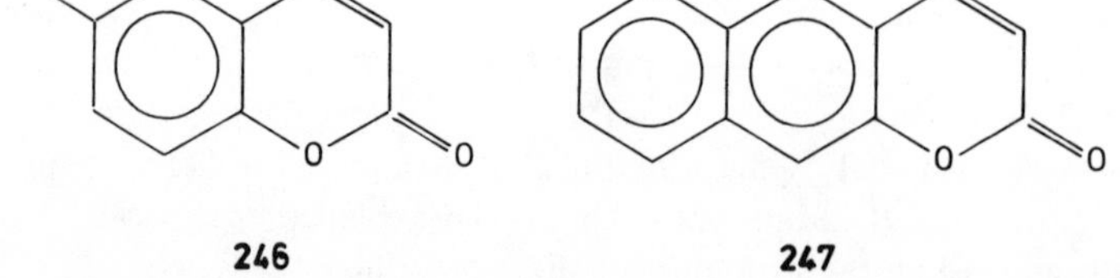

246   **247**

Detection of accidental isochrony — Diastereotopic pairs may often become accidentally isochronous. Anisochrony is easy to prove by means of SRs. The structures of the two, isolated diastereomers of Compound **248** could not be identified by ¹H resonance, since the methylene protons of both compounds produced singlets, although in the case of the *meso* isomer (**a**) they are diastereotopic and potentially anisochronous.[585] In the presence of Eu(DPM)₃, the methylene signal of the compound with the lower melting point splits into an *AB* spectrum (with $J_{AB} = 13$ Hz), while that of the other remains a singlet. Consequently, the former has *meso* (**a**) and the latter has *racemic* (**b**) structure.

a *(meso)*   **b** *(racemic)*

248

Simplification of higher-order spectra into first order — Due to the specific nature of Δ_i, the $J/\Delta\nu$ ratio may often be decreased to such an extent that complicated overlapping higher-order multiplets separate into symmetrical multiplets of first order, enabling the coupling constants to be read directly from the spectrum. The coupling constants determined by computer analysis of the overlapping *AB* part of the *AA′BB′X* multiplet of cyclic phosphorane derivative **249** may also be obtained directly from the spectrum "expanded" by the SR, and they are in good agreement.[1568] From the splitting, the multiplet of the *A* protons is easy to identify (since the higher coupling constants correspond to *diaxial* couplings, $J_{AX} \approx 11$ Hz, and $J_{BX} \approx 4$ Hz), showing that $\Delta\delta A > \Delta\delta B$. This indicates that the compound is the *trans* isomer, **a**, since the oxygen of the dative P → O bond, the most probable coordination center, is closer to proton *A* in this isomer.

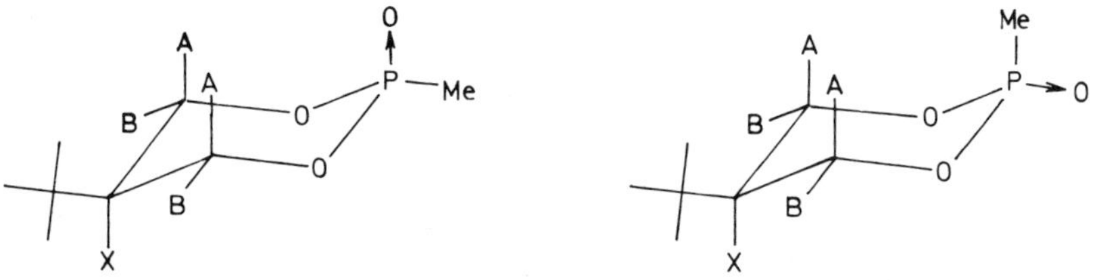

a   **b**

249

Solution of assignment problems — The measurement of the relative magnitude of Δ_i enables one to solve assignment problems, e.g., on the basis of the r^{-3} dependence. The assignment of the three methyl singlets of camphor (**250**) to groups 8, 9, and 10 is unam-

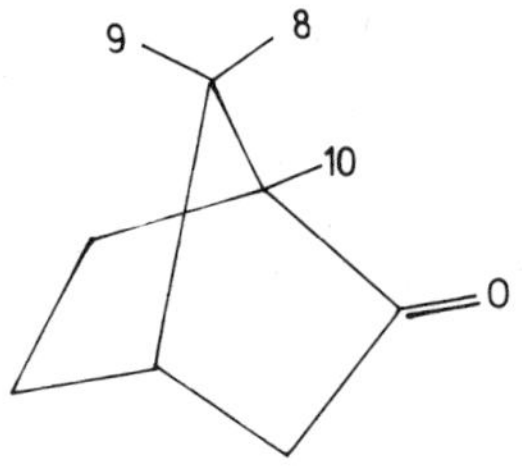

250

biguous on the basis of the predictable sequence, $\Delta_9 < \Delta_8 < \Delta_{10}$, of the shifts caused by the SR.[653] From these shift differences it follows that the original chemical shifts have the sequence $\delta Me_8 < \delta Me_9 < \delta Me_{10}$, as also expected, since the anisotropy of the carbonyl group causes a paramagnetic shift for the nearly coplanar 10-Me group and an opposite shift for 8-Me situated above the plane of the carbonyl bond. The 9-methyl group has a "normal" chemical shift.

Distinction of stereoisomers — The r^{-3} dependence of dipolar shift makes it possible to simply distinguish between the most diverse kinds of stereoisomers, such as cyclic *cis-trans* isomers, geometric (*Z–E*) isomers,[991] s-*cis* and s-*trans* conformers, etc. Their characteristically different Δ_i values or different θ dependences form an easily accessible experimental background. Many examples can be found in monographs[1196,1302] and reviews [33,493-498,673] dealing with SR technique.

As an example, the distinction between isomers **251a** and **251b** is discussed. The signals of H-1 and H-2, 5, 8 may be separated by the decoupling of olefinic protons. In the possession of the assignment, it can be seen that $\Delta\delta H\text{-}2 \gg \Delta\delta H\text{-}5$, and thus the molecule should have Structure **251a**.[786] If in the presence of SR the signals of the isomers are separated sufficiently, the composition of isomer mixtures may also be determined quantitatively.[189,1221]

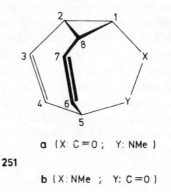

a (X: C=O ; Y: NMe)

251

b (X: NMe ; Y: C=O)

Determination of optical purity, distinction of enantiomers — Perhaps the most fruitful potential application of SR technique is the distinction of enantiomeric nuclei and the enantiomers of optically active substances. The method is principally similar to the use of chiral solvents for the same purpose,* but the chiral medium represented by the optically active SR in the presence of a strong local magnetic field induced by the lanthanides affects the chemical shifts much more than the chiral solvents. Since, in turn, the shift differences between enantiomers in chiral solutions are very small, the best method for the determination of optical purity and the investigation of related problems is just the SR technique. The two main types of chiral shift reagent are the tris {[3-(1-hydroxy-2,2-dimethyl)-propylidene]-*d*-camphor}-europium, **252a**,[1516] its **b** and **c** derivatives,[561,562] and the *d,d*-dicamphoryl-methane

a (R:CMe$_3$; Ln: Eu)   Ln(TBC)$_3$ ≡ Ln(BHC)$_3$

b (R:CF$_3$; Ln: Eu)   Ln(TFC)$_3$ ≡ Ln(FACAM)$_3$

c (R:C$_3$F$_7$; Ln: Eu)   Ln(HFC)$_3$ ≡ Ln(HFBC)$_3$

252

*   See Volume I, p. 41, point 5, and Section 3.7.2 (p. 114-115).

complexes (253)[961] which were originally suggested as the analogue of DPM, but have nowadays been already superseded by SRs 252a to c.

253

The spectrum of enantiomer mixtures separates in the presence of optically active SRs into two components with an intensity ratio corresponding to the relative concentration of antipodes, and all signals corresponding to one of the antipodes shift to a greater extent, i.e., $(\Delta_i)_+ \gtrless (\Delta_i)_-$ for all signals. This was interpreted by assuming that the dissociation constants of the enantiomers are different.[1516] The shift difference $\Delta\Delta_\pm$ measured for the aromatic protons of the enantiomers of 2-phenyl-butan-2-ol (EtPhMeCOH) increases upon the effect of SR 252c up to a mole ratio of $\rho = 0.5$, decreases to $\rho = 1.2$, and then increases again.[562] This shows that the reason for $\Delta\Delta_\pm \neq 0$ is not the differences in dissociation constant, but the different complex stoichiometries of the antipodes (i.e., the relative amounts of 1:1 and 1:2 complexes). Even without this, however, there are obvious differences between the chemical environments, owing to the connection of the chiral centers of substrate and SR (SRLR and SSLR complex species are formed). This phenomenon may be utilized in distinguishing not only enantiomer forms but also *meso-racemic* diastereomeric pairs. The enantiotopic nucleus pairs of the *meso* isomers become anisochronous in the presence of chiral SRs, whereas the chemically equivalent nuclei of *racemic* compounds remain isochronous even in chiral environment. This was utilized in the determination of the configurations of the *meso* (**a**) and *racemic* (**b**) isomers of Compounds 254 and 255, in which the

a *(meso)*   b *(racemic)*

254

   * $\delta A \neq \delta B$ is valid only in chiral environment.

labeled hydrogens give *AB* (**a**) or A_2 (**b**) signal in the presence of Eu(TFC)$_3$.[9,560] The signal of the enantiotopic methylene protons of benzyl-alcohol separates upon the effect of Pr(HFC)$_3$ SR.[510] The two methyl signals of DMSO split in the same manner ($\Delta\Delta\pm = 0.17$ ppm). This indicates that the separation of the signals of the two antipodes is due, instead of the difference between dissociation constants, to the fixed geometry of the complexes (θ-dependence!). The proof for this is that the methylene protons of prochiral benzyl alcohol are anisochronous in the presence of an achiral shift reagent, Eu(FOD)$_3$, too: they give an *AB* multiplet.[550] It can also be proved by partial deuteration that the shift in the signal of the proton in the pro-R configuration is higher. Consequently, this method is applicable for the determination of absolute configurations!

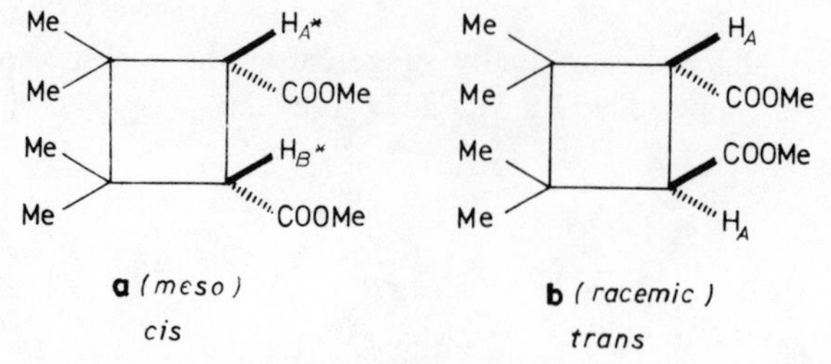

a *(meso)*
cis

b *(racemic)*
trans

255

* $\delta A \neq \delta B$ is valid only in chiral environment.

The application of SRs for studying molecular motions — By combination of SR technique and DNMR method, the determination of the activation parameters of the exchange phenomena of groups giving accidentally isochronous or nearly coincident signals at constant temperature and measuring frequency is possible.[1359] The two *N*-methyl signals of trimethyl carbamate (**256**) are separated at $-23°C$ in $CDCl_3$ solution, but only by 0.03 ppm.[888] Therefore, the activation parameters may be evaluated only very inaccurately. In the presence of $Eu(FOD)_3 \cdot (H_2O)_n$, the *N*-methyl signals separate already at room temperature, to increasing extent with increasing concentration of SR (see Figure 164). Keeping the rate of exchange process constant ($\kappa = 1/\tau$ = constant), i.e., working at constant temperature, the NMR time scale may be varied continuously by means of the SR technique, since the condition of coalescence, $\tau = \sqrt{2}/\pi\Delta\nu$, may be fulfilled at constant temperature and measuring frequency by varying $\Delta\nu_{c,t}$ (the chemical shift difference, in hertz, of the two *N*-methyl signals).

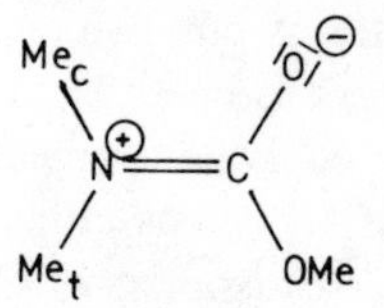

256

Indeed, with gradually increasing SR concentration, first a considerable line broadening occurs, too, with the increase of $\Delta\nu_{c,t}$, but then the diverging *N*-methyl signals become increasingly sharp again. This is illustrated well by a plot of the half-band widths of the *N*- and *O*-methyl signals as a function of SR concentration (more particularly of ρ). One may observe the usual, gradual but weak broadening of the methoxy signal and the sharpening of suddenly broadened *N*-methyl signals, until the signal width becomes equal to that of the methoxy signal and both signals broaden in parallel, at a much lower rate (see Figure 165).

3.7.4. Chemically Induced Dynamic Nuclear Polarization: CIDNP[284,512,530,636,831,841,1204]

The products of certain reactions of radical mechanism show special magnetic properties. In the NMR spectrum of the reaction mixture obtained by common techniques, some of the absorption lines appear with very high (sometimes thousandfold) intensities, and emission lines of similar strength also appear. There are no other changes in the spectrum: the number and position of lines remains the same. This phenomenon is chemically induced dynamic nuclear polarization (CIDNP).

3.7.4.1. Dynamic (DNP) and Adiabatic Nuclear Polarization; Theoretical Interpretation of CIDNP Effect

The phenomenon may be attributed to an unusual population of magnetic energy levels

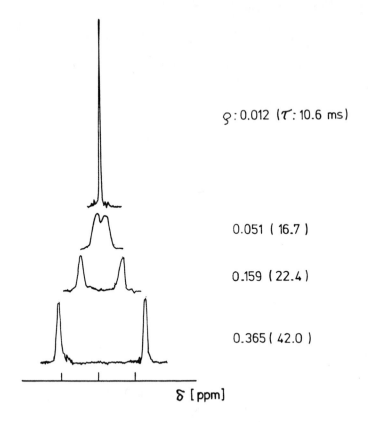

$\varrho : 0.012 \ (\tau : 10.6 \ \mathrm{ms})$

0.051 (16.7)

0.159 (22.4)

0.365 (42.0)

δ [ppm]

FIGURE 164. The *N*-methyl signals of **258** in CCl$_4$, at 27°C, in the presence of varying amounts of SR (as a function of ρ) and the mean lifetime of rotamers as determined by signal shape analysis. (From Sievers, R. E., *Nuclear Magnetic Shift Reagents,* Academic Press, New York, 1971. With permission.)

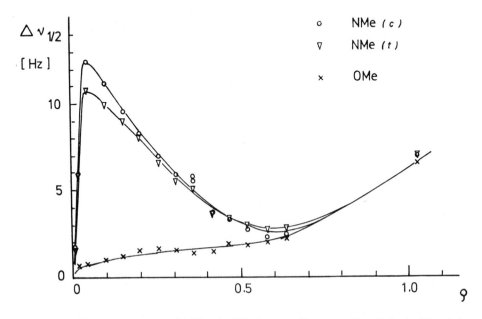

FIGURE 165. The dependence of half-band width ($\Delta\nu_{1/2}$) on SR concentration (ρ) for the ^{1}H methyl signals of **258** in CCl$_4$, in the presence of Eu(FOD)$_3$. (From Sievers, R. E., *Nuclear Magnetic Shift Reagents,* Academic Press, New York, 1971. With permission.)

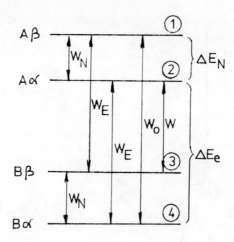

FIGURE 166. Magnetic levels and transitions split by the interaction of an electron spin and a nuclear spin.

(nuclear polarization), which now does not obey the Boltzmann distribution. Nuclear polarization due to electron spin relaxation (dynamic nuclear polarization, DNP), which was regarded as a special case of Overhauser effect, is a well-known phenomenon,[636,1045] and CIDNP was first interpreted in analogous manner.[86,484,829,1473] Its name comes from this interpretation. Accordingly, if in a reaction radicals with infinite spin temperature are formed, they relax in the direction of thermal equilibrium, and during this process nuclear polarization is produced by an exchange of energy which is absorbed by the molecules of the reaction product.

The interaction of an electron spin and a nuclear spin in a polarizing field $\mathbf{B}_0$ produces four energy levels. The level scheme can be seen in Figure 166, where A and B refer to the excited and ground states of the electron spin, and the given sequence of energies is a consequence of the opposite nature of electron and nuclear spins (para- and diamagnetic, respectively). When the free radical is formed, the levels are populated equally and the thermal equilibrium is approached because of the different probabilities of possible transitions:

$$\alpha \leftrightarrow \beta: \quad W_N e^{\pm \Delta E_N};$$

$$A \leftrightarrow B: \quad W_e e^{\pm \Delta E_e};$$

$$A\beta \leftrightarrow B\alpha: \quad W_o e^{\pm(\Delta E_N + \Delta E_e)}$$

and

$$A\alpha \leftrightarrow B\beta: \quad W_o e^{\pm(\Delta E_N - \Delta E_e)} \qquad (300)$$

It can be seen that an irregular nuclear polarization is possible if the probability of combined electron-nuclear spin transitions (W or W_0) is higher than the rate of radical consumption (i.e., the W_R probability of chemical reaction), and the latter is faster than the relaxation of electrons and nuclei: W or $W_0 > W_R > W_e$ and W_N. The sign of nuclear polarization depends on the relative values of W and W_0; when $W > W_0$, the probability of $1 \rightarrow 4$ transitions is low, but that of $2 \rightarrow 3$ is high, therefore levels **1** and **3** become overpopulated, and the probability of $1 \rightarrow 2$ and $3 \rightarrow 4$ nuclear transitions increases: emission (E) of nuclear energy may be observed. When, however, $W < W_0$, levels 2 and 4 become overpopulated (probability of transition $1 \rightarrow 4$ is high and $2 \rightarrow 3$ low); the result is increased absorption (A). The extent of intensity change is a function of the rate of chemical reaction. This has an optimum; since the reaction is too fast ($W_R > W$, W_0), there is no time for the development of irregular polarization, and if it is too slow ($W_R < W_N$, W_e), nuclear polarization is eliminated via relaxations. It can be shown that the maximum increase in nuclear polarization is proportional to the ratio μ_e / μ_N, i.e., it may be $\gamma_e / \gamma_N \approx 660$-fold.

In practice, however, greater absorption enhancement may also be observed, and emission signals thousand times exceeding the intensity of usual absorption signals also occur. The condition $W_R > W_e$ does not hold either, since the relaxations of electrons faster than radical reactions. It also disagrees with the above theory that nuclear polarization occurs not only with radicals but also, and even more frequently, in the spectra of nonradical reaction products. The sign of polarization (A or E) depends on the type of chemical reaction, and in the case of multiplets, the intensities of the individual lines undergo different changes: in one part of the multiplet, emission lines, in the other part, absorption lines of enhanced intensity may be found. These phenomena cannot be interpreted by the DNP model and indicate that the nuclear polarization takes place in the product formation step and not in the radical.

According to the presently accepted theory, the phenomenon stems from the adiabatic interaction of radical pairs in solution (in the solvate cage), which leads to chemical reaction (formation of dimer or unsaturated product) only if both radicals are in singlet electron spin states.[282,748] The interaction of the radical pair gives rise to a singlet (S) and three triplet (T_{-1}, T_0, T_1) electron spin states (spin angular moment is 0 or 1), and the relative transition probabilities between these states are influenced by the nuclear spins. Without any change in nuclear spin states, $S \leftrightarrow T$ transitions may be produced. Therefore, the reactivity (reaction probability) of a given radical pair is a function of the nuclear spin configuration. A nucleus of a given spin quantum number is passed unchanged into the product of radical reaction (from the reagent into the product), and thus the quantum number can be considered as a special sort of labelling of individual atoms. Thus, the intensity changes of the NMR signals reflect the amounts of product with various nuclear configurations.

Although the adiabatic interaction of radical electron spins and the corresponding transitions have low probability (approximately 1%), the resulting change in nuclear polarization is substantial (order of 10^3), since the thermal nuclear polarization is also very low (10^{-5} to 10^{-6}). The interaction of radical pairs is possible with no regard whether the radical pair arises from a common parent molecule (correlated or twin radical pair) or it is produced by the diffusional collision of two radicals formed independently.

The radical pair may react either by "capture" (recombination or disproportionation) or by "escape" (diffusion and reaction with other radical or molecule). Capture reactions may proceed only through the collision of radicals in S state.

If weakly coupled electron spin vectors precess with different velocities, ($v_1 \neq v_2$), $S \leftrightarrow T_0$ transitions are possible. This can be seen from Figure 167 which illustrates the precession of spin vectors.

In state S there is no resulting magnetic moment; in state T_0 the magnetic moment is in the [x, y] plane; the z components of the spin vector cancel one another. States S and T_0 may, therefore, combine only if one of the electron spin vectors rotates around the z axis with respect to the other.[283,830,1472] Since $v = g \, \beta_e B_0 / h$, $\Delta v = v_1 - v_2$ arises in part from the difference in g factors ($g = g_1 - g_2$ is the coupling of electron spin orbitals) and in part from the interaction of the spins of neighboring nuclei and the electron (ESR hyperfine interaction),

$$\Delta v = (\beta_e / h)[\Delta g B_o + \sum_i A_{1i} m_{1i} - \sum_j A_{2j} m_{2j}] \tag{301}$$

where A is the coupling constant characteristic of hyperfine splitting.

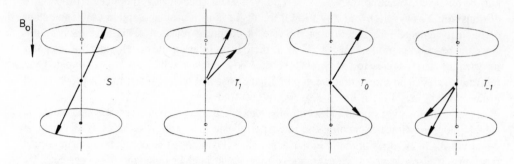

FIGURE 167.   The precession of electron spin vectors in quantum states S, T_{-1}, T_0, and T_1.[1204]

In the free radical $\Delta\nu$ is 10^6 to 10^7 Hz, which makes $S \leftrightarrow T_0$ transitions possible during the lifetime of radical pairs, and thus a significant nuclear polarization may take place. It follows from the above expression that in the product of radical reaction, a given nuclear configuration is enriched. This is easy to see if it is assumed that radical pairs of state S are formed and $A_1 > A_2$. Then the magnitude of $\Delta\nu$ is determined by m_1. If $\Delta g > 0$, then $\Delta\nu$ is lower when $m_1 = -1/2$, i.e., for the case of β-nuclear spins. If, in turn, $\Delta\nu$ is lower, the probability of $S \leftrightarrow T_0$ transitions is also lower and the mean lifetime of the S state of the radical pair becomes higher, and thus the probability of its reaction into "captured" product is higher. Consequently, the nuclei of configuration β are enriched in the disproportionated or recombined product, which leads to the appearance of strong emission line(s) (e.g., see Figure 168).

It is evident that under similar conditions the products of "escape" reactions give stronger NMR absorption signals, since they are enriched in nuclei of α-spin state (in the case of $m = +1/2$, $\Delta\nu$ is higher, the probability of $S \leftrightarrow T_0$ transitions increases, the chances of "capture" reactions decrease, i.e., the probability of diffusion and subsequent "escape" reactions is higher). The nuclear polarization is thus opposite in the products of capture and escape reactions. Nuclear polarization is also opposite when radical pairs of triplet state are formed in the escape product, and emission signals appear in the NMR spectrum. When the radical pairs are produced by diffusional collision (case F), the S and T states have identical statistical weights, and since the S pairs may combine into a product, the probability of T state increases, and the nuclear polarization is the same as in the case of twin radical pairs of T state. Consequently, from the E or A character of NMR signals, conclusions can be drawn on the one hand on the way of product formation (capture or escape) and on the other hand on the electronic state (S or T_0) of the reacting radical pair.

3.7.4.2. The Sign of CIDNP and the Multiplet Effect in CIDNP: Relative Intensities in CIDNP Spectra

On the basis of quantum mechanical considerations, some rules were set up by Kaptein,[747] which can be used to predict the signal of the CIDNP effect. The sign of net nuclear polarization is the product of the signs of the following parameters:

$$P_N = \mu\epsilon\Delta g A_i \tag{302}$$

For the lines of first-order multiplets,

$$P_M = \mu\epsilon A_i A_j J_{ij}\,\sigma_{ij} \tag{303}$$

The positive or negative sign of P_N shows whether total nuclear polarization causes an increase in absorption intensity or the appearance of emission signal, respectively, in the

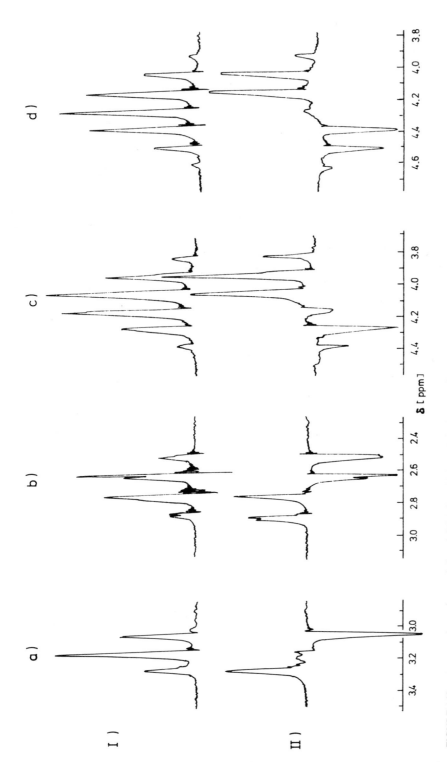

FIGURE 168. A comparison of normal (I) and CIDNP multiplets (II) in the case of (a) an *A/E* triplet, (b) a quartet with fine structure corresponding to higher order splittings, (c) an *E/A* sextet, and (d) an *E/A* septet to illustrate the CIDNP multiplet effect. (From Lepley, A. R. and Closs, G. L., *Chemically Induced Magnetic Polarization*, Wiley-Interscience, New York, 1973. With permission.)

NMR spectrum. The positive or negative sign of P_M is equivalent to the E/A or A/E phase of the lines of the multiplet (E/A corresponds to the case when the downfield lines have emission character, the upfield ones absorption, A/E is the opposite case). μ indicates the state of the parent (precursor) radical pair; it is positive if the precursor is in T_0 state and negative if the precursor is in S state. ϵ pertains to the way of product formation: positive in capture (recombination, disproportionation) reactions and negative in escape reactions. σ_{ij} refers to the interacting nuclei; it is positive if nuclei i and j, participating in the interaction which causes the multiplet, appear in the same radical and negative otherwise. The Δg factors and the hyperfine electron-nucleus coupling constants, A, may be obtained from ESR measurements. It can be shown that gA_i in the expression of P_N has opposite signs for the members of a radical pair. On the other hand, product A_iA_j, important in determining the E/A or A/E phase of the multiplets, depends on the sign of coupling constant J_{ij} only.

Consequently, the structure of the NMR spectra of radical reaction mixtures are predictable, or from experimental spectra the reaction may be characterized. Of course, further facts may also be considered in such cases. It has been shown, for example, that the value of g increases in parallel with the increasing electronegativity of radicals.[483] Thus, in radical rearrangements, where the migrating radical has usually negative polarization, Δg is negative, whereas for the radical pair, $\Delta g > 0$. If, as often occurs, $\Delta g \approx 0$ ($g_1 \approx g_2$), net nuclear polarization is negligible, and a pure multiplet effect may be observed. When, however, the P_N and P_M components are similar in magnitude, the Kaptein rules are inapplicable, and the CIDNP effect corresponding to the experimental spectrum may only be determined by simulation. It is stressed that the Kaptein rules may be used at high B_0 fields only (since at lower B_0 transitions $T_{+1} \leftrightarrow S$ may take place), when the rates of nuclear relaxations are favorable in the radical and the product. In other cases the original quantum mechanical expressions, also derived by Kaptein, must be applied for correct results. In practice, the simplifying conditions are mostly met so that the complicated mathematical treatment may be omitted, and the effect of CIDNP is predictable on the basis of the simple Kaptein rules.

The relative intensities of first-order multiplets subjected to CIDNP (the multiplet effect) are just as easy to obtain as in the case of normal NMR spectra: the binomial coefficients corresponding to the multiplicity of one less are written twice below one another, shifted by one position to right or left, and the coefficients are subtracted.*

Depending on the direction of shift, the CIDNP multiplets of E/A and A/E phase are obtained,[841] e.g., for quintet and sextet signals, we obtain

normal NMR	CIDNP	
	E /A phase	*A /E* phase
1 3 3 1 +1 +3 +3 +1 1 4 6 4 1	1 3 3 1 −1 −3 −3 −1 −1 −2 0 2 1	1 3 3 1 −1 −3 −3 −1 1 2 0 −2 −1
1 4 6 4 1 1 +4 +6 +4 +1 1 5 10 10 5 1	1 4 6 4 1 −1 −4 −6 −4 −1 −1 −3 −2 2 3 1	1 4 6 4 1 −1 −4 −6 −4 −1 1 3 2 −2 −3 −1

It can be seen (see Figure 168) that for odd line number the central line is missing from the CIDNP spectrum.[841] In the presence of higher-order interaction, similar to normal spectra, further splittings occur and the symmetry of intensity distribution may be distorted. If among the nuclei subjected to higher-order coupling, only one is adjacent to the atom bearing the electron with odd spin, the CIDNP effect (emission or enhanced absorption) may also occur

---

* Instead of adding them, see Volume I, p. 48.

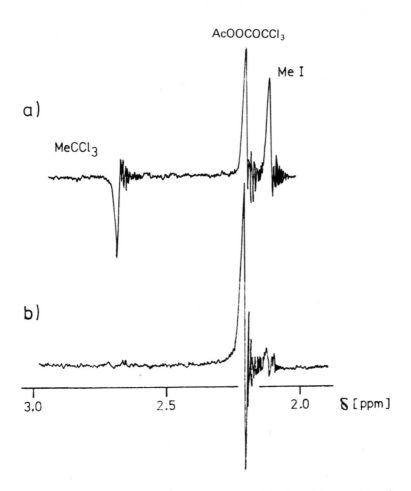

FIGURE 169. The spectrum of the reaction mixture of the thermal decomposition of CH₃COOOCOCCl₃ peroxide, in CCl₄. (a) At 50°C, 5 min after the dissolution of the reactants (peroxide and iodine). (b) The spectrum recorded immediately after cooling the solution to 0°C.[1472]

in the signals of the other nuclei coupled with it. The CIDNP effect is an excellent tool in the determination of the magnetic properties of free radicals and products of radical reactions (the parameters governing the CIDNP effect) in the investigation of radical reactions and studies on radical formation and transformation.

3.7.4.3. Application of CIDNP in Organic Chemistry

Two examples are shown for illustration. The thermal decomposition of acetic acid-trichloroacetic acid mixed peroxide in the presence of iodine yields 1,1,1-trichloro-ethane and methyl iodide (through the formation of methyl and trichloromethyl radicals). The former product is formed via recombination, the latter via "escape":

$$2CH_3-CO-O-O-CO-CCl_3 \xrightarrow{\Delta} 2\dot{C}H_3 + 2\dot{C}Cl_3 + 4CO_2 \xrightarrow{I_2} CH_3CCl_3 + CH_3I + \dot{C}Cl_3 + \dot{I}$$

Theory predicts opposite nuclear polarizations for the products. The experimental spectrum (see Figure 169)[1472] contains, indeed, one emission line at 2.7 ppm, corresponding to tri-chloroethane, one absorption line of about the same intensity at 2.15 ppm, corresponding to methyl iodide, and, of course, the signal of the unreacted peroxide at 2.25 ppm.

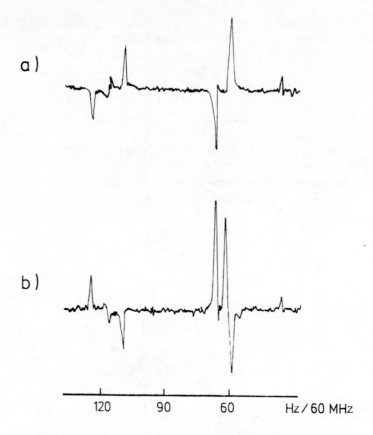

FIGURE 170. The CIDNP spectra of 1,1,2-triphenylethane in toluene solution. (a) Spectrum of Ph_2CH-CH_2Ph produced by a reaction of benzyl and
diphenylmethyl radicals obtained by the thermal decomposition of diphenyl-
methyl-benzyl azide. Spectrum of Ph_2CH-CH_2Ph prepared by a reaction of
triplet carbene $Ph_2\ddot{C}$ and toluene. The zero point of the frequency scale at 60
MHz is the methyl signal of toluene (i.e., 0 Hz corresponds to $\delta = 2.32$ ppm,
see Figure 21). (From Lawler, R. G., in *Progress in NMR Spectroscopy*, Vol.
9, Emsley, J. W., Feeney, J., and Sutcliffe, L. H., Eds., Pergamon Press,
Oxford, 1973, 145. With permission of Pergamon Press, Ltd.)

The second example illustrates the relationship between the multiplicity of electron spin
state and the sign of nuclear polarization. According to the theory, the radicals of S and T
state produce nuclear polarizations of opposite sign. The S and T radical pairs of benzyl
($Ph_2\dot{C}H_2$) and diphenyl-methyl ($Ph_2\dot{C}H$) were prepared.[286,288] The S radical pair was synthesized by the thermal decomposition of benzyl-diphenylmethyl azide ($Ph_2CH-N=N-CH_2Ph$),
and the T analogue by a reaction of triplet diphenylmethylcarbene ($Ph_2\ddot{C}$) and toluene. The
spectra of 1,1,2-triphenyl-ethane (Ph_2CH-CH_2Ph) obtained in the two ways are the mirror
images of one another (see Figure 170), supporting theoretical prediction.

The most common reaction types in which radical mechanism is unusual, and thus the
experimental observation of CIDNP is often probable, are the thermal and photochemical
decomposition of peroxides, the reduction of diazonium salts, rearrangement reactions,
reactions of carbenes, substitution of free radicals onto aromatic molecules, addition of free
radicals, e.g., to nitroso derivatives, photochemical reactions of aldehydes, ketones and
quinones, photooxidation of phenols, decomposition of azo derivatives, biradical reactions,
reaction of metal-organic compounds with aliphatic and aromatic halogen derivatives, radical-
ion reactions, etc.

It is to be noted that the CIDNP effect has been studied not only in proton resonance, several papers report the observation of this effect in ^{13}C, ^{15}N, ^{19}F, ^{31}P, and other resonances.[1204] In ^{13}C NMR spectra, owing to the high gain in intensity, the acquisition of only one transient may be sufficient for obtaining satisfactory signal-to-noise ratio. A further advantage is that nuclear polarization of ^{13}C is much stronger than that of proton, since in carbon the unpaired electron is closer to the nucleus. CIDNP spectra observed in carbon resonance have the same advantages as ^{13}C NMR spectra is general: the proton decoupled singlets are easier to assign in accordance with the relatively larger differences in shift.

^{19}F nuclei, situated far from the lone electron, also have CIDNP effect, i.e., the effect is observable for so large atomic distances where it is does not occur in the ^{1}H NMR spectrum. (This is presumably due to π-p interactions.) The CIDNP effect of ^{31}P NMR nuclei, owing to the high hyperfine coupling constants A, gives a chance for studying the role of this parameter in the CIDNP effect.

Chapter 4

THE RESONANCE SPECTRA OF NUCLEI OTHER THAN HYDROGEN

INTRODUCTION

All nuclei possessing magnetic moments give rise to NMR spectra.* The study of resonance spectra of nuclei other than protons had not been very significant up to the past decade. This was due to the experimental difficulties and extra expenses of observing the NMR spectra of other types of nuclei. The main source of the experimental difficulties** is the small sensitivity of most magnetic nuclei as compared to protons involving an unfavorable signal-to-noise ratio ζ.

At a given temperature the sensitivity is given by:[1216]

$$\zeta = CI (I + 1) \gamma^3 NB_0^2 \tag{304}$$

The factors determining the magnitude of ζ are the following:

The spin quantum number — As the spin of the most important magnetic nuclei (^{13}C, ^{14}N, ^{15}N, ^{17}O, ^{19}F, and ^{31}P) — apart from the isotopes ^{14}N and ^{17}O — is 1/2, the relative sensitivity is not influenced by the spin quantum number, although for ^{14}N and ^{17}O with $I = 1$ and 5/2, the sensitivity increases by about 2.5 and 12, respectively. However, nuclei with $I > 1/2$ always have quadrupole moments, and this leads to too short relaxation times and therefore to broad signals. Line broadening involves a decrease of several orders in relative sensitivity, therefore, larger spin quantum number is rather a disadvantage from the viewpoint of ζ.

The magnetogyric factor — This factor — and thereby the magnetic moments of the various nuclei — varies also within one order (see Table 1), but as it appears on the third power in Equation 304, it has a decisive influence upon sensitivity, which is always smaller than that of protons, especially in the case of the isotopes ^{13}C, ^{14}N, ^{15}N, and ^{17}O. For ^{15}N the sensitivity is reduced by two orders.

The number of nuclei in the sample (concentration) — Because of the small natural abundance (compare Table 1) of certain magnetic nuclei (^{13}C, ^{15}N, ^{17}O), the sensitivity for their resonance drops by several orders.

The magnitude of the polarizing field — The introduction of superconducting magnets has made it possible to reach fields of 12 to 14 T instead of 1.4 to 2.4 T of the CW spectrometers operating with electromagnets. This involves a substantial improvement in the signal-to-noise ratio, due to the B_0^2 term in Equation 304. The signal-to-noise ratio can be improved by increasing the excitation field B_1, but this is limited due to saturation (depending on T_1).***

The relative sensitivities as determined by the above factors are given in Table 1, assuming constant polarizing field, temperature, and the same concentrations, the sensitivity of protons taken as unity. It can be seen, that — except for fluorine — the relative sensitivities are about one to three orders smaller than for protons.

Owing to dramatic developments in the past decade in NMR techniques (FT method and superconducting magnets, etc.), routine spectrometers have appeared on the market for the measurement of nuclei other than proton and fluorine, and primarily ^{13}C NMR spectroscopy is rapidly gaining space; they have become routine methods of chemical structure elucidation.

---

*   Compare Volume I, p. 6.
** Compare Volume I, Table 1, p. 9.
*** Compare Volume I, p. 22 and Equation 60.

In the following sections, a brief discussion of the main achievements of studies on ^{13}C, ^{14}N, ^{15}N, ^{17}O, ^{19}F, and ^{31}P nuclei will be given.

4.1. CARBON RESONANCE SPECTROSCOPY

Since the spin quantum number of the ^{13}C isotope is 1/2, it gives rise to sharp resonance signals. However, its natural abundance is only 1% and its magnetic moment is small as well; the relative sensitivity of the ^{13}C with respect to proton is only 1/6500.[996] Due to these facts, ^{13}C resonance was first observed in only 1957 by Lauterbur,[816] and ^{13}C NMR turned into routine method only at the beginning of the 1970s, with the introduction of FT spectrometers (compare Section 2.2).

The difficulties were first overcome by increasing the ^{13}C content of samples by tiresome and expensive enrichment or by recording the dispersion curve, which is less sensitive to saturation (compare Figure 8), instead of the usual absorption curve. In the majority of cases, the quality of the spectra was very poor. If the sweep* (change in magnetic field** during measurement) is fast enough (fast adiabatic sweep),[817,1227] the dispersion curve is distorted and becomes similar to an absorption curve. By recording the distorted curve with reversed sweep, too, the "negative" spectrum was also measured (see Figure 171a).[820] As chemical shift, the average of the not completely identical maxima observed in the two curves was accepted. This method partly eliminates the error arising from the fast sweep and from the distorted curve. This is now only of historical interest, since at the end of the 1960s the much more sensitive FT spectrometers appeared, immediately solving the above problems. The development is illustrated by a comparison of the CW spectrum of phenol in Figure 171a with the FT spectrum (see Figure 171b).[729]

The next sections deal with the most important results of ^{13}C NMR spectroscopy.[196,729,855,1042,1043,1053,1143,1147a,1372,1410,1492] We restrict the discussion mostly to the recent results because a significant fraction of earlier data has been made obsolete by the development in instrumentation.

Before turning to the detailed discussion of ^{13}C NMR spectrum parameters, it is worth assessing the advantages of carbon resonance studies and pointing out the fields where they give excess information primarily with respect to ^{1}H NMR spectroscopy:

1. The chemical shift range of carbon resonance signals is about 20- to 30-fold of the proton shift range, and thus the signals of individual carbon atoms and the characteristic shift ranges of various carbon types overlap much less frequently, since line widths are about the same.
2. The ^{13}C NMR spectrum gives direct information on the carbon skeleton of the molecule.
3. Several effects (e.g., the inductive effect of substituents) which are very weak in proton resonance owing to the interposed carbon atoms are stronger and observable with less interference.
4. For compounds with many similar protons, ^{13}C NMR spectra are more informative than the featureless ^{1}H NMR spectra, e.g., the ^{1}H NMR spectrum of simpler paraffins and cycloalkanes is often a single, broad absorption maximum, whereas in the ^{13}C NMR spectrum, the signals of the individual carbons are well separated (e.g., compare Figure 173).
5. Functional groups that bear no protons (C=O, C≡N, etc.), on which ^{1}H NMR data give at most only indirect information, may still have ^{13}C NMR signals, e.g., the ^{1}H NMR spectrum of NCCH$_2$COCH$_2$CN molecule is one singlet, whereas there are three lines with characteristic frequencies in the ^{13}C NMR spectrum.

---

*   See Volume I, p. 138.
** In a field of 2.4 T, the ^{13}C resonance line is at 25.1 MHz.

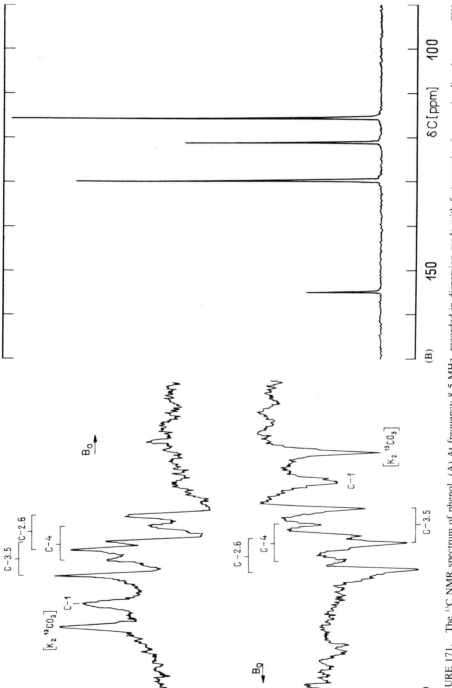

FIGURE 171. The ^{13}C NMR spectrum of phenol. (A) At frequency 8.5 MHz, recorded in dispersion mode with fast scanning in opposite directions on a CW spectrometer.[820] (B) At 25.16 MHz on a VARIAN® XL-100 FT spectrometer. (From Johnson, L. F. and Jankowski, W. C., *C-13 NMR Spectra*, Wiley-Interscience, New York, 1972. With permission.)

6.   Since in the majority of molecules there are fewer carbon atoms than hydrogens, large molecules have generally too complex and thus hardly interpretable ^{1}H NMR spectra, while their ^{13}C NMR spectra are usually simpler.

7.   The carbon resonance spectrum is not complicated by homonuclear couplings, also causing much simpler structure. Owing to the low natural abundance of ^{13}C isotope, it has a very low probability that two neighboring carbons of the same molecule are both ^{13}C isotopes.

8.   The splittings due to C–H couplings may be eliminated simply by means of BBDR (compare Section 2.2.5.5) leading to a further simplification of ^{13}C NMR spectra.

In ^{1}H NMR the three types of information obtained from the spectrum, i.e., the chemical shift, coupling constant, and intensity, are equally important in structure elucidation. In ^{13}C resonance chemical shift is of central importance, since by routine measurements coupling constants may not be determined, and there is no unique relationship, like in ^{1}H NMR, between signal intensity and the corresponding number of equivalent nuclei.*

4.1.1. Theoretical Interpretation of Carbon Chemical Shifts

4.1.1.1. Chemical Shifts in Carbon Resonance Spectra

The ^{13}C NMR spectra of organic compounds embrace a range of about 220 ppm (taking into account extreme cases, too, this range extends to 600 ppm). As reference, nowadays TMS is applied almost exclusively (in aqueous and DMSO-d_6 solutions DSS). With respect to this signal most chemical shifts are positive. Formerly, carbon disulfide was most frequently applied as external reference.** For the recalculation of older literature data, the chemical shifts of some solvents and formerly widespread references[847,1143a] are given on the δ-scale in Table 49. The solvent signal as secondary reference may be employed similarly to the ^{1}H NMR technique. A survey of carbon resonance chemical shifts is given in Table 50, which lists the chemical shift ranges characteristic of various compound types.[191,855,1143,1492]

4.1.1.2. The Effect of the Physical Properties of Carbon Nuclei on Shielding

Shielding constant σ and thus the chemical shift of magnetic nuclei is a resultant of various factors, with widely different relative importance for the different types of nuclei. Consequently, the considerations for protons (compare Section 1.2) may not be transferred without change to carbon nuclei. In order to have a better view of the special properties of carbon resonance shielding phenomena, the theoretical aspects of the problem must be touched briefly.

A quantitative interpretation of nuclear shielding was first attempted by Ramsey,[1164] and his equation is the basis of the various quantum chemical methods developed since then:

$$\sigma_{vw}^{A} = \sigma_{vw}^{d} + \sigma_{vw}^{p} = \frac{1}{2c^2} <\psi_0 \,|\, \sum_{j} (r_j\, r_j^{A}\, \delta_{vw} - r_{vj}\, r_{wj}^{A})\, (r_j^{A})^{-3}\,|\, \psi_0 > -$$

$$-\frac{1}{2c^2} \sum_{k \neq 0}^{\infty} \Delta E_{ok}^{-1}\, [<\psi_0 \,|\, \sum_{j} \mathcal{L}_{vj}\,|\, \psi_k> <\psi_k\,|\, \sum_{l} \mathcal{L}_{wl}^{A}\, (r_l^{A})^{-3}\,|\, \psi_0> +$$

$$+ <\psi_0\,|\, \sum_{j} \mathcal{L}_{vj}^{A}(r_j^{A})^{-3}\,|\, \psi_k> <\psi_k\,|\, \sum_{l} \mathcal{L}_{wl}\,|\, \psi_0 >] \qquad (305)$$

Equation 305 is a modern form of the Ramsey equation[369] in which σ_{vw}^{A} is one of the components of the shielding tensor of carbon nucleus A (v and w represents a pair of Cartesian

---

* Compare Volume I, p. 24 and 142.

** Compare Volume I, p. 43, second footnote.

Table 49
THE ¹³C NMR SHIFTS (PPM) OF COMMONLY USED SOLVENTS AND REFERENCES[847,1143a]

Solvent, reference	Chemical shift		Solvent, reference	Chemical shift	
	Light isotope	Perdeuterio solvent		Light isotope	Perdeuterio solvent
TMS	0	—	Dioxane	67.4	—
CH₃CN	1.6	1.2	CHCl₃	77.2	76.9
CHBr₃	10.2	10.2	CCl₄	96.0	—
C₆H₁₂	27.5	26.1	CH₃CN	117.8	117.8
(CH₃)₂O	30.4	29.2	C₆H₆	128.5	128.0
(CH₃)₂SO	40.5	39.6	CS₂	192.8	—
CH₃OH	49.9	49.0	CS₂	193.7[a]	—
CH₂Cl₂	54.0	53.6	(CH₃)₂CO	205.4	205.7

[a] As external reference in capillary.

Table 50
CHARACTERISTIC ¹³C NMR SHIFT RANGES (PPM) OF VARIOUS FUNCTIONAL GROUPS[190,191,855,1143,1492]

Functional group	Chemical shift range	Functional group	Chemical shift range	Functional group	Chemical shift range
>C< [a](sp³)	0—10	–CH₂N(sp²)	35—75	>C$_{HAr}$–C(sp²)	100—140
CH₃–	0—30	>CHCl	35—85	–SCN	110—120
CH₃S–	5—20	>CHN(sp²)	40—90	–C≡N	115—125
CH₃–C(sp²)	5—30	CH₃O–	45—65	–NCO	115—130
–CH₂–	10—70	C–C(sp²)	45—100	>C$_{HAr}$–N(sp²)	120—170
–CH₂S–	15—60	C–N–(sp²)[b]	50—100	–NCS	125—140
>CH–	15—70	–CH₂O–	55—90	>C=N–	145—170
–CH₂–C(sp²)	20—60	>C=C=C<	65—95	–NC	150—170
>CX / >C–	20—100 / 20—100	>CHO– / CCl[b]	65—100 / 65—110	–COX, –CON –COOR	150—180
CH₃N(sp²)	25—50	C–O–[b]	70—110	–COOH	160—185
>CHS–	25—75	CH₂=C(sp²)	80—135	–CHO	185—210
>CH–C(sp²)	30—70	>C$_{Ar}$H	80—140	–CSO, –CSN	190—210
C(sp³)[b]	30—80	>C(sp²)	80—160	=C=	200—220
>CS–	30—90	–O–C(sp³)–O–	85—110	–CO–	205—220
–CH₂Cl	35—55	>C$_{Ar}$–	90—160	–CS–	220—240

[a] In cyclopropyl ring.
[b] Quaternary carbon atom, Ar: aromatic; HAr: heteroaromatic.

coordinates), ψ_0, ψ_1, . . . ψ_k, ψ_l . . . are the eigenfunctions of electronic states, r_j is the distance from an arbitrary origin, r_j^A and r_{jl} are the distances of electron j from nucleus A or electron l, respectively, δ_{vw} is a component of Krönecker delta,* ΔE_{0k} is the energy difference between the ground state and the kth excited electronic state, $\mathscr{L}_j$ and $\mathscr{L}_l$ are the vector operators of the angular moments of electron orbitals j and l: $\mathscr{L}_j = r_j \times \nabla_j$ and $\mathscr{L}_j^A = r_j^A \times \nabla_j$. It is noted that shielding factor σ is a tensor, i.e., it has nine (six independent) components, but in dilute liquids and gases, owing to the fast molecular motions, only the average of diagonal elements may be observed:

$$\overline{\sigma^A} = (\sigma_{xx}^A + \sigma_{yy}^A + \sigma_{zz}^A)/3 \tag{306}$$

The calculation of carbon resonance shifts on the basis of the Ramsey Equation 305 has two main difficulties. On the one hand, this equation contains the ground and excited state electronic energies of the molecule, and neither they nor the eigenfunctions belonging to these states are known for most of the molecules. On the other hand, the shielding constant, i.e., the chemical shift is a result of a sum of two factors of opposite sign, and as this is a difference of two large, almost equal numbers, large errors may occur.

In order to overcome these difficulties, shielding is decomposed into components by localizing electronic motions:[1231]

$$\sigma_A = \sigma_A^d + \sigma_A^p + \sum_{B \neq A}^{N} \sigma AB \tag{307}$$

where σ_A is the shielding factor of nucleus A in a molecule composed of N atoms, indexes d and p denote dia- and paramagnetic contributions arising from the electrons of atom A, and σ_{AB} is the effect of the electrons of neighboring atoms.

Local diamagnetic shielding arises from the electrons around the nucleus, and in the case of spherically symmetric electron distribution (free atoms), it is the only component of shielding. If the electron density around the nucleus (i.e., shielding) increases, σ^d also increases, and the resonance signal is shifted diamagnetically (toward lower values of δ).

The σ^d contribution, which is described by the Lamb formula (Equation 71) is of decisive importance for protons, whereas for other atoms, including carbon, it is less important. According to approximate theoretical calculations,[1356] a π-electron causes a shielding of 160 ppm for a carbon atom, while the σ^d contribution within this shielding is only 14 ppm. Taking also into account that the $-I$ effect of substituents may cause only about 30% change in the electron density of one electron,[1575] the σ^d component may cause at most 4 ppm change in δC, which is very small regarding the 200 ppm range of carbon chemical shifts. Since, however, the increase in electron density also decreases the σ^p component of opposite sign (see below) and both effects increase shielding, often there is a linear relationship between shielding and electron density.

The σ_{AB} component arising from the anisotropy of the magnetic susceptibility of neighboring groups, the second most important term after σ^d for protons, which also includes the effect of delocalized electrons (e.g., ring currents), is negligible in carbon resonance. The σ_{AB} contribution is a function of molecular geometry only (described by the McConnel Equation 72), and thus the calculation pertaining to protons[1126] may be transferred without change. According to the results, σ_{AB} does not exceed some parts per million. Accordingly, the main contribution of carbon chemical shifts must be the paramagnetic term σ^p, which is negligible in proton resonance.**

---

* Compare Volume I, p. 82.
** Compare Volume I, p. 31.

Except for hydrogen, all nuclei have a considerable shielding contribution from the paramagnetic component due to the mixing of ground and excited electronic states induced by the magnetic field. This represents the restricted electronic motions arising from the molecular environment, i.e., from the chemical bonds (the effect of free electronic motion is represented by the σ^d term).

In the LCAO-MO (linear combinations of atomic orbitals - molecular orbitals) theory,[1127] the simplest expression of σ^p is [756]

$$\sigma^P = \frac{e^2\hbar^2}{2m_e^2c^2\,\Delta E\,<r^3>}\,(Q_{AA} + \sum_{A\neq B} Q_{AB}) \tag{308}$$

where ΔE is the average excitation energy of electronic states. The application of this average eliminates the difficulty connected with Equation 305, i.e., the use of the eigenfunctions of electronic states, but it introduces a new empirical parameter.[969] Factor $<r^3>$ depends on the mean distance $<r>$ of $2p$ orbitals, and the expression in the parentheses represents the elements of the nonperturbed charge density - bond order matrix (in the absence of magnetic field).

The factor $<r^3>$ depends primarily on the effective nuclear charge around A. If the electron density around A increases, $<r^3>$ increases, σ^p decreases, and thus shielding increases, i.e., δC_A decreases linearly with π-electron density.[1066] This is the most important factor of carbon shifts, being influenced by the inductive effects of substituents, the extent of electron delocalization, and steric factors as well.

The decrease in ΔE, through the increase of σ^p, causes deshielding. Therefore, in unsaturated and conjugated systems, δC increases (the energy of $\pi \to \pi^*$ transitions is much lower than ΔE of transitions involving saturated systems). Similarly, the decrease of ΔE is responsible for the parallel change in chemical shift with the order of saturated carbon atoms (the delocalization of σ-electrons increases with the number of substituents, and ΔE of the corresponding $\sigma \to \sigma^*$ transition decreases).

The value of Q_{AA} is 2 if the charge density is unity in all $2p$ orbitals, which is a quite reasonable assumption for hydrocarbons.[1372a] Q_{AB} corresponds to the contribution of multiple bonds, since it is zero for σ-bonds, i.e., its magnitude reflects the relative importance of σ- and π-bonds.

The parameters $<r^{-3}>$, ΔE^{-1}, Q_{AA}, and Q_{AB} are mutually interrelated, too, but for a qualitative estimation of the shielding of carbon atoms, they may be considered separately. The relative shieldings of hydrocarbons calculated theoretically from ΔE and ΣQ_{AB}, for example, are close to the experimental data (within this series $<r^{-3}>$ and Q_{AA} are constant, and thus they may be disregarded).[943,1372b]

Although since the early attempts of Ramsey and Pople quantum chemical methods have become much more refined and large computers enable formerly impossible calculations to perform, the chemical shifts may still not be determined theoretically, except for the simplest or smallest molecules, to an accuracy reasonably approaching that of experimental data. Therefore, carbon resonance shifts will be discussed in terms of structural parameters — substituent effects — familiar to chemists. These terms are suitable for the simple interpretation of shifts, and for the application in structure elucidation. However, in order to understand the basic reasons for the phenomena and thus judge the conditions of applicability of empirical rules, wherever possible, the roles of physical properties discussed here will be pointed out.

4.1.1.3. The Effect of Structural Parameters on the Shielding of Carbon Nuclei

The shielding of carbon nuclei in a given environment, i.e., the chemical shift, may be correlated with certain parameters characteristic of chemical structure. It is worth determining

Table 51
THE ^{13}C NMR SHIFTS
(PPM) OF SOME
257-TYPE
COMPOUNDS[316,904,1496]

	Chemical shift	
X	δCH	δCH$_2$
CH$_2$	—	−2.8
S	—	18.7
NMe	—	28.5
O	—	39.5
CHI	−20.2	10.3
CHBr	13.9	8.9
CHNH$_2$	23.9	7.3
CHCl	27.3	8.9

chemical shifts as a function of these parameters instead of abstract theoretical terms, since thus direct and clear relationships are obtained between shifts and chemical structure which, although theoretically not always strict, may be applied successfully for the estimation of relative chemical shifts, i.e., for the solution of assignment and structure elucidation problems. The next sections will deal with the most important structural properties that determine chemical shifts.

Hybridization state of carbon — As follows from theory, the primary factor determining the chemical shifts of carbon atoms is the hybridization state of carbon.* Like in proton resonance, $\delta C(sp^3) < \delta C(sp) < \delta C(sp^2)$; e.g., the chemical shifts of ethane, acetylene, and ethylene are 5.7, 71.9, and 122.1 ppm, respectively.

Order of carbon atom — In the case of identical substituents (e.g., methyl, chlorine),[882,1354] the chemical shift increases with the order of carbon atom, again in parallel with proton resonance: $\delta CH_3R < \delta CH_2R_2 < \delta CHR_3 < \delta CR_4$, except for Br, I, and CN substituents (see below).

Electron density around carbon, — *I* effect of substituents — Although the resultant shielding around carbon atoms is primarily determined by the paramagnetic contribution,** a drastic reduction in diamagnetic shielding (i.e., in electron density) in carbonium ions of type $R_3C^\oplus$ leads to extremely high (approximately 330 ppm) chemical shifts.[1069] Within a given family of compounds, where the order and bonding character (hybridization state) of the substituted carbon atom are roughly the same and no steric interactions occur, the chemical shift increases when a hydrogen of the carbon atom is replaced by various electronegative substituents or heteroatoms, and the change is approximately proportional to the electronegativity of the substituents or the heteroatom. The substituent effect proportional to the electronegativity (−*I* effect), which may also be observed in proton resonance spectroscopy, is illustrated by the data of Table 51[316,904,1496] (compare also Problems **68, 70, 71, 72, 75,** and **76**), which show the ^{13}C chemical shifts of some three-membered cyclic compounds (**257**). The correlation between δC and the electronegativities of the substituents is the basis

CH$_2$———CH$_2$
\\ X /

257

---

*   Compare Equation 308.
**   Compare p. 146.

Table 52

SUBSTITUENT CONSTANTS (PPM) DERIVED FROM ^{13}C NMR CHEMICAL SHIFTS OF MONOSUBSTITUTED ALKANES FOR ESTIMATION OF CHEMICAL SHIFTS OF POLYSUBSTITUTED DERIVATIVES[427,1101,1143b,1492a]

Substituents	ρ_α	ρ_β	ρ_γ	ρ_δ	Substituents	ρ_α	ρ_β	ρ_γ	ρ_δ
I	−7.0	11.0	−1.5	−1.0	NH₂	29.0	11.5	−5.0	0
CN	3.0	2.5	−3.0	−0.5	COR	30.0	3.0	−3.0	0
C(sp)	4.5	5.5	−3.5	−0.5	CHO	31.0	0	−3.0	0
C(sp³)	9.0	9.5	−2.5	0.5	Cl	31.0	11.0	−5.0	−0.5
SH	11.5	12.0	−3.0	0.5	NC	31.5	7.5	−3.0	0
Br	19.0	11.0	−4.0	−1.0	COCl	33.0	2.5	−3.5	0
C(sp²)	19.5	7.0	−2.0	0.5	NHR	37.0	8.0	−4.0	
COOH	20.0	2.0	−3.0	0	SOR, SO₂R	42.0	−0.5	−3.0	0.5
SR	21.0	6.5	−3.0	0.5	NR₂	42.0	6.0	−3.0	
CON(sp³)	22.0	2.5	−3.0	−0.5	OH	49.0	10.0	−6.0	0
Ph	22.0	9.5	−2.5	0.5	OAcil	54.0	6.0	−5.0	0
COOR	22.5	2.0	−3.0	0	SO₂Cl	55.0	3.5	−3.0	0
SCN	23.0	9.5	−3.0	0	OR	58.0	8.0	−4.0	
COO⊖	25.0	3.5	−2.5	0	NO₂	63.0	4.0	−4.5	−1.0
NH⊕₃	26.0	7.5	−4.5	0	F	70.0	8.0	−6.0	0

of several empirical rules used for the approximate calculation of carbon resonance chemical shifts.[220,855,912,1156,1242,1410,1492,1567] Of course, the substituents have the strongest effect on the signal of directly adjacent carbon atom C_α. The effect is also substantial on the next carbon, C_β, occasionally commensurable with the α-effect, and, if the latter is small, even exceeds it (like in the first five rows of Table 52, see above). The effect on nuclei in the γ- and δ-positions is much smaller. In addition to the role of inductive effect, it may also be seen from the data of Table 51 that, again like in ^{1}H NMR, three-membered rings show much smaller chemical shifts than the corresponding open-chained analogous (compare Table 50 and Problem 78) or larger rings. Substituent effects are additive in carbon resonance, too. Thus, for example, comparing the chemical shifts of monosubstituted normal alkanes with those of unsubstituted analogues, substituent parameters may be derived, which can be used then to estimate the shifts of polysubstituted derivatives with acceptable (5 to 10%) accuracy:

$$\delta C = -2,3 + \Sigma\rho_i - \xi_{ij} \tag{309}$$

where the values of ρ_i can be found in Table 52 and ξ_{ij} are steric correction factors,[1143c] in which i is the order of the carbon atom in question and j is the highest order of the adjacent carbon atoms. For secondary carbons, with the exception of Cl, Br, I, N⊕H₃, CN, and SH substituents, the values of ρ_α and ρ_β should be decreased by 6 and 2 ppm, respectively. The value of ρ_γ changes by 2.0, −4.0, or −1.0 for fixed anti and syn periplanar or for gauche conformation, respectively. The inclusion of ξ_{ij} steric correction is necessary for substituents O−, N=, ⊕N(sp³), S−, SO−, C(sp³), and C(sp²). They are $\xi_{1,3} = -1.0$, $\xi_{1,4} = -3.5$ $\xi_{2,3} = -2.5$, and $\xi_{2,4} = -7.5$ for primary ($i = 1$) and secondary ($i = 2$) carbons. The shift of a tertiary carbon is already decreased by a secondary neighbor: $\xi_{3,2} = -3.5$, $\xi_{3,3} = -9.5$, and $\xi_{3,4} = -15.0$, and, finally, for quaternary carbons $\xi_{4,1} = -1.5$, $\xi_{4,2} = -8.5$, $\xi_{4,3} = -15.0$, and $\xi_{4,4} = -25.0$ ppm.

It can be seen from Table 52 that the analogous I, Br, Cl, and F derivatives or the derivatives substituted with S, N, or O atoms have gradually increasing δC_α chemical shifts.[1354] It is, however, also clear that only the δC_α values are proportional to the $-I$ effect; the change in shift on the β- and γ-carbons is practically constant, i.e., neither the theory assuming a gradually decreasing positive polarization nor that assuming alternating polarization, as a consequence of $-I$ effect* are in agreement with the experimental results. The δC_β and δC_γ shifts are, therefore, caused by different effects (β- and γ-effect).

Field effect (steric compression shift) — The theoretical interpretation of the β-effect discussed in the previous section and shown in Table 52 is not yet clear, it is probably a result of several contributions. However, the γ-effect, most frequently cited as *field effect* (or steric compression shift), is well interpretable and provides the key for the solution of stereochemical problems by means of ^{13}C NMR (compare Problems **67, 74, 75, 76,** and **77**). The field effect substantially decreases, even by 20 ppm,[327,579] the chemical shifts of carbon atoms attached to atoms or groups in steric proximity. Field effect arises in part from the interaction of electrons of adjacent groups and in part from a minor distortion in molecular geometry, bond lengths and bond angles. Qualitatively, the field effect may be interpreted as the polarization of C–H bond leading to an increase in electron density around the carbon atom (orbital expansion increases shielding.) Field effect may always be expected when two carbon atoms also bearing hydrogens are in γ-*gauche* (**258a**) position. When a carbon takes part simultaneously in more such interactions, the field effects are additive. The effect was first reported and interpreted by Grant et al.,[328] who determined the field effect for some rigid systems and found that a 2- to 6.5-ppm increase in shielding corresponds to one

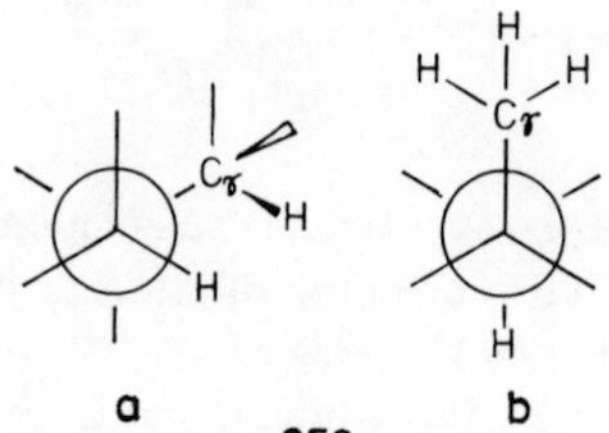

258

interaction. The range is relatively broad since the molecules may avoid this steric compression to various extents. In *trans* 1,2-dimethylcyclohexane the mutual field effects of the two *equatorial* methyl groups decrease the shift of the methyl carbons by 2.6 ppm. In the case of three adjacent *equatorial* methyl groups (**259**), the signal of the middle carbon atom shifts by 6.2 ppm (thus, one interaction corresponds to a shift of 3.1 ppm). On the terminal methyl

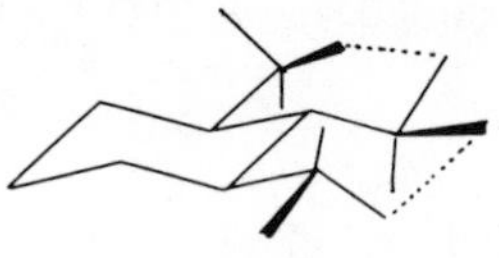

259

signals, the effect is 2.1 ppm, slightly less than for *trans* dimethylcyclohexane. The *axial* methyl groups of methylcyclohexanes are close to two (H-3 and H-5) *syn-axial* hydrogen atoms (**260**): their signal is shielded by 4 ppm (the fraction of field effect corresponding to

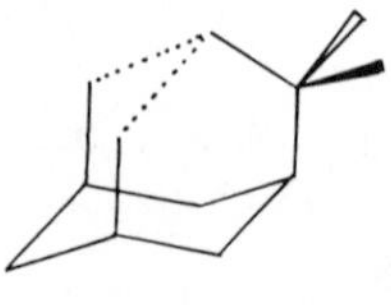

260

---

* Compare p. 172.

one interaction is therefore 2 ppm). The C-3,5 signals show much stronger effect, -6.3 ppm (due probably to the fact that they can not avoid steric compression). On the methyl signal of *cis* 1,2- and 1,1-dimethylcyclohexane the field effect is 11.4 and 8.4 ppm, respectively (3.8 and 4.2 ppm corresponding to one interaction), and a same effect is observable on the C-3,5 atoms (4.2 ppm). In *cis-cis* 1,2,3-trimethylcyclohexane (**261**), the signal of

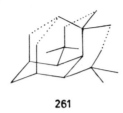

261

the middle, *axial* methyl group is shielded by 17.8 ppm, owing to the four, superimposed interactions. The proximity of the *endo-axial* hydrogens of *cis* anellated fused cyclohexane derivatives (**262a**) leads to a field effect of 8.5 ppm in rigid systems. The above data pertain to *rigid* systems. Field effect may be observed for flexible systems as well, but the measured chemical shifts are the weighted averages of larger unperturbed and smaller perturbed shifts; the higher the probability of sterically compressed conformation, the closer the measured shift to the minimum characteristic of the rigid, sterically compressed structure; e.g., in *cis* decaline, the carbons compressed in one of the structures (**262a**, full circles) get by ring inversion into noncompressed positions (**262b**). Therefore, the field effect observed for the decaline isomers (5.2 ppm) is much lower than the effect (8.5 ppm) characteristic of rigid systems mentioned above.* If the hydrogen is replaced by a heteroatom, the field effect causes 1 to 3 ppm higher increase in shielding;[1026] e.g., in the case of 4-*t*-butyl-cyclohexanol or 4-*t*-butyl-chlorocyclohexane the effect increases to approximately 6 from 4 ppm characteristic of the methyl analogue.

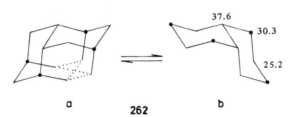

a   b
262

Field effect has an important role in carbon resonance spectroscopy. The ^{13}C NMR shifts are extremely sensitive to steric changes in the molecule, and thus the interpretation or field effects is of basic importance in the solution of the most diverse structure elucidation problems. This will be illustrated by many examples;** e.g., in 1-haloalkanes δC_γ increases in the order Cl, Br, I (the substituent effects are -4.6, -3.1, and -1.0 ppm), since the probability of γ-*gauche* conformation decreases in the same order (higher atomic radius causes stronger steric compression), together with the field effect which would decrease the chemical shift.[1372]

Hyperconjugation — Similar to the field effect, the heteroatoms may exert shielding not only in the γ-*gauche* but also in the *trans-periplanar* (**258b**) conformation.[423] This shielding is $1.5 - 7.5$ ppm in the case of N, O, and F atoms, but small (<0.5 ppm) for S and Cl atoms. The shifts are attributed to the parallel orientation, and thus large overlap, of the C_α

---

* Compare Problem **75**.

** Compare e.g., p. 161, 164, 165, 166, 167—169, 171, 177, 179, 181, 184, 187, 188, 195, 197, 200 and 203 and Problems **67, 74, 75, 76,** and **77**.

- C_β σ-bond and the π-orbitals of the heteroatom (**263**), i.e., to an increase in electron density around C_γ due to hyperconjugation.[423] (Inductive effects would cause opposite shifts.)

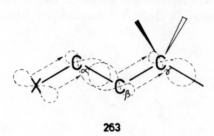

263

Electric field effect — The electric field of polar or electrically charged substituents may affect the electron distribution of the molecule and thus the shielding of its carbons. According to the Buckingham equation:[224]

$$\Delta\delta_E = -AE_z - BE^2 \tag{310}$$

where E_z is the z component of the axially symmetric electric field **E** formed around the X – Y bond, and A and B are parameters characteristic of the X – Y bond and depending on substituent X which produces field **E**. $\Delta\delta_E$ is very sensitive of steric structure,[91] and thus its magnitude may be utilized in conformation analysis. Since field **E** and gradient $\partial E/\partial r$ simultaneously determine the value of $\Delta\delta_E$ and since their effects on shielding[90] are opposite, $\Delta\delta C_\alpha \leq \Delta\delta C_\beta$ because the gradient is proportional to r^{-3}, whereas the field to r^{-2}, and thus the former, effect manifests only in the immediate neighborhood of substituent X producing the field. The deshielding effect of *syn-axial* heteroatoms[594] may be interpreted similarly (by the electric field of the substituent); e.g., in 11β-hydroxysteroids, the 18- and 19-Me groups are 1.7 and 2.6 ppm more deshielded than in the 11α-epimers.[595]

Heavy atom effect — Although the chemical shifts of halomethanes or halogenated *c*-propane or *c*-hexane derivatives increase in the order of I, Br, Cl, and F (compare Tables 51, 59, and 60), the hydrogen → iodo exchange causes shielding, and the shift corresponding to bromosubstitution is much less than expected from ¹H NMR analogy. If the hydrogens of methane are exchanged successively for halogen, shielding gradually occurs in the case of iodine, but for bromosubstitution the mono and dibromo derivatives show increasing shifts, and the shift decreases for bromoform and turns into negative for tetrabromo methane (compare Table 60). For chloro and fluoro derivatives, the shifts monotonously increase, quite evenly for the former, but at a suddenly slowing rate for the latter. The effect is obviously a result of two opposite terms.

This phenomenon is called heavy atom effect (compare Problems **69** and **74**).* It was interpreted originally as anisotropic effect,[1356] but it transpired later that the anomal behavior is due to other factors: bond contraction, bond polarization, and a special bonding structure for methyl halides.[882,1249] It is interesting to note that the chemical shift of tetraiodomethane, – 292.5 ppm[882] is the smallest ¹³C chemical shift observed so far. Concerning the effect of halogen substitution, there is no analogy between ¹H and ¹³C spectroscopy.

On the basis of the heavy atom effect, however, alkyl halide analogues which could not be distinguished on the basis of UV, IR, and ¹H NMR spectra are easy to distinguish. Figure 172 shows the ¹³C NMR spectra[729] of 1-chloro- and 1-bromo-3-methylbutane (Me$_2$CH–CH$_2$–CH$_2$X, X = Cl, Br). The shifts are in the order of carbons 4-3-2-1: 22.0,

---

* See p. 176 and 179.

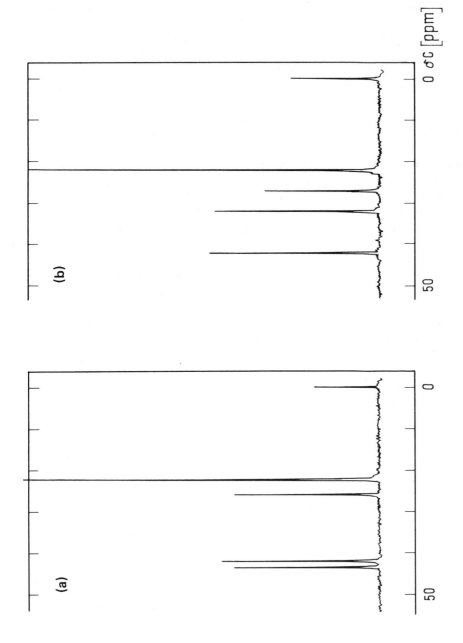

FIGURE 172. The CMR spectra of (a) 1-chloro- and (b) 1-bromo-3-methyl-butane. (From Johnson, L. F. and Jankowski, W. C., *C-13 NMR Spectra*, Wiley-Interscience, New York, 1972. With permission.)

25.7, 41.6, and 43.1 ppm and 21.8, 26.8, 41.7, and 31.7 ppm, respectively.[729] This sequence of shifts arises from the fact that the values corresponding to the order of carbon are increased by the substituent for C-1 and by the adjacent tertiary carbon for C-2. The former effect is much stronger for the chloro derivative, since the increase in chemical shift due to the $-I$ effect of bromine is compensated in part by the heavy atom effect of opposite sign. The assignment of C-3 and C-4 to the corresponding signals is unambiguous on the basis of the multiplicities (quartet for C-3 and doublet for C-4 because of the C–H couplings, whereas the signals of C-1 and C-2 are both triplets).

Anisotropic neighbor group effect — This effect of basic significance in ¹H NMR hardly influences ¹³C NMR shifts,[269] since it has the same absolute magnitude (usually <1 ppm) as in ¹H NMR.[1118] Therefore, in ¹³C NMR, this effect could be detected in a few cases only,* where it was due almost exclusively to aromatic ring currents, since their anisotropic effects are greater than usual. The "middle" methylene signals of [12]-paracyclophane (**6**, $n = 12$) are by about 8 ppm more shielded than the ones adjacent to the aromatic ring, but their shifts are only some tenth of ppm less than in cyclohexane.[842] The effect of ring currents is usually masked by other effects.[611,842,1363]

Mesomeric effects — In conjugated unsaturated systems the mesomeric electron shifts cause great differences in shielding, since the effective nuclear charges of carbons influence the paramagnetic shielding contribution (compare Problem **70**). In the case of conjugated olefins, cumulenes, sp-bonded systems, and aromatic and conjugated carbonyl derivatives, mesomery has a decisive role in net shielding. Numerous examples will be given for this.** The terminal carbon of allenes, ketenes, and keteneimines is gradually more shielded[481,482] corresponding to the increasing weight of mesomeric structure $> \bar{C}^{\ominus} - C \equiv X^{\oplus}$, i.e., to the mesomeric effect increasing in the order of X = $CH_2 < O < N$. The strong deshielding of carbonyl carbons is also due to mesomeric effects, since the polarization of carbonyl bond (the preference of $> C^{\oplus} - \bar{O}^{\ominus}$ mesomeric structure) decreases the electron density around the carbon atom.*** It is a further consequence of the role of mesomery in the shielding of carbon atoms that the shifts of aromatic carbons, primarily those of *para* and, with slightly greater deviation, of *meta* position,[1030] are proportional to Hammett's σ.

Solvent effects, van der Waals' interactions, isotope effect — Further, less significant effects may also influence ¹³C chemical shifts. *Solvent effect* is usually small, arising from the position of carbon atoms inside of the molecule, although its magnitude is higher than in proton resonance. In certain cases, however, very large solvent effects occur. This may be expected primarily in the case of carbonyl derivatives, where the formation of hydrogen bond or protonation may increase chemical shifts to a great extent, e.g., the carbonyl signal of acetone measured in 26 solvents[903] changes over an interval of approximately 40 ppm.† Deshielding occurs in polar and protic solvents upon the effect of hydrogen bonds.[817] However, solvent effects are generally much weaker, not exceeding some parts per million.[72,862,1030] The chemical shift of chloroform changes, e.g., over 4.3 ppm.[862] The highest solvent effects measured on carbons 1 to 4 of phenol are approximately 3.7, 3.5, 1.8, and 4.4 ppm.[901] A slight change in shift may also occur upon dilution. Of this effect, the strongest has been observed for methyl iodide in cyclohexane (about 7 ppm).[1356] The solvent effect is specific on the individual carbons of a molecule (e.g., Reference 194), and thus it may be utilized to solve assignment problems.

Van der Waals' interactions may occur between substituents getting into steric proximity. They increase chemical shifts (deshielding), and the effect is proportional to the polarizability and ionization potential of the interacting group or atom.[54] These intramolecular dispersion

---

*   Compare p. 187.
**   Compare p. 175 (point 2), 176 (point 4), 179, 180 (point 1), 181, 187, 197, etc.
***Compare p. 187.
†   Compare p. 183.

forces are short-range interactions, proportional to r^{-6}, where r is the distance between the carbon and the perturbing substituent. *Isotope effect* may be observed when an atom is replaced by its heavier isotope. In the case of H $\rightarrow$ D exchange, e.g., shielding of the carbons separated by one or two bonds increases[111,569,1432,1491] except in some cases[469,639,899,1004] (compare Table 49). The extent of shielding depends on the hybridization state of the carbon and on the relative electron density around it.[193] Thus, e.g., by deuterating the –CONH– groups of a peptide, the carbonyl signals of the peptide bond may be distinguished from those of acids and amides.[469] The splitting in carbon resonance signals upon OH $\rightarrow$ OD exchange (corresponding to the molecules containing OH and OD groups, respectively) is suitable for the identification of the C_α and C_β signals of alcohols and phenols,[1034] for the assignment of the ^{13}C NMR spectra of carbohydrates,[268,535] and for the solution of structure elucidation problems.* Isotope labeling has important roles in the study of reaction mechanisms and biosyntheses (compare Section 4.1.5).

4.1.2. The Chemical Shifts of Various Compounds in Carbon Resonance Spectra

4.1.2.1. Alkanes

The carbon signals of saturated hydrocarbons appear in the -3- to $+60$-ppm range.[581,1143d,1354] The shifts of the simplest alkanes may be found in Table 53.

The chemical shifts of normal alkanes may be determined to an accuracy of ± 0.2 ppm by means of the additivity rule:

$$\delta C_i = 3.85 + \Sigma\, n_{ij}\rho_j \qquad (311)$$

where n_{ij} is the number of carbon atoms in position j (α, β, γ, etc.) with respect to C_i, and ρ_j is the additivity parameter of carbon in position j and $\rho_\alpha = 9.1$, $\rho_\beta = 9.4$, $\rho_\gamma = -2.5$, and $\rho_\delta = 0.35$.[1097] For branching paraffins one may estimate the carbon shifts with slightly larger uncertainty (± 0.8 ppm) by means of[867]

$$\delta C = A_a + \Sigma_i\, \beta_{bi}B_{ab} + \gamma C_a + \delta D_a \qquad (312)$$

where a is the number of adjacent carbon atoms (the order of the atom under study), b is the order of adjacent atoms (the number of carbon atom attached thereto), β_{bi} is the number of carbons of the same order, γ and δ are the number of carbons in three or four bond distances, and A, B, C, and D are constants. Their values for different a and b are given in Table 54.

The ^{13}C NMR investigation of alkanes gives much more information than the ^{1}H NMR spectra, since the latter are usually featureless,** owing to the many almost similar protons. As an illustration, the two spectra of *racemic* and *meso* 2,4,6,8-tetramethylnonane [$CH_2(CHMe–CH_2–CHMe_2)_2$] are given in Figure 173.[719] The carbon signals may be assigned on the basis of the shifts calculated from Equation 312: $\delta C\text{-}4',6' = 20.61$, $\delta C\text{-}1,2'$, $8',9 = 22.62$, $\delta C\text{-}2,8 = 25.92$, $\delta C\text{-}4,6 = 28.38$, $\delta C\text{-}5 = 44.36$, and $\delta C\text{-}3,7 = 46.55$ ppm. The chemical shifts of the C-3,7 atom pair and of the middle C-5 atom are, therefore, much higher than the others. It can be seen from Figure 173 that two signals are really separated downfield from the rest, and of these, since it corresponds to a single carbon, the upfield one has lower intensity.

Carbons 1,2' and 8,9', respectively, are diastereotopic and thus chemically nonequivalent. The probable reason for the significant shift difference is steric hindrance which hinders the rotation of methyl groups.[216,987] With respect to molecular symmetry and chemical equivalence, the principles set up for ^{1}H NMR are, of course, valid in carbon resonance, too;

---

*   Compare p. 91 and 232.

** Compare p. 9.

Table 53
¹³C NMR CHEMICAL SHIFTS (PPM) OF SOME SIMPLE ALKANES[581,1097,1354]

-2.3
CH_4

5.7
$CH_3—CH_3$

15.4 15.9
$CH_3—CH_2—CH_3$

13.1 24.9
$CH_3—CH_2—CH_2—CH_3$

13.6 22.4 34.4
$CH_3—CH_2—CH_2—CH_2—CH_3$

13.7 22.8 31.9
$CH_3—CH_2—CH_2—CH_2—CH_2—CH_3$

25.0 24.3
$CH(CH_3)_3$

11.4 31.7 29.7 21.9
$CH_3—CH_2—CH(CH_3)_2$

8.5 36.5 30.2 28.7
$CH_3—CH_2—C(CH_3)_3$

27.4 31.4
$C(CH_3)_4$

14.0 20.5 41.6 27.6 22.4
$CH_3—CH_2—CH_2—CH(CH_3)_2$

18.4 36.5 29.1 11.1
$CH_3—CH(CH_2—CH_3)_2$

19.1 33.9
$(CH_3)_2CH—CH(CH_3)_2$

Table 54
CONSTANTS (PPM) IN FORMULA 312[867]

Constant		Order of the carbon atom			
		$a = 1$	$a = 2$	$a = 3$	$a = 4$
A		6.80	15.34	23.46	27.77
	$b = 2$	9.56	9.75	6.60	2.26
B	$b = 3$	17.83	16.70	11.14	3.96
	$b = 4$	25.48	21.43	14.70	7.35
C		-2.99	-2.69	-2.07	0.68
D		0.49	0.25	—	—

Table 55
^{13}C NMR CHEMICAL SHIFTS (PPM) OF
CYCLOALKANES AS COMPARED WITH
THE SHIFT OF THE MIDDLE CARBON IN
THE OPEN CHAIN ALKANE CONTAINING
THE SAME NUMBER OF CARBON
ATOMS[230,965]

Size of the ring (n)	δC		Δδ
	Cycloalkane	Open chain	
3	−2.8	15.9	32.8[a]
4	22.1	24.9	12.6[a]
5	25.3	34.4	9.1
6	26.6	31.9	5.3
7	28.2	29.3	1.1
8	26.6	29.5	2.9
9	25.8	30.0	4.2
10	25.0	30.1	5.1

[a]   Values given considering also the difference in the order
    of the neighboring carbons (instead of two and one primary
    carbons, there are secondary neighbors in the case of c-
    propane and c-butane, respectively) between cyclic and
    open-chain analogues by constant 9.75 ppm of formula
    312.

diastereotopic groups may have separate signals.[785] This cannot be taken into account by Equation 312. The difference between the two isomers due to the field effect cannot be interpreted either by this equation; in the sterically more compressed *racemic* isomer, some signals show upfield shifts.

From the constants given in Table 54 some general conclusions may be drawn on the effect of neighboring groups. The monotonous increase of A with a (order) means that the chemical shift increases with the order of the carbon atom. Adjacent carbons increase the shift, to an extent proportional with their order. The effect of β-carbon is the same in direction and magnitude, whereas the γ-carbon atoms cause smaller and opposite shifts. These facts are in harmony with the conclusions drawn in Section 4.1.1.2 on the relationships between structural parameters and chemical shifts. Equation 312 and similar relationships[417,581,867] arise from the additivity of the effect of adjacent groups on chemical shift[581,1097] and are very helpful in the assignment of signals.

The substituent effects may be estimated to a first approximation on the basis of the parameters of Table 52. Several empirical rules are known, by means of which the shifts of alcohols,[415,1213] aliphatic amines,[414] or nitroalkanes[416] may be estimated.

4.1.2.2. Cycloalkanes

The chemical shifts of the first few members of the homologous series of cycloalkanes are given in Table 55, in comparison with the shift of the middle methylene group of the open chain analogue.[230,965] It can be seen that for n > 4, the shift hardly changes, being in the 23- to 28-ppm range up to n = 20. Owing to the field effect, the carbons of cyclic methylene groups are by about 3.5 ppm more shielded (n > 5) than those of aliphatic analogues.

The carbon atoms of cyclopropane ring have extreme shielding similar to its protons (compare Tables 50 and 51), which is attributed to electron delocalization and the resulting

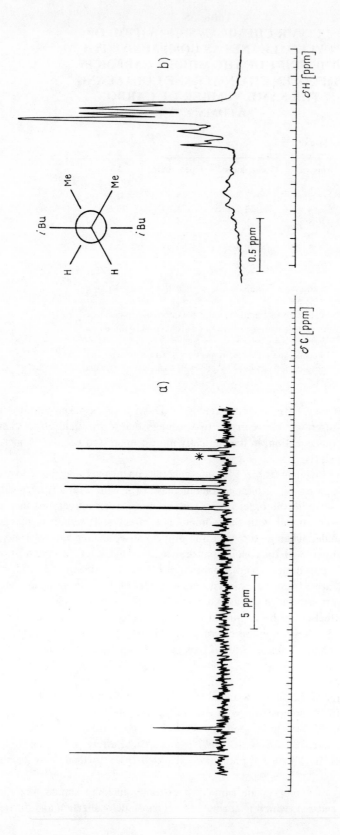

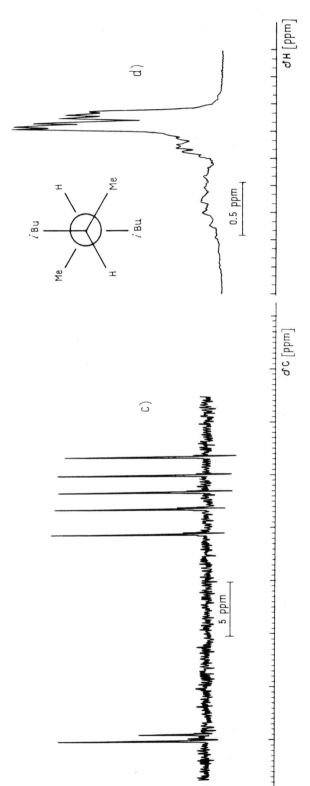

FIGURE 173. The ^{13}C and ^{1}H NMR spectra of (a) and (b) *racemic* and (c) and (d) *meso* 2,4,6,8-tetramethylnonane. (From *JEOL (Jpn. Electron Opt. Lab.) News*, 7c, 16, 1970. With permission.)

Table 56

¹³C NMR CHEMICAL SHIFTS (PPM) OF SOME CONDENSED AND SPIRO BICYCLOALKANES[598,876,878,1161]

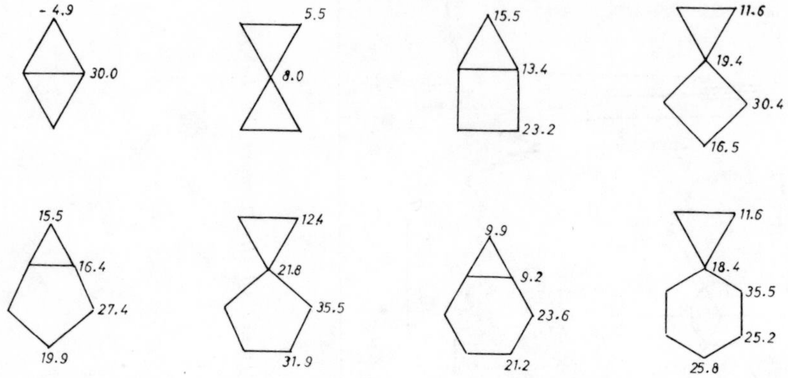

Structures (top row):
- −4.9, 30.0
- 5.5, 8.0
- 15.5, 13.4, 23.2
- 11.6, 19.4, 30.4, 16.5

Structures (second row):
- 15.5, 16.4, 27.4, 19.9
- 12.4, 21.8, 35.5, 31.9
- 9.9, 9.2, 23.6, 21.2
- 11.6, 18.4, 35.5, 25.2, 25.8

ring currents.[1161] Strong shielding may also be observed in fused and spiro structures, as indicated by the data of Table 56. In fused systems with larger rings, chemical shifts are relatively constant, but the various isomers of rigid or conformationally homogeneous rings with different ring anellation or position of substituents are easy to distinguish by the field effect.

The methylene carbons of adamantane (**105**) are deshielded by 11.6 ppm compared to

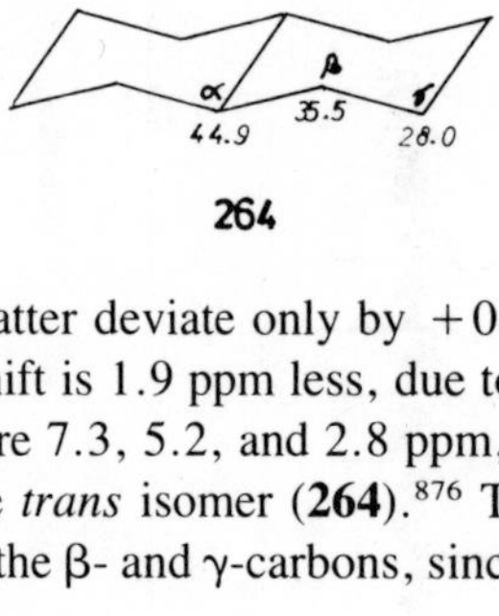

α 44.9   β 35.5   γ 28.0

264

those of cyclohexane,[878] which latter deviate only by +0.4 ppm from the value calculated by Equation 312. The methine shift is 1.9 ppm less, due to the field effect. The α-, β-, and γ-carbons of *cis* decaline (**262**) are 7.3, 5.2, and 2.8 ppm, respectively, more shielded than the corresponding carbons of the *trans* isomer (**264**).[876] This is a consequence of the field effect, which is much weaker for the β- and γ-carbons, since they alternate between sterically

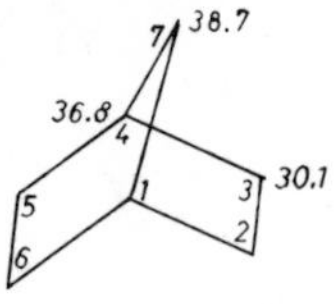

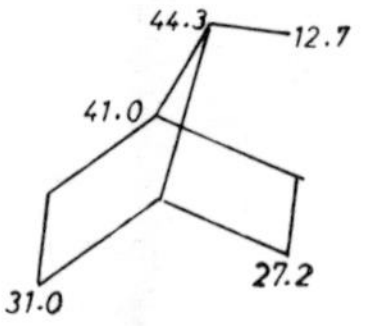

265   7: 38.7, 36.8, 4, 30.1, 5, 6, 1, 3, 2

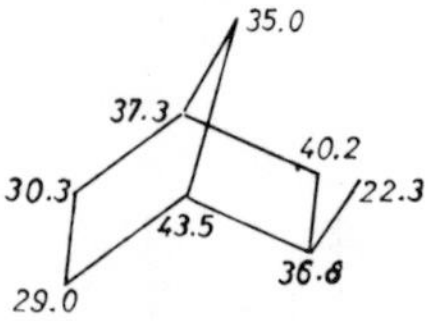

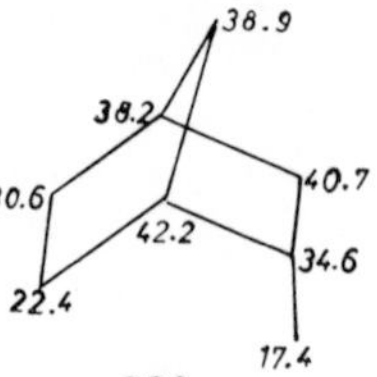

266   35.0, 37.3, 40.2, 43.5, 22.3, 30.3, 36.8, 29.0

267   44.3, 12.7, 41.0, 31.0, 27.2

268   38.9, 38.2, 40.7, 30.6, 42.2, 34.6, 22.4, 17.4

compressed and noncompressed positions during inversion,* and the measured effect is an average of those corresponding to the two structures. Consequently, for rigid or conformationally homogeneous analogues, much higher field effects may be expected and observed, too (compare Problem **75**). The field effect is illustrated by the methyl-substituted norbornane derivatives **265** to **268**, indicating at the same time how this phenomenon may be utilized in the determination of the position and orientation of substituents attached to a rigid skeleton.[598]

The *equatorial* (*endo*) 2-methyl group of Compound **266** decreases the shift of C-7 due to the field effect, whereas it increases the C-1,2,3 shifts through the α- and β-effects. The

---

* Compare p. 155.

Table 57
¹³C NMR CHEMICAL SHIFTS (PPM) OF MONOMETHYL AND 1,2-DIMETHYL CYCLOPENTANES AND CYCLOHEXANES[278,327]

Cyclohexane structures:

Structure 1: 22,8 Me; 33.1; 26.4; 35.8; 26.6

Structure 2: 15,8 Me; 34,5; 23.7; 31.6; Me

Structure 3: 20,3 Me; 39.6; 26.9; Me; 36.1

Cyclopentane structures:

Structure 4: 20.2 Me; 34.6; 34.6; 25.2

Structure 5: Me; 37.4; 23.0; 33.0; 14.9 Me

Structure 6: Me; 42.5; 23.1; 13.5 Me; 34.8

shifts of C-4,5,6 atoms hardly change with respect to the precursor **265**. The increasing chemical shifts of C-1,2 signals are compensated in part by the field effect. The field effect of the 2-*axial* (*exo*) substituent (**268**) may also be observed with respect to the *endo* isomer on the C-1,2,6 and methyl signals; it is the highest, of course, for C-6. In contrast, C-7 is deshielded since there is no steric compression like in the *endo* isomer. In the 7-methyl analogue (**267**), the two methylene carbons of *syn* position show field effect.

In the investigation of configuration and conformation problems, ¹³C NMR spectra are especially important whenever the ¹H NMR spectrum is not informative, e.g., with cycloh-exanols the conformation is easy to determine from the position and splitting of the δCH(OH) ¹H NMR signal. With methylcyclohexane this is already impossible, since the signals of *axial* and *equatorial* methyl groups do not separate sufficiently, and the very weak long-range couplings of type $^4J(H,H)$ with the *vicinal* ring hydrogens do not cause significant splitting. On the other hand, in the ¹³C NMR spectrum, the signal of the cyclic carbons of methylcyclohexanes is shifted depending on the orientation of the methyl group. Higher δ-values may be observed for *equatorial* methyl substitution,[579] both on the *geminal* and on the α- and β-carbons ($\Delta_{e,a}\delta C \approx$ 4.5, 3.5, and 5.5 ppm). Table 57 shows the ¹³C chemical shifts of *cis* and *trans* 1,2-dimethylcyclohexane and -pentane in comparison with the mon-omethyl derivative.

For the analogous carbon atoms of isomer pairs, $\delta C(cis) < \delta C(trans)$ without exception, and the field effect is, obviously, the highest for the substituted and methyl-carbons. Ac-cordingly, the net shift of the carbon atoms of the *cis* isomers is much less than that of the *trans* derivatives.[1107] With dimethylcyclohexanes $\Sigma\Delta\delta C(trans, cis)$ is 34.6 ppm, for the cyclopentane analogues it is 21.1 ppm. The magnitude of the field effect measured for the substituted carbon signals is the same for cyclohexane and cyclopentane derivatives, whereas in the case of methyl signals it is higher for the six-membered ring, since the methyl groups are closer to one another. Field effect occurs already for the monosubstituted derivatives, too, and thus the measured chemical shifts are much smaller than the ones calculated from the substituent constants applied for alkanes. It is interesting that with methyl substituted cyclobutanes a γ-effect similar in sign and magnitude to that of cyclohexanes was observed,[425] and even for the certainly planar methyloxethanes, the signal of the γ-carbon atom is shifted diamagnetically.

From the systematic investigation of methylcyclohexanes,[327,1107] substituent increments were derived for the carbon atoms of the ring (see Table 58). By adding these values to 26.6 ppm of cyclohexane, the shifts of cyclic carbon atoms may be determined. The shift of the methyl groups may be estimated in a similar way (Table 58).[1143e]

The highest shielding (16.3 ppm) is measured in *cis* 1,2-dimethyl-cyclohexane, and the lowest (34.3 ppm) is measured in 1,1,3-trimethylcyclohexane, both for the *equatorial* 1-

Table 58
SUBSTITUENT CONSTANTS (PPM) FOR ESTIMATION OF CHEMICAL SHIFTS OF RING CARBONS AND METHYL CARBON ATOMS IN POLYMETHYL SUBSTITUTED CYCLOHEXANES[327,1107,1143e]

Substituent	C_s	C_α	C_β	C_γ	Orientation of substituent		Me(a)	Me(e)
Me(a)	1.9	5.9	−5.9	0.4	δMe		18.8	23.1
Me(e)	6.5	9.5	0.5	0.2				
Me₂(gem)	−3.3	−0.8	2.5		Substitution	gem	6.4	10.4
	−2.9 (a)	2.1 (a)				2a	−2.8	−2.8
Me₂(vic)ᶜ	−2.5 (e)	−0.3 (e)				2e	−6.8	−2.8
Me₂(vic)ᵗ	−2.0					3e	2.0	0
						3a,4a,e	0	0

Note: a, axial; e, equatorial; gem, geminal; vic, vicinal; C_s, substituted carbon; c, cis; t, trans.

Table 59
¹³C NMR CHEMICAL SHIFT (PPM) OF MONOSUBSTITUTED CYCLOHEXANES[1143g]

Substituent	δC-1	δC-2	δC-3	δC-4
CN	28.3	30.1	24.6	25.8
I	31.8	39.8	27.4	25.5
Me	33.4	36.0	27.1	27.0
SH	38.5	38.5	26.8	25.9
COOMe	43.4	29.6	26.0	26.4
Ph	45.1	34.9	27.4	26.7
COO⊖	47.2	30.9	26.9	26.9
NH₂	51.1	37.7	25.8	26.5
N⊕H₃	51.5	33.4	25.6	26.0
Ac	51.5	29.0	26.6	26.3
Br	52.6	37.9	26.1	25.6
Cl	59.8	37.2	25.2	25.6
OH	70.0	36.0	25.0	26.4
OAc	72.3	32.2	24.4	26.1
OMe	78.6	32.3	24.3	26.7
NO₂	84.6	31.4	24.7	25.5
F	90.5	33.1	23.5	26.0

methyl group.[327] From the data of Table 58, the former may be calculated exactly, the latter may be calculated with a deviation of −0.8 ppm.

Owing to the field effect, for substituted cyclohexanes and their heteroanalogues, the $\delta CR_a < \delta CR_e$ relationship is applicable for a great variety of substituents.[327] By means of low-temperature measurements, this has been shown for methylcyclohexane[45] and t-butyl-substituted cyclohexanes, too, where $\Delta_{a,e}\delta C \simeq 5$ ppm.[1004]

The substituted derivatives of cycloalkanes show similar effects as their open chain counterparts (see the chemical shifts of monosubstituted cyclohexane derivatives collected in Table 59). By means of the substituent parameters derived from these shifts, the chemical shifts of polysubstituted derivatives may also be calculated if the field effects are taken into account.

Table 60
^{13}C NMR CHEMICAL SHIFTS
(PPM) OF HALOGENATED
METHANE
DERIVATIVES[688,882,1143f]

X	F	Cl	Br	I
CH$_3$X	75.2	25.1	10.2	− 20.5
CH$_2$X$_2$	109.0	54.2	21.6	− 53.8
CHX$_3$	116.4	77.7	12.3	− 139.7
CX$_4$	118.5	96.7	− 28.5	− 292.3

4.1.2.3. Halogen Derivatives

The ^{13}C NMR chemical shifts of halomethanes are summarized in Table 60. In the rows of the table, the chemical shift, increasing from right to left, varies with the −I effect of substituents, in analogous manner with proton resonance. A similar trend is shown by the first two columns: the shift increases with the number of fluorines or chlorines. However, with the bromo-derivatives the direction changes at bromoform, and with iodo-derivatives the sequence is opposite. This cannot be explained by the anisotropic effect of halogens as formerly assumed,[1354] since it is negligible.[595] According to theoretical interpretation[882] this irregularity, known as "heavy atom effect", is due to bond polarization and bond length changes.*

A wealth of data may be found on haloalkanes in the ^{13}C NMR literature.[246,874,982,985] The shifts of conformationally homogeneous (4-t-butyl) halogenated cyclohexanes are temperature dependent, owing to the stretching of the C–X bond at higher temperatures.[1258]

4.1.2.4. Alcohols, Ethers, and Acyloxy Derivatives

The carbon atom of alcohols attached to the oxygen atom has 32 to 52 ppm higher chemical shift than that of the corresponding alkane, due to the strong −I effect of oxygen. The chemical shift of the adjacent carbon atom increases by 5 to 10 ppm.[1213] With α,ω-diols and ω-haloalkanols, similar to the α,ω-dihalo-derivatives, the substituent effects are additive if at least three carbon atoms separate the substituents.[1372d] In ethers, the chemical shifts were found to change with respect to alcohols by further 9.5 ± 0.5 and -1.7 ± 0.7 ppm,[875] indicating that the oxygen atom "transmits" the β- and γ-effects similarly as the carbon atoms.

In the acetoxy derivatives of primary alcohols, α-effect is 1.5 to 4.0 ppm; in those of secondary alcohols, it is about 10 ppm. The β-effect has opposite sign in both cases, and

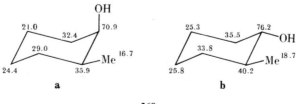

269

the extent of shielding is 1 to 5 ppm.[278] These facts are illustrated by the data of Table 61.

In the investigation of cyclic alcohols and their derivatives, a special role is played by cyclohexanol derivatives and their heteroanalogues, since the sensitivity of ^{13}C NMR spectra on stereochemistry can be utilized in the solution of configuration and conformation problems.

*   Compare p. 156.

Table 61
^{13}C NMR CHEMICAL SHIFTS (PPM) OF SOME SIMPLE OPEN-CHAIN ALIPHATIC ALCOHOLS AND OF THEIR ACETOXY AND ETHER DERIVATIVES[278,437,875,1007,1213,1230]

49.0
CH_3OH

17.6　57.0
$CH_3{-}CH_2OH$

10.0　25.8　63.6
$CH_3{-}CH_2{-}CH_2OH$

25.1　63.4
$(CH_3)_2CHOH$

54.8　97.9
$(CH_3O)_2CH_2$

50.7　170.7　19.6
$CH_3{-}OOC{-}CH_3$

13.8　59.8　170.0　20.0
$CH_3{-}CH_2{-}OOC{-}CH_3$

10.5　22.4　66.1　170.7　20.7
$CH_3{-}CH_2{-}CH_2{-}OOC{-}CH_3$

20.4　66.8　169.5　21.4
$(CH_3)_2CH{-}OOC{-}CH_3$

51.1　115.0
$(CH_3O)_3CH$

59.7
$(CH_3)_2O$

14.7　67.7
$(CH_3{-}CH_2)_2O$

11.1　24.0　73.2
$(CH_3{-}CH_2{-}CH_2)_2O$

23.0　68.5
$[(CH_3)_2CH]_2O$

125.6
$(CH_3O)_4C$

FIGURE 180.　The ^{13}C NMR spectrum of 2-methyl-4-hydroxy-chlorobenzene (281).　(From Johnson, L. F. and Jankowski, W. C., *C-13 NMR Spectra*, Wiley-Interscience, New York, 1972. With permission.)

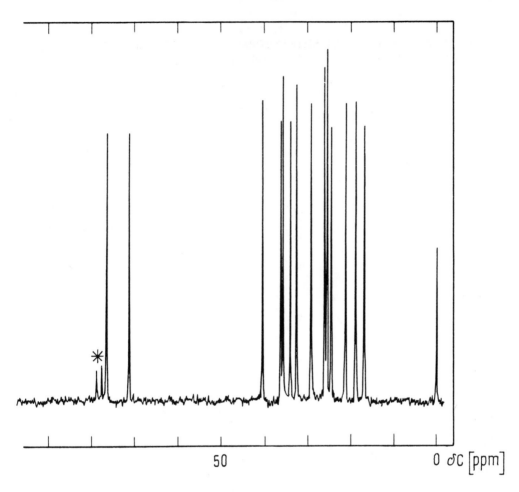

FIGURE 174. The ^{13}C NMR spectrum of the mixture of *cis* and *trans* 2-methylcyclohexanol (**269**)[1213]. (From Johnson L. F. and Jankowski, W. C., *C-13 NMR Spectra*, Wiley-Interscience, New York, 1972. With permission.)

The analogous signals of *cis* and *trans* 2-methylcyclohexanol (**269a,b**) have different δ-values. The differences in chemical shift can be attributed to the field effect, as a consequence of which all carbons of the *cis* isomer are more shielded. Figure 174 shows the spectrum of an isomeric mixture. On the formulas the individual shifts[1213] are shown. The assignment is aided by the quartet and doublet splitting of methyl and of C-1,2 signals, respectively, owing to the C–H couplings. The two latter signals may also be identified on the basis of their downfield shifts (substituent effect). Moreover, δC-3,6 > δC-4,5 (C-3,6 have tertiary neighbors!), and δC-6 > δC-3, owing to the −*I* effect of the hydroxy group. Finally, in isomer **a**, δC-5 < δC-4 arising from the field effect (1,3-*diaxial* interaction between the *axial* C-5 and the hydroxy group). For isomer **b** the analogous difference is very low, but the two signals still appear separately (Δδ = 0.5 ppm). In the solution of **269a,** there is a conformational equilibrium, and the observed shifts are the weighted averages of the values corresponding to the ring inverses.

By using the data of conformationally homogeneous (e.g., *t*-butyl-substituted) analogues, the ratio of conformers may be estimated for the cyclohexanol derivatives. Thus, comparing the chemical shifts of *cis* and *trans* 4-*t* -butylcyclohexanol (Table 62) to those of *cis* and *trans* 3-methyl-cyclohexanol, it can be calculated that in the latter, the equilibrium ratio of the inverse-containiing *equatorial* hydroxy group is about 10%.[1213]

Table 62

^{13}C NMR CHEMICAL SHIFTS (PPM) OF ALKYLSUBSTITUTED CYCLOHEXANOL AND CYCLOPENTANOL DERIVATIVES[278,1213]

Table 62 also contains the data of *cis* and *trans* 2- and 3-methyl-cyclopentanols. It can be seen that in the more flexible five-membered analogues there is no field effect for 1,3-disubstitution: $\Delta\delta C(t,c)$ = 0, 0.2, −1.1, 0.3, −0.2, and −0.4 ppm, for C-1 through C-5 and the methyl carbon, respectively, and even in the case of 1,2-dimethylsubstitution (see Table 62) it is smaller than with cyclohexanes, since the substituents are more distant and there is a free pseudorotation. The shift differences $\Delta\delta C(t,c)$ are in the above sequence 4.6, 2.4, −0.6, −0.6, 0.8, and 4.6 ppm,[278] i.e., the difference is significant for the substituted and methyl carbons only.

The δC-1 shift of the acetoxy cyclohexanols increases by about 2 or 3 ppm in the case of *equatorial* or *axial* acetoxy groups, respectively, whereas for methyl ethers a deshielding of about 9 ppm may be observed regardless of the steric position.[223] It is interesting to note that in the case of 1-methylcyclohexanol acetylation causes a downfield shift of 11.6 ppm in the signal of quaternary C-1, from 69.6 ppm to 81.2 ppm.[1372e] This proves the steric origin of the deshielding effect of acetoxy substitution (a *geminal* methyl group hinders the free rotation of the acetoxy substituent). A further proof is furnished by the δC-1 shifts of 1-methyl cyclopentanol and its acetoxy derivative (79.2 and 88.9 ppm), where the analogous effect is slightly smaller (9.7 ppm).[278,1213]

Dioxane derivatives are also favored subjects of stereochemical studies.[763,1207] This may be illustrated, e.g., by the data of *cis* and *trans* 4,6-dimethyl-1,3-dioxane (see Table 62). Owing to the identical inverses of the *trans* isomer, the field effect is halved, but the isomers have still high $\Delta\delta C(t,c)$ shift differences: 6.5, 4.9, 3.3, and 2.5 ppm for the C-2, C-4,6, C-5, and methyl signals.[197] Similarly, in 2-methyl-1,3-dioxane, the difference $\Delta\delta C$-2(e,a) is approximately 7 ppm; with respect to the unsubstituted ring, the *equatorial* 2-methyl substituent causes downfield shifts of 3 to 5 ppm, whereas the *axial* substituent causes approximately 1.7-ppm upfield shift on the C-2 signal (see Table 62). The corresponding changes in the C-4,6 signal are 0.1 and −7.3 ppm, respectively.[763] This is shown by the *trans* and *cis* 2,4,4,6-tetramethyl-1,3-dioxane isomer pair, where the former shows an average effect owing to the conformational equilibrium: $\Delta\delta C$-2 = 3.4 (4.2) and $\Delta\delta C$-6 = 3.6 (4.0) ppm. The measured values in parentheses are in good agreement with the calculated ones. Of course, for the C-4 signal no significant difference may be expected; the 1,3-*diaxial* interaction, and thus the field effect, is present in both isomers.

4.1.2.5 Carbohydrates

The principles laid down in connection with cyclohexanols and dioxanes may also be utilized in the stereochemical investigation of carbohydrates with pyranose rings. The systematic investigation of glycosides has shown that the signals of the anomers have strictly

270

parallel relative shifts, and the anomeric carbon atom of methyl glycosides is always more shielded in the isomer which contains the *axial* methoxy group.[197] This is illustrated by the chemical shifts of α- and β-methyl-D-arabopyranose (**270a** and **b**). The methoxy signal is shifted, too, in the same sense, whereas the signals of C-3 and C-5 are shielded in the spectrum of the β-anomer, owing to the field effect.

The carbon atoms bearing *axial* hydroxy group is more shielded by approximately 2 ppm in most cases[1454] than that of the *equatorially* substituted analogues.[382] The signal of the carbons attached to *syn-axial* hydrogens is upfield shifted, too.[195,197,382,1105] The δC-1 signal of methyl glycosides shifts downfield by approximately 7 ppm with respect to the hydroxy derivative, like in the case of alcohols and ethers.[192,195,377,382,383,1105,1451]

^{13}C NMR spectroscopy is a very powerful method in the investigation of carbohydrates in other respects, too. The presence of many similar functional groups makes other carbohydrate spectra (such as IR and ^{1}H NMR) often so complex that little information may be obtained from them on structure. In contrast, in the range of acetalic (O–C–O) ^{13}C chemical shifts (approximately 80 to 110 ppm), there is practically no signal from other groups, and thus these shifts are particularly characteristic. Symmetric and asymmetric isomers are simple to distinguish by the number of signals[1453] (compare Problems **67** and **74**).

In aqueous solutions, aldoses are present as a four-component mixture of furanose and pyranose epimers, and in principle the open aldehyde form may also occur. The ^{13}C NMR spectrum makes it possible to detect this equilibrium and to estimate the relative amounts of components.* Figure 175 shows the 22.53-MHz spectrum of D-ribose in D$_2$O solution,[192] which illustrate that the anomeric carbon atom of pyranoses is more shielded than that of furanoses and that δC-1 (α) < δC-1 (β) due to the steric hindrance (field effect) which occurs with furanoses between the adjacent 1,2-hydroxy groups, whereas with pyranoses between the *axial* 1-hydroxy group and the *syn-axial* 3-hydroxy group or H-5.

Peracetylation causes upfield shift, except for the C-6 signal.[383] The ^{13}C NMR investigation of carbohydrates, including di- and polysaccharides, has a very rich literature; in addition to those cited so far, see References 231 and 1452, too.

4.1.2.6. Amines and Their Salts

The electronegative NH$_2$ substituent causes an approximate 30-ppm downfield shift in the signals of the substituted carbon atoms of amines. The substituent parameters are additive for α,ω-diamines. The ^{13}C NMR shifts measured for methyl, ethyl, and trimethylamine are

* Compare p. 40—41.

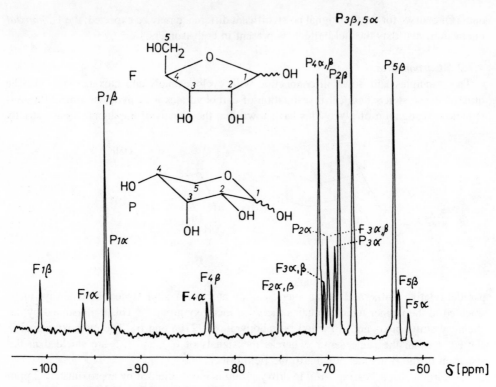

FIGURE 175.   The ^{13}C NMR spectrum of D-ribose at 22.63 MHz in D$_2$O solution.[192]

26.9, 35.9,[697] and 47.5 ppm.[875] In amine salts there is a diamagnetic shift,[1372f] opposite to that expected on the basis of increasing $-I$ effect.

In the C–C$_\alpha$–C$_\beta$–C$_\gamma$–C$_\delta$ chain of *n*-butyl, *c*-hexyl- and 1-adamantyl amine (**271**), and of piperidine, the shifts change on protonation alternately.[1002] Consequently, instead of the classical picture of inductive effect,[702]

$$\underset{R_{(-I)}}{\delta-} \longleftarrow \underset{C_\alpha}{\delta+} \longleftarrow \underset{C_\beta}{\delta\delta+} \longleftarrow \underset{C_\gamma}{\delta\delta\delta+} \longleftarrow \underset{C_\delta}{\delta\delta\delta\delta+} \dots$$

the actual charge distribution is probably more accurately described by assuming alternating partial charges along the carbon chain[1130]

$$\underset{R_{(-I)}}{\delta-} \longleftarrow \underset{C_\alpha}{\delta-} \longleftarrow \underset{C_\beta}{\delta\delta+} \longleftarrow \underset{C_\gamma}{\delta\delta\delta-} \longleftarrow \underset{C_\delta}{\delta\delta\delta\delta+} \dots$$

$$\underset{N^\oplus}{\delta\delta-} \longleftarrow \underset{C}{\delta\delta-} \longleftarrow \underset{H}{\delta+}$$

Consequently, on protonation, the electron density increases around the α-carbon: the amino substituent polarizes the C–H bonds

271

and the extent of this effect changes along the carbon chain.

4.1.2.7. Saturated Heterocycles

The substituent effects observed with ethers and amines are slightly modified in saturated heterocycles; a survey is given in Table 63. The ^{13}C NMR spectroscopy of this family of compounds is reviewed by Eliel and Pietrusiewicz.[424] Some more recent data are available, among others, on oxa-, dioxa-,[426] and thiocyclopentanes,[81] as well as on dioxaphospholanes.[113,127,1139,1202]

4.1.2.8. Functional Groups with Unsaturated Nitrogen

According to the $-I$ effects of unsaturated functional groups containing sp^2 and sp nitrogens, the adjacent carbons are deshielded. If the carbon atom is attached to the nitrogen by an unsaturated bond, its shift increases characteristically. Of the simplest molecules in this group, acetonitrile (MeCN), the analogous isonitrile (MeNC), isocyanate (MeNCO), and isothiocyanate (MeNCS) have the following shifts on the methyl and unsaturated carbon atoms, respectively: 0.3, 117.7; 29.3, 158.5; 26.1, 121.3; 29.1 and 128.5 ppm. The shifts of the corresponding CN$^\ominus$, NCO$^\ominus$, and NCS$^\ominus$ ions are 168.6, 129.8, and 134.4 ppm.[897] Scattered data on further shifts may be found in the literature (e.g., in Reference 1266 on aliphatic methyl groups).

It is striking that alkyl isocyanides (the signal of the sp carbons of ethyl and cyclohexyl derivatives are at 156.6 and 156.5 ppm)[897] and the cyanide anion have much higher shifts, similar to that of carbon monoxide (182.2 ppm),[456] than the other compounds mentioned. A plausible explanation may be found in the analogous mesomeric structures:

$$|C^\ominus \equiv O|^\oplus \leftrightarrow |C = \overset{\frown}{O} \leftrightarrow |C^\oplus - \overline{O}|^\ominus$$

$$|C^\ominus \equiv N^\oplus - R \leftrightarrow |C = \overline{N} - R \leftrightarrow |C^\oplus - \underline{\overline{N}}^\ominus - R$$

In contrast, the shifts of cyanides and isothiocyanates are close to that of carbon dioxide (124.2 ppm),[456] the mesomeric structures of which contain the carbon in analogous position:

$$|O^\oplus \equiv C - \underline{\overline{O}}^\ominus \leftrightarrow \overset{\frown}{O} = C = \overset{\frown}{O} \leftrightarrow |\underline{\overline{O}}^\ominus - C \equiv O|^\oplus$$

$$\overset{\frown}{O} = C = \overline{N} - R \leftrightarrow |\underline{\overline{O}}^\ominus - C \equiv N^\oplus - R$$

$$\overset{\frown}{S} = C = \overline{N} - R \leftrightarrow |\underline{\overline{S}}^\ominus - C \equiv N^\oplus - R$$

$$R^\oplus \ldots C = \overset{\frown}{N}^\ominus \leftrightarrow R - C \equiv \overline{N}$$

The nitro derivatives have extreme characteristics because of the strong substituent effect of nitro group; except for fluorine, the nitro group causes the largest deshielding of the adjacent carbon. The signal of nitromethane is at 57.3 ppm[1354] and that of the substituted carbon of nitrocyclohexane is at 85.1 ppm.[1100]

The ^{13}C NMR chemical shifts of nitroso derivatives are sensitive of $Z - E$ isomerism,[1140] as indicated by the chemical shifts of N,N-dimethyl- and N,N-diethyl-nitrosamine (**272,273**) given on the formulae.

272   **273**

Table 63
¹³C NMR CHEMICAL SHIFTS (PPM) OF SATURATED
HETEROCYCLES[82,430,463,466,532,582,658,734,810,850,904,980,1112,1273,1275,1372g,1384,1428]

Table 64

^{13}C NMR CHEMICAL SHIFTS (PPM) OF SOME SIMPLE OLEFINS AND CYCLOOLEFINS[379,609,1083,1143b,1372h]

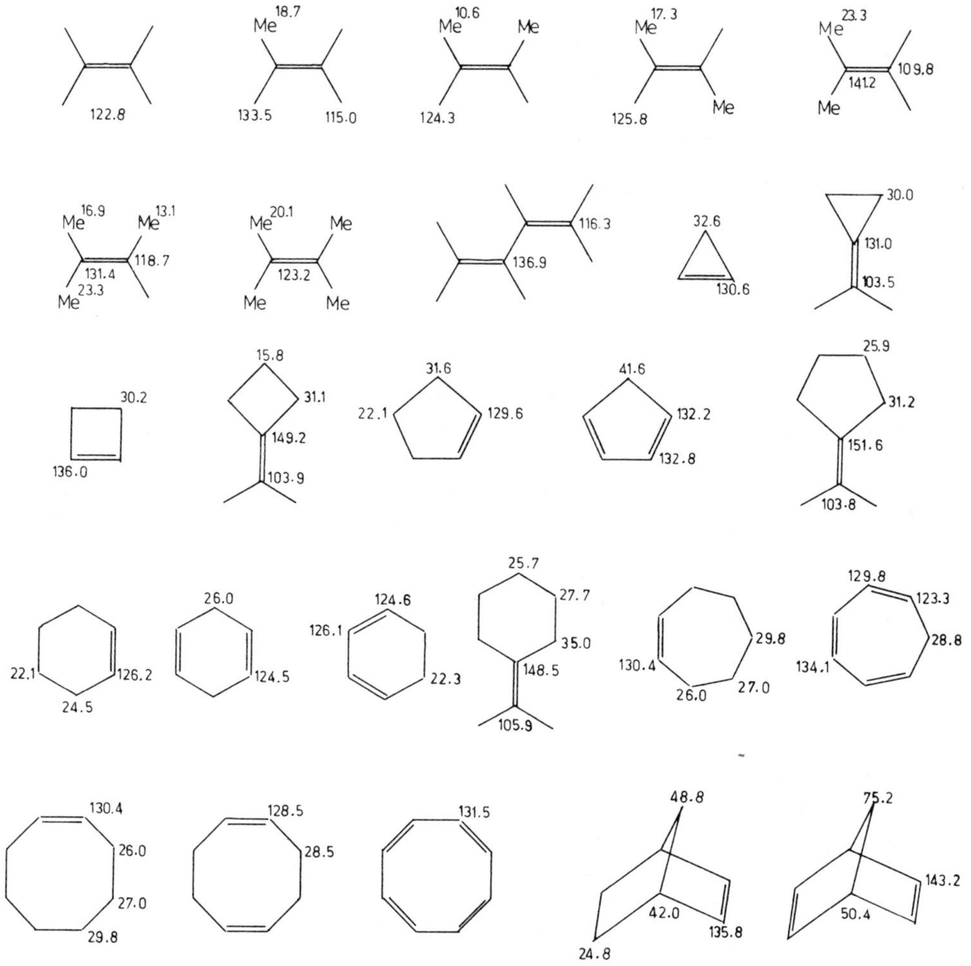

18.7 Me / 10.6 Me Me / 17.3 Me / 23.3 Me

122.8 133.5 115.0 124.3 125.8 Me Me

141.2 109.8

16.9 Me 13.1 Me / 20.1 Me Me / 116.3 / 32.6 / 30.0

131.4 118.7 123.2 136.9 130.6 131.0 103.5

23.3 Me Me Me

25.9

15.8 31.6 41.6

30.2 31.1 22.1 129.6 132.2 31.2

149.2 132.8 151.6

136.0 103.9 103.8

25.7

26.0 124.6 27.7 129.8 123.3

126.1 35.0 29.8 28.8

22.1 126.2 124.5 22.3 148.5 130.4 134.1

24.5 105.9 26.0 27.0

130.4 128.5 131.5 48.8 75.2

26.0 28.5 143.2

27.0 42.0 50.4

29.8 24.8 135.8

In amine oxides (RR′R″N → O), a 9.5 to 16.5 ppm deshielding may be observed for the α-carbon (β-effect), while the opposite shift of C_β signal (γ-effect) is approximately 4.8 to 6.6 ppm.[24,1372g]

4.1.2.9. Alkenes

The carbon signals of olefins appear in the 80- to 165-ppm range; the terminal carbon is more shielded by 10 to 40 ppm than its *C*-substituted analogue.[1141] The ethylidene carbon atom of the *exo* double bond attached to rings absorbs in the 100- to 110-ppm region.[1109] The shifts of the carbon atoms of an alkyl chain attached to an olefinic bond are, except for C_α, equal within 1 ppm to those of the corresponding alkanes. For C_α the difference varies between +4.5 and −5.5 ppm.[379] The data of Table 64 suggest some general conclusions:

1. The β-effect is opposite to that of alkanes; alkyl substitution decreases the shift of the olefinic carbon in β-position.
2. Neither conjugation nor the ring size of cycloalkenes has significant effect on the chemical shifts.

Table 65

**SUBSTITUENT PARAMETERS (PPM) FOR
ESTIMATION OF ^{13}C NMR CHEMICAL SHIFTS OF
OLEFINIC CARBON[379]**

			Substitution			
ρ_i	Generally	Mono	1,1-di	Z -1,2-di	E -1,2-di	tri
ρ_α	10.6	12.6	8.8	2.2	2.6	6.4
ρ_β	7.2	4.6	5.9	6.5	6.5	6.5
ρ_γ	−1.5	−1.2	−1.1	−0.7	−0.7	−0.7
$\rho_{\alpha'}$	−7.9	−8.0	−6.6	−3.0	−2.6	−5.1
$\rho_{\beta'}$	−1.8	−2.0	−1.9	−1.8	−1.8	−1.2
$\delta_{\gamma'}$	1.5	1.4	2.0	0.9	0.9	1.2

3.   An olefinic bond as substituent increases the chemical shift of C_α in the case of open chains and less than six-membered rings and decreases it for six-membered or larger rings.

4.   With conjugated cycloalkanes, similar to the ^{1}H NMR data, the shifts of the "inner" carbon atoms are higher, e.g., the sp^2 carbons C-1,2 and 3 of cycloheptatriene absorb at 123.3, 129.8, and 134.1 ppm.[610]

5.   From the data of simple unsaturated hydrocarbons, additive bond parameters may be derived for the sp^2 carbons (see Table 65), which approximately reproduce the chemical shifts by the formula

$$\delta C_{olefin} = \delta_0 + \Sigma \rho^i \qquad (313)$$

where δ_o = 123.3 and 122.2 ppm for E and Z configurations, respectively. Originally, further correction factors were also suggested,[379] but they do not improve the accuracy of calculations significantly (for *geminal* α,α- or α,α'-disubstitution and for one or more β-substituents, δ_o is, in turn, 118.5, 125.8, or 125.6 ppm). The ρ^i parameters of olefins substituted in different ways are significantly different, and the calculation is more accurate if these special values are taken into account instead of the general parameters. The agreement is rather poor for cycloolefins.

The chemical shifts of polysubstituted olefins may be estimated by using the data of vinyl derivatives[895,1240] (see Table 66) as substituent parameters:

$$\delta C_{\alpha,\beta} = 123.3 + \Sigma \rho^i_{\alpha,\beta} \qquad (314)$$

It can be seen from the values of ρ^i that the neighborhood of heteroatoms increases the shift of α-carbons, whereas the shift of β-carbons decreases, as follows from the mesomeric equilibrium

$$\overline{X}-CH{=}CH_2 \quad \longleftrightarrow \quad \overset{\delta+}{X} \overset{\cdots\cdots}{-\!-\!-} CH -\!\!-\!\!- \overset{\delta-}{\overline{C}H_2}$$

With halogen substituents this effect is compensated in part (chlorine) or overcompensated (bromine and, particularly, iodine) by the heavy atom effect. In nitrile and carbonyl derivatives an opposite change takes place:

Table 66

SUBSTITUENT CONSTANTS (PPM) DERIVED FROM ^{13}C NMR CHEMICAL SHIFTS OF VINYL DERIVATIVES FOR ESTIMATION OF CHEMICAL SHIFTS OF POLYSUBSTITUTED ETHYLENES[895,1143i,1240]

R	ρ_α	ρ_β	R	ρ_α	ρ_β	R	ρ_α	ρ_β
I	− 38.0	7.1	COOEt	6.4	7.1	SiMe$_3$	17.0	6.8
CN	− 15.0	14.3	Me	10.7	− 7.8	OAc	18.5	− 26.6
Br	− 7.8	− 1.4	Ph	12.6	− 11.0	N$^\oplus$Me$_3$	20.0	− 26.6
NC	− 3.8	− 2.6	CHO	13.2	12.8	NO$_2$	22.4	− 0.8
Cl	2.7	− 6.0	SO$_2$–CH=CH$_2$	14.4	8.0	F	25.0	− 34.2
COOH	4.3	9.0	Ac	15.1	5.9	OMe	29.5	− 38.8

Table 67

^{13}C NMR CHEMICAL SHIFTS (PPM) OF E AND Z 1,2-DISUBSTITUTED ETHYLENES AND THE SHIFT DIFFERENCES OF ANALOGUE CARBONS IN THE GEOMETRIC ISOMERS[95,379,900,985,1240]

$R_\alpha = R_\beta$	$\delta C(Z)$	$\delta C(E)$	$\Delta^{E,Z}\delta$	$R_\alpha = R_\beta$	$\delta C(Z)$	$\delta C(E)$	$\Delta^{E,Z}\delta$
Me	124.3	125.8	1.5	CN	120.8	120.2	− 0.6
Et	131.2	131.3	0.1	Cl	119.3	121.1	1.8
COOMe	128.7	132.4	3.7	Br	116.4	109.4	− 7.0
COOEt	130.5	134.1	3.6	I	96.5	79.4	− 17.1

R_α	R_β	$\delta C_\alpha(Z)$	$\delta C_\alpha(E)$	$\Delta^{E,Z}\delta C_\alpha$	$\delta C_\beta(Z)$	$\delta C_\beta E)$	$\Delta^{E,Z}\delta C_\beta$
Me	Cl	126.2	128.9	2.7	119.6	117.2	2.4
Me	Br	130.4	134.2	3.8	110.9	105.2	− 5.7
Me	CN	151.0	150.8	− 0.2	102.2	102.1	− 0.1
nBu	I	141.7	147.2	5.5	83.7	76.1	− 7.6
Me	Et	122.0	122.5	0.5	131.5	132.1	0.6
Me	i Pr	122.8	123.9	1.1	129.7	130.6	0.9

$$\overline{N}{\equiv}C{-}CH{=}CH_2 \longleftarrow \overset{\delta-\quad\delta+}{\overline{N}{=\!=\!=}C{-}\overset{\ominus}{C}H{-}CH_2^\oplus}$$

$$\overset{}{\underset{}{O}}{=}C{-}CH{=}CH_2 \longleftarrow \overset{\delta+\;\;\delta++\qquad\delta+++}{|\overline{O}^\ominus{-}C{=}CH{\cdots\cdots}CH_2}$$

This causes δC_β to increase in both cases, whereas δC_α decreases in nitriles and increases in carbonyl compounds (electron density increases on C_α or on the oxygen, respectively). Thus, it can be seen, that the substituent effect is more complex than with the alkanes, and a simple downfield shift proportional to the − I effects of substituents cannot be expected. The C-1 and C-2 signals, e.g., of vinyl acetate appear at 141.8 and 96.8 ppm; the analogous shifts of vinyl-methyl ether are 153.2 and 84.1 ppm.[729]

It is useful to observe that there is a linear relationship between the chemical shifts of aromatic and olefinic carbon atoms bearing the same substituent.[895] Likewise, there is a linearity between the size of the R substituents of RO–CH = CH$_2$ vinyl ethers and the chemical shifts of the terminal carbon.[634]

Generally downfield shifts may be observed for the Z isomers of 1,2-disubstituted olefins owing to the field effect. The difference is, however, small and may be opposite in some cases (see Table 67).

The spectrum of 1-bromopropene-1 (CHBr = CHMe) is a good example for the application of ^{13}C NMR chemical shifts in the distinction of Z − E isomers (see Figure 176). The figure

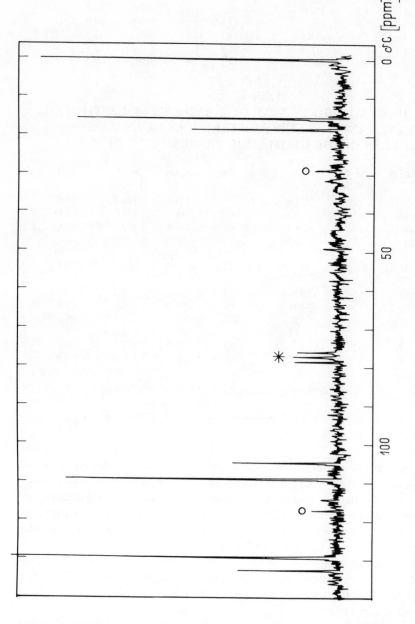

FIGURE 176.   The CMR spectrum of 1-bromopropene-1.   (From Johnson, L. F. and Jankowski, W. C., *C-13 NMR Spectra*. Wiley-Interscience, New York, 1972. With permission.)

Table 68
^{13}C NMR CHEMICAL SHIFTS (PPM) OF MONOSUBSTITUTED ALLENES[1369]

R	δC_α	δC_β	δC_γ
H	74.8	213.5	74.8
Me	84.4	210.4	74.1
Br	72.7	207.6	83.8
CN	80.5	218.7	67.2
OMe	123.1	202.0	90.3

shows the spectrum of a sample rich in the Z isomer.[1240] The signals of the isomers may be identified on the basis of the intensities. The upfield shift of the olefinic carbon adjacent to bromine is due to heavy atom effect* and the stronger shielding for C-2,3 of the Z isomer[379] to the field effect. The heavy atom effect is stronger for the E isomers of 1,2-disubstituted ethylenes, probably owing to the more extensive delocalization of the electrons, which decreases the energy of excited electronic states.

Comparing the shifts of 1,1,2-trichloroethylene (δC_α = 125.0, δC_β = 117.2 ppm) and tetrachloroethylene (121.4 ppm),[1097] it can be seen that the effect of chlorination (δC_α increases, δC_β decreases, (compare Table 66) remains the same for polychloroethylenes: it is additive. Accordingly, the shift difference $\Delta\delta C_{\alpha\beta}$ is high for 1,1-dichloroethylene: δC_α, 127.1 and δC_β, 113.3 ppm.[983]

^{13}C NMR is particularly important in the investigation of tri- and tetrasubstituted ethylene derivatives where the IR and ^{1}H NMR spectra are not informative.

4.1.2.10. Allene and Acetylene Derivatives

The signal of the central carbon of allene derivatives appears in the 200- to 220-ppm region, and it is thus easy to identify since only the oxo and thioxo groups absorb in this region. The signal of the terminal carbon atoms, like that of acetylene derivatives,[95] appears between the signals of olefins and saturated hydrocarbons, in the 65- to 95-ppm region (compare Table 50), in analogy to ^{1}H NMR experiences. The chemical shifts of some monosubstituted allene derivatives (R–CH=C=CH$_2$) are shown in Table 68; some further data can be found in the literature.[1191,1369]

Comparing these data to the changes found with olefins, parallel features may be found in the δC_α and δC_β shifts, but only in direction, since the actual magnitudes are rather different. For nitrile substitution, even the direction is opposite. Of course, the α-carbon atom is the most sensitive to substituent effects, whereas the range of δC_β is the narrowest. This shows that mesomeric effects have a decisive role in the substituent effect here. The shifts of tetramethyl and tetraphenyl allene (δC_α = 92.6 and 113.6; δC_β = 200.2 and 209.5 ppm)[1369] reflected the additivity of substituent effect.

The ^{13}C NMR chemical shift of acetylene is 71.9 ppm[1143j] and substituted derivatives absorb in the 60- to 95-ppm region. These shifts are much smaller than expected on the basis of the anisotropy of the triple bond.[380] The carbons bearing the substituent are deshielded and thus, for monosubstituted derivatives, $\delta C_\alpha > \delta C_\beta$.

The signal of terminal carbon atoms (C_β) can be identified not only by its upfield shift, but also on the basis of its doublet nature in proton decoupled or off-resonance spectra.[1424] The shifts of some acetylene derivatives are given in Table 69, and further data may be found in the literature.[1224]

---

* Compare p. 156.

Table 69
¹³C NMR CHEMICAL SHIFTS (PPM) OF
ACETYLENE DERIVATIVES[368,528,729,1223,1424,1514]

R_α	R_β	δC_α	δC_β	R_α	R_β	δC_α	δC_β
Me	H	79.2	66.9	Cl	nBu	56.7	68.8
Et	H	85.0	67.3	Br	nBu	38.4	79.8
nBu	H	84.5	69.1	I	nBu	−3.3	96.8
OEt	H	89.4	23.2	Ac	nBu	87.0	97.4
SEt	H	72.6	81.4	Me	Ph	79.8	85.7
Ph	H	83.3	77.3	Me	Me	73.9	73.9
F	H	86.7	12.2	Ph	Ph	89.9	89.9

Table 70
¹³C NMR CHEMICAL SHIFTS (PPM) OF SOME SIMPLE CARBONYL
COMPOUNDS[358,362,499,583,616,705,855a,898,928,929,1063,1143k,1374]

R	RCOMe	RCHO	RCOO$^\ominus$	RCOOH	RCONH$_2$	RCOCl	RCOOMe	(RCO)$_2$O
H	199.3	194.1	171.4	166.0	165.5		161.6	
Me	203.8	199.3	181.4	176.9	172.7	169.0	170.7	166.1
Et	206.3	201.5	184.8	180.1	177.2	174.7	173.3	170.8
CMe$_3$	210.4		188.6	185.9	180.9	180.3	178.9	173.9
CH$_2$Cl	200.7	193.3	174.7	173.8	168.3	168.8	167.8	162.1
CCl$_3$	186.3	175.9		167.1		162.6	161.0	154.1
CH=CH$_2$	196.9	192.1	179.3	170.4	168.3	165.6	165.5	
Ph	195.7	190.7	175.6	173.5	169.7	168.7	165.6	162.8

Methyl substitution increases δC_α and decreases δC_β by about 5 ppm. The effect of phenyl substituent is twofold on the C_α signal and the same but opposite on the signal of terminal carbon. Consequently, the sign of the substituent effect depends on the nature of transmitting electrons: the β-effect is transmitted by σ-electrons in methylacetylene and by π-electrons in phenylacetylene.

The effect of ethoxy groups may be interpreted by the importance of mesomeric structure $H\overline{C}^\ominus = C = \overline{O}^\oplus R$, that of ethylthio groups by $HC^\oplus = C = \overline{S}R^\ominus$. Consequently, in the former, a $+M$ and in the latter a $-I$ effect plays a dominant role in shielding. The opposite polarizations explain the extremely large difference of shift $\Delta\delta C_{\alpha,\beta}$ (in the case of iodohexine it is 100 ppm!). The ethinyl group as substituent increases shielding, by 5 to 15 ppm on the adjacent α-carbon and approximately 5 ppm on the β-carbon.

4.1.2.11. Carbonyl Compounds

Table 70 shows the carbonyl carbon shifts of some simple carbonyl derivatives. The carbonyl signals are easy to identify, since owing to the strong deshielding they seldom overlap with signals of different origin. The large shift is connected to the polarization of carbonyl bond ($C=O\rangle \leftrightarrow C^\oplus-O^\ominus|$) causing a reduction in electron density around the carbonyl carbon. When a hetero atom is attached to the carbonyl carbon, the shift decreases, which may have two reasons.

1. The lone electron pairs of the hetero atom compensate, through their mesomeric effect, the decrease in electron density around the carbon, i.e., the center of positive charge, shifts to the heteroatom.
2. The $-I$ effect of heteroatom suppresses the polarization of the carbonyl group and thus the decrease in electron density around the carbon.

Therefore, the signals with the largest shift are produced by the aliphatic ketones; they may be expected to occur in the 203- to 218-ppm region.

The shift increases in parallel with the order of the carbon in α-position.[705] Thus, the most deshielded carbonyl group is that of *t*-butyl-*i*-propyl ketone[705] (217.0 ppm); the carbonyl signal of di-*t*-butyl ketone is slightly more shielded (215.2 ppm) owing to field effect. (This anomaly also occurs in the IR and UV spectra.) The analogous signal of cyclic ketones shows no gradual change with ring size,[1241] the signals of cyclobutanone, cyclopentanone, cyclohexanone, and cycloheptanone appear at 208.2, 213.9, 208.8, and 211.7 ppm,[1496] respectively.

α-Halogen substitution or a conjugated, α,β-unsaturated group decreases the shift (compare Table 70). In the former case, the $-I$ effect of the halogen suppresses the polarization of the carbonyl bond, whereas in the latter the positive center of charge is shifted, owing to conjugation, to the substituent:

$$\overset{|}{\underset{\diagup}{\overset{\diagdown}{C}}}=\overset{|}{C}-\overset{|}{C}=O \longleftrightarrow \overset{\oplus}{\underset{\diagup}{\overset{\diagdown}{C}}}-\overset{|}{C}=\overset{|}{C}-\underline{\overline{O}}\,|^{\ominus}$$

Accordingly, the signal of the sp^2 carbon in β-position is more deshielded in the case of α,β-unsaturated derivatives than the carbon adjacent to the carbonyl group.*

The olefinic signals of this group of compounds appear in the 120- to 160 -ppm region, being more deshielded than in common olefins. The carbonyl signals of aldehydes may be assigned unanimously in proton-decoupled or off-resonance spectra on the basis of doublet splitting.[191]

Carbonyl substitution causes a significant increase in the shift of the α-carbon, the extent decreases with the order of carbon. The γ-carbon becomes slightly more shielded, while the β-effect is small and varying in direction. The methyl signals of acetone, acetaldehyde, acetate anion, acetic acid, acetyl chloride, and methyl acetate appear at 28.1, 31.2, 24.0, 21.1, 32.7, and 19.6 ppm, respectively;[616,705,875,1372i] the shifts of methyl ethyl (**274**), methyl isopropyl (**275**), and methyl *t*-butyl (**276**) ketones are shown in their formulas.[637]

274   **275**   **276**

The methyl signal of acetophenone appears at 24.9 ppm,[359] i.e., it is much more shielded than that of acetone (30.2 ppm). Here, conjugation decreases the positive polarization of carbonyl carbon (see above). The methyl shift increases in the case of *ortho* substitution,[359] the planar arrangement becomes impossible, and the extent of conjugation decreases. Accordingly, the substitution of the ring of acetophenone has significant effect on the carbonyl shift whenever the substituent is in *ortho* position (hindering planar configuration).[358]

Therefore, the effect of conjugation on electron distribution may be used to distinguish the planar and nonplanar conformations of α,β-unsaturated ketones by ^{13}C NMR.[928] The ranges of δC_α, δC_β and $\delta C=O$ chemical shifts characteristic of planar and nonplanar conformations are 122.5 to 142.5 and 120.5 to 136.5; 127.5 to 178.5 and 136.5 to 142.5; and 195.0 to 197.5 and 198.0 to 205.2 ppm, respectively.

---

* Compare the analogous phenomenon in ^{1}H NMR, p. 52.

Table 71
¹³C NMR CHEMICAL SHIFTS (PPM) OF CYCLIC CARBONYL
COMPOUNDS[115,344,377,777,928,1007,1230,1491,1496]

For similar reasons, the carbonyl signal of benzoic acid methyl ester derivatives is more deshielded and has a broader range (164 to 171 ppm) for *ortho* substitution than for the same substituent in *meta* or *para* position[362] (165 to 167 ppm). There is a linear relationship between the carbonyl shifts of *meta*-substituted benzaldehydes and the Hammett constants.[947]

The geometric isomers of acrylic acid derivatives cannot be distinguished with certainty on the basis of the carbonyl shift, e.g., the carbonyl signals of *Z* and *E* β-chloro- and β-bromo-acrylic acid are uniformly at 165.9 and 166.3 ppm, respectively, and those of the corresponding methyl ester isomers are at 163.9 and 164.1 ppm and 163.6 and 163.9 ppm, respectively.[877]

Table 71 is a collection of cyclic carbonyl derivatives. It is added to the data that lactones with six-membered or larger rings absorb in the 165- to 177-ppm region, the characteristic

Table 72
THE CARBONYL ^{13}C NMR CHEMICAL SHIFT (PPM) OF ACETONE IN VARIOUS SOLVENTS[903]

Solvent	$\delta C = O$	Solvent	$\delta C = O$
Cyclohexane	202.8	MeCN	207.3
Et$_2$O	203.2	CHCl$_3$	207.5
CCl$_4$	203.9	MeOH	208.9
C$_6$H$_6$	204.4	MeCOOH	211.4
Pure liquid	205.2	TFA	219.3
Dioxane	205.2	H$_2$SO$_4$	244.4

range of quinones is 170 to 190 and lactams, 165 to 180 ppm, and those of dialkyl and diaryl carbonates are approximately 155 and 152 ppm, respectively.

^{13}C NMR signals are usually insensitive of solvent and concentration, except when hydrogen bond formation is possible. Therefore, significant solvent effects have been observed most frequently in connection with carbonyl compounds.

As an illustration, Table 72 shows the shifts of the carbonyl signal of acetone in various solvents.[903] Another example is the variation of the chemical shift of the carboxy carbon of acetic acid with concentration in various solvents.[905]

In polar acetone, the shift increases (by about 6.5 ppm) with gradually decreasing concentrations, indicating that cyclic dimer association is replaced by stronger solvent-solute hydrogen bonds. In the case of nonpolar cyclohexane, most of the molecules remain associated over the entire concentration range; the slight variation in the shift (<1 ppm) is a consequence of the solvation of dimers for a small portion of molecules. In weak proton donor, chloroform, the shift first decreases (the proton of chloroform, associating with the carboxy oxygen of the dimer acid, loosens the hydrogen bond of the latter without breaking it, and thus increases the electron density around the carboxy carbon), and then increases (dimers are split on further dilution and solvent-solute hydrogen bonds become dominant). The shift of the carbonyl signal increases more slowly and to a smaller extent than in acetone (weaker hydrogen bond), but faster and more, by approximately 3 ppm, than in dioxane. By decreasing the concentration in aqueous solution, chemical shift first increases and then, after a flat maximum, it becomes constant. Water molecules associate both with the carbonyl and hydroxy groups of the carboxy moiety, through their hydrogen and oxygen atoms, and thus in dimer association at high concentrations, too, there is about the same electron distribution around the acid groups after dissociation of dimers as in pure liquid (see Figure 177).

The carbonyl signal of N-substituted carboxylic amides splits due to hindered rotation. The signals of N-methylformamide are at 163.4 and 166.7 ppm.[856] A similar splitting occurs with ketoximes owing to the *syn-anti* isomerism. Thus, the signals of methyl ethyl ketoxime (**277a,b**) are doubled (see Figure 178), and their relative intensities make the assignment easy.

277

The geometric isomers of keteneimines (**278a** and **b**) are also easy to distinguish. When R=Me, R'=H, the ratio of E and Z isomers is 93:7, and for the R=R'=Me case, the E isomer

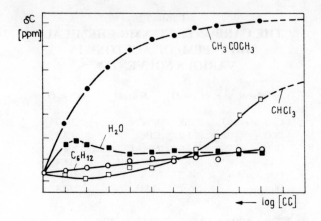

FIGURE 177.   The concentration and solvent dependence of the ^{13}C NMR chemical shift of the carbonyl-carbon atom in acetic acid.[905]

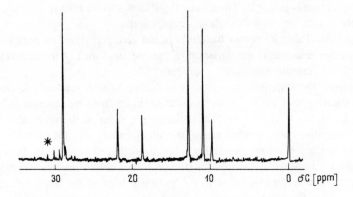

FIGURE 178.   The ^{13}C NMR spectrum of the mixture of (a) *syn*- and (b) *anti*- methyl-ethyl-ketoximes (**277a** > **277b**).[856] (From Levy, G. C. and Nelson, G. L., *Carbon-13 Nuclear Magnetic Resonance for Chemists*, Wiley-Interscience, New York, 1972. With permission.)

is less favored owing to the mutual steric hindrance of methyl groups, and thus the *E:Z* ratio is 24:76.[139] For similar reasons, the compound R=Ph, R′=H occurs exclusively in the *E* form, whereas with the R=Ph, R′=Me analogue the *E:Z* ratio is 2:1.[222] The ^{13}C NMR

278

a (*E*)   **b** (*Z*)

shifts show that the substituted phenyl ring is not coplanar with the C=N bond, since the signal of methyl substituent R′ hardly changes in the isomers: δCH_3 (2′) = 19.7 (*E*) and 20.3 (*Z*). At the same time, the signal of the methyl carbon attached to the C=N bond is by 8.2 ppm more shielded for the *E* isomer (21.2 ppm) than for the *Z* isomer (29.4 ppm), owing to the steric proximity of the phenyl (R) and methyl groups. The ^{13}C NMR investigation

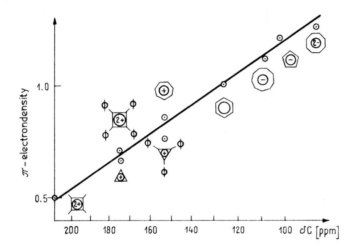

FIGURE 179.   Correlation betwee π-electron density and ¹³C NMR chemical
shifts of various ionic aromatic systems.[1066]

of similar geometric isomers is feasible in much more complex cases, e.g., for compounds
of type $RSO_2-N=CH-NR'-N=CR''R'''$, where eight isomeric structures are, in principle,
possible.[709] The C_α signal of the R group of nitrosamines $(R-N=N \rightarrow O)$ shifts upfield in
the spectrum of the *syn* isomer containing the oxygen in Z position.[92,431]

The signals of carbonate and hydrocarbonate anions appear at 170 and 160 ppm, respectively.[1091] Metal carbonyls are strongly deshielded, e.g., the signal of $Ni(CO)_4$ is at 191.6
and the signal of $Fe(CO)_5$, is at 209.6 ppm.[828,1374]

The carbonyl signals of thiocarbonyl derivatives show downfield shifts of 20 to 30 ppm
with respect to the carbonyl analogues, and thus, disregarding carbonium cations, thioketones
have the most deshielded carbons.[1143m] There is a linear relationship between carbonyl and
analogous thiocarbonyl shifts[744]

$$\delta C=S = 1.45 (\delta C=O) - 46.5 \qquad (315)$$

Using this relationship, the structures of tautomeric systems may be determined[337] from a
comparison of the ¹³C NMR shifts of analogous C=O and C=S derivatives. Thioacetamide
$(MeCSNH_2)$ and thioacetic acid $(MeCSOH)$ absorb at 207.2 and 194.5 ppm, respectively.[897]
The thiocarbonyl signals of γ-thiolactams and five-membered cyclic thioureas are in the 182
to 186 and 173 to 176 ppm regions.[227]

4.1.2.12. Aromatic Compounds

The ¹³C NMR spectra of aromatics were studied intensively already in the early stages
of spread of carbon resonance spectroscopy.[95,358,360,434,816,819,822,1355,1371] The signals of unsubstituted aromatic carbons may be expected in the 123- to 142-ppm range, and the shifts
are determined primarily by the local π-electron density, from which it was concluded that
the shift range of alternating benzene homologues (approximately 10 ppm) is much narrower.[733] A proof is the linear relationship between chemical shift and π-electron density[1066,1356]
that was found for ionic aromatic systems (see Figure 179). The shifts of some simple
aromatic compounds may be seen in Table 73.

Compound **279**, although aromatic on the basis of the Hückel rule (has $14 = 4n + 2\pi$-
electrons), is more reactive and less stable than the dehydro derivative (**280**) which contains
16 π-electrons, i.e., must be "antiaromatic". Carbon resonance spectroscopy shows the
real reason for the anomaly: the "middle" double bond does not take part in the delocalized

Table 73

^{13}C NMR CHEMICAL SHIFTS (PPM) OF SOME SIMPLE AROMATIC COMPOUNDS[23,733,735,736,819]

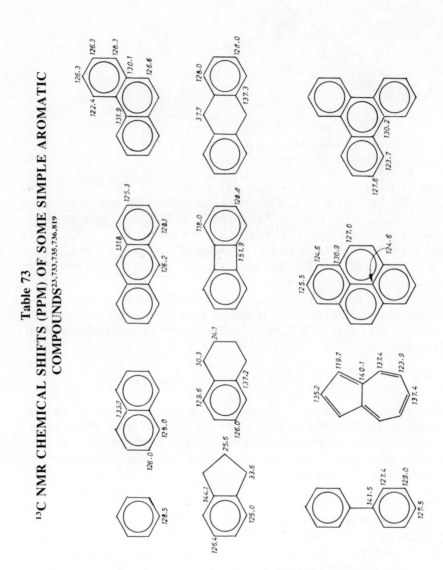

Table 74
¹³C NMR CHEMICAL SHIFT OF MONOSUBSTITUTED BENZENES AS REFERRED TO BENZENE[729,855c,1143n,1355]

Substituent	$\Delta\delta C^s$	$\Delta\delta C^o$	$\Delta\delta C^m$	$\Delta\delta C^p$	Substituent	$\Delta\delta C^s$	$\Delta\delta C^o$	$\Delta\delta C^m$	$\Delta\delta C^p$
I	−34.1	8.7	1.4	−1.6	SMe	9.9	−2.0	0.1	−3.7
CN	−15.4	3.6	0.6	3.9	NHAc	11.1	−9.9	0.2	−5.6
CF₃	−9.0	−2.2	0.3	3.2	SO₂H	15.0	−2.2	1.3	3.8
Br	−5.5	3.4	1.7	−1.6	NH₂	18.0	−13.3	0.9	−9.8
COOH	2.1	1.5	0	5.1	NO₂	20.0	−4.8	0.9	5.8
SH	2.2	0.7	0.4	−3.1	NHMe	21.7	−16.2	0.7	−11.8
COCl	4.6	2.4	0	6.2	NMe₂	23.0	−16.0	2.0	−12.0
CONH₂	5.4	−0.3	−0.9	5.1	OAc	23.0	−6.4	1.3	−2.3
Cl	6.2	0.4	1.3	−1.9	OH	26.9	−12.7	1.4	−7.3
CHO	8.6	1.3	0.6	5.5	OMe	31.4	−14.4	1.0	−7.7
Me	8.9	0.7	−0.1	−2.9	F	34.8	−12.9	1.4	−4.5
Ac	9.1	0.1	0	4.2	NO	37.4	−7.7	0.8	7.0

system. The separated character is proved by the much lower shifts for **280**; the anisotropic effect of the π-electrons localized on the periphery of the ring system increases shielding around the carbon atoms in the middle of the system. A proof for delocalization is that the shift range of peripheral carbons is much narrower ($\Delta\delta$ = 12.4 ppm) than for the dihydro analogue ($\Delta\delta$ = 23.9 ppm).

279   **280**

The substituted carbon atoms of benzene derivatives absorb in the 120 to 170-ppm range; the unsubstituted ones absorb in the 110- to 140-ppm range.

When investigating the chemical shifts of monosubstituted benzene derivatives (Table 74) as a function of the electron-withdrawing effect of substituents, the shift of *para* carbon atom changes over a range of approximately 22 ppm (− 14 to + 8 ppm) roughly linearly (with Hammett's σ),[1355] the signal of the *meta* carbon is not sensitive to substitution, appearing in a very narrow interval close to the benzene signal (between − 1.5 and + 3.5 ppm), whereas the shift of the *ortho* carbon shows larger and unsystematic shifts between − 18 and + 11 ppm. The reason is the field effect, which has the strongest influence on the shielding of the *ortho* carbons, whereas in *para* position only the inductive and mesomeric effects prevail. The substituent effect has an even stronger impact on the substituted carbon atom, and thus the signal of the latter shows the largest and most unsystematic change (between − 35 and + 35 ppm). It is worth noting that the deviation of the chemical shift of substituted and *ortho* carbons from that of benzene are usually opposite in sign (compare Table 74). For most substituents, $\delta C^o < \delta C^p$ (except for R, X, and O$^\ominus$), and $\Delta\delta C^{o,p}$ is approximately 5 ppm for O, 4 ppm for C=O, and approximately 3.5 ppm for N.

The carbon shifts of polysubstituted benzenes can be estimated similarly as ¹H NMR shifts, and more significant deviations occur only for adjacent substituents.

$$\delta C_{Ar} = 128.5 + \sum \rho^i \qquad\qquad (316)$$

where the ρ^i constants are the $\Delta\delta C$ values given in Table 74.

As an example, for the assignment of aromatic ^{13}C NMR signals, the spectrum of 2-methyl-4-hydroxy-chlorobenzene (**281**) is shown (see Figure 180). The upfield signal corresponds of course, to the methyl carbon. The expected shifts of the aromatic carbons may be calculated by Equation 316. The result can be found in Table 75. On this basis the assignment is straightforward. The deviation between calculated and observed values is ≤0.4 ppm, except for the atoms C-1,2 ($\Delta\delta$ = 1.7 and 1.8 ppm). The larger discrepancy is due obviously to the steric proximity between the adjacent groups, i.e., the field effect. It is noted that the signals of substituted carbons may also be identified on the basis of their lower intensities.*

Cl

Me

OH

281

Field effect was also observed, e.g., for 2-substituted toluenes, where it causes an upfield shift in the methyl signal.[1557] Therefore, the C-1 and C-2 shifts of *ortho*-dichlorobenzene are lower (132.6 ppm) than the analogous C-1 and C-3 shifts of the *meta* isomer (135.1), and the deviation from the value calculated by Equation 316 is 2.5 ppm for the former and only 0.9 ppm for the latter.

The additivity of substituent effect has an extensive literature for benzene derivatives,[358-362,443-447,578,820,822-824,903,1239,1396,1507] and similarly for fused benzene derivatives, primarily substituted naphthalenes,[374,443-447,578,631,769,971,1507,1530] and the anomalies connected with *ortho* substituents are in the limelight of interest.

The shifts of saturated groups attached to aromatic rings are in the 10- to 60-ppm range. The 2-methyl signal of 2,3-dimethylanisole is at 10.9 ppm.[1372j] Data on some further hydroaromatic systems may be found in Table 73.[1352]

4.1.2.13. Heteroaromatic Systems

The presence of heteroatoms has the consequence that the ^{13}C NMR chemical shifts of unsubstituted heteroaromatic compounds are spread over a much wider range (100 to 170 ppm) than those of benzene analogues. Within this, the ranges of five- and six-membered cyclic systems are somewhat narrower (104 to 155 and 120 to 170 ppm, respectively), but still rather broad.

The chemical shifts show some characteristics also observed in ^1H NMR. The neighborhood of heteroatom causes a paramagnetic shift in the Cα signal, and thus the effect is the strongest for carbons between two heteroatoms. The upper limit is represented by the 167.5-ppm shift of triazole. The carbons not adjacent to heteroatoms are of course more shielded (e.g., the β-carbon of pyridine), whereas carbons adjacent to one heteroatom represent medium shifts.[825,1148] Table 76 shows the data of some simple five-membered heterocycles.

The increase in aromatic character in the sequence furan - pyrrole - thiophene - selenophene

---

* Compare p. 221 and 228.

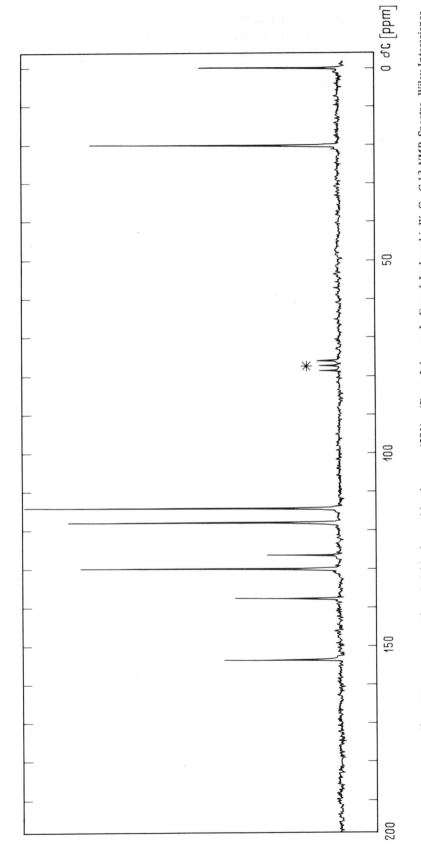

FIGURE 180. The ^{13}C NMR spectrum of 2-methyl-4-hydroxy-chlorobenzene (281). (From Johnson, L. F. and Jankowski, W. C., *C-13 NMR Spectra*, Wiley-Interscience, New York, 1972. With permission.)

Table 75

THE MEASURED AND CALCULATED ^{13}C NMR CHEMICAL SHIFTS (PPM) FOR 2-METHYL-4-HYDROXY-CHLOROBENZENE (281) AND THE ASSIGNMENT DEDUCED FROM THE COMPARISON OF THESE DATA

| Carbon atom | Substituent contributions from Table 74 | | | δC | | |
	Cl	Me	OH	Calculated	Measured (Reference 729)	|ΔδC|
C-1	6.2	0.7	−7.3	128.1	126.3	1.8
C-2	0.4	8.9	1.4	139.2	137.5	1.7
C-3	1.3	0.7	−12.7	117.8	117.8	0
C-4	−1.9	−0.1	26.9	153.4	153.3	0.1
C-5	1.3	−2.9	−12.7	114.2	114.1	0.1
C-6	0.4	0.1	1.4	130.2	129.8	0.4

Table 76

^{13}C NMR CHEMICAL SHIFTS (PPM) OF BASIC FIVE-MEMBERED HETEROAROMATIC COMPOUNDS[537,544,729,783,1080,1148,1149,1390,1494]

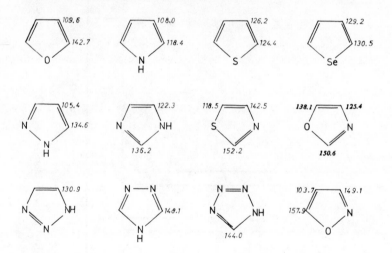

Table 77

SUBSTITUENT CONSTANTS DERIVED FROM ^{13}C NMR CHEMICAL SHIFTS (PPM) OF MONOSUBSTITUTED THIOPHENES FOR ESTIMATION OF CHEMICAL SHIFTS OF POLYSUBSTITUTED DERIVATIVES[1390]

| R ρ_R^i | 2-Substituent ($i = 2$) | | | | 3-Substituent ($i = 3$) | | | |
	C-2	C-3	C-4	C-5	C-2	C-3	C-4	C-5
I	−49.6	11.5	3.5	7.9	3.7	−47.5	9.0	4.8
Br	−11.8	3.8	1.5	2.7	2.5	−15.8	3.9	−1.4
COOMe	8.8	6.6	0.9	7.5	7.5	6.7	0.9	1.0
Me	14.6	−1.5	0.2	−1.8	−4.3	10.6	2.6	0.3
CHO	19.0	10.2	1.9	10.2	12.7	16.4	−1.3	2.9
OMe	42.3	−22.6	−1.7	−13.0	−27.7	32.7	−7.0	0.1

is represented both in the decreasing $\Delta\delta C_{\alpha,\beta}$ values (33.1, 10.4, 1.8, and 1.3 ppm) and in the increasing shifts. There is a proportionality between the proton and carbon shifts,[1080] i.e., the latter are determined primarily by the π-electron density:

$$\delta C = 22.24 \, (\delta H) - 152.5 \tag{317}$$

Hence, the deviation from linearity may be regarded as a measure of σ-contribution to the shielding of carbon atoms. It is interesting that the proportionality constant is practically the same as that found for benzene derivatives.[1356]

Table 77 shows the substituent parameters arising from the chemical shifts of monosubstituted thiophenes,[1390] which, again in parallel to proton resonance, reflect the "insensitivity" of δC-4 and δC-5 shifts (for the 2- and 3-substituted series, respectively) to substitution (they are the analogues of *meta* carbons in benzenes). Of course, the shift of substituted carbon shows the largest change, and δC-5 is strongly influenced by 2-substitution, although this carbon is not adjacent to the substituent. These considerations can be deduced by comparing the shift ranges of the various carbons derivable from the data given in Table 77. With the aid of ρ^i parameters the approximate shifts of polysubstituted derivatives may be estimated in the usual manner:

$$\delta C^k = \delta_o^k + \Sigma \rho_i^k \tag{318}$$

where k refers to the position of the carbon (2, 3, 4, or 5), ρ_o^k is the corresponding shift of thiophene, and ρ_i^k is the contribution of substituent in position i to the shift of atom k. By means of these parameters, the shifts of pyrroles may also be estimated. For substituted pyrroles[6] and thiophenes,[1390] other additivity rules are also known. In the 2-substituted series, δC-3 and δC-5 are in a linear relationship.[1390]

The effect of protonation is analogous in the case of imidazole and pyrrole, but different for pyrazole.[1149] In the series anion - imidazole - protonated imidazole, δC-2 and δC-4,5 equally decrease (by 8.9 and 1.6 ppm for C-2 and 4.5 and 2.2 ppm for C-4,5). For pyrazole δC_α increases with respect to the neutral molecule, both in the anion (5.2 ppm) and in the cation (1.7 ppm), whereas δC_β decreases in the anion (-1.3 ppm), but increases in the cation (4.3 ppm). Comparing these differences to the shift changes expected on the basis of $-I$ effect alone, only the last series is not anomal. The theoretical interpretation of measured shift (e.g., Reference 1149) is not complete yet, but the experimental data are suitable for determining the site of protonation in fused systems containing more nitrogen atoms. Thus, originating from the shifts of imidazole and benzimidazole and their cations it could be determined that in purine N-1 is protonated first.[1150]

The shifts of some simple representatives of six-membered heterocyclic rings are shown in Table 78. The literature contains a large number of additional data, e.g., on the ^{13}C NMR shifts of isoquinolines.[448,722,1376]

On protonation, the shielding of α-carbon of pyridine and its fused derivatives and of heteroaromatic rings with more nitrogens generally increases, whereas that of β- and γ-carbon atoms decreases. As an illustration, Figure 181 shows the shift changes of the various carbons of quinoline as a function of pH.[194] One may observe the upfield shift of C-2,8a and the analogous behavior of C-5 on protonation, the large and small downfield shift, respectively, of δC-4,6,7 and of C-3,4a,8 signals, respectively, from which pK may be determined (see Figure 181b).

An effect similar to protonation may be observed in the aqueous solution of pyridine on dilution, which may be attributed to association with water.[1148] There is a linear relationship between the chemical shifts and π-electron density,[825] although the deviation is quite high, indicating the importance of other contributions, too.

Table 78
¹³C NMR CHEMICAL SHIFTS (PPM) OF SOME SIMPLE SIX-MEMBERED HETEROAROMATIC COMPOUNDS[188,192,825,1149,1151,1391]

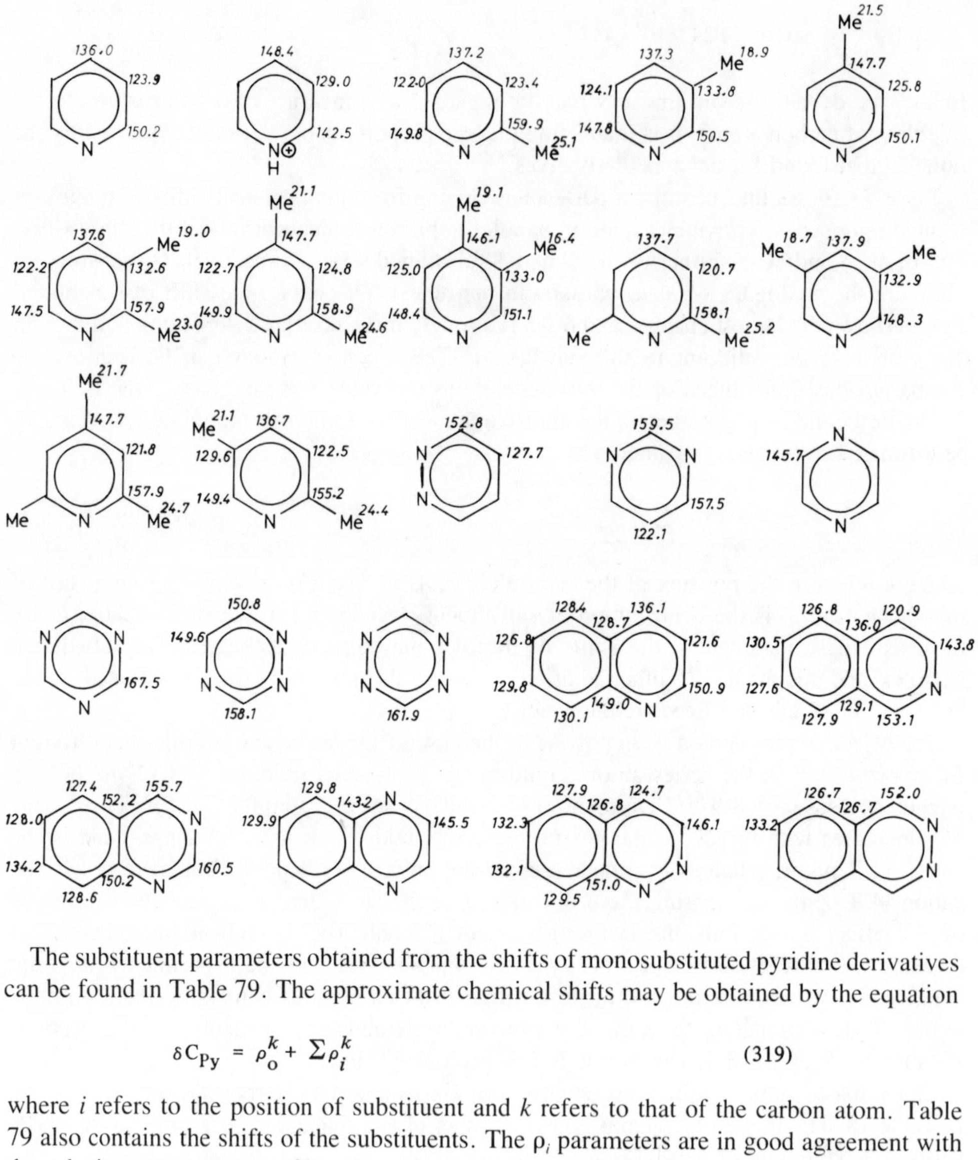

Pyridine: 136.0, 123.9, 150.2

Pyridine-N-oxide (N⊕H): 148.4, 129.0, 142.5

137.2, 122.0, 123.4, 159.9, 149.8, N, Me 25.1

Me 18.9: 137.3, 124.1, 133.8, 147.8, 150.5

Me 21.5: 147.7, 125.8, 150.1

137.6, Me 19.0, 122.2, 132.6, 147.5, 157.9, Me 23.0

Me 21.1: 147.7, 122.7, 124.8, 149.9, 158.9, Me 24.6

Me 19.1: 146.1, Me 16.4, 125.0, 133.0, 148.4, 151.1

137.7, Me 18.7, 120.7, 158.1, Me, Me, N, 25.2

137.9, Me, 132.9, 148.3

Me 21.7: 147.7, 121.8, 157.9, Me, N, Me 24.7

Me 21.1, 136.7, 129.6, 122.5, 149.4, 155.2, Me 24.4

152.8, 127.7 (N, N)

159.5, 157.5, 122.1 (N, N)

145.7 (N, N)

150.8, 149.6, 167.5 (N, N, N)

N, N, 158.1

N, N, 161.9

128.4, 128.7, 136.1, 126.8, 121.6, 129.8, 150.9, 130.1, 149.0, N

126.8, 136.0, 120.9, 130.5, 143.8, 127.6, 129.1, 127.9, 153.1, N

127.4, 152.2, 155.7, 128.0, 134.2, 150.2, 128.6, 160.5, N, N

129.8, 143.2, N, 129.9, 145.5, N

127.9, 124.7, 126.8, 132.3, 146.1, 132.1, 151.0, 129.5, N, N

126.7, 152.0, 126.7, 133.2, N, N

The substituent parameters obtained from the shifts of monosubstituted pyridine derivatives can be found in Table 79. The approximate chemical shifts may be obtained by the equation

$$\delta C_{Py} = \rho_o^k + \Sigma \rho_i^k \qquad (319)$$

where i refers to the position of substituent and k refers to that of the carbon atom. Table 79 also contains the shifts of the substituents. The ρ_i parameters are in good agreement with the substituent parameters of benzene, mainly in the case of β-substitution, but the deviations are below 2.5 ppm for γ-substitution as well. In pyridine-*N*-oxide ($C_5H_5N\rightarrow O$) the C_α and C_γ signals show large diamagnetic shifts (10.5 and 10.3 ppm), whereas C_β shows an opposite shift of 3.4 ppm.[47,324]

Table 80 contains the chemical shifts of some other fused nitrogen heterocycles, which may be used for orientation in the assignment of the signals of other ring systems and substituted derivatives. The structure of the product 3-hydroxy-2-picoline amide (**282b**), formed in the ammonolysis of 3-hydroxy-2-picolinic acid (**282a**), was proved by isotopic labeling,[997] refuting the previously assumed 3-amino-2-picolinic acid structure (**282c**). Half of the mobile protons was exchanged for deuterium by adding three equivalents of D_2O to the anhydrous DMSO solution of the reaction product, whereupon a doublet splitting could

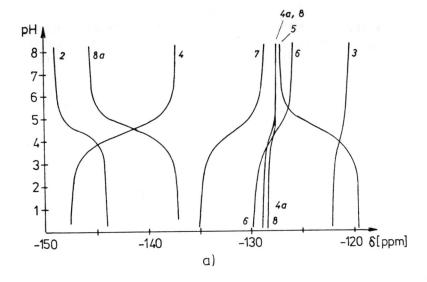

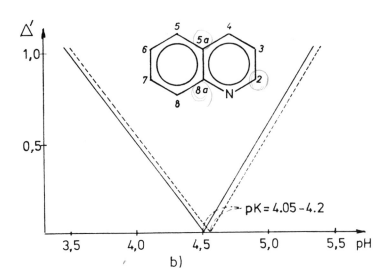

FIGURE 181. The pH-dependence of (a) ^{13}C NMR chemical shifts of quinoline and (b) the $\Delta' = \log (\delta_{max}-\delta)/(\delta-\delta_{min})$ values corresponding to δC-4 (solid line) and δC-5 (broken line) for determination of pK by the relationship[1043b] $pH = pK + \Delta'$.[194]

a: $R = R' = OH$

b: $R = NH_2$, $R' = OH$

c: $R = OH$; $R' = NH_2$

282

be observed in the C-3,4,6 signals owing to the isotope effect.* Since, unlike hydroxy protons, NH$_2$ protons are capable of fast exchange, an NH $\rightarrow$ ND - substitution would not have caused splitting, and even if it did, a triplet splitting would occur, corresponding to the molecules containing NH$_2$, NHD, and ND$_2$ groups.

* Compare p. 158—159.

Table 79

SUBSTITUENT CONSTANTS (PPM) DERIVED FROM ^{13}C NMR CHEMICAL SHIFTS OF MONOSUBSTITUTED PYRIDINES FOR ESTIMATION OF CHEMICAL SHIFTS OF POLYSUBSTITUTED DERIVATIVES AND THE SHIFTS OF THE SUBSTITUENTS[1143p,1190,1192,1461]

Substituent	Site	ρ^2	ρ^3	ρ^4	ρ^5	ρ^6	$\delta C(Me)$	δC^a
CN		−15.9	5.0	1.6	3.6	1.4		118.4
Br		−6.7	4.8	3.3	−0.5	1.4		
Cl		2.3	0.7	3.3	−1.2	0.6		
CHO		3.5	−2.6	1.3	4.1	0.7		193.9
Ac	$i = 2$	4.3	−2.8	0.7	3.0	−0.2	25.0	199.5
Me		8.8	−0.6	0.2	3.0	−0.4	25.1	
NH₂		11.3	−14.7	2.3	−10.6	−0.9		
Et		13.6	−1.8	0.4	−2.9	−0.7	14.2	32.0
F		14.4	−13.1	6.1	−1.5	−1.5		
OMe		15.3	−7.5	2.1	−13.1	−2.2	53.1	
I		7.1	−28.4	9.1	2.4	0.3		
CN		3.6	−13.7	4.4	0.6	4.2		117.5
Br		2.1	−2.6	2.9	1.2	−0.9		
CHO		2.4	7.9	0	0.6	5.4		192.2
Cl		−0.3	8.2	−0.2	0.7	−1.4		
Ac	$i = 3$	3.5	8.6	−0.5	−0.1	0	26.8	197.5
Me		1.3	9.0	0.2	−0.8	−2.3	18.9	
Et		−0.4	15.5	−0.6	−0.4	−2.7	15.4	26.8
NH₂		−11.9	21.5	−14.2	0.9	−10.8		
OMe		−12.9	31.3	−16.0	−0.1	−8.8	55.3	
OH		−10.7	31.4	−12.2	1.3	−8.6		
F		−11.5	36.2	−13.0	0.9	−3.9		
CN		2.1	2.2	−15.7				117.5
Br		3.0	3.4	−3.0				
CHO		1.7	−0.6	5.5				192.7
Ac		1.6	−2.6	6.8			26.9	197.8
Me	$i = 4$	0.5	0.8	10.8			21.5	
Et		−0.1	−0.4	17.0			13.9	28.3
NH₂		0.9	−13.8	19.6				
F		2.7	−11.8	33.0				
OMe		0.5	−14.1	28.9			55.0	

^{a} Shift of carbons in groups CH_2, CN, or C=O.

4.1.2.14. Steroids

Carbon resonance spectroscopy is particularly important in the investigation of complicated macromolecules. The molecular weight limit of the chances of successful structure determination which was about 300 to 500 in ^{1}H NMR, shifted considerably. With compounds containing many similar types of protons, including biologically important carbohydrates, terpenes, and steroids, the ^{1}H NMR limit is close to the lower boundary owing to the featureless, overlapping, and thus rather uninformative signals. In contrast, in the ^{13}C NMR spectrum of even the most complicated steroid molecule, the signals of each carbon appear separately in most cases, and their assignment is also possible without particular difficulties.

The ^{13}C NMR investigation of steroids has ample literature including one pioneering review[1186] and two more recent surveys summarizing tabulated shifts and assignments of

Table 80
^{13}C NMR CHEMICAL SHIFTS (PPM) OF SOME FUSED HETEROAROMATIC COMPOUNDS[418,467,1007,1084,1143r,1150,1152,1230]

Structure 1 (indole): 121.0, 128.5, 102.4, 122.0, 124.9, 120.0, 135.9, 111.6, N–H

Structure 2 (benzimidazole): 115.4, N, 141.5, 122.9, 137.9, N–H

Structure 3 (indazole): 120.4, 122.8, 133.4, 120.1, 125.8, 139.9, 110.0, N–H, N

Structure 4 (benzoxazole): 120.5, 140.1, N, 152.6, 125.4, 124.4, 150.0, 110.8, O

Structure 5 (benzisoxazole): 124.3, 122.2, 147.1, 123.0, 130.6, 162.7, 109.9, O, N

Structure 6 (benzothiazole): 123.1, 153.2, N, 155.2, 125.9, 125.2, 133.7, 122.1, S

Structure 7 (indolizine): 119.6, 99.4, 133.4, 117.2, 114.1, 110.4, 125.6, 113.0, N

Structure 8 (pyrazolo): 117.4, 96.3, 139.5, 141.3, 122.4, 110.8, 128.1, N, N

Structure 9 (imidazopyridine): 118.2, 130.6, 119.9, 119.4, 112.7, 122.8, 128.4, N, N

Structure 10: 117.6, 145.6, N, 134.1, 124.6, 112.2, 127.0, 113.4, N

Structure 11: 151.0, 148.9, N, N, 135.2, 109.4, 136.1, 112.7, N

Structure 12: 144.8, 128.4, N, 147.9, 152.0, N, 154.9, N–H, N

about 400 and 100 steroids, respectively.[157,1314] In the following, some data and spectra are shown to illustrate the possibilities offered by ^{13}C NMR methods for the structure elucidation of steroids.

The assignment of the spectrum of testosterone (see Figure 182), given along with Formula **283**, is based on the following considerations. The smallest shifts may be assigned to the methyl, and the largest may be assigned to the keto-carbon atoms. Of the two, also substantially deshielded olefinic signals, the downfields one can be assigned to the quaternary atom. The signal of C-17 also separates from the others (downfields), owing to the hydroxy substituent. The tertiary carbons (C-8,9,14) are more deshielded than the methylene ones. This, nevertheless, does not apply for C-8 on account of the field effect (1,3-*diaxial* interaction between the two methyl groups and H-8). The signals of quaternary C-10,13 and secondary C-11 atoms appear upfield for similar reasons.

Steroid structure 283 with shifts: OH, 11.0, 36.4, 81.2, 20.6, 42.7, 30.1, 35.6, 17.3, 53.9, 50.4, 23.2, 33.8, 38.6, 35.6, 199.4, 171.4, 31.5, O, 123.6, 32.7

283

As the next example, the 90- and 500-MHz spectra of 3β-acetoxy-cholest-5-ene (**284**) are shown in Figure 183a and b,[218] along with the spectrum of the solid sample obtained by ''magic angle'' spinning method* (see Figure 183c).[219] Figure 183a and b illustrate the sensitivity enhancement effect of magnetic field: the signal-to-noise ratio of the 500-MHz

---

* Compare Volume I, p. 219—220.

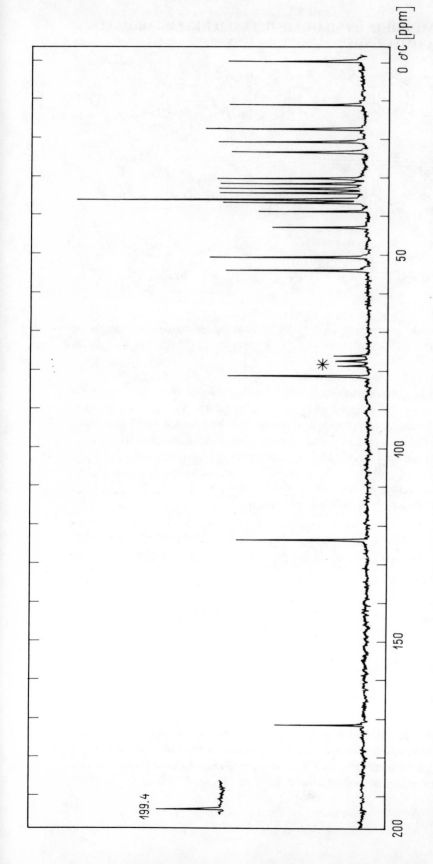

199.4

FIGURE 182.   The ^{13}C NMR spectrum of testosterone (**283**). (From Johnson, L. F. and Jankowski, W. C., *C-13 NMR Spectra*, Wiley-Interscience, New York, 1972. With permission.)

spectrum obtained by a single pulse from a 10% solution is better than that of the 90-MHz spectrum obtained by averaging of 400 scans from a saturated solution. In the spectrum of the solid sample, significant changes are caused by the conformation effects. The splitting of C-3 (carbonyl), C-5,6 (olefinic), and C-18 (methyl) signals is well observable, as well as the shifts, broadenings, or splittings of some of the aliphatic signals.

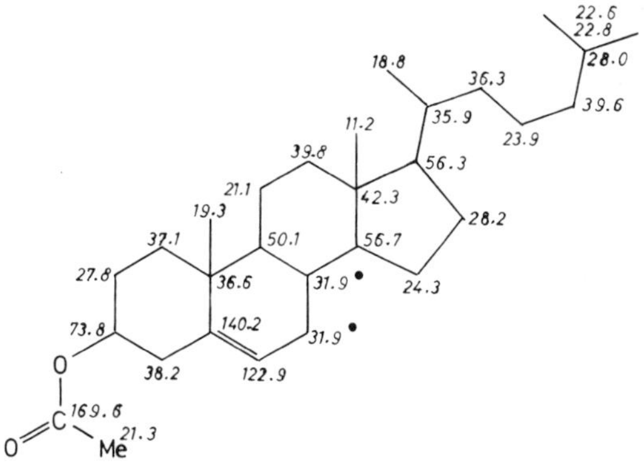

284

As to the shifts, the following facts are worth noting. When compared to testosterone (**283**), the largest differences may be found in the signals of C-3,4,5,6,17, reflecting the most serious changes in structure. δC-3 is much less than δC-17 of testosterone which is also substituted by an sp³ oxygen, since C-3 in the former has two secondary carbon neighbors, whereas in the latter one neighbor of C-17 is a quaternary carbon (C-13). δC-5 is much less than in testosterone since the conjugation with the carbonyl causes strong deshielding. Of course, δC-17 is also smaller owing to the weaker substituent effect of the tertiary carbon replacing the hydroxy substituent. The change in functional groups is also "felt" by the neighbors, e.g., δC-2 decreases and δC-14 increases, the former is due to the C=O → CHOAc change, and the latter is due to the increased flexibility of the B ring, which decreases the field effect which is substantial in **283**. The other shifts are roughly the same, like in the case of the side chain with respect to Compound **286** (see below).

An investigation of the series of androstanones[1372k] has shown that the carbonyl carbons adjacent to the angular methyl groups 19 and 18 (i.e., C-1,12,17) are more strongly deshielded than those of other carbonyl groups. In the six-membered A, B, and C rings, the carbonyl shifts C-2,3,4,6,7,11 are 209.8, 210.0, 211.2, 210.4, 209.9, and 209.4 ppm, whereas δC-1 and δC-12 are 213.2 and 213.0 ppm. In the five-membered D ring, δC-15 and δC-16 are 215.2 and 216.0 ppm, but δC-17 is 219.7 ppm.

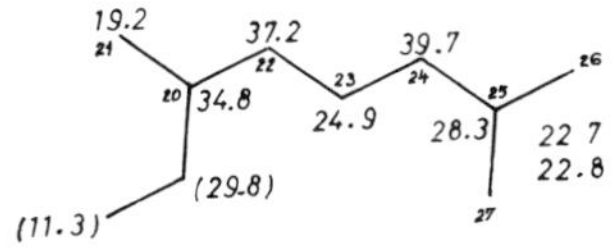

285

It is worth noting the method of assignment generally applicable for complicated molecules, but first used in the identification of the signals of the side chain of cholesterine derivatives.[1372l] As model, the simple molecule corresponding to the side chain, i.e., 2,6-dimethyloctane (**285**) was used assuming that the shifts of this molecule remain essentially unchanged in the steroid (**286**). This is really the case. For atoms C-21, 24, 25, 26, and 27, Δδ ≤ 0.2 ppm with respect to the model compound, and the difference is small for C-22 and 23 as well (0.5 and 0.6 ppm, respectively).

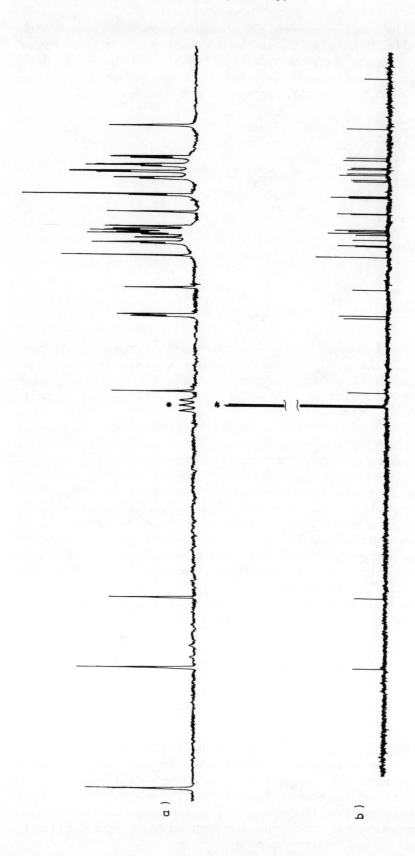

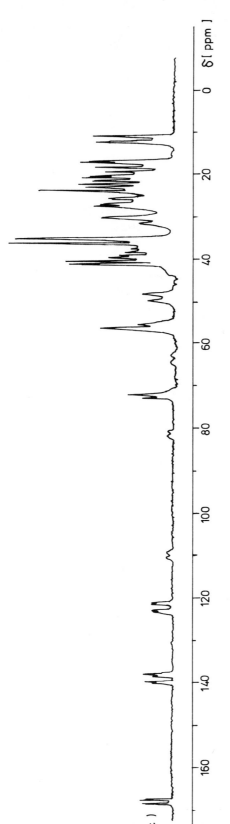

FIGURE 183. The ^{13}C NMR spectrum of 3β-acetoxycolest-5-en (**284**). (a) At 22.63 MHz in saturated CDCl$_3$ solution (400 scans),[218a] (b) at 125.8 MHz in 1% CDCl$_3$ solution (1 scan),[218b] and (c) at 50.3 MHz in solid sample, with magic angle spinning.[219]

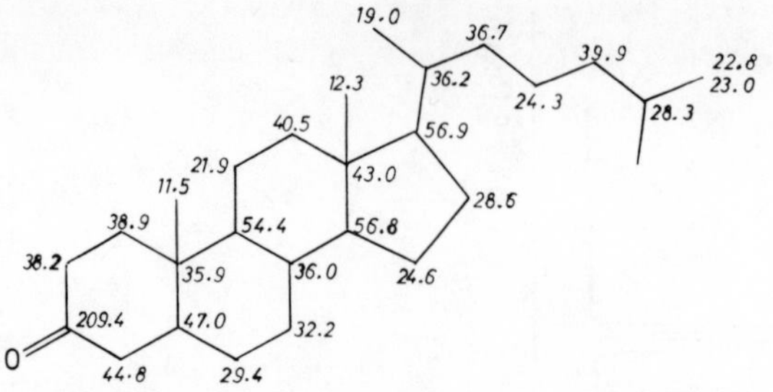

286

Field effect plays a particularly useful role in the investigation of steroids. The *cis* anellation of A and B rings (5β steroids) causes a 11- to 12-ppm paramagnetic shift in the 19-methyl signal with respect to the *trans* (5α) analogues.[574] Whereas in the latter (**287**) five hydrogens may be found in *syn-axial* position with respect to the 19-Me groups (H-2,4,6,8,11), the β-steroids (**288**) contain only three such protons (H-6,8,11). Simultaneously, steric compression occurs between H-2,4 and H-7,9, respectively, and thus the shielding of the corresponding carbons increases: their chemical shifts decrease by approximately 1, 2, 5, and 14 ppm.

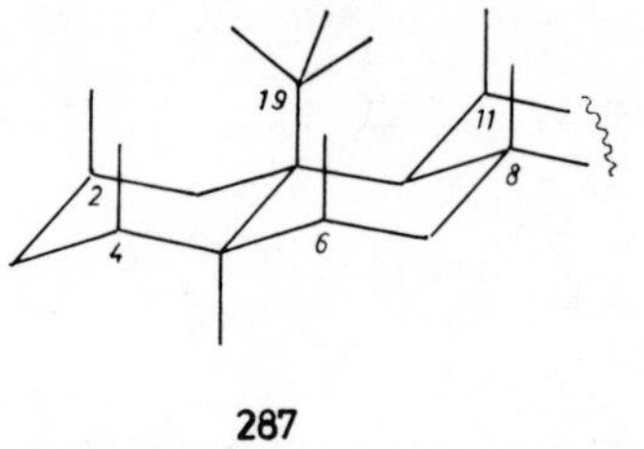

287 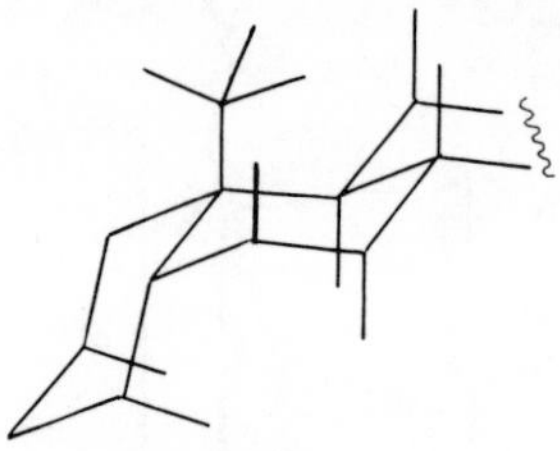   **288**

Consequently, in 19-nor steroids, where the methyl group on C-10 is replaced by hydrogen, the signal of this carbon is not deshielded, as expected, since the field effect causes stronger shielding than the opposite effect of methyl substitution.[1186] In 5-ene derivatives, the chemical shift of 19-Me carbon increases by approximately 7 to 8 ppm, since the *syn-axial* interaction with H-6 is no more present and the remaining interactions (with H-2,4,8,11) are also weaker owing to the more flexible (*twist*) conformation of the B ring (**289**), e.g., for 3β-acetoxy-7α-hydroxy-5α-cholestane, δC-19 = 11.1 ppm, whereas for the cholest-5-ene analogue, δC-19 = 18.2 ppm.[1314a] Of course, the C/D ring anellation is reflected in a similar manner in the shift of 18–Me.[1563,1564]

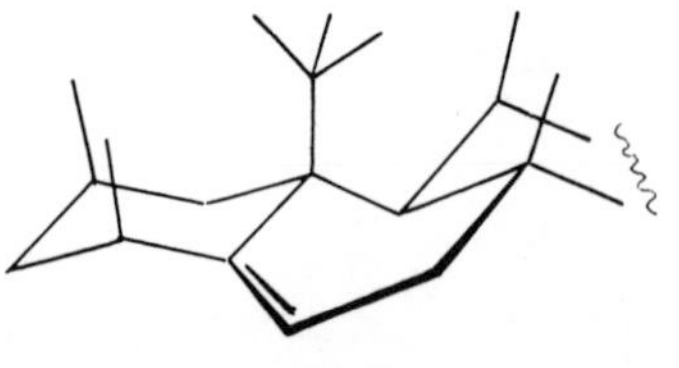

289

Likewise, the δC-5 shift of 2α,3α-epoxy-17β-acetoxy-androstane decreases by 5.3 ppm with respect to the β-epoxy isomer,[1414] owing to the field effect. The substituent effect was investigated in detail[129,695] and empirical additivity parameters were introduced[157,1314b] for the estimation of the shifts of substituted steroids.

As a further example of substituent effect, the ^{13}C NMR shifts of 11α-hydroxyprogesterone (**290**) and 3-acetoxy-estrone (**291**) are shown[574] for the sake of orientation in the assignment of steroid signals.

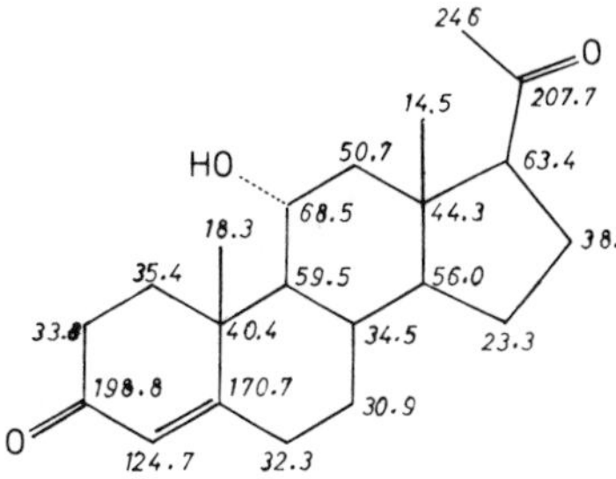

290 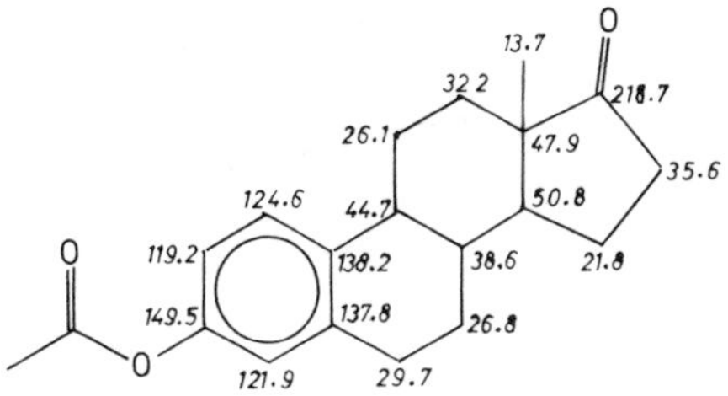   **291**

In acetylated steroids, the signal of the α-carbon shows a 3- to 4-ppm downfield shift; the β-carbon shows a 2- to 3-ppm upfield shift with respect to the hydroxy derivative. If the hydroxy group is *axial*, the signal of the γ-carbon atom also shifts upfield by about 1 ppm.[574]

The assignment of the ^{13}C NMR signals of aza-steroids may be carried out on the basis of the spectra of the corresponding steranes and the analogous cyclohexane-piperidine derivatives.[1446]

The investigation of the carbon resonance spectra of terpenes,[60,109,156,196a,373,714,765,773,885,1203,1274,1372m,1508-1510] poses similar problems than that of steroids. Norbornanes with rigid skeleton (e.g., **265** to **268**) must be mentioned separately, since they are suitable models for the investigation of stereochemical effects (see also References 499, 878, and 416).

4.1.2.15. Alkaloids[310,311]

^{13}C NMR spectroscopy has opened new horizons in the investigation of alkaloids, too. It is illustrated by two protoberberine derivatives (**292** and **293**) how the anellation of saturated rings may be determined. With these compounds a *cis* ⇌ *trans* isomerization may take place via the inversion of the bridge-head nitrogen with a relatively low energy, and thus ring anellation is substituent sensitive. The upfield shifts of C-6,8,13,13a for **293** proves that it contains *cis* anellated skeleton, whereas **292** has a *trans* anellated one.[745,746]

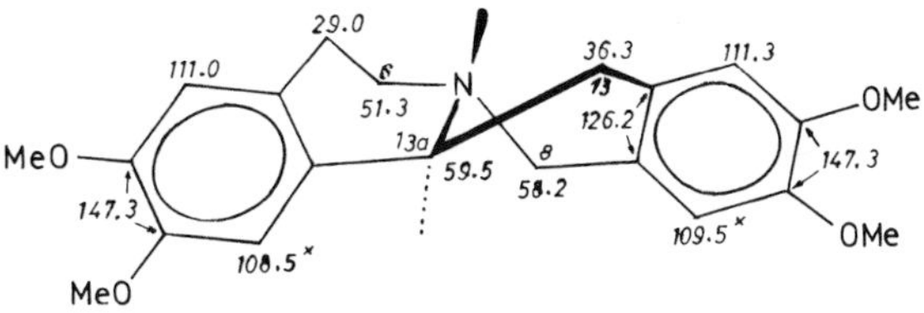

292

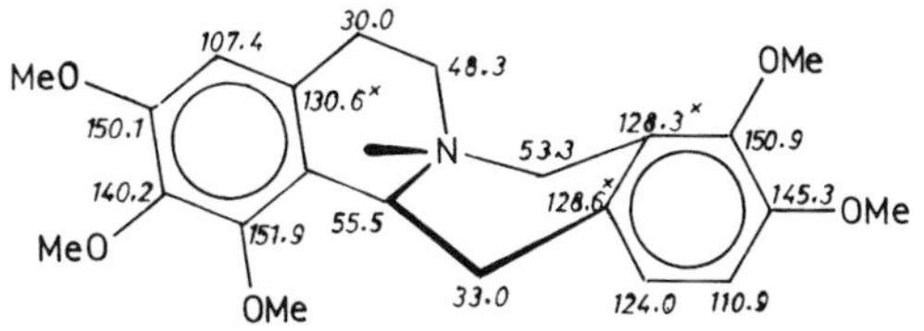

293

Quinolizidine alkaloids **294** and **295** may be distinguished already on the basis of the number of carbon signals, since molecular symmetry requires the coincidence of 7 pairs of signals for the former.[166] Moreover, the C-8 signal is 9.1 ppm more shielded for **295** owing to the steric compression with the lone pair of nitrogen. The same interaction causes a diamagnetic shift of 4.7 ppm in the C-6 signal of isomer **294**.

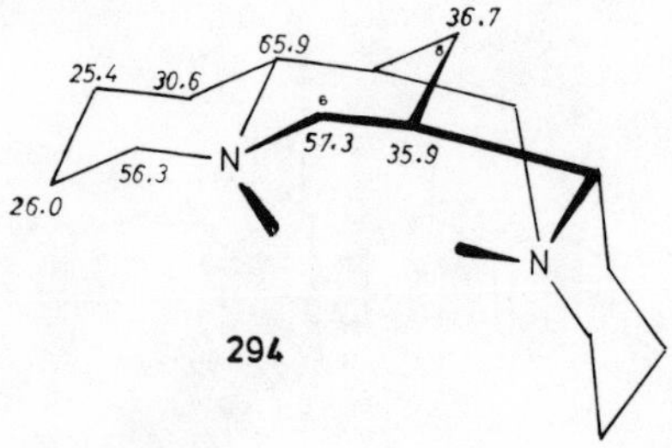

294

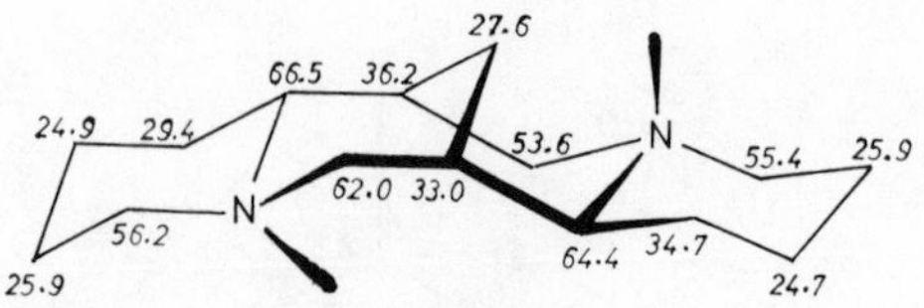

295

The power of ^{13}C NMR spectroscopy is shown, e.g., by the complete assignment of strychnine (**296**).[1360]

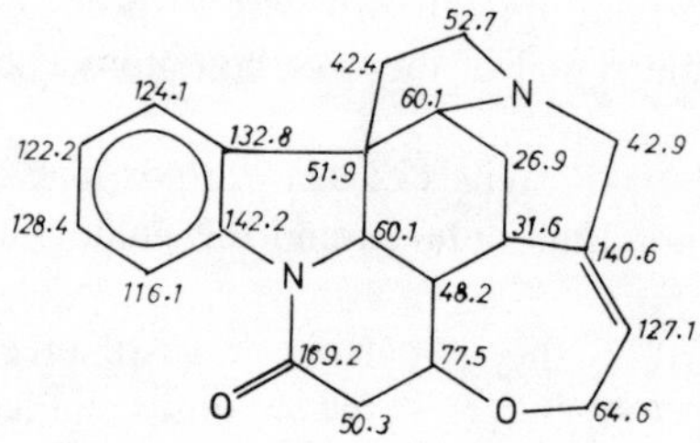

296

4.1.2.16. Nucleosides, Nucleotides

The carbon resonance literature of nucleosides and nucleotides of literally vital importance had become considerable already in the early stages of ^{13}C NMR.[381,738,1456] Here we restrict ourselves to the presentation of the chemical shifts of citidine (**297**), uridine (**298**), anhydro-uridine (**299**), adenosine (**300**), and inosine (**301**) to give a basis for the prediction of the shifts of carbons occurring in the most frequent building units of these compounds. Knowing the already discussed shifts of purine (see Table 80) and aldopyranoses,* the signals of pyrimidine, purine, and pyranose rings may be selected from the spectra, and thus the identification of the remaining signals, arising from the substituents, i.e., structure eluci-dation, becomes much simpler. Significant deviations of the chemical shifts assigned to the

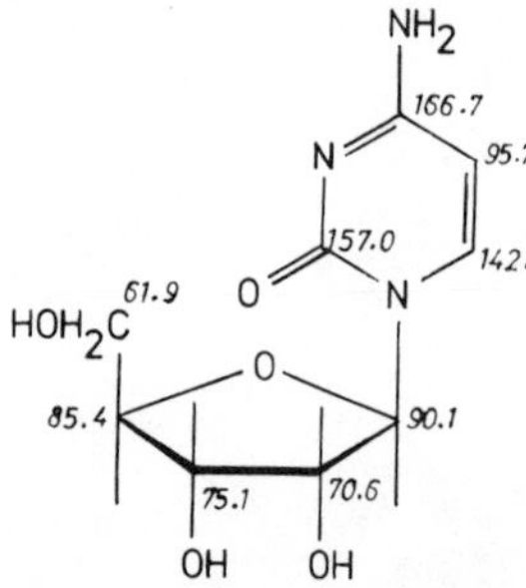

297

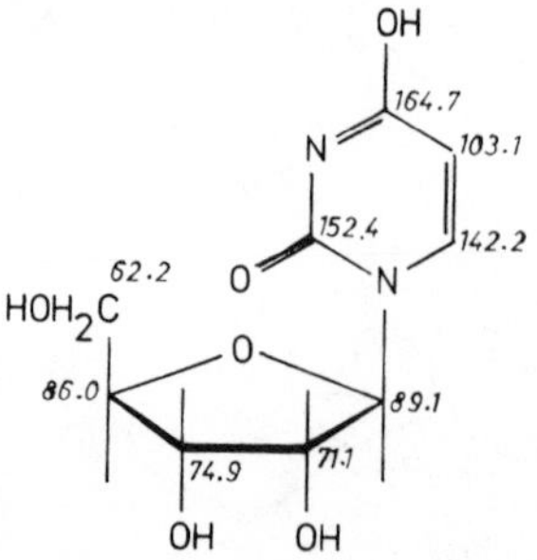

298

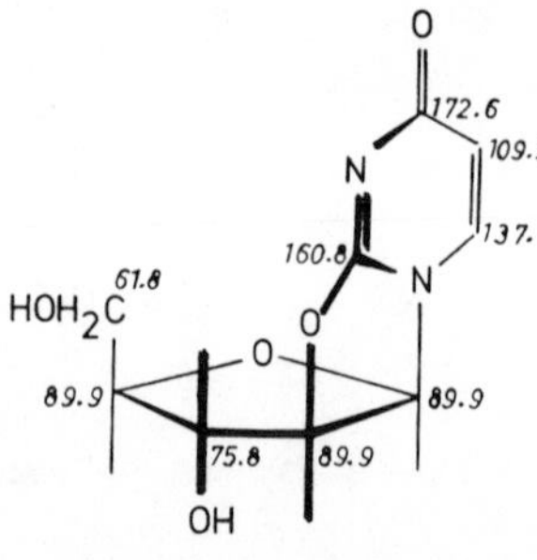

299

* Compare Section 4.1.2.5 (p. 170—171).

carbons of the skeleton from the shifts of the basic ring may give further information for the identification of structure. There is only a very rough proportionality between the π-electron densities calculated theoretically (e.g., Reference 664) and the δC shifts, and the theoretical estimation of chemical shifts for this group of compounds brought so far less promising results than for alkanes or aromatic molecules.

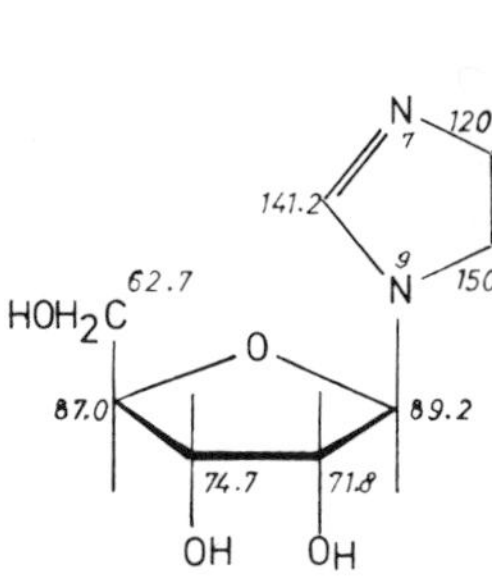

300 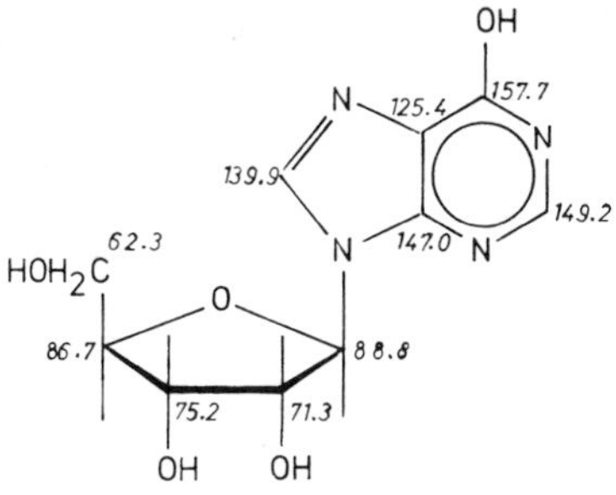   **301**

4.1.2.17. Amino Acids and Peptides[352,966]

The "models" for the ^{13}C NMR investigation of peptides are the amino acids themselves; their chemical shifts became available among the first carbon resonance data.[191,680,681,697,1396] The carbonyl signal appears in the 168- to 183-ppm range; the signal of the α-carbon appears between 40 and 65 ppm. In this case, too, the main advantage is the larger range of shifts over proton resonance, but the line broadening occasionally occurring for larger peptides owing to DD interactions is also smaller than in proton resonance,[681] since in contrast with the protons of peripheral site in the molecule, and thus subjected to DD interactions, the carbon atoms are inside of the molecule. The chemical shifts of the amino acids are shown in Table 81.

The spectra of amino acids and peptides are pH dependent.[488] The deprotonation of NH$_3^\oplus$, COOH, and SH groups causes deshielding of approximately 3 ppm on the α-carbon and 4 to 6 ppm on the β-carbon.

The activation parameters of the hindered rotation of proline peptides around the amide bond are in a range permitting the NMR detection of the rotamers. Thus, the Z and E isomers may be distinguished:[1457] owing to the field effect in the "cis" (E) isomer, the γ and δ atoms, in the "trans" (Z) isomer, the α- and β-atoms of the proline ring have smaller chemical shifts. In the spectrum (see Figure 184) of acetylproline amide (**302**) taken in D$_2$O solution, the lines of the isomers are easy to distinguish on the basis of the ~3:1 Z:E ratio. isomer ratio is solvent dependent, and by means of investigations as a function of temperature

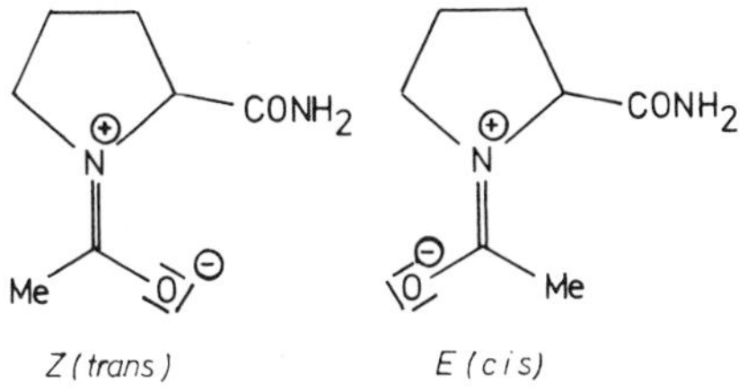

Z (trans)   E (cis)

302

the activation parameters of isomerization may be determined.[908] The difference $\Delta\delta C_{\beta,\gamma}$ is larger for the E isomer, which may be utilized in the conformation analysis of proline peptides.[384] Thus, two conformers could be identified for the cyclohexapeptide

Table 81

^{13}C NMR CHEMICAL SHIFTS (PPM) OF AMINOACIDS[681,1455,1458]

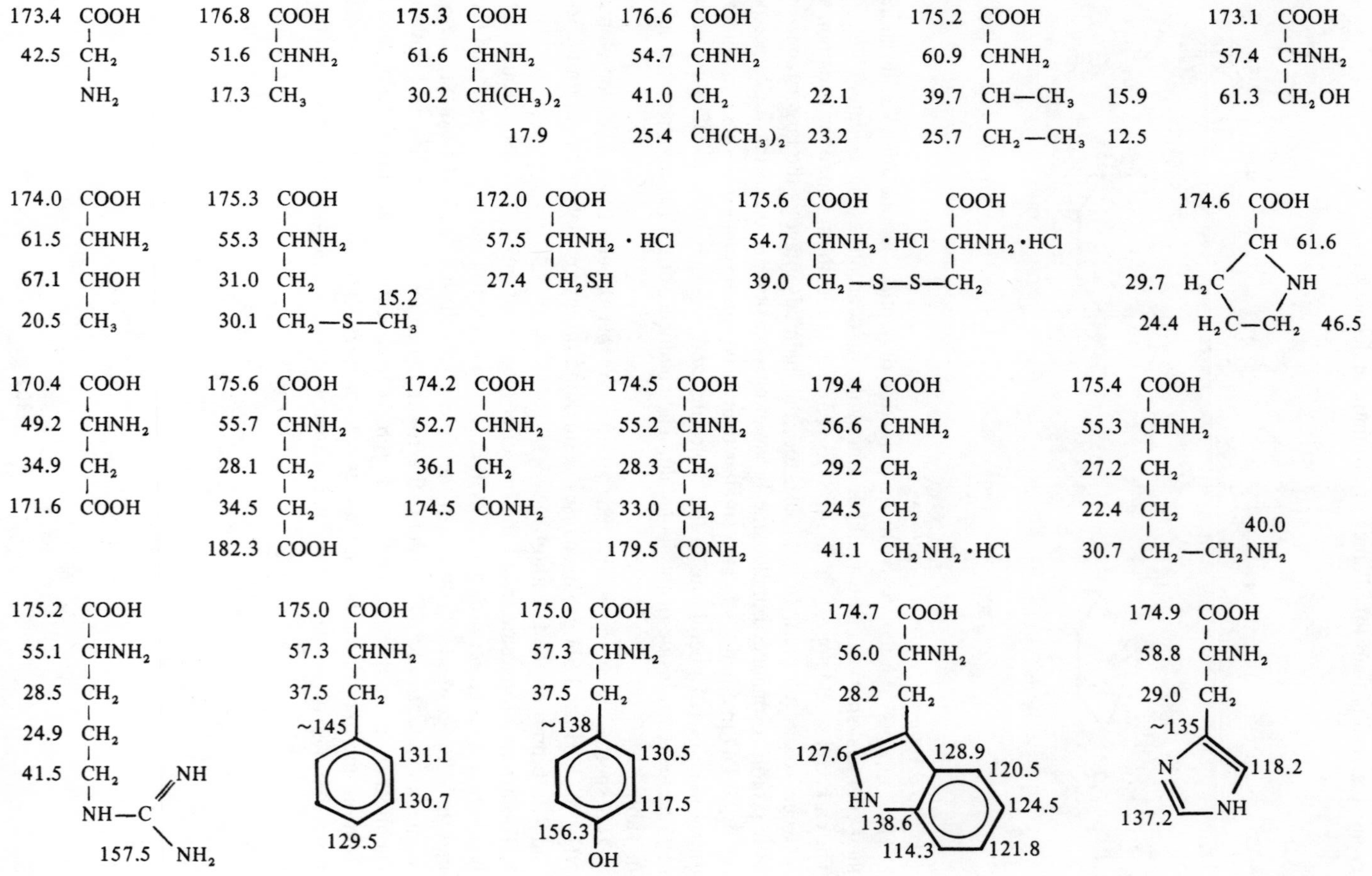

173.4	COOH
42.5	CH₂
	NH₂

176.8	COOH
51.6	CHNH₂
17.3	CH₃

175.3	COOH
61.6	CHNH₂
30.2	CH(CH₃)₂
17.9	

176.6	COOH
54.7	CHNH₂
41.0	CH₂
25.4	CH(CH₃)₂
22.1	23.2

175.2	COOH
60.9	CHNH₂
39.7	CH—CH₃
25.7	CH₂—CH₃
15.9	12.5

173.1	COOH
57.4	CHNH₂
61.3	CH₂OH

174.6 COOH
CH 61.6 — NH
29.7 H₂C
24.4 H₂C—CH₂ 46.5

174.0	COOH
61.5	CHNH₂
67.1	CHOH
20.5	CH₃

175.3	COOH
55.3	CHNH₂
31.0	CH₂
30.1	CH₂—S—CH₃
	15.2

172.0	COOH
57.5	CHNH₂ · HCl
27.4	CH₂SH

175.6	COOH
54.7	CHNH₂ · HCl CHNH₂ · HCl
39.0	CH₂—S—S—CH₂

170.4	COOH
49.2	CHNH₂
34.9	CH₂
171.6	COOH

174.2	COOH
52.7	CHNH₂
36.1	CH₂
174.5	CONH₂

174.5	COOH
55.2	CHNH₂
28.3	CH₂
33.0	CH₂
179.5	CONH₂

175.6	COOH
55.7	CHNH₂
28.1	CH₂
34.5	CH₂
182.3	COOH

179.4	COOH
56.6	CHNH₂
29.2	CH₂
24.5	CH₂
41.1	CH₂NH₂ · HCl

175.4	COOH
55.3	CHNH₂
27.2	CH₂
22.4	CH₂
30.7	CH₂—CH₂NH₂ 40.0

175.0	COOH
57.3	CHNH₂
37.5	CH₂

~138 ring 130.5, 117.5, 156.3 OH

174.7	COOH
56.0	CHNH₂
28.2	CH₂

127.6 HN 138.6, 128.9, 120.5, 124.5, 121.8, 114.3

175.2	COOH
55.1	CHNH₂
28.5	CH₂
24.9	CH₂
41.5	CH₂—NH

157.5 NH—C(=NH)—NH₂

175.0	COOH
57.3	CHNH₂
37.5	CH₂

~145 ring 131.1, 130.7, 129.5

174.9	COOH
58.8	CHNH₂
29.0	CH₂

~135 N 118.2, 137.2 NH

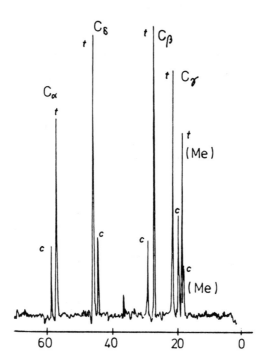

FIGURE 184.   The ^{13}C NMR spectrum of acetylproline
(**302**) in D$_2$O solution.[1561]

(Pro - Pro - Gly)$_2$; in one of them the configuration of one proline-amide bond is different, and in the other conformer all bonds have the same configuration.[670]

The carbonyl and α-carbon signals of helical and random γ-benzylglutamate are different by about 3 ppm,[1082] which makes it possible to analyze the configuration of macropeptides by means of ^{13}C NMR. In connection with the investigation of biopolymers, fibrine proteins, and membranes,* we refer to the literature.[778,1411]

It is worth noting that diastereomeric peptides may be distinguished, too,[182,1520] which may be utilized in determining the extent of racemization in coupling reactions.[356,1500] The possibilities of carbon resonance studies are illustrated by the spectrum of the cyclodeca-peptide **303** of antibiotic effect (see Figure 185).[553] It can be seen that except for the carbonyl

303

---

*   Compare p. 233.

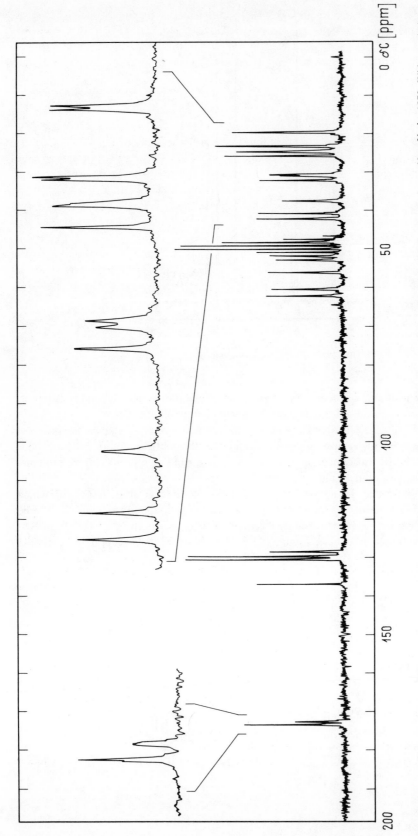

FIGURE 185.   The ^{13}C NMR spectrum of gramicidine-S (**303**). (From Johnson, L. F. and Jankowski, W. C., *C-13 NMR Spectra*, Wiley-Interscience, New York, 1972. With permission.)

signals and three further signal pairs, which have alternative assignments, the signals of all chemically nonequivalent carbons are separated and may be assigned. Thus, carbon resonance supplies information on structural changes at any part of the molecule. Finally, the investigation of the segment motions of biopolymers by the measurement of ^{13}C spin-lattice relaxation times* is mentioned.

4.1.2.18. Polymers[180,986,1245]

In addition to the broader chemical shift range, a further advantage over ^{1}H NMR spectroscopy is that in ^{13}C NMR no complications arise from spin-spin couplings. Similar to ^{1}H NMR spectroscopy, the first attempt for application was the determination of the "aromaticity" (ratio of aromatic components) of mineral oil and gasoline distillation and cracking products by comparing the intensities of aromatic carbons to that of all signals.[529]

The sensitivity of ^{13}C NMR spectra on steric structure makes it possible to compare the structures of natural and artificial rubber products by determining the ratio of 1,4- and 1,2-polybutadiene as well as the Z-E ratio.[397]

In the ^{13}C NMR spectrum of polymethylmethacrylate, the number of lines in the carbonyl region is characteristic of the number of pentades; the intensities reflect the relative occurrences. This fine structure is analogous to the structure of α-methyl signal in the 220-MHz proton spectrum.[728]

In the spectrum[1243] of a maleic acid-ethylene oxide copolymer (**304**) of a molecular weight of approximately 5000, four methylene and carbonyl signals, respectively, may be observed, which arise from the triads AAA, BAA, AAB, and BAB; the intensity ratios indicate a 74 to 80% occurrence of the A triad.

$$\text{+\!\!\left(\!\!\underbrace{CH_2\!-\!CH_2\!-\!0}_{A}\!\!\right)_{\!n}\!\underbrace{CO\!-\!CH\!=\!CH\!-\!COO}_{B}\!\!\right)_{\!m}}$$

A   B

304

The nitrile and methine signals of polyacrylonitrile split into three lines each according to the *iso-*, *syndio-*, and *heterotactic* triades.[1244] Upon diluting the 20 wt% DMSO-d$_6$ solution all signals split into two to three further pentade components, thus, of the nine pentades possible, six may be detected, proving that the polymer is built up from completely random units. Similarly, the methine signals of the three triad of poly-vinyl chloride are also separated.[246]

Instead of further details, we refer to some original papers on the investigation of vinyl polymers,[703,771,950,951,1244] diene-polymers[294] and synthetic biopolymers (poly-amino acids).[1082,1387] In the investigation of polymers, new horizons were opened by the high-resolution investigation of solid samples[1246] realized by magic-angle spinning.**

4.1.2.19. Carbonium Cations, Metal-Organic Compounds

On the carbon resonance chemical shifts of carbonium cations several papers were published already at the beginning of the spread of the ^{13}C NMR method.[1067,1148] The investigation of classical carbonium cations led to the observation of very strongly deshielded carbons, in agreement with theoretical predictions; they represent the largest δ-values observed in carbon resonance. The shifts measured for the ionic carbons of $Me_2C^{\oplus}H$, $Me_3C^{\oplus}$, $Ph_2C^{\oplus}H$, and $Ph_3C^{\oplus}$ cations are 319.6, 330.0, 200.2, and 212.7 ppm[1069] in $SO_2ClF\cdot SbF_5$ and SO_2SbF_5 solution, respectively. (The methyl signals of the first two compounds are at 61.8 and 48.3 ppm.)

---

*   Compare p. 233.
** Compare Volume I, Section 2.2.9.2 (p. 219—220).

In contrast with the above "classical" carbonium ions, "nonclassical" ones may occur in various equivalent forms, in which the positive charge is localized on different carbon atoms. If the interconversion of these forms is not too fast on the NMR time scale, the signals of the nonionic and ionic carbon atoms appear separately, upfields and downfields, respectively. If the process is fast, an average shift may be measured. An example for the latter case is cyclopentane cation (**305**), which has a regular decet in the proton-coupled ^{13}C NMR spectrum[1069] at 99.2 ppm in $SO_2ClF \cdot SbF_5$ solution, at $-70°C$. It must be assumed for the interpretation that the classical **a** to **e** cations are interconverting very rapidly through 1,2-hydride migration (the positive charge is delocalized), and thus the equivalent carbons are spin-spin coupled with nine, also equivalent, hydrogens. The measured shift is very close to the average of the expected individual shifts of one ionic carbon and two pairs of α- and β-carbons: $(317 + 2 \cdot 27 + 2 \cdot 52)/5 = 95$ ppm.

305

The shifts of some Me_nM-type metal-organic compounds are as follows: -15.0 (MeLi), -9.4 (Me_4 Sn), -4.0 (Me_2Zn), -3.2 (Me_4Pb), -0.6 (Me_4Ge), 1.2 (Me_2Cd), and 23.7 Me_2Hg).[967,1498] For comparison the shift of the methyl signal of neopentane is 31.6 ppm and for TMS, it is of course, 0.

The metal complexes of acetylacetone (CH_2Ac_2: ACAC) may be used as "T_1 - reagents" (relaxation accelerators) for the "development" of signals which are weak or not observable at all owing to saturation.* Although for this experiment only traces of the complex are necessary, it is worth knowing where its signals may be expected. In the spectra of approximately 20 different complexes the methyl signal is between 23 and 29 ppm the carbonyl carbon signal is between 189 and 197 ppm, and the olefinic signal (ACAC, with β-dioxo-structure forms complexes in the enolic form) is at 97 to 104 ppm.[624] The studies dealing with the metallotropic rearrangement of σ-cyclopentadienyl derivatives are worth mentioning.[591]

4.1.3. Spin-Spin Interaction in ^{13}C NMR Spectroscopy; $^{13}C - {}^1H$ and $^{13}C - {}^{13}C$ Coupling Constants

In ^{13}C NMR spectroscopy, $^{13}C - {}^1H$ and $^{13}C - {}^{13}C$ couplings and the corresponding splitting are the most important. However, the latter is not observable in routine spectra, since the low natural abundance of ^{13}C isotope makes it very improbable that the same molecule contains two ^{13}C atoms near one another. Thus, the splitting due to $^{13}C - {}^{13}C$ coupling is usually studied in ^{13}C-enriched samples,** although the high sensitivity of modern FT instruments allows the detection of this splitting in the spectra of natural samples, too.[567,896]

---

* Compare p. 233 and Volume I, p. 24 and 142.

** By excitation with specially phased pulse sequences it is now possible "to filter out" the satellite of 1% relative intensity corresponding to $^{13}C - {}^{13}C$ splittings from ^{13}C NMR spectra via suppressing the main line. This is the "double quantum coherence" technique. (Compare Bax, A., Freeman, R., and Kempsell, S. P., *J. Am. Chem. Soc.*, 102, 4849, 1980 and Bax, A. and Freeman, R., *J. Magn. Resonance*, 41, 349, 1980.)

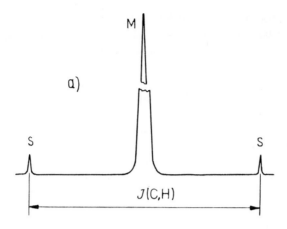

FIGURE 186. ^{13}C satellites in the ^{1}H NMR spectrum in the case of a singlet. M = main signal of the ^{12}CH groups; S = satellite lines of the ^{13}CH groups, with 1% intensity relative to the main signal.

4.1.3.1. Multiplicity of Carbon Resonance Signals: the Measurement of $^{13}C - {^1}H$ and $^{13}C - {^{13}}C$ Coupling Constants

In routine ^{13}C NMR spectra the splittings due to $^{13}C - {^1}H$ coupling may not be observed, either, since these spectra are usually produced by the BB proton decoupling technique (compare Section 2.2.5.5). Of course, proton-coupled carbon resonance spectra can also be measured in which $^{13}C - {^1}H$ couplings are observable. Moreover, the "off-resonance" method (compare Section 2.2.5.1) also permits one to utilize (at least in part) the sensitivity enhancing effect of NOE and to decrease $^{13}C - {^1}H$ splitting to an extent which allows multiplicities to be determined but separates the signals of adjacent multiplet. Of course, the splittings observed in this case may not be used to determine the true value of J (^{13}C,^{1}H) coupling constants, but the multiplicities of the signals give the order of the carbon atom in question unambiguously: in the proton-coupled or off-resonance spectrum the multiplicity of the signal is $n + 1$, where n is the number of protons attached to the carbon.

Figure 99 shows the off-resonance spectra of 1,3-butanediol.[191] The multiplicities directly give the assignment of the methyl (quartet) and methine signals (doublet), whereas the two triplets can be assigned on the basis of the relative chemical shifts or by graphical cross-correlation of the ^{1}H and ^{13}C NMR spectra (compare Figure 99). From the splittings due to $^{13}C - {^1}H$ coupling, it is often difficult to determine the coupling constants in the proton-coupled spectra because of overlap of the multiplets. In this case one may investigate the spectra of selectively deuterated model compounds. In their BBDR spectra, the $^{13}C - {^1}H$ interactions are eliminated and only the signals of deuterated carbons are split. From the magnitude of $^{13}C - {^2}H$ splittings, the analogous $J(^{13}C, {^1}H)$ coupling constant is easy to determine (compare Equations 78 and 324).*

4.1.3.2. Satellite Spectroscopy[567]

Before the spread of FT spectrometers, the poor quality of carbon resonance spectra made it impossible to determine $J(^{13}C,{^1}H)$ coupling constants, which could be determined, however, from the proton resonance spectra. The couplings, owing to the low natural abundance of ^{13}C nucleus (1%), appear as symmetrical "side bands" or satellites on the two sides of ^{1}H NMR lines with relative intensities of 1%. The distance of these lines gives the magnitude of J(C,H) coupling constants (see Figure 186). Since these couplings are in the order of

---

* The up-to-date method for separating overlapped multiplets and to decode complex split spectra is two-dimensional spectroscopy, 2DFTS (compare Volume I, Chapter 2, Section 2.2.9.4).

120 to 320 Hz, the weak lines of the satellite spectrum occur at 60 to 160 Hz distance from the main spectrum in both directions. If the main signals is a multiplet, all lines of it are repeated with an 0.01-fold intensity. A condition of the measurement of satellite spectra is that there is no other absorption in this region, since it would suppress the weak satellite lines.

Of course, the main source of $J(C,H)$ coupling constants is nowadays the proton-coupled carbon resonance spectrum, but satellite spectroscopy still has two applications where it supplies excess information over other NMR methods.

Determination of the relative signs of $^{13}C–H$ coupling constants — $^{13}C - {}^1H$ and $^{13}C - {}^{13}C$ coupling constants, like $J(H,H)$ may be positive or negative depending on whether the lower or the higher of the two new energy levels of the coupled spin pair corresponds to the antiparallel alignment of the two spins. When at least three spins interact, the relative (but not the absolute) sign of the coupling constants may be determined from the spectrum.* They may be determined empirically (e.g., on the basis of solvent dependence)[110] or by DR.**

Thus, e.g., by the partial decoupling of the satellite spectrum of acetaldehyde, the sign of the $^1J(C,H)$ and $^2J(C,H)$ coupling constants could be determined. The molecules containing ^{13}C in the methyl group may be regarded as an A_3MX spin system. In the ^1H NMR spectrum (see Figure 187) on the two sides of the methyl doublet at a distance of $^1J(C,H)/2$ and on the sides of the quartet of aldehyde protons of a distance of $^2J(C,H)/2$, satellite lines may be observed.[1103] From the former, $^1J(C,H) = 127$ Hz and from the latter $^2J(C,H) = 26.6$ Hz. By decoupling the satellite doublet at the upfield side (irradiating the sample with its frequency) the upfield satellite quartet, by decoupling the downfield doublet, the downfield quartet coalesces into a singlet, indicating*** that the signs of $^1J(C,H)$ and $^2J(C,H)$ are the same.[1229]

Determination of couplings between chemically equivalent protons — This is another application of satellite spectroscopy. This method was used first to determine the $^1J(C,H)$ coupling constants of Z and E dichloroethylene, symmetrically 1,2-disubstituted ethanes, and allene.[289,1288,1513] The $J(H,H)$ coupling constants of ethane, ethylene, acetylene,[894] and benzene[1173] were determined by this method in ^{13}C enriched samples.

From the singlet ^1H NMR spectra of the above symmetric molecules, the $J(H,H)$ coupling constant cannot be determined. However, the $H^{12}C \equiv {}^{13}CH$ molecule may be regarded as an ABX spin system, since ^{12}C is magnetically inactive. As the chemical shift difference of ^1H and ^{13}C isotopes is several orders greater than $J(C,H)$, the first-order treatment of $^{13}C - {}^1H$ coupling is correct. The hydrogens are nonequivalent owing to the isotope effect. The long-range isotope effects are stronger than those for the hydrogens immediately attached to the isotopes, and thus the chemical shifts of the hydrogens are slightly different, and the satellite spectrum is asymmetric with respect to the original signal. Of course, in the ^1H NMR spectrum only the AB part of the ABX spectrum may be observed, and it consists of two quartets. The coupling constants may be read directly from the spectrum, as shown in Figure 188.

4.1.3.3. On the Theory of Spin-Spin Interactions

The deeper reasons for the relationship between chemical structure and coupling constants may be understood only on the basis of the theory of spin-spin interactions. Thus, very briefly, this topic must also be discussed here.

---

* Compare Volume I, p. 55.
** Compare Volume I, Sections 2.1.6.2 to 2.1.6.4 (p. 149—158).
***Compare Volume I, p. 149.

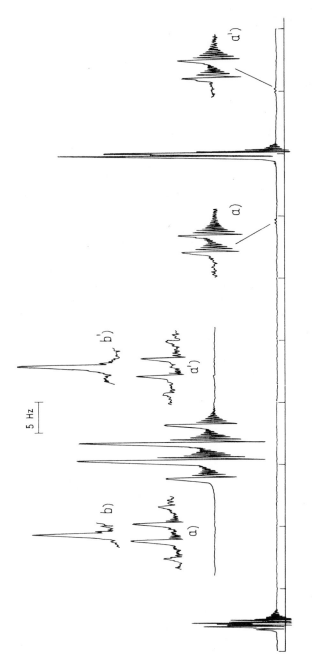

FIGURE 187. The ¹H NMR spectrum of acetaldehyde. (a) With the satellite signals of the ¹³CH₃CHO molecules and (b) the change of the satellite quartets by DR, irradiating with the frequency of the doublets.[1103]

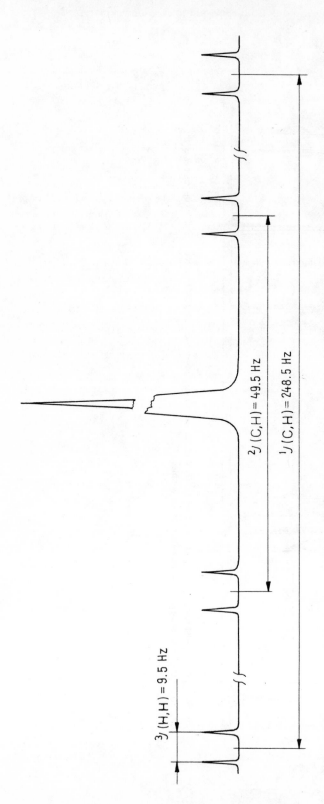

FIGURE 188.   Schematic ^{1}H NMR spectrum of ^{13}C-labeled acetylene (H^{12}C≡^{13}CH) with the ^{13}C satellite lines for the illustration of determination of coupling constants ^3J(H,H), ^1J(C,H), and ^2J(C,H).

For the interpretation of spin-spin interactions, Ramsey developed a general perturbation theory, which is still valid.[1165] Accordingly, the spin-spin interaction energy is

$$E' = hJ(A,B) I_A I_B \qquad (320)$$

where I_A and I_B are the quantum numbers of interacting nuclei (compare Equations 88, and 89, and 132). Taking into account that the z component of the nuclear magnetic moment is $\mu = \gamma hI$, one obtains

$$E' = \frac{2\pi}{\hbar} \frac{J(A,B)}{\gamma_A \gamma_B} \mu_A \mu_B = K(A,B) \mu_A \mu_B \qquad (321)$$

where $K(A,B)$ is the reduced coupling constant.*

The spin-spin interaction may be transmitted by the electrons in three ways: (1) direct nuclear spin - electron spin interactions or Fermi-contact interaction, (2) DD interactions between the nuclear magnetic moment and electronic orbitals, and (3) DD interaction between nuclear and electron spins. In the Fermi-contact interaction the nuclear and electron spins as well as the bonding electrons force one another into antiparallel alignment, and thus for nuclei attached by odd number of valences the parallel alignment of nuclear spins is less favorable energetically than the antiparallel one.** In proton-proton interactions the other two terms may be neglected,[1012] and thus the coupling constant is proportional to the product of s-electron densities around the nuclei coupled. The Fermi-contact interaction has a dominant role in the coupling of protons with many other magnetic nuclei (including carbon) as well.[711,1012,1410a,1423]

The importance of orbital and dipole-dipole interactions increases with the number of p electrons[145-147,712,965,1410a] or when the coupled magnetic nuclei are joined by multiple bonds.[147,445,1263] For the theoretical calculation of coupling constants, many quantum chemical methods have been developed.[83,659,754,957,1170,1229] Pople and Santry[1131] published an equation for the calculation of coupling constant from atom-atom polarizability.

4.1.3.4. The $^1J(C,H)$ Coupling Constant

Most of the "direct", $^1J(C,H)$ coupling constants fall into the 120- to 320-Hz range[434b,965,966a] and are positive.[122,225,1353] The data of Table 82, containing $^1J(C,H)$ for some simple compounds show that the magnitude of coupling depends primarily on the hybridization state of the carbons atom, increasing proportionally to the s-character in agreement with theory. Accordingly, the sp^3, sp^2, and sp carbons may be distinguished unanimously on the basis of the direct $J(C,H)$ coupling constant. Related to the change in s-character, the coupling constant decreases when the ring size of cycloalkanes increases from 3 to 6. Thus, the carbons of cyclopropane may be distinguished on the basis of $^1J(C,H)$ from other saturated carbons (compare Problem **78**). Several examples may be found[158,477,555,909,1124,1228,1270] on the linear relationship between $^1J(C,H)$ and s-character.

On the basis of this relationship, the chemical properties of various fused, rigid systems with strained rings could be predicted.[121,285] The coupling constants supported experimentally the theoretical assumption on the sp^2 character of cyclopropane carbons. The sensitivity of $^1J(C,H)$ on hybridization state is illustrated by the coupling constants of bicyclobutane (**306**).[1562]

The couplings determined experimentally are in excellent agreement with the hybridization states of the various bonds: the theoretical s-character of C–H bonds for sp^3, sp^2, and sp carbons is 0.25, 0.33, and 0.50. Using the mentioned relationship, the $^1J(C,H)$ coupling

---

* Compare p. 251 (Equation 346).
** Compare Volume I, p. 46—47.

Table 82

^1J(C,H) COUPLING CONSTANT (Hz) OF SOME SIMPLE MOLECULES[492,543,576,650,902,973,1009,1480,1513]

Molecule	^1J(C,H)	Molecule	^1J(C,H)
CH_3-CH_3	125.0	Cyclohexane	125.0
$CH_2=CH_2$	156.2	Cyclopentane	128.0
Benzene	159	Cyclobutane	136
$H_2C=C=CH_2$	168	Cyclobutene (C-3)	140
$H_2C=O$	172	Cyclopropane	160.5
$HC\equiv CH$	248.7	Cyclobutene (C-1)	170

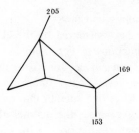

306

constants of ethane, ethylene, and acetylene yield 0.25, 0.31, and 0.50 for the s-characters of the C–H bonds, i.e., the only discrepancy is 0.02 units for ethylene.

Table 83 shows the ^1J(C,H) coupling constants of methyl compounds, from which follows that it is proportional to the $-I$ effect of substituents, again because of the changes in s-character. There are numerous analogous experiences for the most diverse compound types,[887,1570] including the substituted carbons of monosubstituted cyclopropanes (see Table 84)[318] and the methylene carbons of their heteroanalogues.[871] This explains the extremely large coupling constant (230 Hz) measured for the olefinic carbons of cyclopropenone.[198] For similar reasons, ^1J(C,H) of methyl amine (133 Hz) is 8 Hz higher than that of ethane,[902] and that of HCN is 20 Hz larger (269 Hz) with respect to acetylene. The correlation holds, therefore, for unsaturated compounds, too. The electron-withdrawing effect of the substituents is reflected in the linear relationship between the values of ^1J(C,H) coupling constants of benzene derivatives and the Hammett or Taft constants.[249,511,647,696,996,1018,1306,1459,1490,1571] Since ^{13}C NMR chemical shifts are proportional to the $-I$ effect, correlation is often found between δC and ^1J(C,H), e.g., for acetylene[1224] and hydrocarbon derivatives.[309] For halogenated methanes (see Table 85), in addition to the influence of $-I$ effect, the ^1J(C,H) coupling constant increases in parallel with the order of carbon atom.

The substituent effects are additive, and thus the coupling constants of polysubstituted aliphatic compounds, may be predicted on the basis of substituent parameters determined from the data of monosubstituted analogues:[910]

$$^1J(C,H) = 125 + \Sigma\, \rho^i \tag{322}$$

The ρ^i coefficients for some substituents may be found in Table 86.

There is a much better agreement between observed and calculated data (deviation ≤ 3 Hz) if the interaction of substituents is also taken into account:[385,910]

$$^1J(C,H) = J_0 + \sum_{i=1}^{3} \rho^i + \sum\sum \rho^{i,j} \tag{323}$$

where $J_o = J(CH_4)$.

Table 83

$^1J(C,H)$ COUPLING CONSTANTS (Hz) OF SOME COMPOUNDS CONTAINING METHYL GROUPS[110,386,387,393,395,396,583,584,614,645,686,1010,1062,1479,1481]

Compound	$^1J(C,H)$	Compound	$^1J(C,H)$	Compound	$^1J(C,H)$	Compound	$^1J(C,H)$
MeLi	98	Me$_2$CO	127	Me$_2$CN	136	Me$_4$N$^\oplus$	145
Me$_2$Mg	105.5	Me$_3$P, Me$_3$P $\rightarrow$ O	129	Me$_2$S, Me$_2$S $\rightarrow$ O	138	Me$_3$S$^\oplus$	146
Me$_4$Si	118.4	NMe$_3$	132	Me$_2$SO$_2$	139	MeNO$_2$	147
Me$_4$Ge	124.4	MeCOCl	133	Me$_2$O	140	MeF	149
MePh	126	MeCOBr, MeCOI, Me$_4$P$^\oplus$	134	Me$_3$N $\rightarrow$ O	143	MeCl	150
						MeBr, MeI	151

Table 84

$^1J(C,H)$ COUPLING CONSTANTS (Hz) FOR SOME CYCLOPROPANES AND FOR THE HETEROANALOGUES OF CYCLOPROPANE[318,871]

X	$^1J(C,H)$			X	$^1J(C,H)$
	C^s	C^c	C^t		
CHNH$_2$	169.7	160.5	159.9	S	170.5
CHI	187.7	165.4	162.7	NH	172.0
CHBr	192.6	165.0	162.3		
CHCl	192.1	164.5	162.6	O	175.5

Note: c, cis; t, trans.

Table 85

$^1J(C,H)$ COUPLING CONSTANTS (Hz) FOR THE HALOGEN DERIVATIVES OF METHANE[505,1479,1481]

X	CH$_3$X	CH$_2$X$_2$	CHX$_3$
F	149.1	184.5	239.1
Cl	148.6	176.5	208.1
Br	150.5	177.7	204.3
I	150.3	171.9	

Table 86

SUBSTITUENT CONSTANTS (Hz) FOR ESTIMATION OF $^1J(C,H)$ COUPLING CONSTANT OF ALIPHATIC COMPOUNDS[910]

Substituent	i	Substituent	i
CMe$_3$	-3	NH$_2$	8
Ac	-1	CCl$_3$	9
Me, Ph	1	CN	11
CHO	2	SOMe	13
CH$_2$Cl, CH$_2$Br	3	OH, OPh	18
CHCl$_2$, NMe$_2$	6	F	24
CH$_2$I, NHMe, C(sp)	7	Cl, Br, I	27

The relation $^1J(C,H_a) < {}^1J(C,H_e)$ equally holds for cyclohexanes,[244] dioxanes,[160] and methyl glycosides.[159,964] For the latter, the characteristic regions are $^1J(C,H_a) = 158$ to 162 and $^1J(C,H_e) = 169$ to 171 Hz.

Molecular geometry affects direct couplings in other cases, too. This may be seen from the data of Table 84, where $^1J(C,H)^c > {}^1J(C,H)^t$ for monosubstituted cyclopropanes.[318] The same may be observed, e.g., for oxaziridines,[716] where the difference is larger (approximately 6 Hz) and arises presumably from the different relative positions of unpaired electrons.

The $^1J(C,H)$ coupling constants of the carbon atoms of quinolizidines adjacent to the bridge-head nitrogen are characteristic of the anellation of the rings, although the assumptions on the assignment are contradictory.[32,137,1392]

For vinyl halogenides $^1J(C,H)_Z < {}^1J(C,H)_E$, except for fluoroethylene.[956] The difference is 1.7 to 4.9 Hz. The $^1J(C,H)$ coupling of ethylene derivatives substituted with more and bulkier substituents decreases with increasing steric compression.[13,1572]

In benzene derivatives carrying electron-withdrawing substituents, the relation $^1J(C,H^p) > {}^1J(C,H^o) > {}^1J(C,H^m)$ was stated.[1307] The $^1J(C,H)$ coupling constants of aromatic and heteroaromatic systems are reviewed in Table 87.

In the vicinity of hetero atoms the coupling constant substantially increases, being determined by the lone pairs transmitter of coupling and the $-I$ effect. Therefore, the substituent position may, in principle, be determined from the $^1J(C,H)$ coupling constants of unsubstituted carbons.

On comparing the couplings of furan, pyrrole, and thiophene, it can be seen that there is no simple correlation with the aromatic character. A neighboring oxygen causes extremely large increase, which is also shown by the coupling between the carbonyl carbon and the "aldehyde" proton of aldehyde derivatives, where $^1J(C,H)$ is much greater than for olefins and further increases upon substitution, but not proportionally to its $-I$ effect. For derivatives with R=H, Ph, NMe$_2$, CCl$_3$, OMe and F, $^1J(C,H) = 172$, 173.7, 191.2, 207.2, 226.2, and 267 Hz were measured.[816,911,1008] $^1J(C,H)$ also increases on protonation, and, owing to the effect of lone electron pair, $^1J(C,H_Z) = 210 > {}^1J(C,H_E) = 198$ Hz.[1069] Similarly, on protonation, the coupling constant of HCN increases from 269 to 320 Hz (the latter is the largest $^{13}C - H$ coupling constant observed so far).[1069]

With acetaldoxime (**162**), the coupling of the olefinic carbon is higher for the Z isomer.[1573] The dependence of $^1J(C,H)$ on solvent, temperature, and phase[213,387,1479] is in relation with bond polarization, change in association structure, and complex formation.[1388]

On the basis of Equation 321, $^{13}C - H$ and $^{13}C - D$ couplings are proportional:[193,291]

$$^nJ(C,H) = (\gamma_H/\gamma_D) J(C,D) \approx 6.55 {}^nJ(C,D) \tag{324}$$

The change in ^{13}C chemical shift on deuteration is only 1 to 2 ppm.[902] The $^1J(C,D)$ couplings are in the 18- to 24-Hz range.[193] It is noted that partial deuteration of CH$_n$ group decreases the $^1J(C,H)$ coupling constants by approximately 1 Hz[1276] by decreasing C–H bond length in CH$_{n-1}$D.

4.1.3.5. The $^2J(C,H)$ Coupling Constant[461]

The $^2J(C,H)$ coupling constants are in the 0- to 60-Hz range.[996] Therefore, in the proton-coupled and off-resonance mode (mainly in the case of aldehydes, acetylenes, and olefins), $^2J(C,H)$ couplings cause well observable splittings, even in routine spectra. The signs of these couplings are variable, for saturated compounds are generally negative and lower in absolute value than the analogous ^3J couplings.[316,317,1396] The $^2J(C,H)$ coupling constants of ethane, ethylene, and acetylene are -4.8, -2.4, and 49.7 Hz.[576] The regions characteristic of their derivatives and of aromatic compounds, aldehydes, and carbonyl compounds are $^2J(C–C–H) = 1$ to 6, $^2J(C=C–H) = 0$ to 16, $^2J(C≡C–H) = 40$ to 60, $^2J(C_{Ar}–C_{Ar}–H) = 1$

Table 87
**¹*J*(C,H) COUPLING CONSTANTS (Hz) FOR AROMATIC AND
HETEROAROMATIC COMPOUNDS**[433,1143s,1405,1417,1418,1449]

to 4, $^2JC_{\alpha}CO–H)$ = 20 to 50, and $^2J(O{=}C{-}C_{\alpha}{-}H)$ = 5 to 8 Hz.[1143r] It is clear that, similar to ^1J(C,H), these coupling constants are also proportional to the s-character of carbon atoms. There is a similar relationship with the $-I$ effect of substituents and the order of the carbon. Thus, the ^2J(C,H) coupling constants measured for iodo-, bromo-, chloro-, and fluoroacetylene are 51.5, 56.0, 60.5 and 65.5 Hz,[887] and those of acetaldehyde and its mono-, di-, and trichloro derivatives are 26.7, 32.5, 35.3, and 46.3 Hz, respectively.[1566] The analogous ^2J(C,H) and ^2J(H,H) values are proportional, the latter being approximately 2.5 times higher.[749]

The ^2J(C,H)coupling constant is sensitive of molecular geometry: for 1,2-dichloroethylene $^2J(C,H)_Z$ = 16 ≫ $^2J(C,H)_E$ = 0.8 Hz.[996] Similarly, the anomers of carbohydrates are easy to distinguish on the basis of coupling; $^2J(C_1H_{2a})$ = 6 Hz for β-D-glucose, whereas $^2J(C_1,H_{2a})$ < 1 Hz for the α-anomer.[196b]

The two ^{13}C–O–H couplings of acetic acid ^{13}C labeled on the carbonyl group and protonated at $-65°C$ by FSO$_3$H [H$_3$C^{13}C$^{\oplus}$(OH)$_2$] are 7.5 and <0.5 Hz! This indicates the localization of double bond and charge, respectively, and thus the fixed (different) steric positions of the two hydroxy groups (*syn* and *anti*).[1068]

In benzene, ^2J(C,H) = 1 Hz,[1493] in its derivatives, 1 to 4 Hz. Couplings involving the α-hydrogen of pyridine are much higher: 7 to 9 Hz,[1393] whereas the analogous interactions of β- and γ-hydrogens are the same as in benzene derivatives. In the case of furan and thiophene, the relationship ^2J(C,H) < ^3J(C,H) is reversed.[1350]

The ^2J(C,H) couplings of thiethane dioxide (**307**) and, in Table 88, those of **190** five-membered heteroaromatic rings are shown.[850,1494] The data of the former compound illustrate how the α- or β-position of electron-withdrawing SO$_2$ group influences the H–C$_\alpha$–C$_\beta$ coupling and how great is the effect of the relative position of the double bond: the difference between H–C(sp^2)–C(sp^3) and H–C(sp^3)–C(sp^2) couplings. The data of five-membered heterocycles indicate that C$_3$–C$_2$–H$_2$ and C$_3$–C$_4$–H$_4$ couplings vary in parallel with the aromatic character of the ring. The same does not hold on the C$_2$–C$_3$–H$_3$ interaction, due evidently to the vicinity of the heteroatom.

Table 88
^2J(C,H) COUPLING
CONSTANTS (Hz) OF FIVE-
MEMBERED
HETEROAROMATIC
RINGS (190)[1494]

X	^2J(C,H) (Hz)		
	$C_2–H_3$	$C_3–H_2$	$C_3–H_4$
O	7.0	14.0	4.0
NH	7.6	7.8	4.6
S	7.4	4.7	5.9
Se	7.0	4.5	6.0

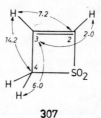

307

4.1.3.6. ^3J(C,H) Couplings

^3J(C,H) coupling constants are in the 0- to 10-Hz range. In alkanes and benzene derivatives ^3J(C,H) > ^2J(C,H), allowing the signals arising from the *ortho* and *meta* carbon atoms of monosubstituted benzene derivatives to be distinguished. Their chemical shifts are often hardly different, whereas ^3J(C,H) = 7 to 10 Hz[693,1397] while ^2J(C,H) < 4 Hz. Therefore, under medium resolution only the splittings corresponding to ^3J(C,H) occur, and thus the carbons in *meta* position give double doublet, and those in *ortho* (and in *para*) position have triple doublet signals. In pyridine derivatives both ^2J(C,H) and ^3J(C,H) are larger; the coupling constants of interactions involving nitrogen i.e., ^3J(C,H) of C–N–C–H or C–C–N–H type are 12 to 13 Hz, whereas those of type C–C–C–H are 3 to 5 Hz.[1393]

It was observed that the ^3J(C,H) couplings depend on the dihedral angle in a similar way as the analogous *vicinal* proton-proton interactions.[12,346,393,532,837,1106,1477] Accordingly, linear relationship was found between ^3J(C,H) and ^3J(H,H) of olefins: the former are lower by a factor of 0.6.[1462] The situation for the analogous quinolines and methylquinoline derivatives is similar.[280]

According to the Karplus relation, the *trans* periplanar conformation is thus characterized by larger ^3J(C,H) coupling constants than the *gauche* rotamer for the most diverse compound types. The characteristic ranges of alkanes are 4.5 to 3.5 and 0 to 3 Hz, respectively. In olefins, where the C–C–C–H chain is replaced by a C–C=C–H one, the values of ^3J approximately doubled.[1265] The ^3J(C,H) coupling constants of amino acids (HC$_\beta$–C–C=O), ethers (C–O–C–H), α,β-dicarboxylic acids (H–C–C–COOH), and dinitriles (H–C–C–CN) are 9 to 12 or 1 to 3 Hz, respectively.[378,455,1469] These relationships may be used for the determination of the conformation of peptides.[232]

Using the modified Karplus equations, (Equation 80a)[1043a] the empirical values of the constants were determined for couplings of type H–C$_\alpha$–N–C=O and H–N–C$_\alpha$–H. Constants A', B', and C' are −0.8, −4.4, and 9.0 Hz for the former[233] and 0.4, −1.1, and 9.4 Hz for the latter.[234]

Substituent effects were studied only sporadically.[630] For dihalobenzenes linear correlation was found between ^3J(C,H) and the −I effect of halogens.[1397]

Table 89
$^1J(C,C)$ COUPLING CONSTANTS (Hz) FOR SOME SIMPLE COMPOUNDS[126,526,576,584,881,894,1372n,1497]

Molecule	$^1J(C,C)$	Molecule	$^1J(C,C)$
CH_3-CH_3	34.6	$(CH_3)_4C$	36.9
C_6H_6	57.0	$(CH_3)_3CCl$	40.0
$H_2C=CH_2$	67.6	$(CH_3)_2C=O$	40.6
$H_2C=C=CH_2$	98.7	CH_3Ph	44.2
$HC≡CH$	171.5	CH_3CN	56.5

$^4J(C,H)$ coupling constants are less than 2 Hz[984] and negative for most of the cases investigated so far.[1397] However, for dimethylacetylene (MeC≡CMe) 1.6 Hz was measured.[640] $^nJ(C,H)$ couplings with $n > 4$ have no significance; they are negligibly small.

4.1.3.7. $^{13}C - ^{13}C$ and $^{13}C - X$ (X ≠ C,H) Couplings

Couplings between carbon atoms, since they do not occur in routine spectra and may be studied simply in ^{13}C-enriched samples, have only theoretical significance. The range of $^1J(^{13}C,^{13}C)$ coupling constants is 10 to 185 Hz. Some characteristic data may be found in Table 89.

There is a linear correlation between $^1J(C,C)$ and $^1J(C,H)$ couplings,[1497] as can be seen from a comparison of the data of Tables 82 and 89, consequently, the magnitude of $^1J(C,C)$ is also proportional to the s-character:[1037]

$$^1J(C_A,C_B) = 0.06S_AS_B - 10.2 \text{ Hz} \qquad (325)$$

From this equation a negative and very low coupling constant is obtained for bicyclobutane (**306**), which had also been proved experimentally ($J = -5.4$ Hz for the bridge-head atoms of the bicyclobutane derivative **308a**).[1121,1122] The smallest couplings are characteristic of cyclopropanes and their fused derivatives, in agreement with the p-character of C–C bonds.

$$308 \quad \begin{array}{l} \textbf{a}: R = Me, \ R'= COOEt \\ \textbf{b}: R = CN \quad R'= H \end{array}$$

Substituted cyclopropanes have small $^1J(C,C)$ constants, too: the $C_1 - C_2$ spin-spin coupling constants of iodo-, bromo-, and chlorocyclopropane are 12.9, 13.3, and 13.9 Hz,[535] and in the case of dicyclopropyl ketone ($C_3H_5 - CO - C_3H_5$) it decreases to 10.2 Hz.

For the cyano derivative **308b** $J_{1,2} = 22$ and $J_{1,3} = 16$ Hz.[1123] In methinyl cyclopropane (**309**) $J_{1,2} = 95.2$ and $J_{2,3} = 23.2$ Hz.[609]

309

The $C_1 - C_2$ couplings of cyclobutanone, cyclopentanone, and cyclohexanone are 29.7, 37.2, and 37.3 Hz, respectively. [535]

The coupling constants of benzene derivatives are in the 53- to 70-Hz range. For nitrobenzene and iodobenzene: $^1J(C_1,C_2) = 55.4$ and 60.4; $^1J(C_2,C_3) = 56.3$ and 53.4, $^1J(C_3,C_4) = 55.8$ and 58.0 Hz, respectively, i.e., the substituent effect is not very large. This topic is reviewed in the literature. [129] For ethylene derivatives $^1J(C,C)$ increases with the number of substituents. [89,107,896] The coupling of the sp carbons of diphenylacetylene, 185 Hz, is the highest C–C coupling constant measured so far. [629]

In the spectra of amino acids, [1426,1427] $^1J(C,C)$ coupling constants were found to depend on pH. The magnitude of $^1J(C,C=O)$ decreases in the anione of carboxylic acids and increases in their esters. These coupling constants of acetate ion, acetic acid, and ethyl acetate are 51.6, 56.7, and 58.8 Hz. [583]

The $^nJ(C,C)$ couplings of carbons separated by two or more bonds ($n \geq 2$) are generally small. [934] In saturated compounds $^2J(C,C) < 3$ Hz[934] and $^3J(C,C) < 6$ Hz. The latter depends on the dihedral angle in a way roughly analogous to the Karplus relation. [371,933] To the dihedral angles 0, 65, and 180°, $^3J(C,C)$ coupling constants of approximately 2, 0, and 4 to 6 Hz correspond, but the deviations of measured data are much higher than in the case of H – H and C – H couplings. [371,933]

In benzene derivatives, $^2J(C,C) = 3$ to 4 Hz. [931] In unsaturated compounds the dependence on dihedral angle may not be used for conformation analysis, since around 90°, where the σ-contribution is minimum, there is a maximum overlap of π-electrons, and thus the two components of interaction balance each other in the entire range of dihedral angles. [930]

Extremely high coupling constants were measured for propionitrile, acetone, and diphenylacetylene: $^2J(C,C) = 33$, 16 and 13.1 Hz. [394,584,629] In unsaturated and aromatic compounds $^2J(C,C) < 2.5$ Hz. [107,628,630] In anthracene derivatives the *cisoid* $^3J(C,C)$ coupling is about twofold of the *transoid* one, [932] which may be attributed to the fact that the former interaction may be realized through two paths.* With these compounds linear correlation was observed between the π-bond order and the magnitudes of ^2J and $^3J(C,C)$ couplings.

Of the $^1J(C,X)$ couplings, the ones for X = N are the smallest (0 to 20 Hz), and those for X = Hg, are the largest (approximately 2 kHz). Additional data may be found in Section 4.2. and Table 5 and, of course, in the literature. [883,954,994,1043,1372p,1476]

4.1.4. Spin-Lattice Relaxation in Carbon Resonance[855d,892,1489,1492e]

4.1.4.1. Carbon Resonance Intensities

The main information sources in ^1H NMR spectroscopy are the chemical shifts, coupling constants, and signal intensities. While the latter are strictly proportional to the number of protons,** in ^{13}C NMR there is not such proportionality, owing to the long relaxation of carbon nuclei and the NOE.

For most organic compounds the spin-lattice relaxation times, T_1, of carbon nuclei are higher than 0.1 to 0.2 s, and even the values around 100 s are not rare. With CW technique the measurement time is at least 1 min, which is sufficient for restoring the thermal equilibrium perturbed by resonance. The RF pulses of FT spectrometers follow one another, however, with a spacing of 0.1 to 1 s, commensurable with T_1, and thus for some of the nuclei partial saturation occurs. The extent of saturation may be reduced by certain technical tricks,*** but in routine measurements the partial saturation of nuclei and thus intensity reduction, must always be accounted for. Since the different nuclei of the same molecule have different relaxation times, the relative intensity reduction may be different for the chemical equivalent nuclei. Partly for this reason, there is no proportionality between the number of nuclei and signal intensity.

---

* Compare Volume I, p. 65.
** Compare Volume I, p. 142.
***Compare Volume I, p. 180 and 218—219.

NOE is a consequence of BBDR. The adjacent excited protons decrease the relaxation times of ^{13}C nuclei, and thus they may absorb more RF power without saturation, i.e., the corresponding signal intensities increase.* NOE may increase intensities by a factor of at most 3. In routine measurements proton decoupling is applied in order to simplify the spectrum and to increase sensitivity (not only through NOE but also by collapse of the lines of the multiplets). Intensity enhancement increases with the number of hydrogens attached to the carbon in question, but it is not strictly proportional to it, and even if the same number of hydrogens is attached to chemically different ^{13}C atoms, the enhancements are not necessarily the same. NOE may act not only between directly bonded atoms, but also between nonbonded ones in steric proximity. Consequently, for larger, mainly rigid molecules the intensity of quaternary carbons may often increase as well and even maximum NOE may occur. This supports further that the intensities of ^{13}C NMR signals give no direct information on the number of absorbing carbon atoms. Qualitatively, the intensity of the signals depends on the order of the carbon atom, and the signals of the carbons with the same order are, although not proportional, characteristic of the number of equivalent nuclei.

Let us reconsider already discussed spectra. In Figure 171b, the signal of the substituted carbon of phenol at 154.9 ppm is much weaker than the other signals,** and the relative intensities of the other three signals indicate (although not quantitatively) that there is only one carbon in *para* position (121.0 ppm) and two each in *meta* and *ortho* positions (the signals at 129.7 and 115.4 ppm are stronger).

The approximately twofold intensity of the two equivalent methyl groups, moreover that the signal of the methine carbon is weaker (less NOE) for the bromo derivative (but only for this), is well observable in Figure 156. The lower intensity of substituted carbon signals is also observable in Figure 180. For the same reason, the carbonyl signals of the spectrum in Figure 185 are also weaker.

Figure 173 is an example that, occasionally, the ^{13}C intensity may also reflect the relative number of different carbons correctly. The signal corresponding to the single, C-5 carbon is approximately half of the other signals, which have approximately identical intensities. This is obviously related to the very similar character of carbon atoms in this paraffin.

The intensity enhancement of NOE is easy to measure experimentally,*** and the extent of saturation may be influenced by the parameters of measurement. Since spin-lattice relaxation times depend on molecular structure, they may be used as a valuable source of information in structure elucidation, and the interpretation of spectra.

4.1.4.2. Spin-Lattice Relaxation of Carbon Nuclei

The fact that in the first period of NMR investigations, mostly restricted to ^{1}H measurements, little attention was paid to the measurement of spin-lattice relaxation times was due partly to technical difficulties and partly to the lack of theoretical interpretation. The technical problems have been eliminated by the FT method, with which the routine measurement of T_1 is possible.† On the other hand, for carbon nuclei, the correlations between spin-lattice relaxation times and chemical structure are much simpler and easy to interpret.

There are no complications from homonuclear couplings due to the low abundance of ^{13}C isotope. Such interactions may really cause difficulties, as shown by the example of diethyl malonate and its labeled derivative EtO–C*O–C*H$_2$–COOEt.[998] Relaxation time T_{1C} of the methylene carbon is the same in both compounds but the relaxation of labeled carbonyl group is faster. The reason is that the former relaxes through C—H dipolar interactions, which are the same for the labeled and unlabeled molecule, whereas in the relaxation of

---

* Compare Volume I, Section 2.2.5.3 (p. 192—194).

** Compare p. 188 and 228.

***Compare Volume I, p. 194—197 and 208.

† Compare Volume I, Sections 2.2.7.1—2.2.7.4 (p. 205—208).

carbonyl carbon, the dipole-dipole interaction $C^*(O)–C^*(H_2)$ is also significant in the labeled molecule.

As a consequence of BBDR, the intensity of singlet ^{13}C signals changes purely exponentially,[790] which makes the determination of T_1 simple and accurate. Since, unlike peripheral hydrogens, the carbons are in the interior of the molecule, intermolecular relaxation processes are negligible. As the carbon signals are distributed over a very broad shift range, all signals of even complicated molecules are usually separated, and their intensities may also give information about the motions varying over the molecule (segmental motions). The phenomenon of relaxation, its theoretical interpretation, various relaxation mechanisms (compare Section 2.2.6), and its measurement (see Section 2.2.7) are not discussed here; only some characteristics of the relaxation of carbon nuclei will be added to the general principles.

4.1.4.3. Dipole-Dipole Relaxation

The spin-lattice relaxation of carbons with time constant T_{1C} takes place in solution primarily according to dipole-dipole (DD) interactions, caused by the actual field defined by Equation 248 induced by the magnetic moments of neighboring protons.[248a]

T_{1C} may be expressed as a function of correlation time τ_c,[1346] and taking into account the effect of the field induced by a single proton

$$(T_{1C}^{DD})^{-1} \equiv R_{1C}^{DD} = \frac{\hbar^2 \gamma_C^2 \gamma_H^2}{r_{CH}^6} \left[\frac{\tau_c}{1 + (\omega_H - \omega_C)^2 \tau_c^2} + \frac{3\tau_c}{1 + \omega_C^2 \tau_c^2} + \frac{6\tau_c}{1 + (\omega_H + \omega_C)^2 \tau_c^2} \right] \qquad (326)$$

which is analogous to Equation 249 given for the case of nuclei H and C and shows that for longer correlation times (slower molecular motions, i.e., large molecules and viscous liquids) and high measuring frequencies, T_1 is frequency dependent, which has also been proved experimentally (see Figure 189).[892a] It can be seen from Equation 326 that T_1 is minimum under the condition $(\omega_H + \omega_C) \tau_c \approx 1$, and that by increasing the measuring frequency, the minimum is shifted towards smaller τ_c values. If the extreme line narrowing condition*[248b] is met, i.e., $(\omega_H + \omega_C)\tau_c \ll 1$, there is not frequency dependence and one obtains the simpler form

$$R_1^{DD} \equiv (T_1^{DD})^{-1} = \sum_{i=1}^{N} \frac{\hbar^2 \gamma_C^2 \gamma_H^2}{r_{CH_i}^6} \cdot \tau_c^{eff} \qquad (327)$$

if the local fields of N hydrogen atoms are taken into account at distances r_{CH_i} from the carbon atom, and it is assumed that they have the same isotropic effective correlation time, τ_c^{eff}. This condition is met strictly for rigid molecules of spherical symmetry only, in which neither the molecule nor its groups have distinguished axes of rotation.

Regarding that the effect of hydrogens is inversely proportional to the sixth power of their distance from the carbon atoms to a first approximation, it is sufficient to take into account n_H protons attached directly to the carbon in question:

$$R_1^{DD} = n_H \hbar^2 \gamma_C^2 \gamma_H^2 \tau_c^{eff} r_{CH}^{-6} \qquad (328)$$

i.e., R_{1C}^{DD} is inversely proportional to the order of carbon atom.

Using the reasonable assumption[1a] that the relaxation mechanisms are independent, the reciprocals of the corresponding carbon relaxation times, i.e., the relaxation rates, are additive[892b]

---

*   Compare Volume I, p. 201.

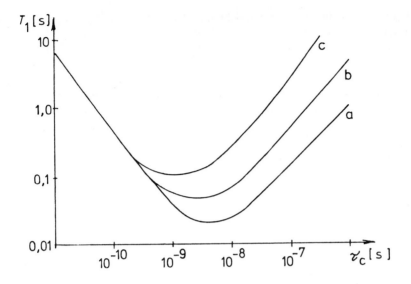

FIGURE 189. The τ_c-dependence of relaxation time T_1 at (a) 15.1, (b) 25.2, and (c) 55.4 MHz. (From Lyerla, J. R., Jr. and Levy, G. C., in *Topics in Carbon-13 NMR Spectroscopy*, Vol. 1, Levy, G. C., Ed., Wiley-Interscience, New York, 1979, 88. With permission.)

$$R^{obs}_{1C} \equiv (T^{obs}_{1C})^{-1} = \Sigma R^i_{1C} = \Sigma (T^i_{1C})^{-1} \qquad (329)$$

As the experimental value of T_{1C} is, in principle and often in practice, the sum of the contribution of various mechanisms, it is important to know the principles of their separation.

We have shown* that the intensity enhancement due to NOE is a consequence of the BBDR of protons, and thus the measuremen of η characteristic of it allows the determination of T^{DD}_{1C}.

The transition probabilities used in Equation 242 to define η give the change in distribution at thermal equilibrium between the $A(\alpha)$ and $A(\beta)$ levels of an AX spin system due to the equilibration of the populations of $X(\alpha)$ and $X(\beta)$ levels caused by the excitation of X. The non-Boltzmann distribution involves great difference between the populations of $A(\alpha)$ and $A(\beta)$ levels ($W_2 > W_o$, W_1), and if A is a carbon and X a hydrogen attached to it, the phenomenon increases the intensity of the carbon resonance signal. In this case the probability of excitation of the carbon ($\beta \rightarrow \alpha$ transition) is W_1. Whereas W_2 and W_o, depend on the carbon-hydrogen interaction only, influencing only the value of T^{DD}_{1C}, W_{1C} depends on all relaxation mechanisms.

From the quantum mechanical expression of transition probabilities[1346] used in Equation 242, it follows that for nucleus A of an AX spin system represented by C and H atoms, the change of the z component of magnetization as a function of time is described by two time constants:

$$\dot{M}^A_z = -\langle M^A_z \rangle T^{-1}_{1A} - M^A_e - M^X_z T^{-1}_{1AX} - M^X_e \qquad (330)$$

where M_e is the equilibrium value of the M_z component of the two spins, and

$$T^{-1}_{1A} = W_o + 2W_{1A} + W_2 \quad \text{and} \quad T^{-1}_{1AX} = W_2 - W_o \qquad (331)$$

* Compare Volume I, p. 193.

i.e., they are the numerator and denominator, respectively, of η in Equation 242, which may be thus written in the following form, too:

$$\eta = (T_{1C}/T_{1CH})\,(\gamma_H/\gamma_C) \tag{332}$$

with the substitution $A = C$ and $X = H$. This expression is the solution of differential Equation 330 for steady state under the $M_z^X = M_z^H = 0$, i.e., for the case of BB excitation of protons. This is the reason for the simple exponential relaxation of carbon signals, since according to Equation 330, without the BB excitation of protons (when $< M_z^X > \ne 0$), M_z^A is a function of two time constants (T_{1A} and T_{1AX}), and thus in the general case its change is not a simple exponential.

Since the maximum NOE[1] $\eta_o = \gamma_H/2\gamma_C$ appears when the carbon relaxations proceed exclusively through DD mechanism (i.e., $T_{1C}^{obs} = T_{1C}^{DD}$), according to Equation 332, $T_{1CH} = 2T_{1C}^{DD}$. Substituting this into Equation 332 and rearranging, one obtains

$$T_{1C}^{obs} = (\eta/\eta_o)\,T_{1C}^{DD} \tag{333}$$

where $\eta_o = 1.988$ is the maximal NOE.*

T_{1C}^{obs}, on the other hand, involves the contributions of all possible mechanisms, thus

$$T_{1C}^{obs} = T_{1C}^{DD} + T_{1C}^{o} \tag{334}$$

where T_{1C}^{o} represents the contribution of the other terms of Equation 329, i.e., the ones other than T_{1C}^{DD}, to T_{1C}^{obs}. Therefore,

$$T_{1C}^{o} = T_{1C}^{DD}(\eta_o/\eta - 1) \tag{335}$$

It is stressed that the above considerations and Equation 242 pertain only to the extreme line narrowing condition, i.e., if $\omega^2\tau_c^2 \ll 1$,[248b] in other cases NOE is frequency dependent,** as shown in Figure 190.[372]

The presumable consequences of the further increase of measuring frequency (and magnetic field) are mentioned here. The introduction of superconducting magnets has made it possible to increase the magnetic fields and thus to construct commercially available 500- to 600-MHz instruments. Although this improves the sensitivity (ζ) of instruments and the spread of signals substantially (although this latter is significant only in proton-coupled carbon resonance spectra), the NOE is maximum only for the faster motions (at 60 MHz $\tau_c = 10^{-9}$ s is sufficient for η_o to occur; at 360 MHz this limit is already $2 \cdot 10^{-10}$s), and T_{1C} becomes frequency dependent: it increases with measuring frequency and thus ζ decreases (saturation). The extreme line narrowing condition ($\omega_o^2\tau_c^2 \ll 1$) is met only for faster motions, i.e., at higher temperatures or in dilute solutions or for smaller molecules. This hinders the interpretation of the relationships between T_1 and molecular structure or dynamics, e.g., the relationship between T_{1C}^{DD} and τ_c is described by the more complex Equation 326, and the much simpler Equation 327 may not be used.

4.1.4.4. Other Carbon Relaxation Mechanisms

The other most important relaxation mechanism of carbon atoms, *spin rotation* (SR) interaction, is significant only for small and symmetric molecules,*** particularly at higher

---

*   Compare Volume I, p. 193.
** Compare Equation 326 and Volume I, p. 201.
***Compare Volume I, p. 201.

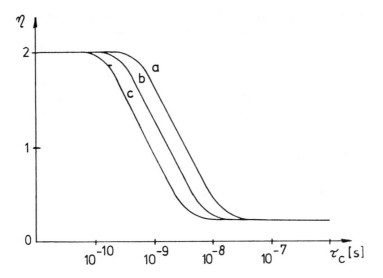

FIGURE 190. The τ_c-dependence of the NOE (factor η) at (a) 15.1, (b) 25.2, and (c) 55.4 MHz.[372]

temperatures, since T_{1C}^{SR} decreases (τ_J increases) while the efficiency of other relaxations decreases (τ_c shortens). Thus, the temperature dependence of T_{1C}^{obs} is a proof for the significant contribution of SR mechanism.[355]

In diamagnetic solutions generally the DD and SR mechanisms are significant, thus T_{1C}^{v} $\approx T_{1C}^{SR}$. The SR term is important for quaternary carbons and for methyl groups; in the case of free rotation, otherwise in large and asymmetric molecules, particularly for protonated carbons, the DD term is almost exclusive. The greater contribution of SR mechanism to the relaxation of quaternary carbons has been proved experimentally, e.g., for substituted benzenes.[849]

The T_{1C}^{SA} term arising from *chemical shift anisotropy* may be proved by its frequency dependence.* Since the other terms of T_{1C}^{obs} are frequency independent,** determining it at two measuring frequencies, the difference is, according to Equation 257,

$$\Delta^{1,2} R_{1C}^{obs} = (8/15)\pi^2(\sigma_\| - \sigma_\perp)^2\tau_c\Delta^{1,2}\nu_0^2 = (2/15)(\sigma_\| - \sigma_\perp)^2\tau_c\gamma_C^2\Delta B_0^2 \tag{336}$$

where $\Delta^{1,2}\nu_0 = (\nu_0)_1 - (\nu_0)_2$ is the difference between the measuring frequencies and τ_c is the correlation time identical with that of the DD mechanism. This measurement also gives the anisotropy of the chemical shift, $\sigma_\| - \sigma_\perp$. According to theoretical calculations, the SA term is significant only when the field exceeds 2.5 T (frequency >100 MHz for protons), but with the increasing use of superconducting magnets its effect must be taken into consideration.***

The relaxation of 1,4-diphenylbutadiine molecule (Ph–C≡C–C≡C–Ph) is one of those rare cases where SA contribution was detected.[849,857] The "inner" (β) acetylenic carbons (isolated from hydrogens and thus free of DD interaction) have strongly frequency-dependent T_1: 135 s at 2.35 T and approximately 340 s near zero field. The molecule is long, its fast rotation is improbable, and thus the SR mechanisms may not be effective. The situation is opposite

* Compare Volume I, Equations 257 and 258 (p. 202 and 203).
** As a possible exception is the very rare T_1^{SC} term, which is, however, not necessarily frequency dependent.
***At 6 T the SA term is already commensurable with DD; compare Volume I, p. 203.

for CS_2, where a high SA term was assumed formerly,[522,1255] but later it turned out that even at 6.5 T the SR mechanism dominates,[1357] and the SA mechanism is preferred only at high fields and low temperatures (below $-60°C$) when spin rotation becomes hindered by increasing viscosity.*

Scalar mechanism is active in the spin-lattice relaxation of carbons if $\Delta\omega^2\tau_{SC}^2 \approx 1$, i.e., the extreme line narrowing condition is not fulfilled (in this case T_{1C}^{obs} is frequency dependent), or if it is fulfilled, but $\Delta\omega^2$ is small and τ_{SC}^2 is not particularly large, either. This is the case with C–Br bonds** when the SC mechanism is active,*** but not frequency dependent.[519,890] In bromoform $R_{1C}^{SC} \gg R_{1C}^{DD}$, and from T_{1C}^{SC} one can determine T_1 (^{79}Br) = 1.5 μs and ^2J(Br,H) $\gg 41$ Hz.[207,465] The relaxation studies of bromo-derivatives are complicated by the different relaxation times of the two magnetic bromine isotopes, due to which the intensity change of carbon signal is not exponential.[843,849]

4.1.4.5. Nuclear-Electron Relaxation

In the presence of paramagnetic species, the analogous *nuclear-electron* interactions are competing with the C–H dipolar interactions, and in sufficient concentrations they become the main mechanism of spin-lattice relaxation because $\mu_e \gg \mu_H$,† despite of the greater nucleus-electron distance. If there is no interaction between the paramagnetic substance and the solute molecules, an equation can be derived for the rate of nuclear-electron DD relaxation under the assumption of rigid spherical shells, from which R_1^e is proportional to the concentration of the species and does not depend explicitly on the nuclear-electron distance:[1b]

$$R_{1C}^e \equiv (T_{1C}^e)^{-1} = (16/15) \pi^2 N_e \hbar^2 J(J+1) \gamma_e^2 \gamma_C^2 \eta/(kT) \qquad (337)$$

where N_e is the density of paramagnetic species, η is viscosity, and the $\gamma_e^2 \hbar {}^2J(J+1)$ factor is the square of the magnetic moment of the paramagnetic species. The relation of this equation with the distance-dependent expression(262) of R_1^e is obvious if Equation 246 of τ_c is substituted into the latter.

The validity of Equation 337 is proved experimentally by the fact that the "(R_{1C}^e)" values related to unit viscosity (1 cP) and concentration (1 mol of the paramagnetic species) and measured at the same temperature[848] are practically the same for *all* carbon atoms of the dissolved molecule. The relaxation times T_{1C}^e of 2,2,4-trimethylpentane measured in the presence of Cr(ACAC)$_3$ are the same, including those of the "inner" quaternary carbon atoms.[848]

For various substances, "inert" with respect to Cr(ACAC)$_3$, $(R_{1C}^e) \approx 30$, i.e., this is a measure of the "efficiency" of the paramagnetic substance. The R_{1C}^e term may be obtained as the difference of R_{1C} measured in the presence of paramagnetic substance and R_{1C}' measured in pure diamagnetic solution

$$R_{1C}^e \equiv (T_{1C}^e)^{-1} = (T_{1C})^{-1} - (T_{1C}')^{-1} \equiv R_{1C} - R_{1C}' \qquad (338)$$

If the concentration of paramagnetic additive is sufficient for R_{1C}^e to be commensurable with R_{1C}^{DD}, NOE decreases proportionally, and at a sufficiently high paramagnetic concentration, it is completely eliminated.[524] If the NOE factors measured with and without paramagnetic substance are η^* and η,

$$T_{1C} = (\eta^*/\eta)T_{1C}^{DD} \qquad (339)$$

similar to Equation 333.

---

* Compare Volume I, p. 202.
** Compare Volume I, p. 203.
*** This is true for ^{79}Br isotope only since ^{81}Br relaxes almost exclusively according to the DD mechanism.[465,843,890]
† Compare Volume I, p. 5 and 204.

The paramagnetic additives may be used for the determination of T_{1C}^{DD}, for the elimination of NOE (e.g., to "develop" the signals suppressed by the negative NOE of nitrogen owing to $\gamma_N < 0$)* and for the intensification of weak or not observable quaternary carbon signals by the acceleration of relaxations[524,538] (in the presence of these "T_1 reagents", ζ of FT spectra is independent of t_{aq} and θ, i.e., pulse angle).

If the paramagnetic substance forms a specific complex with the solute molecules, i.e., the latter have functional groups capable of coordination, the acceleration of relaxation, proportional to r^{-6}, may be interpreted by Equation 262 instead of Equation 337.[462,524] Then, the components T_{1C}^e measured for the different atoms of the molecule may be used in the solution of assignment and stereochemical problems.[852] It is interesting that T_1^e is not angle dependent, unlike the lanthanide-induced shifts, and its magnitude does not depend on the order of carbon, either,[805,852] for "inert" substances. No quantitative relationship can be found between T_{1C}^e and the bond length in the complex, since owing to the fast exchange phenomena and eventually the formation of many types of complexes with different equilibrium constants, the measured value of T_{1C}^e can be regarded as a weighted average.

Besides the most widespread relaxation reagents,[524,538] Fe(ACAC)$_3$ and Cr(ACAC)$_3$, an r^{-6} dependent $(T_{1C}^e)^{-1}$ contribution is produced by the Gd(DPM)$_3$ lanthanide complex,[462] without causing a simultaneous change in chemical shift. Thus, e.g., in n-butanol, T_1 (C-1) $< T_1$ (C-2) $< T_1$ (C-3) $< T_1$ (C-4).[462] This is advantageous since the application of normal SRs in ^{13}C NMR spectroscopy is hindered by the substantial contact term due to which there is no pure r^{-3} dependence.**

An important practical aspect of nuclear-electron relaxations is the acceleration effect of paramagnetic molecular oxygen absorbed in solvents. The corresponding value of T_{1C}^e is approximately 100 s,[849] and this contribution may cause substantial errors if $T_{1C} > 10$ s. Therefore, the removal of dissolved oxygen from the sample solution (most simply by bubbling nitrogen through it or by boiling) is of utmost importance when measuring ^{13}C relaxation times longer than 10 s. Of course, the presence of other paramagnetic substances (transition metals and rare earth metals) in traces also leads to false results.

4.1.4.6. Application of Dipolar Spin-Lattice Relaxation Times in Structure Analysis

The ^{13}C $-$ ^{1}H DD mechanism has a decisive role in ^{13}C relaxation, and as we have shown, this is easy to separate from other mechanisms by NOE measurement. As the ^{13}C $-$ ^{1}H DD interactions rapidly decrease with internuclear distance (r_{CH}^{-6} dependence) to a first approximation, it is sufficient to take into account the hydrogens directly attached to the carbon, and in this case a simple inverse proportionality exists between T_{1C}^{DD} and τ_c (compare Equations 326 to 328). Under these conditions, the measured T_1 data and intensities are useful in structure elucidation: in the study of assignment, molecular motions, and intermolecular interactions.

On the basis of the r^{-6} dependence of dipolar relaxations and the direct proportionality between relaxation efficiency and the number of neighboring or directly attached hydrogens, quaternary and protonated carbons are easy to distinguish, and of the latter, chemically similar secondary and tertiary carbons may also be distinguished (compare Figures 156, 171b, 172, and 180 discussed in the introduction).

For adamantane, $T_1(CH) \approx 2T_1(CH_2)$. The measured values are 20.5 and 11.4 s.[790] The ratio (1.8) is in good agreement with the one expected when the effect of the further hydrogens (two *geminal* for the methylene carbons and six for the methine carbons) is taken into account. When only the directly attached hydrogens are considered, the theoretical ratio is exactly $T_1(CH)/T_1(CH_2) = 2$ on the basis of Equation 329.

---

* Compare p. 259.
** Compare Section 3.7.3.2, p. 120—126.

Taking into account the T_1 data of the carbons of 3β-chloro-cholest-5-ene (**310**),[27] the signals of methine and methylene carbons are easy to distinguish, since the former have about double T_1 relaxation times. For the methine carbons C-3, C-9, C-14, C-17, and C-20, T_1 = 0.50, 0.51, 0.44, 0.51, and 0.49 s, respectively, and for the methylene carbons C-1, C-2, C-4, C-11, C-12, C-15, and C-16, T_1 = 0.25, 0.25, 0.27, 0.25, 0.31, 0.23, and 0.23 s.* This is particularly important since the similar carbons of steroids often give close, sometimes overlapping signals, and thus the undecoupled or even the off-resonance spectra are too complex for the determination of multiplicities and, therefore, the order of the carbon for all signals. Thus, the T_1 values represent the only aid in assignment.

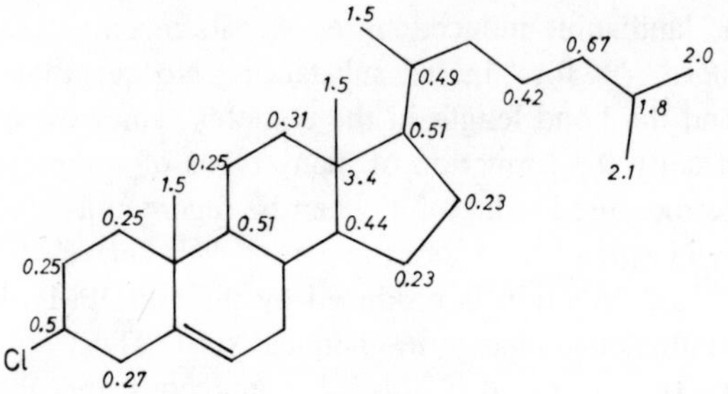

310

The agreement of ratios T_1 (CH)/T_1(CH$_2$) with theory also indicates that in steroids, like in large molecules generally, the dipolar relaxation mechanism is dominant.[27,53,851] Of course, quaternary carbons may be distinguished from the protonated ones on the basis of relaxation times, generally an order higher, i.e., the partly or completely saturated signals. Thus, already on the basis of intensities, the lines of quaternary carbons may be identified in the ^{13}C NMR spectrum. (The relative intensity of the quaternary carbon signals can be increased, at the expense of measuring time or by increasing t_{pi} or decreasing pulse angle θ.** For the protonated C-2,6, C-3,5, and C-4 carbons of benzonitrile, T_1 = 9.3, 9.3, and 6.0 s,[1489a] whereas for C-1 and CN carbons T_1 = 86 and 80 s, respectively.

Although relaxation of quaternary carbons has a contribution from the DD interaction of *geminal* and more distant hydrogens, NOE does not reach the maximum of protonated carbons ($η_o ≈ 2$). In this case the relative magnitude of NOE ($η/η_o$) and the number of neighboring hydrogens may be the basis of assignment and thus structure determination (compare Equation 327).

The different T_1 values of the substituted C-3,5, C-4, and C-1 nuclei of **311** may be interpreted by the number of the adjacent (*geminal*) hydrogens. C-4 relaxes most slowly, since it has no hydrogen neighbor. C-3,5 may be characterized by medium relaxation times corresponding to the one hydrogen neighbor, whereas T_1 measured for C-1 *geminal* to two hydrogens is the shortest.[846]

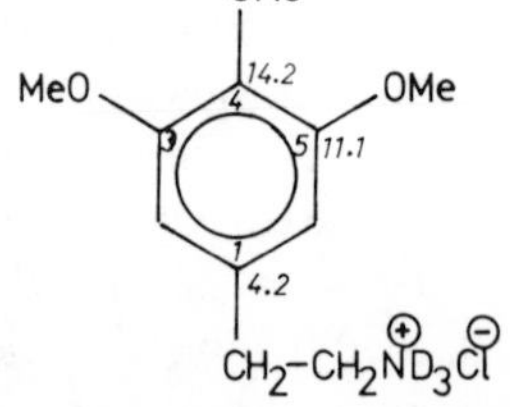

311

---

* The data, obtained by inversion method, are the first spin-lattice relaxation times measured for steroids.
** Compare Volume I, p. 178 and Section 2.2.4.4. (p. 188).

Of the five quaternary carbons of codeine (**312**), the signal of C-13 can also be distinguished on the basis of its upfields position; the assignment of C-3, C-4, C-11, and C-12 is made unambiguous, however, only by the T_1 relaxation times.[1487] They are, in the sequence of the numbering of quaternary carbons, 5.2, 8.9, 1.8, 5.1, and 1.5 s. Presumably T_1 (C-3) $< T_1$ (C-4), since C-3 has one *geminal* hydrogen neighbor, whereas C-4 has none. Similarly, owing to the three *geminal* protons, T_1 (C-11) $\ll T_1$ (C-12), since the latter has no hydrogen neighbor. The fastest relaxation (1.5 s) arises from C-13 adjacent to four protons.

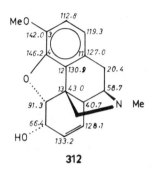

312

The quaternary carbons of Compound **313** may be distinguished in a similar manner, on the basis of the shortening relaxations with the number of *geminal* hydrogens: according to the 0, 2, and 3 *geminal* hydrogens for C-8a, C-4a, and C-1,8, T_1 = 112, 73, and 66 s, respectively.[36] In the case of **313** and other smaller fused aromatic compounds, in which the rotation of the molecule is relatively fast so that the SR term is also significant beside the DD mechanism, the actual ratio of DD term may be determined only from their partial NOE[737] according to Equation 333. For C-8a, C-4a, and C-1,8 of Compound **313**, e.g., η = 0, 0.2, and 0.3 s, indicating the significant SR contribution. Of course, if the DD mechanism dominates in the relaxation processes, T_{1C}^{obs} depends on the number of neighboring hydrogens, and thus the relative values measured for the different carbons are proportional to T_{1C}^{DD}. When, however, another mechanism (e.g., SR) is also significant, the measured T_1 values may prompt false conclusions on the assignment. Thus, e.g., of the 80- and 86-s relaxation times of benzonitrile belonging to the substituted aromatic and nitrile carbons, the former may be assigned to the aromatic and the latter may be assigned to the nitrile signal, considering the two hydrogen neighbors of the aromatic carbon. However, the relative NOE values support the opposite assignment: $T_{1C}^{DD} = \eta_o T_{1C}^{obs}/\eta$ is 250 and 150 s in the above sequence. The reason is that the significant SR term shortens T_{1C}^{obs} for the nitrile carbon atoms so much that it overcompensates the effect of *geminal* hydrogens for the aromatic carbons.[1489a]

313

On deuteration, T_1 changes significantly, not only for the deuterated, but also for the neighboring carbon atoms if their relaxation is affected by adjacent hydrogens. According to the lower values of γ_D (compare Equations 326 to 328), T_{1C}^{DD} (CD) is about tenfold of the corresponding value of T_{1C}^{DD} (CH).

The C-2′ and C-4′ chemical shifts of vitamin B_1 (**314**) (163.8 and 164.2 ppm) and T_{1C}^{obs} relaxation times (4.2 and 4.7 s) are hardly different, and NOE is exactly the same.[412] On exchanging acidic protons to deuterium (using $CD_3OD - D_2O$ mixture as solvent instead of $CH_3OH - H_2O$ system), T_{1C}^{DD} values change as shown on the formula.

314

On the basis of $T_{1C}^{DD}(D) - T_{1C}^{DD}(H)$ differences, not only the C-4' and C-2' signals may be assigned unambiguously (through the much stronger isotope effect for C-4'), but also the conformation given in the formula may be supported on the basis of the strong isotope effect on C-4 (the ND_2 and 4-methyl groups are close). This is only one example for the rapidly recognized fact that T_{1C} and NOE data have equally important roles in conformational analysis.[1372r]

The originally doubtful assignment of the C-4a,5a and C-1a,8a signals of phenantrene (**182**) was proved by deuteration. The slightly lower relaxation time of the former carbons (51 vs. 59 s) already suggested the verified conformation (atoms C-4a,5a have two, and C-1a,8a have only one hydrogen neighbor), but the T_1 values measured after replacing H-1,8 atoms by deuterium (59 and 80 s) made the assignment unambiguous, owing to the much higher isotope effect on the atoms relaxing more slowly also in the original molecule.[1077]

4.1.4.7. Investigation of Molecular and Segmental Motions

Carbon spin-lattice relaxation times are the almost exclusive information source of molecular microdynamical studies. The thermal motion of small molecules is often anisotropic, i.e., for inertial, electrostatic, or steric reasons, the rotation of the molecule about a preferred axis or axes is more probable. It frequently occurs with larger molecules that while the motion of the molecule as a whole is roughly isotropic, some terminal groups move much faster than the molecule. The motion of one end of chain molecules may be hindered for various reasons; in this case, mobility increases gradually along the chain. These phenomena are in direct connection with T_1, e.g., the rotation of monosubstituted benzene derivatives around the symmetry axis is obviously preferred. This is reflected in the slower relaxation of *ortho* and *meta* carbon atoms with respect to the also protonated *para* carbon atom, which lies in the axis. This is illustrated by the T_1 data of toluene and nitrobenzene shown in Formulas **315** and **316**.[26] The local fields which promote relaxation are averaged during rotation around the molecular axis (the efficiency of DD mechanism decreases). Relaxation times much longer than T_1 measured in pure liquids are characteristic of the CCl_4 solution of aniline (values in parentheses in Formula **317**), owing to the faster molecular motions. In contrast, in acetic acid solution very short T_1 values may be measured (**318**), since protonation slows down molecular motions (the molecules are enclosed in a "solvent cage" on the effect of ion pair formation and solvation), and the differences between T_1(C-4), T_1(C-2,6), and T_1(C-3,5) are higher: the molecular symmetry axis is more preferred, i.e., rotation anisotropy increases.[849]

315   **316**   **317**   **318**

Rotation is slowed down by heavy substituents, and thus relaxation is accelerated: T_1 decreases.[849] This can be seen from a comparison of the data of toluene and nitrobenzene, but a more direct proof is given by ferrocene derivatives bearing a substituent on one of the cyclopentadiene rings. Here, in the case of acetyl derivative **319**, the T_1 value measured for the equivalent carbons of the unsubstituted ring rotating faster around the molecular axis is about four times higher than the average relaxation time of the carbons of the substituted ring.[845]

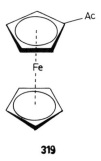

319

From analogous reasons, the relaxation times of cyclic ethers decrease with increasing ring size,[853] whereas those of the methyl groups attached to the ring increase. The reorientational motion of the molecule as a whole slows down (i.e., relaxation time decreases) with increasing molecular weight, while the contribution of the internal rotation of the methyl group to the relaxation of the methyl carbon (see below) increases, increasing thereby $T_1(Me)$.

The T_1 data of salicylic acid methyl ester **320** indicate[1489b] that of the two preferred axes, the rotation around the axis also passing the heavy carbomethoxy group is, logically, the more frequent: $T_1(C-5) > T_1(C-4)$. The assignment of the very close C-3,6 and C-4,5 lines of *ortho*-dichlorobenzene ($\delta C-3,6 = 130.4$ and $\delta C-4,5 = 127.9$ ppm) is possible on the basis of the faster relaxation of the "axial" C-4,5 atoms (6.3 s; for the other atom pair, it is 7.8 s).[518]

320

If there is no preferred conformation owing to steric compression, the dipolar relaxation time of (mostly methyl) groups moving fast as compared with the reorientation of the molecule increases.[21,789] In the case of free rotation, the contribution of SR mechanism may become significant,[21,465,708,789,1256,1470] and thus the increase in T_{1C}^{obs} may be less than approximately threefold as expected.[891] The relatively slow relaxation of methyl groups can be seen in the T_1 data of steroid derivative **310**. Even in the case of isotropic motion, the nonrotating methyl group with three protons would be expected to have three times shorter T_1 than the methine protons, on the basis of Equation 328. Thus, T_{1C}^{DD} may theoretically increase nine times in the case of fast methyl group rotation.

If steric hindrance increases, the rotation of methyl groups becomes faster (there are not preferred conformations). Therefore, in androstane derivatives, the $T_1(C-19) > T_1(C-18)$ relation is general,[53] since the rotation of the 19-Me group is more hindered, owing to the five 1,3-*diaxial* interaction, than that of 18-Me affected only by three such interactions.

For similar reasons, in 1,2,3-trimethylbenzene $T_1(Me-2) > T_1(Me-1,3)$,[21] and in dimethylformamide, $T_1(Me^Z) > T_1(Me^E)$.[856] In the former molecule the rotation of the middle

2–Me group, in the latter of the methyl group *cis* to carbonyl is hindered, and thus, there is no preferred conformation, and the rotation becomes faster.

The polar end of normal alcohols is "fixed" by hydrogen bonds, and segmental motion becomes gradually less hindered toward the nonpolar end of the molecule: the motion becomes faster. In *n*-decanol (**321**), e.g., the relaxation time of the carbon atoms gradually increases along the chain from 0.6 to 3.1 s.[370] The analogous phenomenon observed[856] with *N,N*-di-*n*-butylformamide **322** may also be attributed to association. Hydrogen bond explains that in ethers T_1 is always longer than in the corresponding alcohols.[853]

$$\overset{3.1}{C}H_3-\overset{2.2}{C}H_2-\overset{1.6}{C}H_2-\overset{1.1}{C}H_2-\overbrace{CH_2-CH_2}^{0.84}-\overset{}{CH_2}-\overbrace{CH_2-CH_2}^{0.77}-\overset{0.65}{C}H_2OH$$

321

$$\begin{array}{c}H\\ \diagdown\\ C-N\\ \diagup\diagup\qquad\diagdown\\ O\end{array}\quad\begin{array}{c}\overset{1.1}{C}H_2-\overset{1.6}{C}H_2-\overset{2.3}{C}H_2-\overset{3.1}{C}H_3\\[4pt]CH_2-CH_2-CH_2-CH_3\end{array}$$

322

The role of "anchor" may be played not only by a group capable of association, but also by polar, electrically charged substituents; (in molecules closed into cyclodextrine or crown ether "cage," the enclosed end of the molecule, or the heavy skeleton of molecule in comparison with the side chain.[1031] An example for the latter are steroids, e.g., Compound **310**, in which the T_1 values measured in the side chain are much higher (e.g., with respect to the cyclic analogues in the case of methylene carbons), owing to the greater mobility of the side chain with respect to the molecule as a whole.

The investigation of steroids has also shown[1314c] that the motion of the ring system is not isotropic either; the rotation is preferred along the axis joining C-3 and C-17. Therefore, T_1 for the carbon of axial 3β C–H bond is lower than for the corresponding α C–H bond (e.g., for 3α-ol as compared to 3β-ol). These methods may be used for studying the segmental motions of macromolecules, including biopolymers such as peptides and proteins,[28,29,353,1232] association problems,[22,844] formation of micelles,[854,1214,1524] etc.

4.1.5. Carbon Resonance of ¹³C Labeled Compounds; Investigation of Reaction Mechanisms and Biosyntheses[1373]

The investigation of reaction mechanisms and biosyntheses with the use of the carbon resonance spectra of ¹³C labeled substances was one of the first topics of ¹³C NMR spectroscopy, since the problems connected to sensitivity were smaller due to enrichment. The advantages of ¹³C NMR method were soon discovered,[821] and the first investigations on reaction mechanism and biosynthesis[1201,1394] were reported already at the end of the 1960s.

The concentrations of ¹³C enriched compounds and their reaction products are very simple to determine by comparing the intensities of satellite signals split by J (¹³C,¹³C) coupling and — in the proton decoupled spectra — the singlets arising from the unlabeled molecules.

Similarly, in the proton decoupled spectra of deuterated compounds, the signal of deuterated carbon is split by J (C,D) couplings. Both ¹³C and ²H labeling may be utilized in the assignment of carbon signals. This method was used first in the interpretation of the ¹³C NMR spectra of steroids.[1186]

For the determination of isotope content, of course, mass spectrometry is also suitable, and in case of CH → CD labeling it is more accurate, since the CD signal may be partially saturated owing to the absence of NOE. (For methyl and methylene groups this difficulty does not arise, since for CH₃ and CH₂D or for CH₂ and CHD groups the NOE remains unchanged.[790])

As an example of the study of biosyntheses we discuss the incorporation of tryptophan (**323**) into antibiotic **324**.[939] On adding tryptophan labeled by ^{13}C on its β-carbon to the culture performing the biosynthesis, only C-3 of the product proved to be labeled indicating unambiguously that the side chain of tryptophan is the specific precursor of the pyrrole ring of Compound **324**.

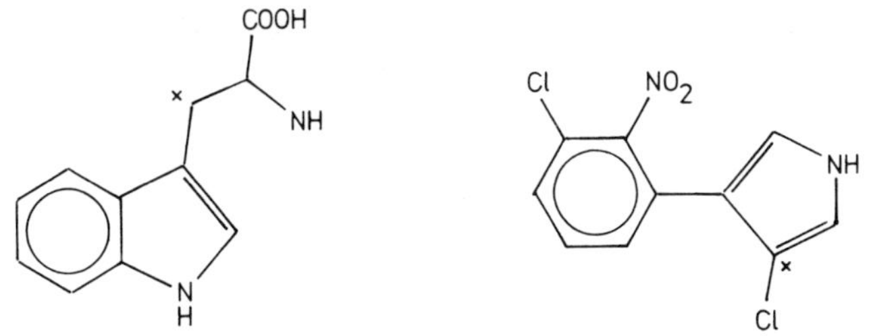

323   **324**

The deuterium labeling of given parts of enzymes or macromolecules enables carbon signals of hardly different shift to be distinguished by DR. In the case of deuterium resonance, for example, the multiplicities of labeled carbons change, and in this manner conformational changes may be followed.[215]

On addition of acetate, ^{13}C labeled at C-1 or C-2 to the bacterium culture producing cephalosporine-C (**325**) 10-, 15-, 18- or 11-, 12-, 13-, 14-, 19- labeled products were formed, indicating the possibilities of acetic acid incorporation.[1033]

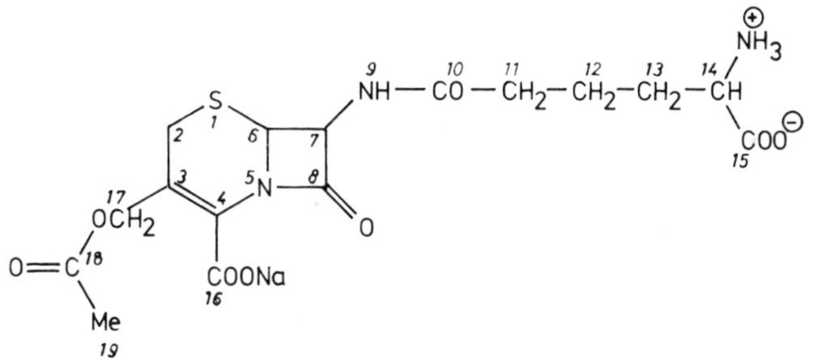

325

^{13}C NMR studies on numerous other biosyntheses can be found in the literature (e.g., References 196b, 966, and 1373a).

The first reaction mechanism investigated by ^{13}C NMR was the formolysis of tosyloxy-methylcyclopentane (**326**).[1200] According to the assumptions, the compound converts into carbonium cation **327**, which stabilizes via a cyclohexylcarbonium cation (**328a-d**), formed by ring expansion, as cyclohexanol (**329**) by simultaneous formation of formyl tosylate.

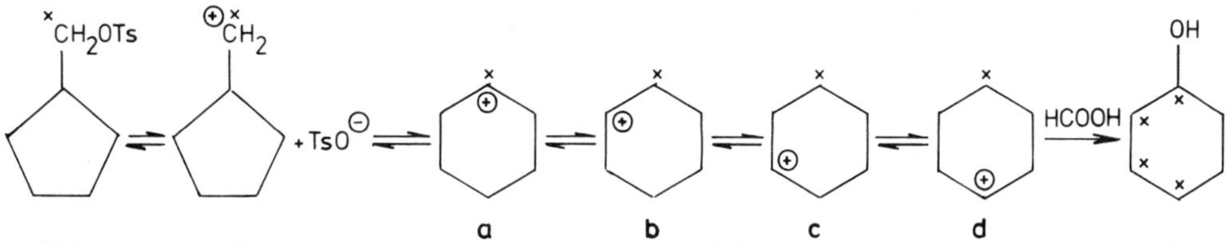

326   **327**   a   b   c   d   **329**
                                           328

On performing the reaction with the starting material labeled on its exocyclic carbon, the

^{13}C NMR spectrum indicated a random (even) distribution of ^{13}C isotope in the cyclohexanol product. This is an unanimous proof for the assumed intermediate of cyclohexyl cation type.

Of the numerous examples available, the structure identification of the dimer of triphenylmethyl is mentioned, which was carried out by the ^{13}C NMR investigation of the products obtained from the monomer labeled at the sp^3 carbon (Ph_3C^*). The two signals at 63.9 and 138.6 ppm, indicating the presence of one labeled sp^2 and one sp^3 carbon atom, exclude the formation of molecule Ph_3C—CPh_3 and prove Structure **330**.[1361]

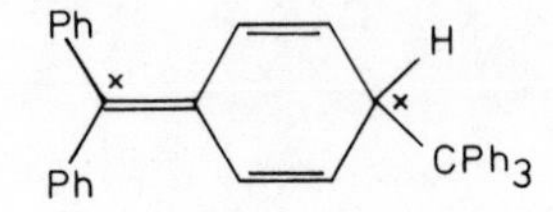

330

4.1.6. Shift Reagents in Carbon Resonance

SR technique is discussed in Section 3.7.3. Since the discussion is restricted to applications in ^{1}H NMR, some special ^{13}C NMR aspects must be mentioned here. The SRs also cause specific shifts in the ^{13}C NMR signals,[204,278,540,1511] promoting therefore the solution of assignment and structure identification problems. The example of isoborneol (**331**) is mentioned, in which the relative magnitudes of carbon and proton shifts (bold numbers) are similar.[34,540] The analogy is, however, not general, since the contact term is much higher for carbon than for hydrogen nuclei.* Therefore, instead of the Eu and Pr complexes most widely used in ^{1}H NMR, in carbon resonance the application of ytterbium-based SRs is recommended,** since with them pseudo-contact interactions are the dominant.[539,723,1360] Formula **332** shows the shifts induced by reagent $Yb(TFC)_3$ on the carbon atoms of 2-bromocholestan-3-one.[1314d] One has to mention the possibility for identification of the C_α signals of oligopeptides on the basis of their extremely high shifts.[196d]

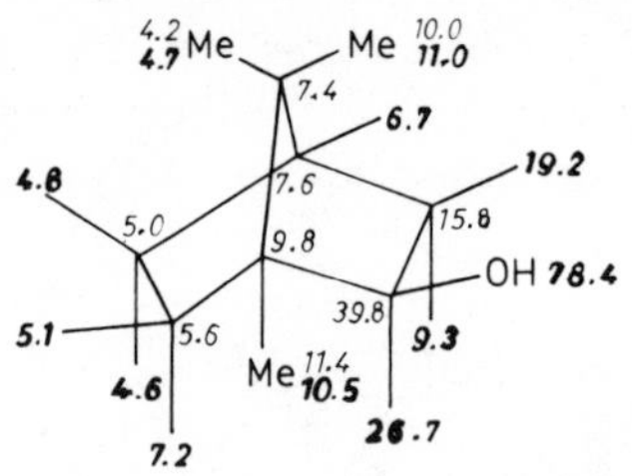

331 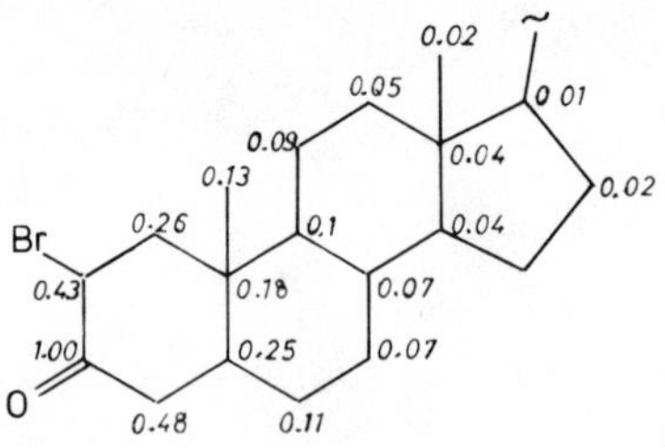   **332**

4.1.7. Dynamic Carbon Resonance Spectroscopy[918]

The investigation of internal molecular motions (interconversion of stable conformers, inversion, tautomery, intramolecular rearrangements, valence isomery, etc.) and intermolecular exchange phenomena by temperature-dependent NMR measurements, i.e., the method of dynamic NMR (DNMR) spectroscopy [134,409,1382,1386] is also applicable in carbon resonance, and even some of the disadvantages occurring in proton resonance are eliminated or reduced in ^{13}C NMR.

1.   Carbon NMR spectra are not complicated by spin-spin coupling. The acceleration of exchange process involves the simplest change in the spectrum, the gradual broadening of two sharp lines, their coalescence into one broad line, and with the further increase

---

*   Compare p. 121—122.
** See p. 127.

of temperature, the sharpening of this single line. Thus, the complicated calculations necessary for the determination of $\Delta G^{\ddagger}$,* which become very time consuming for more than four spins, are superfluous, and there is no error arising from the neglect of coupling (which, even in the case of long-range couplings causing no observable splitting may reach 10% in proton resonance).[391]

2. The number of lines is smaller, their distance is larger, which makes it possible to study faster molecular motions, since the coalescence temperatures are higher,[1373] and the separation of lines may be observed in such cases, too, when the ^1H signals would separate only below the measurable temperatures.

3. Measurements may be more accurate since the coalescence of more signal pairs can be observed separately, and the line width of signals with unchanged environment can be used as reference.

4. From the activation free enthalpies calculated for signal pairs coalescing at different temperatures, the temperature dependence of $\Delta G^{\ddagger}$ may also be determined.

Of the specific applications, some characteristic examples are discussed here. The most frequent field of application of DNMR spectroscopy is conformation analysis and the determination of the activation parameters of conformational motions. The application of carbon resonance spectra for these purposes is important in cases when the ^1H NMR investigation is not informative. In the proton resonance spectrum of dimethylcyclohexane derivatives, e.g., the signals overlap, and the spectrum parameters characteristic of the steric position of substituents, i.e., the chemical shifts and coupling constants of the protons attached to the substituted carbons, may not be determined. In contrast, in the ^{13}C NMR spectrum, for example, one can follow the change of ring inversion rate as a function of temperature, and the broadening and coalescence of the originally separated signals of carbon pairs which exchange environment during the inversion. This is illustrated by Figure 191,[1259] which shows the ^{13}C NMR spectrum of *cis* 1,2-dimethylcyclohexane at -115, -40, and $-20°C$. It can be seen that at $-115°C$ six lines correspond to the six skeletal carbon atoms. With increasing temperature, the signals of the exchanging atom pairs C-1,2, C-3,6, and C-4,5 and of the two methyl carbons first broaden, and then, at $-20°C$, appear as sharp, coinciding lines. From the coalescence temperature ($T_c \approx 220°K$), $\Delta G^{\ddagger} = 41.5$ kJ/mol is obtained for the ring inversion.

The fact of conformation equilibrium, i.e., the presence of more than one conformers, is proved by the coalescence of split signals upon increase of temperature and the repeated separation upon cooling, i.e., the reversible temperature dependence of the spectrum. The reason for the signal splitting of a flavone glycoside observed at room temperature was cleared in this way.[341]

The ligand exchange phenomena of metal-organic compounds may also be studied easily in carbon resonance. The complexes of metal carbonyls with olefins have trigonal bipyramidal structure (**333a**), where two pairs of equivalent carbonyls (*equatorial* and *apical*) may exchange positions via the rotation of double bond ($\mathbf{a} \rightleftharpoons \mathbf{b} \leftrightharpoons \mathbf{c}$).[16] Depending on the rate of ligand permutation, in the carbonyl region of the ^{13}C NMR spectrum the *apical* and *equatorial* carbonyl signals separate or coalesce, respectively. As an example, a spectrum section of dimethyl fumarate iron tetracarbonyl complex is shown in Figure 192 as a function of temperature.[787]

A classical example of valence isomery is bulvalene (**153**), in the spectrum of which one line appears at room temperature, whereas at low temperature the lines corresponding to the four nonequivalent kinds of carbon separate[612,1023,1073] with the expected 1:3:3:3 intensity ratio. Owing to hindered internal rotation,[392,886] the C-2,6 and C-3,5 signals of benzaldehyde

---

* Compare p. 94—95.

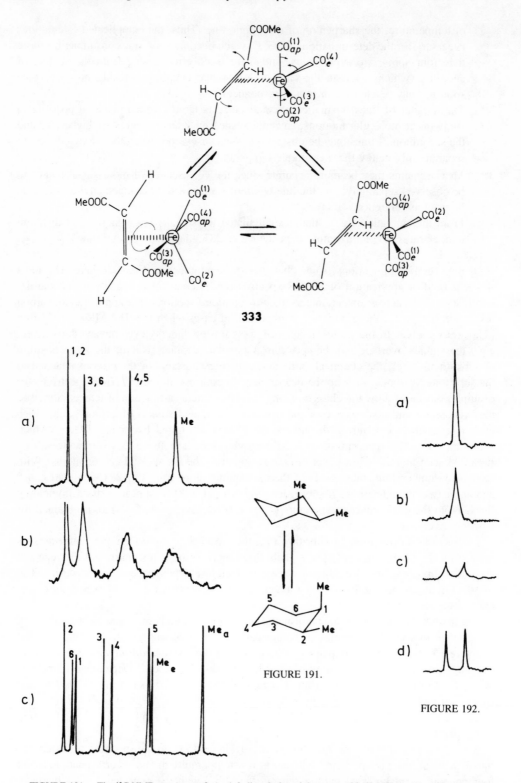

333

FIGURE 191.

FIGURE 192.

FIGURE 191.   The ^{13}C NMR spectrum of *cis*- 1,2-dimethyl-cyclohexane at 22.63 MHz at (a) -20, (b) -40, and (c) $-115°C$.[1259] FIGURE 192. The ^{13}C NMR carbonyl signal of *E*- dimethyl fumarate $Fe(CO)_4$ complex (**333**) at (a) 25, (b) 5, (c) -25, and (d) $-60°C$.[787] (From Mann, B. E., in *Progress in Nuclear Magnetic Resonance Spectroscopy*, Vol. 11, Emsley, J. W., Feeney, F., and Sutcliffe, L. H., Eds., Pergamon Press, London, 1977. With permission.)

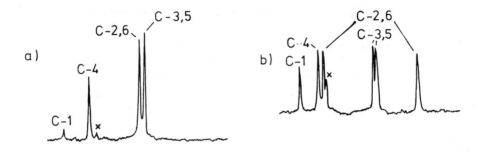

FIGURE 193. The signals of the aromatic carbons in the ^{13}C NMR spectrum of benzaldehyde (334) in dimethyl ether solution at (a) − 60 and (b) − 134°C.[886]

334   **335**

(334a ⇌ b) or *p*-nitroso-*N,N*-dimethylaniline (335) are split at low temperature. Thus, the spectrum of the former contains four lines at room temperature and six at − 134°C in the aromatic region (see Figure 193). For the latter, from the temperature dependence of the spectrum, $\Delta G^{\ddagger}$ = 55.7 ± 2.5 kJ/mol.

For Compound 336 $\Delta G^{\ddagger}$ = 77.8 ± 3.8 kJ/mol was determined from the temperature dependence of nitrogen inversion rate. Corresponding to the chiral *N*-substituent, the ring carbons are not equivalent in case of restricted inversion, but when inversion is fast, their signals coalesce.[590]

336   **337**

As a result of fast tautomery (proton exchange), tropolone (337, R=H) gives four carbon signals at room temperature, arising from the equivalent C-1,2, C-3,7, C-4,6, and C-5 atoms.[1499] The three signals of the acetoxy derivative (R=OAc) split into nine lines at − 70°C. The spectra recorded at different temperatures also show that of the three lines observed at room temperature, one arises from the methyl group, and one arises from the C-1 and C-2 atoms, whereas the third is the coincident signal of C-3,7, C-4,6, and C-5 carbons.[942] From the result of signal shape analysis, $\Delta G^{\ddagger}$ = 45.2 kJ/mol for the acyl migration.

Finally, we call the attention here to the application of CIDNP (compare Section 3.7.4) in ^{13}C NMR.[990]

4.2. NITROGEN RESONANCE[995,1040,1053,1168,1548,1550]

4.2.1. The Basic Problems of Nitrogen Resonance Spectroscopy

Both nitrogen isotopes ^{14}N ($I = 1$) and ^{15}N ($I = 1/2$) are magnetic nuclei. In a polarizing field of 2.35 T their resonance frequencies are 7.225 and 10.135 MHz. The investigation of the resonance spectra of both nitrogen isotopes is hindered by serious experimental difficulties, since the nuclei are insensitive (see Table 1). Therefore, the literature of nitrogen resonance was quite poor[995,1168] and has grown at a higher rate only in the last decade, with the spread of sensitive FT spectrometers.

Both nitrogen isotopes have small magnetic moments; their gyromagnetic ratios with respect to hydrogen (γ_N/γ_H) are 0.072 and -0.101, and therefore the sensitivity in comparison with protons is inherently low. Although the natural abundance of ^{14}N isotope is almost 100%, the line broadening owing to its quadrupole moment ($I > 1/2$), which may reach 800 to 900 Hz, causes a further decrease in ζ.

Nitrogen isotope ^{15}N ($I = 1/2$) always produces sharp signals, but its natural abundance (0.37%) is very low, causing more than two orders lower in ζ. FT instruments, owing to their higher sensitivity, removed the main obstacles in measuring ^{15}N NMR, similar to the measurement of ^{13}C NMR. (Formerly, only samples enriched in ^{15}N could be investigated.) Owing to the fast relaxation processes nitrogen nuclei are incapable of saturation, however, the advantages arising from this do not compensate the much more serious disadvantages caused by the lower magnetic moment and quadrupole line broadening. The difficulties arising from the line broadening caused by short relaxation times in ^{14}N measurements are not eliminated by the FT technique, and problems may be caused for both isotopes by the exchange phenomena and hydrogen bonds, common with NH groups. Therefore, one cannot expect such a rapid development like with ^1H and ^{13}C NMR, although an acceleration could recently be observed in this field, too, particularly in the application of ^{15}N NMR.

It can be seen from the above that ^{14}N and ^{15}N NMR are complementary methods. The former, owing to the high natural abundance, can be applied for the determination of chemical shifts and to study quadrupole relaxation processes using less sensitive CW instruments or dilute solutions. The latter is useful primarily in determining various coupling constants of nitrogen nuclei. The complementary nature of the investigation of the two isotopes also arises from the soon discovered fact (e.g., compare Reference 1168), which was later much debated but now unanimously proved, that there is a linearity for the isotopes both in chemical shifts and in coupling constants, i.e., there is no measurable isotope effect. The ratio ν_{15_N}/ν_{14_N} was studied for molecules with wide range of chemical shifts — directly and indirectly as well. In the direct method the nitrogen resonance frequencies of the isotopes were measured[102] for cases when the signal of ^{14}N resonance is sharp enough and the synthesis of the isotopomer containing ^{15}N nucleus is easy to carry out (preparation of enriched samples). In the indirect method the ^1H NMR spectra of the $NH_4^{\oplus}$ ion and acetonitrile were measured by DR, and the nitrogen isotopes were simultaneously saturated.[1168] The ratio was found to be the same to five decimals. The constant ratio, -1.40275 ± 0.00001, proved to be identical with the $-\gamma_{15_N}/\gamma_{14_N}$ ratio of 1.4027 determined for the bare nucleus by atom-beam methods (e.g., compare Reference 99a). It follows that the shielding constants of nitrogen isotopes for the isotopomers of the same molecule are the same, since

$$\nu^{15}N = \gamma_{15_N} (1 - \sigma_{15_N})B_o \tag{340}$$

and

$$\nu_{14_N} = \gamma_{14_N} (1 - \sigma_{14_N}) B_o \tag{341}$$

and thus if, as shown, the ratio ν_{15_N}/ν_{14_N} is constant, $\sigma_{15_N} = \sigma_{14_N}$ must hold. Likewise, the ratio of the analogous coupling constants of isotopomers is constant, too:[1548]

$$J\,(^{14}N,X) = -0.7129\,J\,(^{15}N,X) \tag{342}$$

which is particularly important since the coupling constants easily measured for the isotopomer containing ^{15}N nucleus can simply be recalculated for the analogous ^{14}N isotopomer, in which the direct measurement of the coupling constants is hindered by quadrupole broadening.

4.2.2. Chemical Shifts in Nitrogen Resonance

Nitrogen chemical shifts embrace a range of approximately 900 ppm. Of the dia- and paramagnetic contributions of shielding factor σ, the latter is dominant. In nitrogen resonance, neighbor effect can be neglected,* since its magnitude, like in proton resonance, is at most 1 to 2 ppm, which is insignificant in nitrogen resonance regarding its two orders wider chemical shift range. For similar reasons, the contribution of solvent shifts (< 10 ppm) are seldom decisive relative to the dia- and paramagnetic terms.

From theoretically determined shielding constants and chemical shifts, the accurate reproduction of experimental data cannot be expected[1548] since these calculations pertain to isolated molecules while the experimental results pertain to solutions, in which solvent effects may cause some parts per million shift. Experimental error may be high, too, particularly in the case of broad ^{14}N resonance signals. Under the experimental conditions, owing to the finite occupation of rotational-vibrational energy levels, the equilibrium molecular geometry is not realized, whereas theoretical calculations are usually based on equilibrium geometry. Despite these inaccuracies, the results of *ab initio* and semiempirical quantum chemical calculations are in qualitative, or in favorable cases quantitative, agreement with the experimental data, and thus they can be utilized in the solution of theoretical, assignment, or structure elucidation problems. Details of this question can be found in the literature.[1053,1548]

4.2.3. Calibration

The accurate determination of the chemical shifts of nitrogen isotopes is difficult. Due to the usually strong solvent effects, the application of external standard is recommended. In this case, however, the eventual errors arising from volume susceptibility** must be taken into account. To improve the small ζ, large sample volume must be applied (sample holders 15 to 25 mm in diameter should be used), in which even in the case of most up to date instruments considerable temperature gradient (2 to 4°C) is developed. Volume susceptibility may cause differences of some tenth of parts per million and they can be restricted to below 0.2 ppm with the application of spherical sample holders. It has also been shown that the temperature gradient causes only insignificant errors: $\Delta\delta \leq 0.1$ ppm/5°C.[20] The variation of chemical shifts with solvent, concentration, and pH, moreover the NOE connected with the application of BBDR for proton decoupling in the FT technique, may, however, cause considerable differences in the position and intensity of signals.[1548] Another significant source of error is the impurity of samples.

In nitrogen resonance there is no inert compound that could be applied as internal reference like TMS in ^1H or ^{13}C NMR. A good external reference is, however, pure liquid nitromethane, which is not too volatile, not hygroscopic unlike other compounds used previously as reference (MeCN, MeNH$_2$, NH$_3$, etc.), its relative "nitrogen content" is high (18.42 M), its ^{14}N signal is relatively sharp, and its ^{15}N signal intensity is insensitive to BBDR (NOE). Therefore, it is increasingly widespread as a reference, replacing other compounds. The chemical shifts of MeNO$_2$ in the most frequently used solvents[1545] — choosing the shift of the pure liquid arbitrarily as zero — are given in Table 90. It can be seen that the solvent

---

* See p. 150.
** Compare Volume I, p. 43 and accompanying footnotes.

Table 90
^{14}N NMR CHEMICAL SHIFT
($\delta N \pm 0.13$ PPM) OF
NITROMETHANE (MeNO$_2$) IN 0.3
M SOLUTIONS, MEASURED AT
30°C, 60 MHz, IN VARIOUS
SOLVENTS[1545]

Solvent	δN	Solvent	δN
DMSO	−2.01	Dioxane	1.82
H$_2$O	−1.98	MeOH	2.01
D$_2$O	−1.94	CH$_2$Cl$_2$	3.21
DMF	−0.69	CHCl$_3$	3.79
−(Pure liquid)	0.00	Et$_2$O	3.91
MeCN	0.20	C$_6$H$_6$	4.38
Me$_2$CO	0.77	CCl$_4$	7.10

effects cause a variation of approximately 9 ppm in the chemical shift of MeNO$_2$, i.e., it must be applied as external reference. A further advantage is that the chemical shifts given in the early literature with respect to nitromethane in nonaqueous solutions and nitrate ion in aqueous solutions as internal reference are comparable without calculation with more recent results. Regarding that for the above reasons the accuracy of the measurements before 1972 allows for a variation of some parts per million, the resulting uncertainty does not increase further owing to the different references. This is also shown by the data of Table 91 (which contains the chemical shifts of some simple nitrogen compounds). The shifts of CH$_3$NO$_2$ in CHCl$_3$ and of nitrate ion in H$_2$O differ only by 3.5 ppm with respect to MeNO$_2$ as external reference.

In early measurements, several other nitrogen derivatives, such as tetranitromethane C(NO$_2$)$_4$, nitric acid, ammonium and tetramethyl-ammonium ion (N$^{\oplus}$Me$_4$), ammonia, methylamine (MeNH$_2$), acetonitrile (MeCN), nitrobenzene (PhNO$_2$), pyridine, DMF, etc., were applied as reference. DMF is particularly useful in the case of nitro derivatives, when signals are expected close to the CH$_3$NO$_2$ line.[1535]

The signal of CH$_3$NO$_2$ is at about the middle of the roughly 900 ppm range of nitrogen chemical shifts, i.e., large negative and large positive δ-values are equally possible. In contrast with the shift scales of ^{1}H and ^{13}C NMR, the chemical shifts decrease downfield, i.e., in the direction of higher frequencies; positive δ-values correspond to higher shielding constants, to more strongly shielded nuclei. This follows the most general convention recommended in the literature.[1040,1548]

4.2.4. Relationships Between Chemical Shifts and Molecular Structure

Nitrogen shifts may be characterized by the following general rules.

1. There is a direct proportionality between ^{14}N and ^{15}N nuclei: from the δ-value determined experimentally for one of the isotopes, it is easy to calculate the chemical shift of the other isotope.
2. The paramagnetic component of shielding factor is dominant, the diamagnetic contribution is of secondary importance, about 10%,[757] and the effect of neighboring electrons is negligible. The paramagnetic contribution is determined by the molecular wavefunctions and the excitation energies,[1550] thus the shifts for a given structure can be predicted after careful considerations only. This can also be seen from the data of Table 91, e.g., the nitrogen is more shielded in ammonia than in the ammonium cation, and the situation is similar for trimethylamine and its tetramethyl-substituted quaternary

Table 91
^{15}N NMR CHEMICAL SHIFTS (PPM) OF SOME SIMPLE MOLECULES

Molecule	Chemical shift	Solvent,[a] counterion	Ref.	Molecule	Chemical shift	Solvent,[a] counterion	Ref.
PhNO	−530	Me$_2$CO (sat.)	1541	N$_2$	70	Liquid (77°K)	186
p-MeOC$_6$H$_4$NO	−428	Et$_2$O (3 M)	1541	CH$_2$NN	95	Et$_2$O (3 M)	864
p-Me$_2$NC$_6$H$_4$NO	−405	Me$_2$CO	646	EtSCN	100	Liquid	646
CF$_3$SNO	−368		589	CN$^{\ominus}$	106	H$_2$O (0.3 M), K$^{\oplus}$	1545
ON–NO$_2$	−302	Liquid (223°K)	37	PhCN	120	Me$_2$CO	646
NO$_2^{\ominus}$	−229	H$_2$O, Na$^{\oplus}$	1545	[NNN]$^{\ominus}$	132	H$_2$O (0.3 M), Na$^{\oplus}$	1545
Me ONO	−182		589	MeCN	136	Liquid	1545
Me$_2$NNO	−157	Me$_2$CO (sat.)	1366	Me$_2$NNO	149	Me$_2$CO (sat.)	1366
PhN=NPh	−130	Et$_2$O	646	NCS$^{\ominus}$	174	H$_2$O (0.3 M), K$^{\oplus}$	1545
Me$_2$CHNO$_2$	−96	MeNO$_2$ (0.05 M)	1537	C$_6$H$_5$N$^{\oplus}$	179	10 M HCl (0.5 M), Cl$^{\ominus}$	1545
ON–NO$_2$	−67	Liquid	944	PhOCN	211	Liquid	1159
MeCH$_2$NO$_2$	−49	MeNO$_2$ (0.05 M)	1537	MeNC	240	Liquid	646
Me$_3$CNO$_2$	−30	MeNO$_2$ (0.6 M)	1549	Me$_2$NCHO	277	Liquid	1545
MeCHClNO$_2$	−23	MeNO$_2$ (0.05 M)	1537	[NNN]$^{\ominus}$	282	H$_2$O (0.3 M), Na$^{\oplus}$	1545
CH$_2$NN	−9	Et$_2$O (3 M)	864	NCO$^{\ominus}$	303	H$_2$O (sat.) K$^{\oplus}$	1545
PhCH$_2$NO$_2$	−7.5	MeNO$_2$ (0.1 M)	1549	PhNH$_2$	325	Liquid	864
MeNO$_2$	0	Liquid	1545	Me$_4$N$^{\oplus}$	338	H$_2$O (0.3 M), Cl$^{\ominus}$	1545
NO$_3^{\ominus}$	3.5	H$_2$O, Na$^{\oplus}$	1545	NH$_4^{\oplus}$	350	10 M HCl (1 M), Cl$^{\ominus}$	1545
PhNO$_2$	9.5	Liquid	1545	NMe$_3$	369	Liquid	1545
HNO$_2$	45	Liquid	1269	MeNH$_2$	378	Liquid	1545
	47	Liquid	272	NH$_3$	382	Liquid	1545
C(NO$_2$)$_4$	47	Liquid	1545	NH$_3$	400	Gas	20
C$_5$H$_5$N	62	Liquid	1545	[Co(NH$_3$)$_6$]$^{3\oplus}$	425	H$_2$O, Cl$^{\ominus}$	646

[a] Concentration or temperature is in parentheses.

Table 92
CHEMICAL SHIFT RANGES (PPM) OF SOME COMPOUND TYPES IN N NMR SPECTRA[814,995,1040,1379,1535,1541,1548,1549,1550]

Type of compound	Chemical shift	Type of compound	Chemical shift
Amines and salts	400—300	Nitriles (RCN)	140—120
Aminoboranes	380—240	Diazoderivatives (RNN)	140—80
Isocyanates (RNCO)	365—330	Azides (RNNN)	135—125
Hydrazines	340—280	Imine-N-oxides	110—75
Azides (RNNN)	320—280	Thiocyanates (RSCN)	~100
Isothiocyanates (RNCS)	290—260	Nitroesters (RONO$_2$)	70—0
Amides	285—240	Nitramines (NNO$_2$)	60— -20
Heteroaromates	250— -20	Oximes	50—0
Thioamides	240—210	Nitroderivatives	50— -120
Isonitriles (RNC)	220—180	Triazines (RNNNR'R'')	40— -80
Cyanates (ROCN)	~200	Diazoderivatives (RNN)	25— -50
Fulminates (RCNO)	180—160	Azoderivatives	-130— -170
Azides (RNNN)	175—140	Nitrosoderivatives	-150— -450

derivative; however, nitrogen is more shielded in pyridinium cation than in pyridine or more shielded in nitrobenzene than in nitrosobenzene. The nitrogen of ammonium cation is 580 and 350 ppm more shielded than the nitrite and nitrate cations, respectively. The π-electron density plays, therefore, a significant part in the resulting chemical shift, and pure diamagnetic effects appear only in the case of sp^3 nitrogens. Consequently, nitrogens with covalent and ionic bonds do not necessarily separate into characteristically different chemical shift categories.

3. When an sp^3 carbon atom is attached to the nitrogen, shielding and thus chemical shift decreases with the order of this carbon. This is the β-effect.

4. If the lone pair of an sp^2 nitrogen is replaced by a covalent bond (quaternerization, protonation, N-oxides), a significant increase in shielding, i.e., chemical shift, occurs. Although less generally, this is also observable for compounds with sp^3 nitrogens.

5. For groups capable of tautomery, isomery, or similar rearrangements involving changes in bond character of the nitrogen, the nitrogen shifts of the isomers may be widely different, even by several hundred parts per million. Therefore, nitrogen resonance may be in most cases an extremely useful tool in the identification of such structures.

6. Solvent effects are most frequently negligible if the nitrogen is inside of the molecule, and the differences between various nonpolar solutions do not exceed 3 to 4 ppm even in the case of peripheral nitrogens. The extreme shifts are usually measurable in CCl$_4$ and H$_2$O, the average difference is 10 ppm,[1548] but occasionally it may be even twice as large (e.g., in the case of six-membered heteroaromatic derivatives).[1545]

The literature of nitrogen chemical shifts is quite rich (e.g., References 589, 646, 808, 811, 860, 944, 945, 995, 1040, 1053, 1168, 1534, 1548, and 1550), but the data, particularly the earlier ones, must be accepted with due criticism. For information, Table 92 shows the chemical shift ranges of some compounds. It is stressed that these data can only be regarded as approximate, since the source of the various shift ranges is quite varied (representing the data of only some, at most 100 to 150 related compounds), the ranges are too wide even for the most frequent types, and they could be further broadened if some extreme shifts were taken into account. Moreover, several very important compound types are missing from the table, a lack of sufficient data. Since the table contains many early data, too, its value is further decreased by eventual experimental errors.

4.2.5. The Characteristic Chemical Shift Ranges of Various Compound Types in Nitrogen Resonance

4.2.5.1. Amines, Hydrazines, and Related Derivatives

Amines have the highest shielding constants. The chemical shifts fall mostly in the 300 to 380-ppm range. Within this shift range shielding decreases with the order of the α-carbon atom,[864,1474] i.e., the β-effect also holds for these compounds. Nitrogen is less shielded in aniline derivatives,[864] shielding decreases with the electronegativity of the substituents, by approximately 16 ppm in the Me, H, I, Br, Cl, NO_2 series.[67,187]

The chemical shift of aziridine (393 ppm)[14] indicates, similar to ^1H and ^{13}C resonance, an extremely high shielding. Shielding also increases upon the effect of silicon in aminosilanes. In molecules of type $R_3SiCH_2NH_2$, δN = 376 ± 1 ppm.[1261] In the silatranes **338**, when a methyl group is attached to silicon instead of hydrogen atom (R = Me), the nitrogen is deshielded,[1542] reminding one of the β-effect. It is therefore interpreted as a proof for the formation of a *trans*-annular N–Si bond.

338 (R = H , Me)

The nitrogen shift of silyl-aminoboranes of type $R_2B–NR'–SiMe_3$ is usually in the 250 to 310 ± 10 ppm range, and R' influences the shift according to the β-effect.[98] The shift range of aminoboranes (RR'B–NR''R''') is very broad, the nitrogen signals appear between approximately 240 and 380 ppm.[98] The β-effect is also observable, e.g., for derivatives $B(NHR)_3$, when R = Me, δN = 356 ppm, whereas when R = *t*-Bu, δN = 297 ppm.[98] Electron-withdrawing substituents have generally a deshielding effect: for the $RB(NMe_2)_2$ type in the R = Me, Ph, F, Cl, Br, I series, δN = 341, 286, 371, 284, and 293 ppm, respectively, within ± 10 ppm. The data indicate that there is a complex relationship between the chemical shift of the nitrogen and the substituent, and by taking into account the $-I$ effect alone, the shifts of the fluoro and bromo (or iodo) derivatives are out of sequence. The nitrogen shift is obviously the result of several competing effects.

The change of the ^{15}N and ^{13}C chemical shifts of $R^{15}NH_2$ and $R^{13}CH_3$ derivatives with substituent R is completely analogous. The higher sensitivity of ^{15}N shifts with respect to R indicates that the determining factor of the paramagnetic contribution of shielding is the mean distance of *p*-electrons from the nucleus.[1537] In the salts of alkylamines, shielding and consequently chemical shift decreases by 5 to 30 ppm.[69,1536] Although the direction of shift depends on the type of compound, the phenomenon may always be used to determine the site of protonation in the molecule when there are more possibilities. Thus, the structures of the salts of nucleotides and nucleosides can be elucidated on the basis of the NMR spectrum.[927]

The nitrogens of hydrazines absorb in the 280 to 340-ppm range.[201] The chemical shift of unsubstituted nitrogen is decreased by about 27 ppm upon methylation, on analogy to the β-effect. On the substituted nitrogen, the effect is only approximately 9 ppm. On substituting both nitrogens, the latter effect, in α-position, is absent, and the signals of the nitrogens in β-position are shifted downfield by approximately 27 ppm. In the case of mono-, di-, and tri-phenylhydrazine, the signal of the substituted N_α atom is shifted, by approximately 45 ppm toward higher frequencies.[865]

4.2.5.2. Carboxylic Amides, Thioamides, Peptides

Carboxylic amides have a signal in the 240- to 285-ppm range. Deshielding occurs in

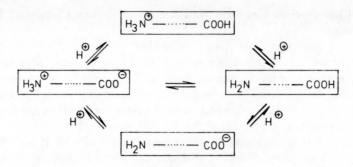

FIGURE 194.   Equilibrium of nonionic, zwitterionic, protonated, and deprotonated forms of amino acids.

thioamides, the signals appear in the 210- to 240-ppm range.[995] The chemical shift of lactams is independent of ring size.[1528] A linear relationship was shown to exist in some amides and thioamides between nitrogen chemical shift and the internal rotation barrier of the C–N bond.[937]

Nitrogen resonance spectroscopy is particularly important in the investigation of amino acids, peptides, and their analogues of fundamental biological importance (e.g., compare Reference 1206). The signal of the amine and ammonium groups is easy to distinguish from the signals of amide groups on the basis of shielding, which is about 100 ppm higher for the former. For some simple dipeptides, $\delta NH_2/NH_3^{\oplus} = 335$ to 365 ppm, whereas $\delta NCO = 250$ to 270 ppm.[1548] Owing to the equilibrium between the nonionic and zwitterionic or the protonated and deprotonated forms, the nitrogen resonance spectra of amino acids and peptides are pH dependent. The signal of the more shielded deprotonated nitrogen ($\delta N = 363$ ppm in glycine)[296] is easy to distinguish from those of the other nitrogens involved in the equilibrium (see Figure 194), at 351 ppm.[296] To a lesser extent, the amide signals of dipeptides are also pH dependent; the amide nitrogen shifts of glycyl glycine and of its deprotonated form are 270 and 264 ppm, respectively.[704]

4.2.5.3. Azides, Compounds with sp Nitrogen

The characteristic chemical shifts of sp nitrogen are given in Table 92. It is noted that the shieldings of $NCO^{\ominus}$ and $NCS^{\ominus}$ anions are averages taken for RNCO, RNCS, ROCN, and RSCN, respectively. When the $-I$ effect of substituent R increases, the shielding of isonitriles (RNC) increases accordingly, whereas that of the analogous nitriles decreases.[1532] The nitrogen of alkyl cyanates and isocyanates is more shielded than of the aryl analogues.[271]

The terminal nitrogen atoms of azide anion are more shielded than the central one ($\delta N_{term.} = 281$ ppm, $\delta N_{centr.} = 132$ ppm), in agreement with the mesomeric structure $N^{\ominus} = N^{\oplus} = N^{\ominus}$. In the spectra of alkyl azides RN_3 three nitrogen signals can be found at approximately 300, 130, and in the 145- to 170-ppm region.[1533,1548] The highest shift is assigned to the nitrogen adjacent to the alkyl group, interpreting the shift changes in the R = Me, Et, *t*Bu, Ph series (320, 305, 286, and 286 pm) as β-effect. The signal with approximately unchanged shift in the series (128 to 134 ppm) belongs obviously to the central nitrogen. Thus, the terminal nitrogen gives the most substituent-sensitive signal appearing at 170.5 for R = Me, 162 for R = *t*Bu, and 144 ppm for R = Ph. The strong deshielding with respect to the azide ion may be attributed to the mesomeric structures $R-\overline{N}=N^{\oplus}=N\!\!>^{\ominus} \leftrightarrow R-\overline{N}^{\ominus}-N^{\oplus} \equiv \overline{N}$, considering that the second mesomeric structure is more favorable in phenylazide than in alkyl azides.

The nitrogen NMR investigations on the azide-tetrazole valence isomerism, carried out with ¹⁵N labeled models, confirmed the results of ¹H and ¹³C NMR studies,* i.e., that in TFA, the azide form, and in DMSO the tetrazole form is preferred.[1400]

* See p. 89.

Table 93

CHEMICAL SHIFTS (PPM) OF SOME FIVE-MEMBERED AROMATIC NITROGEN HETEROCYCLES IN THE NITROGEN RESONANCE IN PURE LIQUIDS[a] OR IN H$_2$O,[b] DMSO,[c] AND Me$_2$CO[d] SOLUTION[76,1233,1252,1474]

4.2.5.4. Azoles

The chemical shifts of the two types of nitrogens occurring in five-membered, aromatic, nitrogen-containing heterocyclic derivatives (azoles), i.e., of nitrogens formally sp^3 and sp^2 in character, are significantly different, the former nitrogens being much more shielded (see Table 93). It can be seen from the table, by comparing imidazole and pyrazole to their *N*-methyl derivatives, that the fast tautomeric equilibrium causes an averaging for the nitrogen shifts. The mean shifts of the nitrogens of *N*-methyl derivatives are (178 + 73)/2 = 125.5 and (218 + 123)/2 = 170.5, which, also taking into account the possible experimental errors, are practically the same as the shifts measured for the unsubstituted derivatives (134 and 176 ppm).

On comparing isoxazole with *N*-methyl-pyrazole and thiazole with *N*-methyl-imidazole, one may observe the deshielding effect of the stronger $-I$ effect of the hetero atom. In this case, the bond character of nitrogen and therefore the paramagnetic contribution of the shielding constant is not very different in the molecules compared, and thus the shifts are determined by the electron density around the nitrogen atom (i.e., by the diamagnetic contribution). Finally, nearly the same shifts of thiazole and benzothiazole indicate that the shielding of nitrogen remains practically unchanged in fused rings.

In rings containing more nitrogen atoms, the mutual shielding effects of nitrogens are additive, which makes the signals easy to assign:

$$\delta N_{sp^3} = \delta N_o + \Sigma \rho_i^k \tag{343}$$

where δN_o is the shift of pyrrole, i refers to the number of nitrogens (second or third nitrogen in the ring), and k refers to the 2- or 5- and 3- or 4-position, respectively, of the second nitrogen. The actual values of ρ are[1543] $\rho_2^{2,5} = -50 \pm 1.5$, $\rho_2^{3,4} = -30 \pm 4$, $\rho_3^{2,5} = -12 \pm 2$, and $\rho_3^{3,4} = +7 \pm 2.5$ ppm. Similar increments may be determined for nitrogens of type sp^2:

$$\delta N_{sp^2} = \delta N_o + \Sigma \rho^k \qquad\qquad (344)$$

where δN_o is the chemical shift of the sp^2 nitrogen of N-methylpyrazole or N-methyl-imidazole, depending on whether the sp^2 nitrogen is in position 2 or 5 and 3 or 4, respectively, k refers to the mutual position of nitrogens, and $\rho^{1,2} = -48 \pm 6$, $\rho^{2,3} = -79 \pm 7$, $\rho^{2,4} = +23 \pm 8$, $\rho^{2,5} = -23 \pm 9$, and $\rho^{3,4} = -51 \pm 9$ ppm.

Accordingly, the shifts calculated for the nitrogens of 1-methyl-1,2,4-triazole are

$$\delta\text{N-1} = 230 - 50 - 30 = 150 \pm 5.5 \text{ ppm} \quad \text{(measured: 160 ppm)}$$

$$\delta\text{N-2} = 73 - 48 + 23 = 48 \pm 14 \text{ ppm} \quad \text{(measured: 46 ppm)}$$

$$\delta\text{N-4} = 123 + 23 \quad = 146 \pm 8 \text{ ppm} \quad \text{(measured: 140 ppm)}$$

As can be seen, the agreement is quite satisfactory, and even if — like in general with similar empirical methods — the calculated shifts do not reproduce the experimental ones, they make the assignment of the signals after all easier. This method was used in the assignment of the four signals of trimethylene-tetrazole (see Table 93). The shifts expected according to the additivity rule are $\delta\text{N-1} = 95$, $\delta\text{N-2} = -7$, $\delta\text{N-3} = 17$, and $\delta\text{N-4} = 129$ ppm.

The shifts calculated from the additivity rule support the pentazole-type ring structure (**339**) of the compounds obtained in the reaction of p-disubstituted benzene diazonium and lithium azide.[1538] Although only one signal could be observed (70 ppm), generally the substituted sp^3 nitrogen of azoles gives the sharpest signal of the ^{14}N resonance spectrum, and thus the only observed signal arises most probably from the nitrogen in position 1. Indeed, its shift is practically identical to the calculated value ($\delta\text{N-1} = 74$, $\delta\text{N-2,5} = -6$, and $\delta\text{N-3,4} = 16$ ppm).

339

Upon the effect of protonation, as usual, the shielding of azol nitrogens strongly increases, e.g., from 177 to 210 ppm in imidazole[1233] and from 56 to 178 ppm in thiazole. Likewise, in the quaternary derivatives, too, the chemical shift increases,[1050] in the case of 1,3-dimethyl- and diethyl-imidazoles (**340a,b**) to 220 and 191 ppm, respectively (β-effect), although the electron density in the ring decreases. This tendency, contrary to the experience in ^{1}H NMR, is a proof for the dominancy of the paramagnetic shielding component. In the azolinium anions resulting from azoles an opposite change, i.e., decrease in chemical shift, occurs: for the lithium salt of pyrrole and pyrazole as well as for the potassium salt of imidazole, the shifts are 205, 115, and 147 ppm, i.e., 26, 19, and 30 ppm lower than for the original rings.[873,1050]

Table 94

CHEMICAL SHIFTS (PPM) OF SOME CHARACTERISTIC SIX-MEMBERED AROMATIC NITROGEN HETEROCYCLES IN NITROGEN RESONANCE IN PURE LIQUIDS[a] OR IN H_2O,[b] Me_2CO,[c] AND Et_2O[d] SOLUTION[1142,1233,1539,1544,1545]

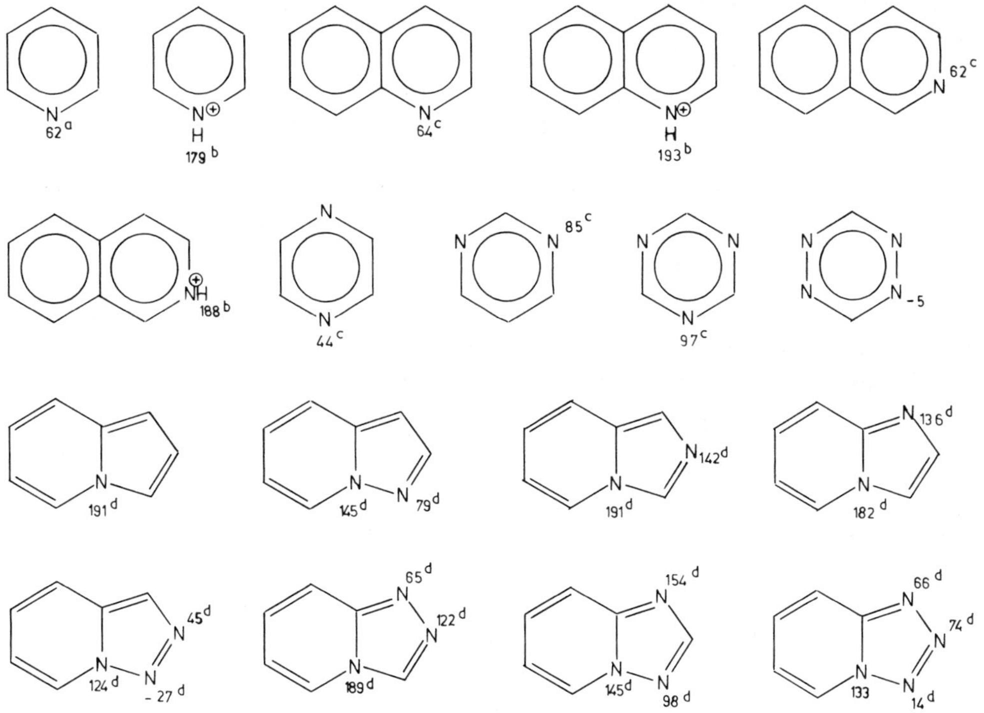

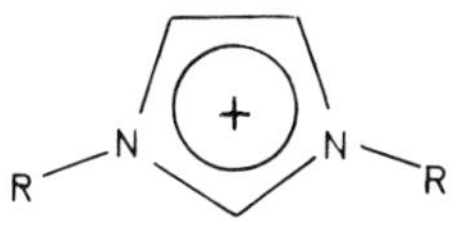

340

a: R = Me

b: R = Et

In agreement with the basic principles given in the introduction, shielding increases when the lone electron pair is replaced with a covalent bond, e.g., on the formation of borane adducts. The chemical shift (176 ppm) of imidazole (compare Table 93) is raised by 33 ppm (to 209 ppm) for BH_3 adduct in CH_2Cl_2 solution. The β-effect is also observable for these compounds[1049] and the shifts of the sp³ and the borane binding sp² nitrogens are 211 and 168 ppm for BMe_3 and 194 and 148 ppm for BEt_3 adducts, respectively. With six-membered nitrogen heterocycles, the increase in shift is even greater, e.g., in the case of pyridine it is +80 ppm.

It is important to note the solvent effects of azoles, which are often very strong, may even exceed 20 ppm. Thus, the shift of pyrazole in acetone is 129 ppm, whereas in DMSO it is 176 ppm.[1233]

4.2.5.5. Azines

The chemical shifts of six-membered nitrogen-containing aromatic rings are reviewed in Table 94. The shifts of the rings containing more than one nitrogen may be estimated from the additivity rule,[1544]

$$\delta N = \delta N_o + \sum \rho^{i,j} \tag{345}$$

where $\delta N_o = \delta N_{pyridine} = 62$ ppm, $\rho^{1,2} = -73 \pm 4$, $\rho^{1,3} = 22 \pm 3$, and $\rho^{1,4} = -20 \pm 6$ ppm. Thus, for pyrimidine $\delta N = 62 + 22 = 84$ ppm (the experimental shift is 85 ppm),[1474] and for tetrazine $\delta N = 62 - 73 + 22 - 20 = -9$ ppm (experimental, -5 ppm).[1539]

It is clear from Table 94 that fusing of rings hardly influences nitrogen shifts, but the mutual position of nitrogens is very important; their effect is additive. In this case, too, protonation increases shielding extremely strongly. The shift also increases in *N*-oxides, and the increase is proportional to the original shift; the chemical shifts of the original ring and its *N*-oxide can therefore be evaluated from one another.[1364,1367]

One of the most efficient applications of nitrogen resonance in structure elucidation is the investigation of tautomery of nitrogen-containing groups.[638,1380] This field will be illustrated by an example of azine derivatives.

With 2- and 8-hydroxyquinolines, tautomeric equilibria may occur (**341 a $\rightleftharpoons$ b** and **342 a $\rightleftharpoons$ b**). However, as known, with the former the oxo-form **b** (possibility of amide-mesomery!), whereas with the latter aromatic form **a** is more stable. Accordingly, the chemical shifts[626] are 238 (**341b**) and 95 ppm (**342a**), i.e., in the regions characteristic of amide and heteroaromatic compounds, respectively (see Table 92).

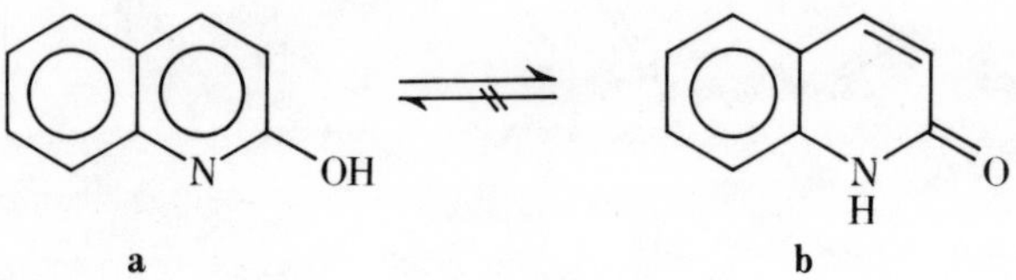

341

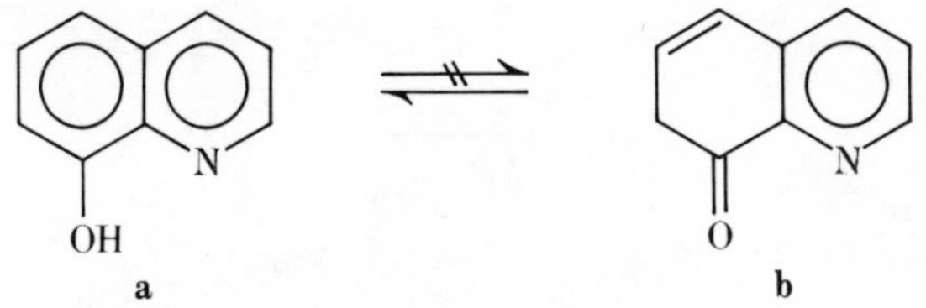

342

In the cases of tautomeric equilibrium, as a result of fast interconversion, a mean shift can be observed in the spectrum.* By comparing the mean shifts with the shift of model compounds of fixed structure, the relative amounts of tautomers taking part in the equilibrium may also be calculated. Thus, the shifts of model Compounds **343** to **345 a, b** (R = Me) are[1365] 215, 181, and 244 ppm for the **a** and 114, 65, and 91 ppm for the **b** isomers, respectively. On the basis of these shifts, from the mean values measured for the nonsubstituted analogues (R = H) the relative amounts of forms **a** and **b** may be determined which are 92:8, 3:97, and 83:17, respectively, in acetone.

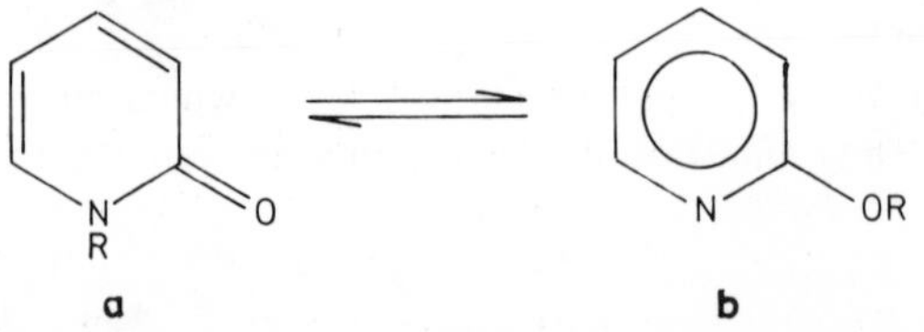

343

---

*   Compare azoles, too; p. 245.

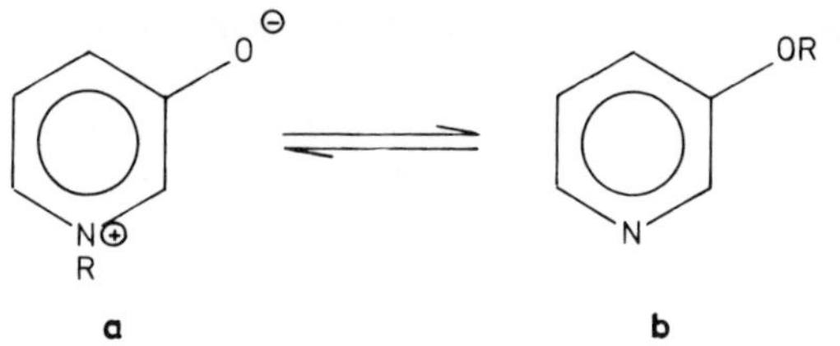

344

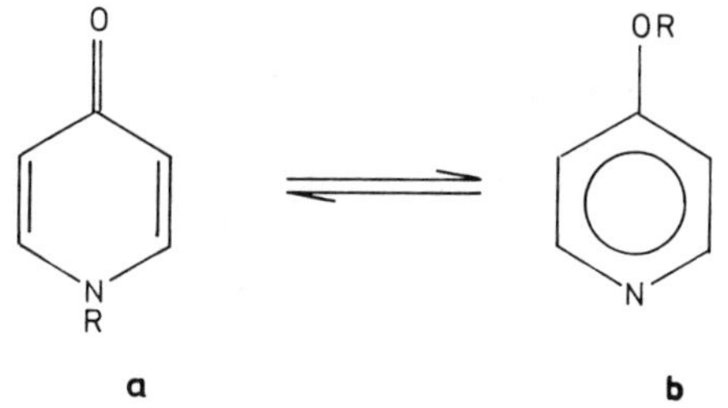

345

The ^{15}N NMR investigation of the thione-thioiminohydrine tautomery of thiourea deriv-atives (Me$_2$N–CS–NMeR $\rightleftharpoons$ Me$_2$N–C(SR) = NMe, R = H) is similar. The δN shifts were determined for the N-methyl derivative (301 ppm) of thione, its SMe-tautomer (300 and 121 ppm, respectively), and for the unsubstituted parent compound (247 ppm), from which the **a:b** (R = H) ratio was found to be 7:3.[1546] From the ratio of tautomers **a** and **b** of urea, it was found that the iminohydrine tautomer does not occur in detectable quantities. This is the reason for the lower chemical shifts of thioureas and thioamides as compared to the analogous ureas and amides. Significant solvent effects can be expected for azines as well. The highest shifts (10 to 20 ppm) may be observed in aqueous solutions, toward the lower frequencies, i.e., higher δ-values.[1545]

4.2.5.6. Oximes and Nitro Derivatives

The oximes and their ethers (R′R″C = N–OR) may be characterized by much lower shielding constants[1541] than the isomeric imine-N-oxides [R′R″C=N(R) → O]. This is a further example for the increase in chemical shift when a lone electron pair is replaced by a covalent bond (compare Table 92). In these derivatives the nitrogen shifts are very useful in the distinction of various geometric isomers,[1027] i.e., the various positions of substituents with respect to the C=N double bond. However, problems of this kind may be solved more convincingly by studying couplings.*

Nitro compounds belong to the derivatives first and most intensively studied by nitrogen resonance. An empirical additivity rule was derived[1537] for nitroalkanes of type RR′R″CNO$_2$, and it was found that δN is proportional to the electronegativities of the substituents.[1134,1537] This is illustrated by the chemical shifts of nitroethane, 1,1-dinitroethane and 1,1,1-trini-troethane (-11, 11, and 26 ppm). The shift of tetranitromethane is 47 ppm. For nitroalkanes R–NO$_2$ the shifts change in a reverse direction in the R = Me, Et, iPr, and tBu series (deshielding), i.e., the β-effect occurs here as well. Conjugation increases the shielding: the chemical shift of nitrobenzene is 10 ppm, that of 1-nitropropene (Me–CH=CH–NO$_2$) is 2 ppm, whereas for 3-nitropropene (CH$_2$=CH–CH$_2$–NO$_2$) -6.5 ppm was measured.[1540] Thus, the chemical shifts of aromatic nitro derivatives (-6 to 22 ppm) are higher than those of aliphatic ones.[1547] Substitution on the benzene ring increases shielding proportionally to the

* Compare p. 255 and 256.

$-I$ effect of the substituent, and the extent of this change is independent of the position (*ortho, meta* or *para*) of the substituent.[1547] The shift increases in nitroamines[138] and nitroesters (covalent) nitrates, too[1533] (compare Table 92). It is worth noting that there is a proportionality between the $v_{as}NO_2$ IR frequency and the frequency of nitrogen resonance:[438] $v_{as}NO_2 = 1573 - 1.147 \, \delta N$ cm^{-1}.

4.2.5.7. Azo- and Diazo- Derivatives

Azo-derivatives (R–N=N–R') absorb at high frequencies (see Table 92), whereas in diazo-derivatives (RR'C=N=N$^\ominus$) shielding is much higher. The corresponding regions are -170 to -130 for the formers and 80 to 140 and -55 to 25 ppm, for the terminal and inner nitrogens of the latters, respectively. The assignment of these signals to the individual nitrogens is debated.[1548] Both the weaker substituent dependence and the comparison with azides make it more probable that the more shielded signals belong to the terminal nitrogens.[946] In contrast, the investigation of the labeled derivatives of Compound **346** led to the reverse assignment.[1548] Since, however, **346** is a mesomeric system, in which mesomeric form **b** may play a significant role, it might occur that under the conditions of investigation (60% sulfuric acid solution, where *O*-protonation may fix the molecule in the tautomeric form corresponding to mesomer **b**) an irregular structure is formed (dissimilar to the normal diazo compounds), and thus the shifts obtained for Compound **346** are not characteristic of other compounds. This is also indicated by the fact that the shift of the terminal nitrogen very similar to the nitrogen of nitriles in mesomeric structure **b** falls into the region characteristic of nitriles in the case of the reverse assignment.

346

Isomeric diazomethane (N$^\ominus$=N$^\oplus$=CH$_2$) and diazirine (N=N) may be distinguished already on the basis of the number of signals, and it is interesting that the shift of the only signal of the latter (52 ppm) is quite close to the average shift of the two signals ($\delta N_{centr.} = -9$, $\delta N_{term.} = 95$ ppm) of the former.[946]

4.2.5.8. Nitroso Derivatives

The highest frequencies of nitrogen NMR spectra are produced by nitroso derivatives. The chemical shifts of nitrosamines (R$_2$NNO) are about -160 ppm[1366] (the shift of NR$_2$ is 149 ppm when R=Me and 124 ppm when R=Et, β-effect!), but in nitrosobenzenes the shift may drop to -500 ppm. In the latter derivatives the chemical shift is extremely sensitive to substitution:[1541] in nitrosobenzene itself it is -536 ppm, whereas the shift of *p* -methoxy derivative is more than 100 ppm different (-428 ppm). There may be a tautomerism between the nitroso and oxime groups (**347 a** ⇌ **b**, R=H), and from the average shift measured for such compounds, similar to previous examples (e.g., Compounds **343** to **345**), the relative amounts of tautomers in the equilibrium can be determined, e.g., **347a** (R=H) is in a 7:1 equilibrium with **347b** in acetone.[1541]

The high frequency signals of *N*-nitroso compounds proved that sidnone imines (**348**) are present in protonated or *N*-acyl form (**350**) only, whereas the base is present in an open

δ N (R = Me) : +5 ppm   δ N (R = Me) : -428 ppm

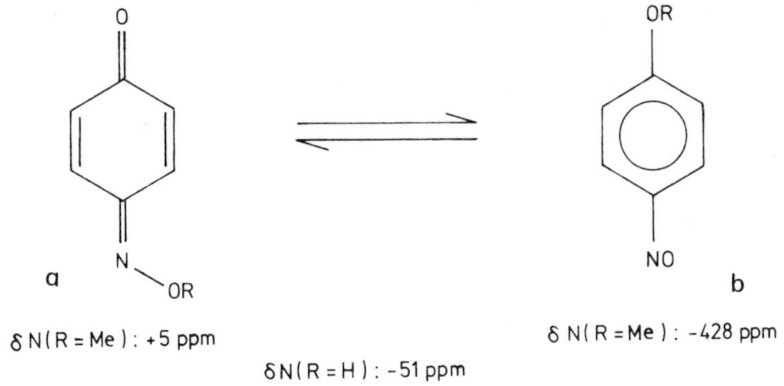

δ N (R = H) : -51 ppm

347

form (**349**), containing *N*-nitroso and nitrile groups. This was shown unanimously by the chemical shifts of all the three nitrogens of the molecule, primarily when they are compared to the shifts of Compound **350**.[1366]

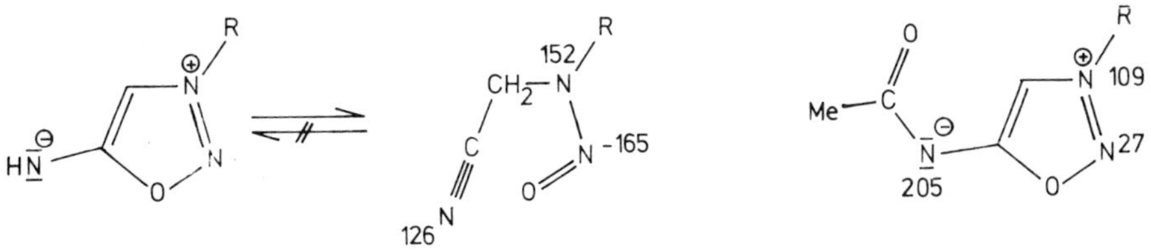

348   **349**   **350**

4.2.6. Spin-Spin Interactions and Coupling Constants of Nitrogen Atoms

Since the coupling constants are directly proportional to the gyromagnetic ratios of the interacting nuclei, the couplings of nitrogen isotopes are different in both sign and magnitude (since $\gamma_{15_N} < 0$).* However, as already mentioned, there is a strict proportionality between them, and from the known ^{15}N–X coupling constant the coupling of ^{14}N–X analogue can be determined, and vice versa.

The splitting due to the spin-spin coupling with the ^{14}N isotope is seldom observable, since the very short relaxation times arising from the quadrupole moment cause signal broadening, and line width is most frequently several orders greater than the coupling constants. Therefore, the main source of nitrogen coupling constants is the ^{15}N resonance spectrum. In these spectra the splittings due to spin-spin coupling are well observable, since the lines are sharp ($\Delta\nu_{1/2} \approx 1$ Hz). Although with CW instruments the multiplets corresponding to the couplings of ^{15}N nuclei could be indentified in the spectra of isotopically enriched samples only, this problem has been eliminated by the spread of the FT technique. On the other hand, by means of CW instrument it is also possible to determine varous ^{15}N–X couplings with samples of natural ^{15}N abundance, e.g., by applying DR or the computer technique of subtraction of satellite spectra.**

In order to overcome the problems arising from the negative sign of the coupling constants of ^{15}N isotope (and of other nuclei with negative gyromagnetic constant, e.g., ^{29}Si), the concept of reduced coupling constant has been introduced:[1162]

$$K_{ij} = J_{if}h\,\gamma_i\,\gamma_j \tag{346}$$

and

---

*   Compare p. 238.
** Compare p. 259.

$$K \, (^{15}\text{N,H}) = -0.822 \, J \, (^{15}\text{N,H}) \text{ Hz} \qquad (347)$$

(for $^{15}\text{N}-^{13}\text{C}$ and $^{15}\text{N}-^{15}\text{N}$ couplings the proportionality constants are -3.26 and 8.10, respectively).[1055] These reduced constants enable one to compare the couplings between widely different nuclei.

4.2.6.1. $^1J(N,H)$ Couplings

The $^1J(^{15}\text{N,H})$ coupling, since $\gamma_{^{15}\text{N}} < 0$, is negative, which has been proved in several cases (e.g., References 206, 1028, and 1183). In the following discussions, the negative sign will not be used owing to the lack of practical significance. However, in theoretical considerations, primarily when comparing this coupling with other interactions, it should always be taken into account.

According to the theory,[755,1071,1165,1166] to a satisfactory approximation, it may be assumed for N–H coupling, similar to ^{13}C–H interactions, that J (N,H) is determined by the Fermi contact term, which is thus proportional to the s-electron density about the nuclei

$$J(\text{N,H}) \sim S_{\text{H}} S_{\text{N}} / \Delta E \qquad (348)$$

where ΔE is the average excitation energy.

Since S_{H} and ΔE are not significantly different in the various compounds, a linear correlation may be expected to hold between S_{N}, i.e., the hybridization state of nitrogen percentage s-character, and $J(\text{N,H})$.[135] For the determination of s-character ($S_{\%}$) from $J(\text{N,H})$, and vice versa, several empirical formulas have been proposed, with the following common form

$$S_{\%} = K^1 J \, (^{15}\text{N,H}) - C \qquad (349)$$

where K and C are constants of values $K = 0.43$, $C = 6$,[135] or $K = 0.34$, $C = 0$;[177] or $K = 0.215$, $C = 3$.[135]

Typical sp^3, sp^2, and sp N–H bonds occur in $\text{NH}_4^\oplus$, $\text{C}_5\text{H}_5\text{NH}^\oplus$, and $\text{HCNH}^\oplus$ cations, for which $^1J(^{15}\text{N,H})$ is 74, 99, and 134 Hz. The former two values were calculated from the couplings of 53^{995} and 70 Hz[1313] measured for the ^{14}N isotope; the latter was measured directly for ^{15}N.[1064] These values give s-characters of 26, 37, and 52% (calculated with $K = 0.46$ and $C = 6$), very close to the theoretical 25, 33, and 50%. There is good agreement for other models too; the first investigations in this field were performed on the protonated forms of aniline, isoquinoline, and benzonitrile resulting $^1J(\text{N,H}) = 77$, 96, and 136 Hz.[135] Protonated nitriles, in agreement with the 50% s-character, have presumably the highest $^1J(\text{N,H})$ and $^nJ(\text{N,H})$ coupling constant. The coupling constant of protonated acetonitrile and benzonitrile, 136 Hz,[135,1064] is the highest measured so far between ^{15}N and H atoms. Direct ^{15}N–H couplings are most frequently between 70 and 95 Hz. The values of $^1J(^{15}\text{N,H}) = 51$ Hz, measured for diphenylketimine ($\text{Ph}_2\text{C=NH}$), represents the opposite limit.[135] If this is not due to experimental error or to the fast exchange between two structures (like with Compounds **343-345**), it is a good example for the exceptional case when, in addition to Fermi contact term, the interaction of nuclear spin with adjacent electronic orbitals also has a determining role in coupling.

When the coupling partner of nitrogen is not a proton, this is always so, e.g., as has been shown for the case of $^1J(^{15}\text{N},^{13}\text{C})$ coupling.[279] If the Fermi contact term has a dominant contribution, the magnitude of $^{15}\text{N}-^{13}\text{C}$ coupling can be determined from the s-characters of the orbitals forming the C–N bond:[135]

$$^1J(^{15}\text{N}, \,^{13}\text{C}) = 0.0125 S_{\text{N}} S_{\text{C}} \qquad (350)$$

However, the extremely high coupling constants (77.5 Hz)[279] measured for the *N*-oxide of trimethyl benzonitrile (**351**) and for other representatives of fulminates are more than twice the possible theoretical maximum! Nevertheless, on the basis of the simple relationship between s-character and couplings given by Equation 349 and 350 one can solve structural problems of wide variety.

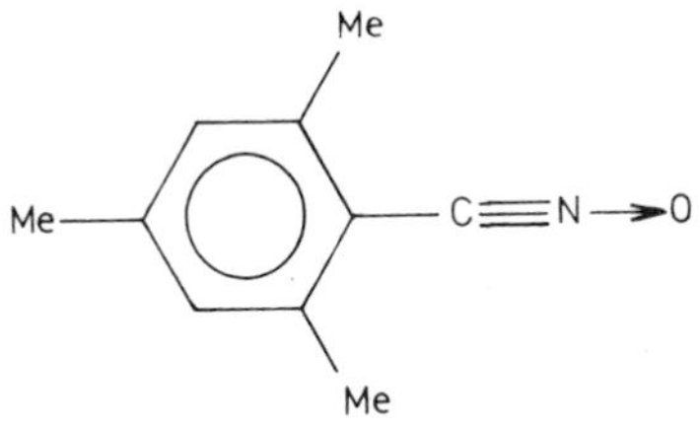

351

The coupling constants of 91 and 97 Hz[1209] obtained for uracil (**352**) prove the planar diketo structure of the molecule, excluding the other possible tautomers **b** to **d**.

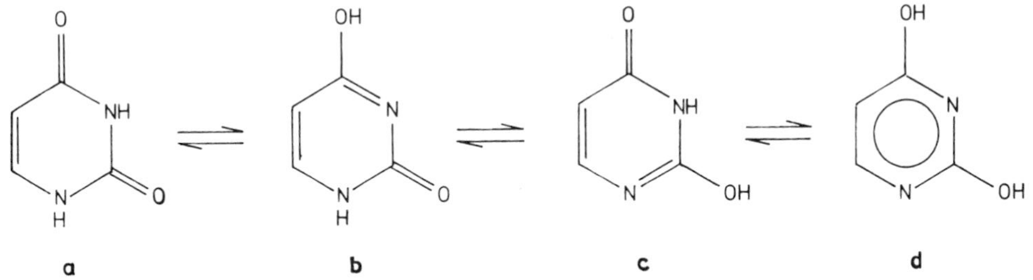

352

The $^1J(^{15}N,H)$ couplings of 88 to 94 Hz of carboxylic amides and thioamides[1168] prove the sp^2 character of amide nitrogen (mesomery). The coupling constant of hydrogen *cis* to the carbonyl group is always lower owing[1381] to the weaker s-character of the longer bond: $^1J(^{15}N,H_c) = 88.0$ to 88.7 Hz; $^1J(^{15}N,H_t) = 88.6$ to 90.0 Hz.[1053]

The amide groups of dipeptides have a characteristic direct ^{15}N–H coupling range of 89 to 95 Hz.[924] The direct ^{15}N–H couplings of ammonia, methylamine (MeNH$_2$), and dimethylamine (Me$_2$NH), 61, 64.5, and 67 Hz,[20] and the couplings of aniline (PhNH$_2$) and its N-methyl derivative (PhNHMe), 82.5 and 87 Hz, show significant substituent effect. The direct NH coupling constant of aniline derivatives increases with the $-I$ effect of the substituents[66,68,187,1168] (which presumably promotes the sp^2 hybridization of nitrogen, i.e., increases its s-character) from 79 Hz (p -dimethyl-aminoaniline) to 92.5 Hz (2,4-dinitroaniline). There is a linear relationship between the direct coupling constants and the Hammett constants.[187]

The increase of direct ^{15}N–H coupling in parallel with the $-I$ effect of substituents was observed in many cases (e.g., Reference 305), among others for pyrroles.[767] However, no solvent, temperature, and concentration dependence could be detected. The direct coupling constant measured for pyrrole itself (96.5 Hz) is characteristic of the sp^2 hybridization state of nitrogen, i.e., to the aromatic character of the ring.[206] The absence of solvent effect does not pertain to the aniline derivatives, where $^1J(^{14}N,H)$ in DMSO-d$_6$ is 4 to 6 Hz higher than in CDCl$_3$.[68,1081]

The N–H couplings are influenced by the mutual position of the hydrogen and the lone pair of nitrogen.[1478] The conformation dependence of direct ^{15}N–H coupling has also been shown[421] for N-acetyl-phenylhydrazine (PhNHNHAc).

The direct ^{15}N–H coupling constant may be used directly in chemical structure elucidation, e.g., the site of protonation may be determined from the magnitude of coupling constant and the multiplicity of the signals. In strong acid media, urea is present in di-protonated state, and the quartet and triplet nitrogen signals with $^1J(^{15}N,H) = 77$ and 97 Hz prove unanimously the presence of $H_3N^\oplus$ (sp^3)and $H_2N^\oplus$ (sp^2) groups, i.e., the *N,O*-diprotonated $H_2N^\oplus = C(OH) - N^\oplus H_3$ structure.[1474]

Compound **353** may occur in two tautomeric forms (**a** and **b**). The value of 89 Hz[399] for the coupling constant unanimously supports structure **a**. The tautomery of Compound **354** (**a** $\rightleftharpoons$ **b**) is fast on the NMR time scale,* and thus the spectrum shows the average of the direct ^{15}N–H coupling constants corresponding to forms **a** and **b**. Thus, the extremely low value of $^1J(N,H)$, 33 Hz,[399] is a convincing proof for tautomery in the case of R=Ph, whereas the coupling of 79 Hz measured for the R=Me analogue indicates the predominance of form **a**. The direct determination of the ratio of forms **a** and **b** from the value of $^1J(N,H)$ is, however, based on the not proved assumption that there is no coupling between the hydrogen and nitrogen in form **b** through the hydrogen bond.

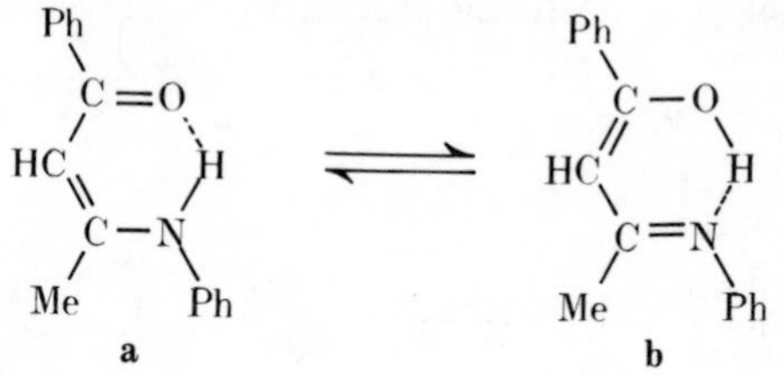

353

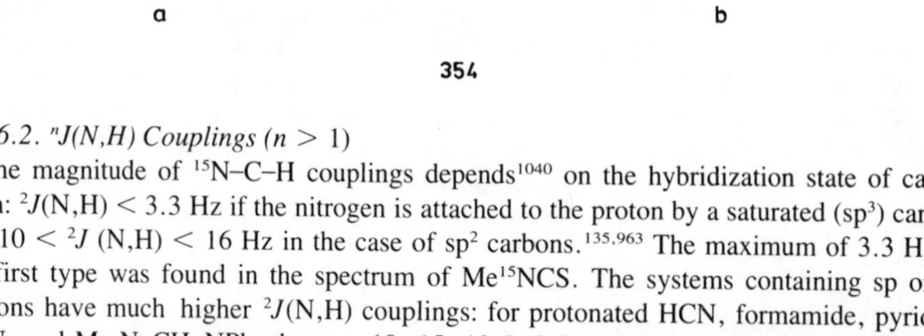

354

4.2.6.2. $^nJ(N,H)$ Couplings (n > 1)

The magnitude of ^{15}N–C–H couplings depends[1040] on the hybridization state of carbon atom: $^2J(N,H) < 3.3$ Hz if the nitrogen is attached to the proton by a saturated (sp^3) carbon, and $10 < ^2J (N,H) < 16$ Hz in the case of sp^2 carbons.[135,963] The maximum of 3.3 Hz for the first type was found in the spectrum of Me^{15}NCS. The systems containing sp or sp^2 carbons have much higher $^2J(N,H)$ couplings: for protonated HCN, formamide, pyridine, HCN, and Me$_2$N–CH=NPh, they are 19, 15, 10.5, 8.7, and 8.5 Hz, respectively.[995,1168]

The absolute value of $^3J(^{15}N,H)$ is approximately 1.5 to 6 Hz. Theoretical considerations suggest that $^2J(N,H)$ and $^3J(N,H)$ may change with molecular geometry. For the latter a formula similar to the Karplus relation was recommended in the case of peptides.[84,233,1399] If in the N–C–C–H chain both carbons are saturated (sp^3), $^3J(N,H)$ is characteristic of the *trans* (approximately 4.8 Hz) and *gauche* (approximately 1.8 Hz) positions.[1317] An example

---

*   Compare p. 93—94.

for the dependence on dihedral angle is 3(*N*)-methyl-1,3-oxazine (**355**), where, of the ^2J(N,H) couplings of 4-methylene protons, only the *equatorial* one causes measurable splitting:[1208] ^2J(N,H$_e$) = 1.5 Hz > ^2J(N,H$_a$) ≈ 0.

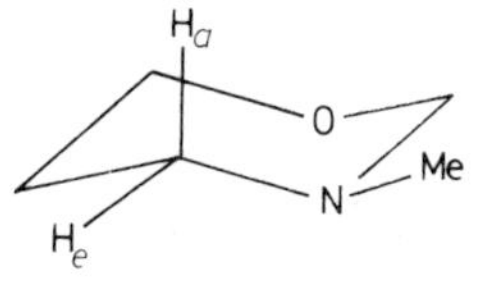

355

All couplings are sensitive to the orientation of the lone electron pair and to the configuration of geometric isomers. In the geometric isomers **356 a** and **b**, the values of ^3J(NNCH) coupling are 5.5 and 10 Hz, respectively.[1027] A similar difference can be seen with the *syn-anti* isomers of imines of type **166** (R″=H), where ^2J(N,H)^{syn} = −9.8 to −10.4 and ^2J(N,H)^{anti} = 3.4 to 3.7 Hz, i.e., even the signs are different.[1569] In other cases there may be larger differences in the values (2 to 4 and 10 to 16 Hz).[1208] However, the difference is large only when one of the substituents is hydrogen.

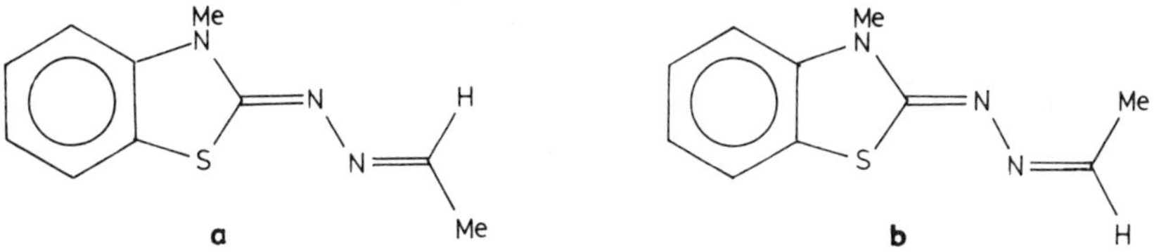

356

The rotamers of amides have different ^1J(N,H) and ^2J(N,H) couplings, e.g., for Compound **357** ^1J(N,H) = 92 (**a**) or 90 (**b**) Hz; ^2J(N,H) = 15 (**a**) or 14 (**b**) Hz.[1217] The conformation sensitivity of the ^1J(N,H) coupling of PhNHNHAc has already been mentioned.* This can be supplemented with the values of ^2J(NNH), which are 1.1 and 1.2 Hz for the Z rotamer and 5.5 and 0.4 Hz for the E rotamer.[421]

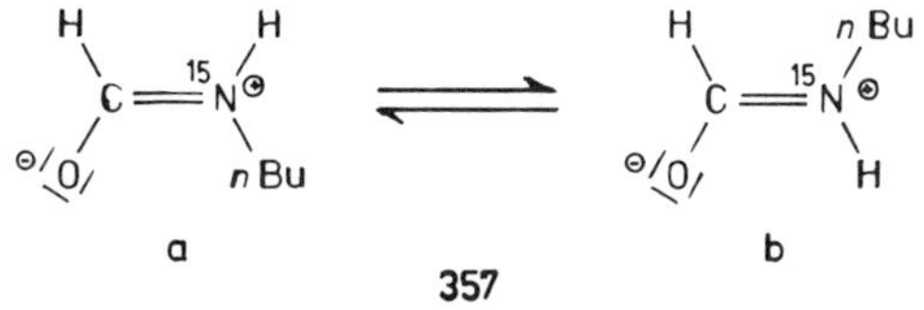

357

In the N–N–C–H chains of nitrosamine derivatives (RR′N–NO), nitroso-hydrazides (R′R″N–NR–NO) and *N*-nitroso-hydroxylamines (R′O–NR–NO) and similarly in the N–O–C–H chains of alkyl nitrites (RO–NO) the ^3J(N,H) couplings are characteristically different in the Z-E isomers: ^3J(N,H$_Z$) < ^3J(N,H$_E$).[64,65,70] For the NNCH chain the couplings are 0 to 1 (Z) and 1.5 to 3.0 Hz (E), whereas for NOCH they are 1.5 to 2.0 (Z) and 2.0 to 2.5 Hz (E).

In α-naphthyl-aziridine (**358**), ^2J(N,H$_C$) > ^2J(N,H$_B$), and since ^2J(N,H) < 0, this means that the electron-withdrawing α-naphthyl substituent decreases coupling.[1058] The absolute value of ^3J(^{15}N,H) coupling increases from −1.5 to approximately −4.5 Hz in heteroaromatic systems pyridine, quinoline.[1168,1420] Similarly, quaternerization or *N*-oxidation also

---

* Compare p. 254.

increases the absolute value of $^3J(N,H)$ coupling.[995] In contrast, the $^2J(N,H)$ coupling constant decreases in absolute value, from -10 to -3 Hz upon the protonation of pyridine, which is attributed to an increase in the $n - \pi^*$ excitation energy.[863] The $^4J(N,H)$ couplings are less than 1 Hz, but in unsaturated systems they often cause measurable splittings.[1548]

358

4.2.6.3. Nitrogen Couplings with Atoms Other than Protons

There are only sparse data on these couplings in the literature. The direct $^{15}N-^{15}N$ couplings — following from the low natural abundance of the ^{15}N isotope — may be studied exclusively on samples enriched in ^{15}N. Their magnitudes are between 4 and 20 Hz.[63,229,1264,1548]

$^1J(^{15}N,^{13}C)$ couplings are in the 1- to 15-Hz range (the extremely high value obtained for the fulminate derivative **351** has already been mentioned).* Some direct $^{15}N-^{13}C$ couplings, together with $^1J(N,H)$ ones, are given in tabular form.[863]

The $^{15}N-^{13}C$ couplings, too, are sensitive to the steric position of the lone pair. The *syn-anti* isomers of acetaldoxime (**162 a,b**) have the following coupling constants: $^1J(N,C) = 2.3$ (**a**) and 4 Hz (**b**); $^2J(N,C) = 1.8$ (**a**) and 9.0 Hz (**b**).[861] It can be seen that the favorable position of the lone pair significantly enhances the generally low value of $^2J(N,C)$. On this basis for the oximes **359**, the values of $^2J(N,CH_3) < 2$ Hz and $^2J(N,C_{Ar}\text{-}1) = 9$ to 10 Hz[221] support the presence of Z configuration. In saturated systems $^3J(N,C)$ is commensurable in absolute value with the direct couplings.[1548]

359

It was shown that the main component of $^1J(N,C)$ couplings is the Fermi contact term, and there is a direct proportionality between its magnitude and the s-character. Its sign proved to be positive for cyanides and negative for isocyanides.[1001] In pyridine hydrochloride $^2J(^{15}N,^{13}C) < ^3J(^{15}N,^{13}C)$.[1]

In quinoline an unusual sequence of values of $^{15}N-^{13}C$ coupling constants occurs: $^1J(N,C) = 0.6$ to $1.4 < ^2J(N,C) = 1.0$ to $2.7 < ^3J(N,C) = 0 - 4.6$ Hz. However, the $^2J(N,C\text{-}8)$ coupling constant is higher, due probably to the favorable orientation of the lone pair. For the same reason, the coupling is solvent dependent.[1142]

4.2.6.4. The Quadrupole Relaxation of ^{14}N Isotope; Technical Problems of the Measurement of ^{14}N Coupling Constants and of Nitrogen Resonance Spectroscopy

According to the spin quantum number $I = 1$, the ^{14}N nucleus has electric quadrupole moment ($Q/10^{-28}$ m$^2 = 1.6 \cdot 10^{-2}$).[832] The central problem of ^{14}N resonance measurements is the signal broadening caused by the short relaxation times due to this quadrupole moment.

---

* Compare p. 252—253.

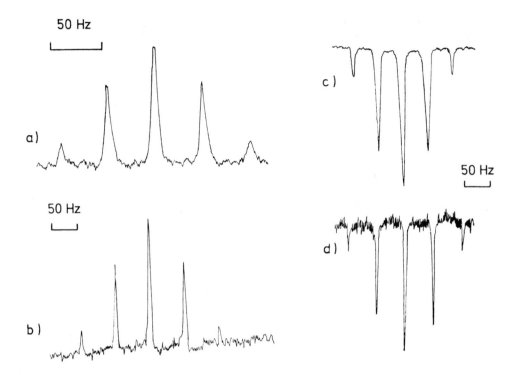

50 Hz

a)

b)

50 Hz

c)

50 Hz

d)

FIGURE 195. The nitrogen resonance spectrum of $NH^{\oplus}_4$ ion for (a) ^{14}N at 3.9 MHz and (b) ^{15}N at 6.1 MHz, in the latter case using the 5 M HCl solution of ammonium chloride enriched to 95% in ^{15}N; and the INDOR spectra of $^{14}NH^{\oplus}_4$ and $^{15}NH^{\oplus}_4$ ions (B_0 and $\nu_1 = \nu H$ constant; ν_2 is varied in the frequency range corresponding to the nitrogen nuclei). (From Randall, E. W. and Gilles, D. G., in *Progress in Nuclear Magnetic Resonance Spectroscopy*, Vol.6, Emsley, J. W., Feeney, J., and Sutcliffe, L. H., Eds., Pergamon Press, Oxford, 1971. With permission.)

The relaxation times T^Q are inversely proportional to the square of the field gradient eq and the quadrupole moment eQ. Consequently, the signal width of a quadrupole nucleus is inversely proportional to the time constant of the quadrupole relaxation:

$$1/T^Q = \pi \Delta \nu_{1/2} N \tag{351}$$

where $\Delta \nu_{1/2} N$ is the half-band width of the absorption signal. Quadrupole relaxation phenomena may decrease the value of T^Q even to 10^{-7} s, which may give rise to several hundred or even thousand hertz line widths if the reasonable value of $\tau_c = 10^{-11}$ s is assumed for the dilute solutions of small molecules.[1053] Owing to Equation 246, τ_c and thus line width increases with increasing viscosity (i.e., increasing concentration and decreasing temperature) and sample volume. When the rotation of the molecule is commensurable with the rotation of one of its groups, the value of T^Q supplies information, via line shape analysis, on the internal rotation.[1464,1465]

When the environment of the ^{14}N nucleus (or the molecule) is symmetrical, η and q (compare Equation 261) decrease, i.e., relaxation slows down (T^Q is higher) and signal width decreases. The more symmetrical the field gradient around the ^{14}N nucleus, i.e., the closer the symmetry to spherical, the sharper the ^{14}N resonance signals. Therefore, in the case of ammonium ions and the quaternary, symmetrically tetrasubstituted alkylammonium ions, the ^{14}N–H spin-spin splitting is well observable. This is illustrated (see Figure 195) by the ^{15}N and ^{14}N resonance spectra of ammonium chloride (the former was obtained by CAT technique with enriched sample.[1168]

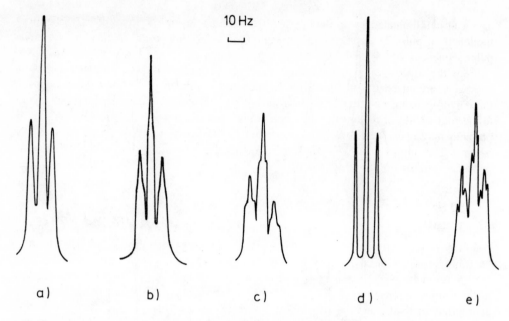

10 Hz

a)   b)   c)   d)   e)

FIGURE 196.   The methyl signal of *N*-methylpyridinium iodide in the ¹H NMR spectrum recorded in D₂O at 60 MHz at (a) 50, (b) 80, and (d) 25°C and the methyl signals of (c) 4-acetamido and (e) 4-trifluoromethyl derivatives at 25°C.[133] (From Mooney, E. F. and Winson, P. H., in *Annual Review of NMR Spectroscopy*, Vol. 2, Mooney, E. F., Ed., Academic Press, New York, 1968. With permission.)

The $^1J(^{14}N,H)$ coupling constants of ammonia and ammonium nitrate are 44 and 53 Hz, respectively.[995] In quaternary *N*-ethyl, *N*-isopropyl, and *N*-*t*-butyl derivatives, the magnitude of $^3J(^{14}N,H)$ couplings of N–C–C–H type is 1.5 to 2.2 Hz.[547] The $^2J(^{14}N,H)$ couplings through the analogous N–C–H path do not cause detectable splitting, since $^2J(N,H) < 0.2$ Hz.[133,1169] It has been shown that ^{14}N–H couplings vary with the dihedral angle.[1399] There are a few coupling constants of ^{14}N with other nuclei reported; e.g., values of 130 to 330 Hz were measured for ^{19}F–^{14}N couplings.[995]

Exceptionally, $J(^{14}N,H)$ couplings may also be observed in ¹H NMR spectra. An example is the direct ^{14}N–H coupling observed in the ¹H NMR spectrum of *N*-ethylpyridinium iodide (see Figure 196), which causes a further 1:1:1 triplet splitting of all lines of the methyl triplet[133] at higher temperatures or even at room temperature if there is an electron-with-drawing substituent in *para* position on the aromatic ring. The reason is that relaxation times increase with temperature owing to the faster thermal motion. Electron-withdrawing sub-stituent in *para* position suppresses the negative polarization of the nitrogen atom, leading to an electronic structure closer to spherical symmetry, thereby decreasing quadrupole mo-ment and thus increasing relaxation time.

The FT spectrometers boosted up primarily the application of ^{15}N resonance, but brought little change in the field of ^{14}N NMR spectroscopy. The main reason is the already discussed line broadening of ^{14}N signals (from 1 Hz to some kilohertz), which may not be eliminated by the FT technique.

The half-band widths of 1 Hz for ^{15}N isotope permit[768] one to enjoy the maximum improvement of ζ with FT technique (compare Equation 235). Therefore, it is possible to get usable spectra from samples with natural isotope abundance, in dilute, 0.1 to 0.001 mol/ℓ solutions, under relatively low t_{aq} acquisition times.[600,1138]

At room temperature, the relaxation times of ^{15}N nuclei are long, 20 to 10^3 s,[879] and thus the saturation of signals is quite frequent, particularly when no hydrogens are attached to the nitrogen. (When the pulse spacing is less than $6T_1$, partial or complete saturation occurs.)

This can be eliminated by commonly applied methods: using "T_1 reagents",[*] shortening the length of pulse,[**] decreasing pulse angle (below 90°), or by the application of various pulse sequences.[***]

From the aspects of avoiding saturation, it is advantageous, as mentioned above, if hydrogens are attached to the nitrogen. In such cases, similar to ^{13}C NMR spectroscopy,[†] these hydrogens may be decoupled by means of BBDR, and therefore the ζ increases owing to the coalescence of the lines of the original multiplet and to the NOE. However, a special problem may occur here, which is connected to the negative gyromagnetic ratio of ^{15}N isotope. The improvement in intensity $\zeta_N = \zeta_{BBDR} - \zeta_{H-coupled}$ arising from NOE is proportional to the ratio of γ_H and γ_{15N}:[1053]

$$\zeta_N \sim 1 + \gamma_H / \gamma_{15N} \tag{352}$$

Since $\gamma_{15N} < 0$, ζ_N may assume negative or zero values. For the extreme narrowing condition,[‡] the limits of ζ_N are $+1$ and -3.93. Within this region, ζ_N depends on the contribution of ^{15}N–H DD term in the relaxation mechanisms. If this is approximately 20%, the ^{15}N resonance signal completely disappears.[1006] This must always be taken into account in the routine applications of FT technique in ^{15}N resonance. (NOE can be eliminated by appropriate methods, and then the signals cancelled by it reappear in the spectrum. In this case, of course, ζ_N is not improved for the signals not adversely influenced by NOE, either.)

Finally let us draw the attention to two special measuring techniques in N NMR spectroscopy. The essence of *differential saturation technique* is the audio-frequency modulation of the resonance frequency, whereby side-bands are produced in the CW spectrum such that the saturation of the side-bands is about 0.1 to 10% of the main band. The joint band shape analysis of the main and side bands by computer enables the νN resonance frequency and T_1 spin-lattice relaxation time to be determined very accurately.[625]

By means of the *difference satellite technique* the ^{15}N–H coupling constants may be measured in the ^1H NMR spectra. The basic principle is that "pure" satellite spectra may be produced as a difference of the coupled and nitrogen decoupled ^1H NMR FT spectra (by "subtracting" the satellite signals from the full proton spectrum), and these spectra usually permit the various ^{15}N–H coupling constants to be read directly.[924,1050]

4.3. OXYGEN RESONANCE[434,1215,1303,1379]

The low natural abundance $3.7 \cdot 10^{-2}$ and small magnetogyric ratio ($-3.6266 \cdot 10^7$ rad $T^{-1}s^{-1}$) of ^{17}O make the detection of its resonance signals very difficult. The relative sensitivity (receptivity, compare Table 1) of ^{17}O isotope is about five orders smaller than that of hydrogen. Due to the nonzero quadrupole moment ($I = 5/2$), signals are extremely wide, the line width is 10 to 10^2 Hz for small molecules, but in case of larger molecule and viscous solutions, it can be as much as 1 kHz which is equivalent to 125 ppm at 8.13 MHz and becomes easily saturated because of the small relaxation times. In order to circumvent this difficulty, the dispersion signal in derivative presentation is recorded instead of the absorption one.[‡]

Since signal broadening as a consequence of fast quadrupole relaxations is not remedied by FT techniques, no fast progress can be expected in oxygen resonance spectroscopy.[632] Nevertheless, the first report on ^{17}O resonance[19] was published as early as in 1951.

---

* Compare Volume I, p. 205 and 219.
** Compare Volume I, p. 178 and 219.
***Compare Volume I, p. 219.
† Compare p. 221 and Volume I, p. 197—198.
‡ Compare Volume I, p. 201.
‡ Compare p. 146.

Table 95
CHEMICAL SHIFTS (PPM) AND SIGNAL WIDTHS (HZ) IN OXYGEN RESONANCE[275,277]

Compound	Chemical shift	Signal width	Compound	Chemical shift	Signal width
MeOH	−37	100	Furan	241	90
D_2O	−3	60	HCOOH	254	80
H_2O	0	60	MeCOOCOMe	259	170
EtOH	6	160	$MeCOCH_2COMe$	269	140
$MeOCH_2OMe$	8	50	(In enolic form)		
$(MeO)_2S$	12	130	$HCONH_2$	304	90
MeSOMe	13	120	$MeCONMe_2$	324	50
EtOEt	15	150	MeCOOMe	355	120
$(MeO)_3PO$	22	80	HCOOMe	361	50
$CH(OMe)_3$	28	160	$MeCOOCH=CH_2$	371	120
$(MeO)_3P$	46	50	EtCOF	373	60
$(MeO)_3PO$	75	20	MeCOOCOMe	393	170
$EtOCH=CH_2$	88	45	$Me(CH_2)_2ONO$	455	60
$(MeO)_2SO_2$	102	120	MeCOCl	507	45
$(MeO)_2SO$	115	60	MeCOBr	536	45
MeCOO Me	137	120	Cyclohexanone	559	140
HCOOMe	139	60	MeCOMe	572	45
$(MeO)_2SO_2$	150	70	MeCHO	595	50
H_2O_2	174	200	$EtNO_2$	600	120
$(MeO)_2SO$	176	35	Et_2NNO	683	190
$MeCOOCH=CH_2$	204	120	$Me(CH_2)_2ONO$	803	60

The chemical shift range is about 800 ppm and the shift ranges for the different types of ^{17}O nuclei rarely overlap. This fact partly compensates for the inaccuracies of the measured shift values.

Table 95 gives summary of the chemical shifts reported by Christ and co-workers more than two decades ago.[275-277] More recent[1215] measurements, however, have not brought about any significant improvement in the accuracy. Because of the sparseness of data, to give characteristic shift intervals is not possible. As reference substance water is applied, its signal is found at high field due to the partially ionic character of the O–H bonds. Smallest chemical shifts are obtained for alcohols. Branching in the aliphatic chain is revealed in a downfield shift enabling one to distinguish primary, secondary, and tertiary alcohols. The signals due to the sp^3 oxygen atoms of ethers, esters, acid anhydrides, and acids are at gradually increasing δ-values. The shift of the carbonyl oxygens is much greater than that of the single-bonded oxygens, and the δ-values increase in the sequence: acids, amides, esters, acid anhydrides, acid halides, ketones, and aldehydes. The greatest chemical shifts is shown by nitro and nitroso derivatives (compare Table 95 and References 872, 873, and 907). The shift of the carbonyl signal upfield is greater, the larger is the extent of the negative polarization of the oxygen atom.

In the spectra of acid anhydrides, two signals appear corresponding to the sp^3 and sp^2 oxygens: assignment is simple on the basis of the intensity ratio 1:2. The equivalence of the sp^3 and sp^2 oxygens of the acids indicates the fast exchange of proton between the oxygens (**360 a** ⇌ **b**) in cyclic dimeric association. This explains also that the signal of the acids is found at a field intermediate between that of the sp^3 and sp^2 oxygens.

The position of the oxygen resonance signal may inform about the enolization. Thus the signal of diacetyl (MeCO–COMe) is found at 571 ppm, while the singlet of the enolized acetyl-acetone (**218**, R=Me: two equivalent oxygen atoms) appears at 260 ppm. Oxygen resonance then is an excellent tool for study of keto-enol equilibria.[571] The large difference

360

between the chemical shifts of single- and double-bonded oxygens is related to the $+M$ effect of the substituents, indicating the importance of the paramagnetic contribution to the resulting shielding. By the same argument, it can be explained that the δ-value of oxygens connected to atoms other than hydrogen or carbon is very variable, because the electronic structure of the oxygen atom in these bonds depends strongly on the nature of the hetero atom. The oxygens in SO and PO double bonds are more shielded than the carbonyl oxygens (compare Table 95). Due to line broadening, the data on the spin-spin couplings involving ^{17}O are sporadic. $^1J(O,H)$ in water[243,1199] and ethanol[1448] is 83 and 81 Hz, respectively. $^2J(O,H)$ for HCOOMe is 38 Hz.[243]

4.4. FLUORINE RESONANCE

Although fluorine compounds do not play a central role in organic chemistry, the literature of ^{19}F NMR spectroscopy is very ample.[257,402,435,436,475,476,739,740,919,993,994,1052]

Due to the 100% natural abundance of the ^{19}F isotope, its 1/2 spin quantum number and the advantageous magnitude of the relaxation times, the relative sensitivity of the fluorine nucleus is not much lower (0.83) than that of proton. The somewhat lower sensitivity is due to the smaller magnetic moment. In addition, an important reason making ^{19}F NMR spectroscopy popular is that the chemical shifts and the coupling constants are about one or two orders greater than in proton resonance.* Consequently many important phenomena in NMR spectroscopy were discovered on fluorine compounds. Examples are the establishment of the optimal conditions for long-range coupling** or the method (use of chiral solvents) for NMR determination of optical purity.***

It should be noted that CW method is often more advantageous than FT in ^{19}F NMR spectroscopy. High ζ is not so essential due to high receptivity of fluorine nucleus; on the other hand serious problems can arise from the spectral width of about 800 ppm $\triangle$ 75 kHz at 2.35 T, which requires very fast ADC† to provide a reasonable solution. Another drawback is that DDBR is often not applicable, on a consequence of the proximity of the vH and vF resonances and of the relatively large $^2J(F,H)$ couplings.[919]

4.4.1. ^{19}F NMR Chemical Shifts

Nowadays, $CFCl_3$ (TFM) is a generally accepted reference substance in ^{19}F NMR spectroscopy. Previously, various compounds, most often TFA, less frequently octafluoro-cyclobutane (C_4F_8), fluoro-, and hexafluorobenzene, CF_4, etc., were used as well. TFM can be simultaneously applied as internal standard and solvent, it is inert and very volatile, but its signal is not sharp, being split into three components of 3:3:1 intensity, corresponding to $CF^{35}Cl_3$, $CF^{35}Cl_2{}^{37}Cl$ and $CF^{35}Cl^{37}C_2$ molecules, due to isotope effect of nuclei ^{35}Cl and ^{37}Cl.[255,740] Further drawbacks of TFM reference are the solvent and temperature dependence of its signal,[436] as well as the inconvenience owing to the low field position of TFM signal

---

* Compare Volume I, p. 26, 68, and Table 5 (p. 55—58).
** Compare Volume I, p. 68.
***Compare Volume I, p. 41 and Volume II, p. 114—115 and 132—133.
† Compare Volume I, p. 182—183.

Table 96

**CHEMICAL SHIFTS (PPM), AS REFERRED TO TFM
STANDARD, IN FLUORINE RESONANCE[402,435]**

Compound	Chemical shift	Compound	Chemical shift	Compound	Chemical shift
FCl	−448	$F_3C–CO–CF_3$	−84	PhCOF	17
$(CF_3)_2CFSF$	−362	F_3CH	−79	FH	40
FCH_3	−272	TFA	−77	$PhSO_2F$	66
$FSiH_3$	−217	$FCCl_2–CFCl_2$	−69	FN=NF (Z)	95
C_6F_6	−163	$F_3C–CS–CF_3$	−68	F_3CSNF_2	103
F_3SiH_2	−151	$PhCF_3$	−64	F_2CS	108
F_2CH_2	−144	CF_4, $(F_3C)_4C$	−63	FN=NF (E)	134
C_4F_8	−136	F_3CCl	−33	NF_3	145
C_5F_{10}	−133	F_3CBr	−21	F_2O	249
$F^\ominus$ ($K^\oplus$)	−125	F_2CCl_2	−8	$FClO_3$	287
PhF	−116	$FCCl_3$(TFM)	0	F_2	429
F_3SiH	−109	$FCBr_3$, F_2CBr_2	7	F_2O_2	865

resulting in mostly negative shifts. To give a basis for assessing the magnitude of ^{19}F NMR shifts, Table 96 contains the δF values of a few simple compounds.

In converting older shift values, one should also consider whether external standards were used. Shifts referring to external standard TFA are smaller by about 2 ppm, since in this case $δ_{TFA} = −79$ ppm.[994]

The solvent effects are significantly greater than in ^1H NMR spectroscopy. The signal of CF_4 is shifted downfield by 5, 7, 9, and 12 ppm relative to the gas phase, if the spectra are recorded in dilute trifluoromethyl-cyclohexane, *n*-heptane, CCl_4 and MeI solutions.[457]

The signal of the fluoride ion is 550 ppm upfield from the F_2 resonance. This shift difference, greater by an order than expectable considering only the diamagnetic contribution, proves the dominance of the paramagnetic term in the resulting shielding. Hence, chemical shifts give information about the covalent or ionic character of a given Q–F bonding. In fluorides of type QF_n, shielding is inversely proportional to the electronegativity of Q.[303] Similarly, in CHF_3, CH_2F_2, and CH_3F, shielding is increased by 16, 81, and 209 ppm, as referred to CF_4 (compare Table 96). A similar change would be expected in the series CF_4, CF_3Cl, CF_2Cl_2, and $CFCl_3$; in this case, however (see Table 96), the smallest shielding of zero parts per million is observed for trichloro-fluoromethane (TFM), and the fluorine in CF_2Cl_2 and CF_3Cl is also less shielded ($δ = −8$ and $−33$ ppm). The reason for this is probably that the fluorine atom has a greater $+M$ effect than chlorine atom, and therefore electronic density about the fluorine atom is smallest in TFM.

Correlation between electronegativity and the chemical shift was found, e.g., for CHFCl-CF_2R–type compounds. The three fluorine and the proton form an *ABMX* spin system, J_{AB} and $\Delta δAB$ of which are inversely proportional, while the midpoint of the *AB* multiplet is directly proportional to electronegativity of substituent R.[408] In Table 97 chemical shift ranges for fluorine compounds are compiled, based on a rather limited number of data which cannot be, therefore, regarded as general.

4.4.2. ^{19}F NMR Parameters of Saturated Open Chain Compounds

Table 98 contains the chemical shifts of a few simple poly- and perfluoroalkanes and their substituted derivatives. The relation $δCH_3 < δCH_2 < δCH$, generally valid in ^1H NMR spectroscopy, does not apply for CF_n groups ($n = 1,2,3$). In fact, the opposite tendency is found (e.g., see compounds number 11, 18, and 23 or 10, 21, and 22 in Table 98).

Also there is not a straightforward correlation between the chemical shifts and $−I$ effect of the substituents, since the shieldings are mainly controlled by the paramagnetic and

Table 97

THE ¹⁹F NMR SHIFT RANGES (PPM) OF VARIOUS FLUORINE CONTAINING GROUPS[435,1379]

Functional group	Chemical shift range	Functional group	Chemical shift range	Functional group	Chemical shift range
$-CH_2F$	$-250--200$	$-CF_2H$	$-150--110$	$(CH_3)FC=$	$-90--85$
CHF^a	$-240--210$	$-OCF_3$	$-145--45$	$CF_3C(sp^3)$	$-90--50$
CHF	$-230--60$	$-CF_2(CF_3)$	$-140--90$	$CF_3N(sp^3)$	$-85--40$
$CF(CH_3)$	$-225--130$	$F_2C=$	$-135--60$	$-CF_2X$	$-85--35$
$CF(CF_3)$	$-225--110$	CFX, CF_2	$-135--10$	CF_3Ar	$-80--60$
$CF-^b$	$-205--80$	$-COCF_2-$	$-130--110$	CF_3X	$-65-0$
$-CF=$	$-205--45$	$-CF_2N(sp^3)$	$-125--60$	$CF_3-metal$	$-60-10$
$(CF_3)FC=$	$-200--100$	$-FN(sp^3)$	$-125--5$	$F-CO-$	$-60-70$
$HFC=$	$-190--100$	$-CF_2Ar$	$-120--100$	$-SF$	$-55-90$
F^c	$-190--20$	$-SCF_3$	$-115--20$	$=NF$	$-55-135$
$CF-$	$-185--40$	$-CF_2S-$	$-110--65$	CF_2X_2	$-10-10$
$-FSi(sp^3)$	$-180--125$	$-CF=N-$	$-105-0$	$-NF_2$	$10-70$
ArF	$-180--90$	$-CF_2(CH_3)$	$-100--85$	$-SO_2F$	$35-70$
PyF^d	$-175--60$	$Py^{\oplus}F^c$	$-100--15$	$-OF$	$45-250$
$CF-^a$	$-170--115$	$-CF_2O-$	$-95--65$	$F-CS-$	$105-165$

Note: ᵃ, in cyclopropane ring; ᵇ, cycloalkanes, except for cyclopropane; ᶜ, in heteroaromatic rings with more heteroatoms; ᵈ, in pyridines; and ᵉ, in pyridine salts.

Table 98

THE ¹⁹F NMR SHIFTS (PPM) OF SOME POLY- AND PERFLUOROALKANES AND THEIR SUBSTITUTED DERIVATIVES[402,435]

-84	18.5	-64		-88	-140	-82	-98	139
1. CF_3-NF_2		9. $CF_3CH_2CF_3$		17. CF_3-CF_2H		25. CF_3-CF_2-OF	

-79	-64	-110	-81 -90
2. CF_3H	10. CF_3Ph	18. CF_2HCH_3	26. $(CF_3-CF_2)_2O$

-79	-62	-76 -190	-132
3. CF_3CHBrI	11. CF_3CH_3	19. $(CF_3)_3CF$	27. $(CH_3)_3CF$

-76	-62	-78 -147	-81 -126 -70
4. $CF_3CHBrCl$	12. $(CF_3)_2O$	20. $(CF_3)_2CFI$	28. $CF_3-CF_2-CF_2Cl$

-76 15	-59	-86 -116	-85 -65
5. CF_3-COF	13. $(CF_3)_3N$	21. CF_3-CF_2Ph	29. CF_3-CF_2I

-72 148	-79 -241	-76 -183	-80 -118 -60.5
6. CF_3-OF	14. CF_3-CFH_2	22. $(CF_3)_2CFPh$	30. $CF_3-CF_2-CF_2I$

-71	-77 -217	-213	-45
7. CF_3CH_2Cl	15. $(CF_3)_2CFH$	23. CH_3CH_2F	31. CF_2ClCH_3

-68	-39	-78 -144	-37
8. CF_3CH_2Br	16. $(CF_3)_2S$	24. $(CF_3)_2CFBr$	32. CF_2BrCH_3

anisotropic terms. This is shown by the smaller shifts of compounds 3-5, 7-10 in contrast with the less shielded compound 11. The order of the δCF_3 values for compounds 12, 13, and 16 is the opposite as for the analogous δCH_3 shifts in ¹H NMR spectra. Similarly, in the halogen derivatives, the shielding decreases in the order Cl, Br, and I (e.g., see the pairs 7 and 8, 28 and 30, or 31 and 32. Examples for the opposites case are the pairs 3 and 4 or 20 and 24).

As mentioned due to the larger shift differences and coupling constants, fluorine resonance is suited for the study of conformers of open chain saturated compounds, too. It may become possible to determine also the spectral parameters of rotamers.

Thus, e.g., from ^{19}F NMR spectrum recorded at $-150°C$ for the rotamers **a** (*trans*) and **b** (*gauche*) of **361**,[1495] $\delta F = -61, -73, -70, -74$ and 61 ppm for fluorine a-e and ^2J $(F,F) = 124, 116,$ and 105 Hz for atom pairs c-e, d-e, and c-d, respectively.

361

The long-range coupling ^5J $(F_c,F_c) = 75$ Hz indicates the presence of a through-space interaction.* Indeed, it can be seen from Formula **361b** that the fluorines c are very close in space. This was the first example for observation of coupling between *geminal* fluorines of a trifluoromethyl group and for the freezing in of rotamers of this group in a saturated compound.

The fluorine spectra of diastereomers are also different. Thus, *threo* and *erithro* isomers of 2-fluoro-3-halogen-n-butanes ($CH_3CHFCHXCH_3$; $X = Cl, Br, I$) can be easily distinguished, e.g., by their $^3J(H,F)$ coupling which is 16 to 19 Hz for the *threo*, whereas it is 9.7 to 10.6 Hz for the *erithro* isomer.[1061] The two rotamers of $CFBr_2$–CF_2Br molecule can be frozen at $-130°C$[1036] and the shifts and ratio of the formers can be determined.

For HF molecule $^1J(F,H) = 530$ Hz.[758] The coupling constants ^nJ (F,H) are generally of the magnitude 45 to 55 ($n = 2$) and 5 to 25 ($n = 3$) Hz, respectively, and often ^4J or ^5J through space couplings can be well detected (>1 Hz), too. The value of $J(F,H)$ depends on the angle and distance of the interacting atoms ($n = 2$) and on the dihedral angle ($n = 3$), respectively, following in the latter case a Karplus-like relationship.[575] Consequently, $^3J(F,H)$ and $^3J(H,H)$ couplings are[1527] roughly proportional: $^3J(H,H) \approx 0.33 \ ^3J(F,H)$. The $^nJ(F,F)$ couplings are naturally larger. Values greater than 200 Hz are quite frequent for $n = 2$. In case of $n = 3$ and 4, they are about 10 Hz and the former are usually smaller than the latter because of the through-space mechanism contributing to ^4J couplings. For CF_3–$CFCl$–CF_2I, the following coupling values were measured: $^3J(CF_3CFCl) = 6.7$ Hz; $^4J(CF_3CF_2I) = 11.9$ and 10.1 Hz.[79] Note that the two diastereotopic and therefore anisochronic fluorine nuclei of the group $-CF_2I$ have different ^4J (F,F) values. Among the $n = 4$ couplings of saturated groups CF–C–CF , those of eclipsed nuclei have the greatest values (~ 26 Hz), the ones in *gauche* position are smaller (~ 16 Hz), and are those of fluorines in *trans* position (''W arrangement''**!) are still smaller (~ 10 Hz). Unlike in proton-proton interactions,** the coupling corresponding to the W arrangement is smaller than those of eclipsed fluorine nuclei. This is due to the importance of through-space F–F interactions.

* Compare Volume I, p. 68.
** Compare Volume I, p. 67.

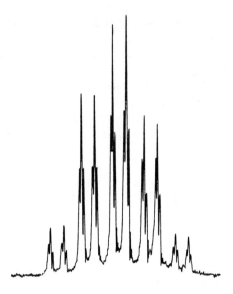

FIGURE 197. The CF_3 signal in the ^{19}F NMR spectrum of perfluoromethyl cyclohexane at frequency 56.5 MHz.[1102]

4.4.3. ^{19}F NMR Parameters of Saturated Cyclic Compounds

Perfluoro-cyclohexane gives at room temperature a singlet at -133 ppm, which is separated to two broad maxima at temperature $-60°C$; $\Delta_{e,a}\delta F = 18.2$ ppm[1407] and in contrast to protons $\delta F_e < \delta F_a$. From the spectra of *cis* and *trans* 4H-perfluoromethyl-cyclohexane, both conformatively homogeneous with *equatorial* CF_3 group, the following chemical shifts and coupling constant could be determined:[674,1099]

$\delta F\text{-}1a = -189$; $\delta F\text{-}2a,6a = -118$;
$\delta F\text{-}2e,6e = -133$ (c) and -129 (t);
$\delta F\text{-}3,5 = -127$ (c), -130 (t,e), and -120 (t,a);
$\delta F\text{-}4 = -236$ (c,e) and -212 (t,a) ppm;

further

^2J (F,F) $= 290$ to 305 Hz; ^3J ($F_e,F_{e,a}$) ≈ 14;
^3J (F_a,F_a) < 3; ^4J (F_e,F_e) ≈ 9;
^4J (F_e,F_a) ≈ 1; ^4J (F_a,F_a) ≈ 27;
^2J (F,H) ≈ 45; ^3J (F_a,H_a) ≈ 17;
^3J ($F_{a,e},H_{e,a}$) ≈ 6; ^4J (F,H) ≈ 4 Hz.

Similar values were obtained from temperature-dependent studies of perfluorocyclohexane.[731] Note that ^4J (F_a,F_a) is rather large owing to the through-space interactions of sterically close 1,3-*diaxial* fluorine atoms. For the *cis* and *trans* isomers of 4-*t*-butyl-trifluoromethyl-cyclohexane, $\delta(CF_3)_a = -67.5 > \delta(CF_3)_e = -75$ ppm.[347]

The effect of substituting fluorine at various positions for protons or for CF_3 groups upon the shifts of the remaining fluorines has been evaluated by comparison with perfluoro-cyclohexane.[674] Distinction can be made between cyclohexanols containing *axial* and *equatorial* hydroxyl groups by the trifluoroacetyl derivatives.[913] Similarly, the configuration and conformation of fluorine-containing sugars can also be determined (e.g., Reference 323, 620). For ^3J(F,H) couplings similar though somewhat larger values are characteristic as for fluorocyclohexanes: ^3J (F_a,H_a) ≈ 24 to 38 (a,a), $^3J(F_e,H_e) \approx 9.5$ to 12 (e,e), and ^3J (F_a,H_e)

≈ 1.5 to 5.5 (a,e), as obtained from studies on a few 1-fluoro-tetraacetyl-hexapyranoses.[621] The CF_3 signal (see Figure 197) is a double quintet due to the *geminal* and the four *vicinal* isochronic and isogamic (fast ring inversion!) fluorines. These ten lines are split further into triplets as a consequence of long-range ($n > 4$) coupling.[1102] Note that due to through-space interactions the relation $^4J(F,F) > \,^3J(F,F)$ holds in this case as well.

The chemical shifts of perfluoro-cycloalkanes of smaller rings are not significantly different ($C_5F_{10} = -133$, $C_4F_8 = -135$ ppm), except for perfluorocyclopropane (-151 ppm) in which, similar to protons in cyclopropane, fluorines are more shielded.[546]

4.4.4. ^{19}F NMR Parameters of Unsaturated Compounds

δF values for $H_2C=CHF$, $F_2C=CF_2$, $FC\equiv CF$, and FCN molecules are -114, -135, -95, and -156 ppm.[402,435] The shifts of fluorine *geminal*, *cis*, and *trans* vicinal to the proton of trifluoroethylene ($CF_2 = CFH$) are -185, -102, and -129 ppm.[435] It has to be noted that the anomaly caused by the anisotropy of the triple bond observed in ^{1}H NMR shifts of acetylene and ethylene is not encountered in fluorine resonance; the natural order of the shifts corresponds to the relative electron density. In the case of $FC\equiv CCF_3$, however, $\delta(\equiv CF)$ $= -203$ ppm and $\delta CF_3 = -51$ ppm.[77]

Contrary to the ^{1}H NMR experience, the chemical shifts of *trans* fluorine signals in $RFC=CFR$-type molecules are generally smaller than those in *cis*, independently of the substituent R (e.g., for R=H, Br, and Cl, $\delta F(Z)$ is -165, -95, and -105, whereas $\delta F(E)$ is -183, -113, and -120 ppm, respectively).[87,435]

The coupling constants are, of course, also informative. $^5J(F,F)$ for the Z and E isomers of $F_3C-CH=CF-CF_3$ molecule are 10 and 2 Hz, respectively (through-space interaction in the Z isomer!), which allows the unequivocal assignment of the isomers.[264] Chemical shifts are also significantly different for all kinds of fluorine atoms: δF-1 $= -57$ and -60, δF-3 $= -113$ and -117, and δF-4 $= -70$ and -75 ppm, respectively. *Vicinal* F–H coupling constant can also be put to use structural work, since similar to the olefinic couplings ^3J (H,H) it holds for $^3J(F,H)$, that $J^Z < J^E$. Thus, e.g., from the ^{19}F NMR spectra of the Z and E isomers of $PhCF=CHMe$, it follows that $^3J(F,H)^Z = 22$ Hz and $^3J(F,H)^E = 36$ Hz.[976]

In $CF_3CF=CF-COF$, the fluorine atom of the acidhalogenide group is subject to very large shifts, $\delta COF = +26.3$ (Z) and $+25.9$ (E) ppm,[78] because of the similarity of the environment of these fluorines to that of the aldehydic protons. The other types of fluorine nuclei are more sensitive towards configuration: $\delta CF_3 = -68$ and -72, $F^\beta \equiv \delta CF(CF_3) = -129.5$ and -147, and $F^\alpha \equiv \delta CF(COF) = -143$ and -159 ppm. Thus, similar to proton resonance, $\delta F^\beta > \delta F^\alpha$. Coupling constants between substituents and the fluorines connected to the same olefinic carbon atom are insensitive towards configuration, thus $^3J(CFCF_3) = 7.3$ and 7.7 Hz and $^3J(CFCOF) = 26.3$ and 21.5 Hz. Coupling constants through the double bond, $^5J(CF_3COF) = 12.5$ and 0.3, $^4J(CFCOF) = 22.3$ and 55.6, furthermore $^4J(CF_3COF) = 11.9$ and 21.1 Hz, reflect, however, the geometrical arrangement. The dramatic change of $^3J(CFCF)$, 9.1 Hz for the Z isomer and 138.2 Hz for its E pair, proves that the ^{1}H NMR rule according to which $^3J(H,H)^E > \,^3J(H,H)^Z$ across olefinic bonds is even more valid in ^{19}F NMR spectroscopy. The same applies to conjugated systems: for perfluoro-butadiene $^3J_{1,2}(E) = 119$ and $^3J_{1,2}(Z) = 32$ Hz. The value of $^2J^{gem} = 51$ Hz is, however, not parallel to the ^{1}H NMR experience, where for olefines ^2J (H,H) < 2 Hz, thus $^2J(H,H) \ll\, ^3J(H,H)^Z$. As it is expected, the coupling between the close nuclei 1 and 3 and 2 and 3 is also large: $^4J_{1,3E} = 14$ and $^3J_{2,3} = 30$ Hz. The other couplings are small, <12 Hz.[914]

The shifts of allenic fluorine nuclei in $F_2C=C=CF-CF_2-CF_3$ are $\delta CF_2 = -75$ and δCF $= -100$ ppm, and the greatest coupling exists between these nuclei, $^4J = 38$ Hz.[79] The δF value of the allene derivatives of type $>RR'C=C=CFH$ is much lower; e.g., for $(F_3C)_2C=C=CFH$, $\delta CF = -185$ ppm.[342]

Table 99

^{19}F NMR CHEMICAL SHIFTS (PPM) OF FLUOROBENZENE AND FLUORINE SUBSTITUTED SIX-MEMBERED NITROGEN HETEROCYCLES[435]

4.4.5. ^{19}F NMR Parameters of Aromatic and Heteroaromatic Compounds

The signal of C_6F_6 is at -163 ppm (see Table 96). The aromatic fluorine nuclei are therefore, in contrast to ^{1}H NMR, more shielded than those of perfluoro-ethylene. Shielding usually increases by substitution. When not all of the hydrogens are substituted by fluorine, the shift usually increases (see Table 99). Coupling constants provide limited structural information as J(F,F) for *ortho*, *meta*, and *para* fluorines are rather similar: ~ 20, 0 to 20, and 5 to 18 Hz, respectively.[7]

Linear relationship was found among the δF-4 and δC-4 shifts of differently substituted fluorobenzenes on the one hand and between the values of δF-4 and J^m (F,F) in the latter series on the other. A similar proportionality exists among δF, J(F,F) and J(F,H) in the various fluorobenzenes and the Hammett, Taft, etc. substituent constants or electro-negativities, etc.[435,739,740,994]

Table 100
THE CHEMICAL SHIFTS (PPM) OF A FEW SIMPLE PHOSPHOROUS COMPOUNDS IN PHOSPHORUS RESONANCE[434,603,1379]

Compound (P^{III})	δP	Compound (P^{III})	δP	Compound (P^V)	δP	Compound (P^V)	δP
P$_4$	-450	P(SMe)$_3$	126	OPClF$_2$	-15	OP(Et)$_3$	48
PH$_3$	-241	P(OPh)$_3$	128	OP(OMe)$_3$	-2	SP(OPh)$_3$	53
H$_2$PMe	-164	P(SPh)$_3$	131	OPCl$_2$F	0	SPMe$_3$	59
HPMe$_2$	-98	P(OMe)$_3$	141	OPCl$_3$	5	OP(SEt)$_3$	68
PMe$_3$	-62	PI$_3$	178	OP(Ph)$_3$	27	SP(OMe)$_3$	73
P(Ph)$_3$	-8	PCl$_3$	215	SPCl$_3$	31	SP(SPh)$_3$	92
PF$_3$	97	PBr$_3$	222	SP(Ph)$_3$	43	SP(SEt)$_3$	93

Peri interactions involving large coupling constants between nearby fluorine nuclei in condensed and polyaromatic compounds and their heteroaromatic analogues are important in structure determinations, too. In this respect, Compound **362** is quoted (compare also Compounds **34** already discussed) where the value of the coupling constants ^5J(F,F) is ~170 Hz.[1144] The coupling constant $J_{1,2}$ of perfluoronaphthalenes and isoquinolines are likewise 60 to 70 Hz.[264,1277]

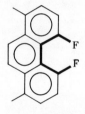

362

4.5. PHOSPHORUS RESONANCE[321,480,920,954,1041]

The 100% natural abundance of the ^{31}P isotope, its convenient relaxation times, and the fact that phosphorus has no quadrupole moment ($I = 1/2$) result in sharp resonance signals of high ζ, consequently relatively small concentrations suffice for obtaining good quality spectra, even on CW spectrometers. Phosphorus resonance is equally important in inorganic and biochemistry.

The phosphorus resonance frequency in a polarizing field of 1.4 T is 24.29 MHz (compare Table 1). The chemical shift values extend over a region of about 400 ppm.[434,1379] Table 100 surveys the chemical shifts and coupling constants of some representatives of various types of phosphorus compounds.[1102]

A common reference substance is 85% phosphorus acid (H_3PO_4) which can be used as external standard only and has a rather broad signal. For this reason shifts referred to PBr$_3$,P(OPh)$_3$ and P$_4$O$_6$[266] are also used. Compounds containing P^{III} show larger chemical shifts and in a much broader range (460 to 250 ppm) than P^V derivatives (-50 to 100 ppm). This may be explained by greater variations of the valence bond directions and angles (hybridization states) of P^{III}.[434]

Substitution increases the chemical shift. Phosphorus is most shielded in P$_4$ and PH$_3$ and most deshielded in PBr$_3$. The anomalous increase of shifts in the order F, I, Cl, Br in PX$_3$-type molecules can be explained taking into account the relative contributions of π-bonding character, valence angle, and the inductive and mesomer effects of the substituent, simultaneously.[606]

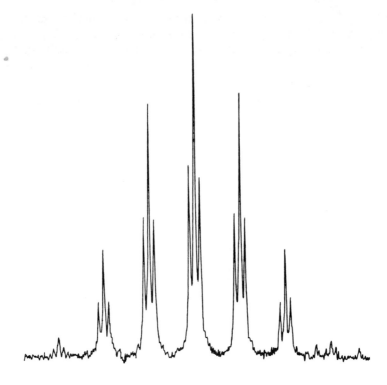

FIGURE 198. The phosphorus resonance spectrum of ethyl-di-(trifluoromethyl)-phosphinite, $(CF_3)_2POEt$, at 24.3 MHz.[1102]

The pairs PR_3 and XPR_3 (X = O or S) are exceptions from the rule $\delta P^{III} > \delta P^V$; pairs $P(OR)_3$ and $XP(OR)_3$, however, behave normally. Within the P^V series the rule $\delta OP < \delta SP$ and in both series $\delta XP(OR)_3 < \delta XP(SR)_3$ apply.

Due to the large chemical shift differences in ^{31}P NMR the solvent effects which never exceed 5 to 10 ppm can be safely neglected.[1379] There are many attempts in the literature to set up additivity rules (e.g., References 587 and 1440), but in practice these fail even for a rough assessment of the chemical shifts.[1257]

The coupling constants between phosphorus and other magnetic nuclei are generally very large and $^1J(P,F)$ values may exceed even 1400 Hz (e.g., for PF_3 J = 1440 Hz). Coupling constants increase with increasing $-I$ effect of the substituents. $^nJ(P,H)$ couplings are characteristically different for P^{III} and P^V compounds:[480] for n = 1, values of 180 to 225 (P^{III}) and 400 to 1100 Hz (P^V) and for n = 2, values of 0 to 12 (P^{III}) and 7 to 10 Hz (P^V) were measured, whereas P–F interactions are insensitive to the oxidation number of phosphorus.[920]

In cyclic compounds anomalous coupling constants can be observed, due to through-space interactions. For Compound **363**, $^4J(P,H)$ is 7 Hz.[1447] Unsaturated compounds have larger coupling constants; for P^{III}–CH=CH$_2$-type groups, ^2J (P,H), ^3J (P,H)^Z and ^3J (P,H)^E values are 15 to 14 and 40 to 70 Hz; for the corresponding ^3J and ^4J (*cisoid* and *transoid*) couplings in OP^V–O–CH=CH$_2$-type compounds, values are ~7, ~1, and ~3 Hz.[480]

$$
\begin{array}{ccc}
 & O\!\!-\!\!-CH_2 & \\
/ & & \backslash \\
P\!-\!O\!-\!CH_2\!-\!CH & \\
\backslash & & / \\
 & O\!-\!CH_2 &
\end{array}
$$

Further coupling constants of ^{31}P nuclei are given in the literature:[480,883] $^1J(P,C)$ couplings are generally negative and in the range -25 to 0 Hz for P^{III}, whereas they are in the range of 45 to 100 Hz for P^V compounds.[480] Figure 198 shows the phosphorus resonance spectrum of $(CF_3)_2POEt$.[1102] The coupling of phosphorus and the six equivalent fluorine nuclei gives rise to a symmetrical septet, each line of which is further split into triplets due to $^3J(P,H)$ coupling.

ABBREVIATIONS IN NMR THEORY AND METHODOLOGY

ADC	analog to digital converter
AF	audio frequency
ASIS	aromatic solvent induced shift
BBDR	broad band double resonance
CAT	computer of average transients
CIDEP	chemically induced dynamic electron polarization
CIDNP	chemically induced dynamic nuclear polarization
CW	continuous wave
DAC	digital to analog converter
DD	dipole-dipole, dipolar, relaxation mechanism or interaction
DEFT	driven-equilibrium FT
2DFTS	two-dimensional FT spectroscopy
DNMR	dynamic NMR
DNP	dynamic nuclear polarization
DRDS	double resonance difference spectroscopy
ESR	electron spin resonance
e.s.u.	electrostatic unit
FFT	fast Fourier transform
FID	free induction decay
FONAR	field focusing nuclear magnetic resonance
FT	Fourier transformation
HSP	homogeneity-spoiling pulse
INDOR	internuclear double resonance
IR	infrared
LCAO	linear combination of atomic orbitals (in quantum theory)
MO	— molecular orbital (in quantum theory)
	— master oscillator
NMR	nuclear magnetic resonance
NOE	nuclear Overhauser effect
NQR	nuclear quadrupole resonance
PFT	pulse Fourier transformation
ppm	part per million
RF	radio frequency
SA	chemical shift anisotropy (CSA), relaxation mechanism
SC	scalar, relaxation mechanism or interaction
SEFT	spin-echo FT
SPI	selective population inversion
SR	— spin rotation, relaxation mechanism or interaction
	— shift reagent
VCO	voltage controlled oscillator
WEFT	water-eliminated FT

CHEMICAL ABBREVIATIONS — COMPOUNDS AND FUNCTIONAL GROUPS

Ac	acetyl, CH_3CO-
ACAC	acetylacetone, $CH_3COCH_2COCH_3$
Ar	aryl (aromatic ring containing groups)
BHC	3-*t*-butylhydroxymethylene-*d*-camphor $\equiv$ TBC

*n*Bu	*normal* butyl, $CH_3CH_2CH_2CH_2-$
*i*Bu	*iso*-butyl, $(CH_3)_2CHCH_2-$
*s*Bu	*sec.*-butyl, $CH_3CH_2CH(CH_3)-$
*t*Bu	*tert.*-butyl, $(CH_3)_3C-$
Bz	benzoyl, C_6H_5-CO-
Bzl	benzyl, $C_6H_5-CH_2-$
DMF, DMF-d$_7$	dimethyl-formamide, heptadeuterio-DMF
DMSO, DMSO-d$_6$	dimethyl-sulfoxide, hexadeuterio-DMSO
DPM	dipivaloyl methane, $(CH_3)_3C-CO-CH_2-CO-C(CH_3)_3$
DSS	2,2-dimethyl-2-silapentane-5-sulfonic acid sodium, $(CH_3)_3Si(CH_2)_3SO_3Na$
Et	ethyl, CH_3CH_2-
FACAM	3-trifluoroacetyl-*d*-camphor
FOD	1,1,1,2,2,3,3-heptafluoro-7,7-dimethyl-octane-4,6-dione
HFBC	3-heptafluorobutyryl-*d*-camphor ≡ HFC
HFC	see HFBC
HMPA (HMPT)	hexamethylphosphoramide
LIS	lanthanide (shift reagent) induced shift
LS	shift reagent — substrate complex
LSR	lanthanide shift reagent
Me	methyl, CH_3-
Ms	mesyl, CH_3SO_2-
Ph	phenyl, C_6H_5-
*n*Pr	*normal* propyl, $CH_3CH_2CH_2-$
*i*Pr	*iso*-propyl, $(CH_3)_2CH-$
Py, αPy, βPy, γPy	pyridyl, α-, β- and γ-pyridyl, C_5H_5N-
PVC	poly(vinyl chloride), $[CH_2Cl-CHCl_2]_n$
TBC	see BHC
TFA, TFA-d	trifluoro-acetic acid (CF_3COOH), deuterio-TFA (CF_3COOD)
TFC	see FACAM
TFM	trichloro fluoromethane, $CFCl_3$
THF	tetrahydrofurane, C_4H_8O
TMHD	2,2,6,6-tetramethylheptane-3,5-dione ≡ DPM
TMS	tetramethylsilane, $(CH_3)_4Si$
Ts	tosyl, $pCH_3-C_6H_4-SO_2-$

LIST OF NOTATIONS

Basic Physical Quantities

c	$(2.997925 \pm 0.000003) \cdot 10^8$ [m s^{-1}], velocity of light
e	$4.8024 \cdot 10^{-10}$ [e.s.u., CGS] $= 1.60210 \cdot 10^{-19}$ [A s], [C], charge of electron
h	$6.6256 \cdot 10^{-34}$ [J s], Planck's constant
$\hbar$	$h/2\pi = 1.0545 \cdot 10^{-34}$ [J s]
k	$1.38054 \cdot 10^{-23}$ [J K^{-1}], Boltzmann's constant
m_e	$(9.1091 \pm 0.0004) \cdot 10^{-31}$ [kg], static electron mass
m_H	$(1.67252 \pm 0.00008) \cdot 10^{-27}$ [kg], static mass of proton
N	$6.02252 \cdot 10^{23}$ [M^{-1}], Avogadro's number
R	8.3143 [J K^{-1} M^{-1}], Rydberg's constant (universal molar gas constant)
β	$9.2732 \cdot 10^{-24}$ [J T^{-1}], Bohr magneton
β_N	$5.0505 \cdot 10^{-27}$ [J T^{-1}], nuclear magneton
μ_H	$2.67519 \cdot 10^8$ [s^{-1} T^{-1}], magnetogyric (gyromagnetic) ratio of proton
H	$1.41049 \cdot 10^{-26}$ [J T^{-1}], magnetic moment of proton

Symbols and Their Meanings

a	antisymmetric spin state, basic function transition, etc.
a, a'	*axial, quasi axial*
A	— signal intensity
	— hyperfine electron-nucleus interaction constant
$\mathbf{B}$	magnetic induction field (magnetic flux density), T [m kg s^{-2} A^{-2}], [m^{-2} s V]
$\mathbf{B}_0$	static external (polarizing) field of the spectrometer
$\mathbf{B}_l$	local magnetic field
$\mathbf{B}_\ell$	induced magnetic field
$\mathbf{B}_{1,2...}$	first, second, etc. exciting (RF) magnetic fields associated with ν_1, ν_2, etc.
C	constant, linear combination constant
$C_\parallel, C_\perp$	spin rotational coupling constants
D	diffusion coefficient
e, e'	*equatorial, quasi equatorial*
E	— energy, N [m^2 kg s^{-2}]
	— electronegativity
$\mathbf{E}$	electric field
E_z	z component of the electric field
$\mathscr{F}$	Fourier transformation operator
f	mole fraction
g	nuclear or electronic Lande- (g-) factor
$\mathscr{H}$	Hamilton operator (in energy or frequency unit)
$\mathscr{H}^0, \mathscr{H}'$	the spin-external field and spin-spin interaction terms in the Hamiltonian
$\mathbf{H}_{ij}$	element of matrix representation of $\mathscr{H}$
H	Hamilton function
$\mathscr{L}$	spin vector operator
$\mathscr{L}_w$	components of the spin vector operator $\mathscr{L}$

$\mathscr{J}^+$	raising ("absorption") spin operator
$\mathscr{J}^-$	lowering ("emission") spin operator
$\mathscr{L}^2$	square of the spin vector operator
I_w	eigenvalue of $\mathscr{J}_w$
I^2	eigenvalue of $\mathscr{L}^2$
$\mathbf{I}_{ij}$	elements of matrix representation of $\mathscr{L}^2$
i	$\sqrt{-1}$
I	— current intensity [A]
	— moment of inertia
	— nuclear spin quantum number
$\pm I$	inductive effect of atoms and functional groups
J	nuclear spin-spin coupling constant [Hz]
$^nJ(X,Y)$ or $^nJ_{XY}$	coupling constant between X and Y nuclei through n bonds [Hz]. Subscripts are given only as algebraic symbols for any interacting pairs of nuclei. Brackets are used for indicating the coupled nuclei, e.g., $J(^{19}F, {}^1H)$ or $J(F, H)$ or the coupling path, e.g., $J(NCCH) \equiv {}^3J(N, H)$.
J^*	reduced coupling constant (in off-resonance)
J	— atomic quantum number representing the total angular moment
	— rotational quantum number
$^nK(X,Y)$	reduced nuclear spin-spin coupling constant [$m^{-3} A^{-2} N$]
K	— combination transition
	— constant, equilibrium constant of conformational motions and other exchange processes
$\mathscr{L}$	angular moment vector operator of electron orbitals
ℓ	— distance
ℓ	— line of force
$\mathbf{L}$	torque, N [m^2 kg s^{-2}]
m	— mass of particles (electrons, protons, etc.) [kg]
	— eigenvalue of I_z of a nucleus
m_T	total magnetic quantum number for a spin system: eigenvalues of I_{zT}
M	mole
$\pm M$	mesomeric effect of atoms and functional groups
$\mathbf{M}$	macroscopic magnetization of a spin system, T
$\mathbf{M}_0$	macroscopic magnetization in field $\mathbf{B}_0$
M_e	equilibrium macroscopic magnetization
M_w	components of M
n	population difference of magnetic quantum states
n_e	equilibrium population difference
N	— Newton
	— nucleus
N	number of magnetic nuclei
$\mathbb{O}$	operator, representing any physical parameters
$\mathscr{P}$	permutation operator
p	— p-orbital
	— mole fraction fractional population of rotamers, etc.
$\mathbf{P}$	angular impulse moment [kg m^2 s^{-1}]
P	nuclear polarization
q	electric field gradient

Q	— nuclear quadrupole moment
	— various quantities and line distances in spectra of different spin systems, consisting of chemical shifts and coupling constants of the system
	— contributions of σ- and π-electrons to shielding component σ^P in quantum theoretical calculations of the shielding
R_w	element of matrix representation of transition moment
r	— polar coordinate
	— distance of atoms and groups, atomic radius, bond length, radius of ring currents, etc. [m], [Å]
R	rate of relaxations
R_1, R_2, R^{DD}, R^{SC}, R^{SA}, R^{SR}, R^Q, see at T_1, etc.	
s	s-orbital
s	symmetric spin state, basic function, transition, etc.
S	electron density around a nucleus
S	singlet electron state
t	time [s]
t_p	pulse time
t_{aq}	acquisition time
t_d	delay time
t_{pi}	period of repetitive pulses
t_{si}	period of repetitive pulse sequences
T	— forbidden transition
	— triplet electron state
	— temperature [°C, °K]
T_c	coalescence temperature
T_s	spin temperature
T_1, T_{1N}	spin-lattice relaxation times (for nucleus N), [s]
T_2	spin-spin relaxation times [s]
$T_{1\rho}$, $T_{2\rho}$	spin-lattice and spin-spin relaxation times in the rotating frame
T_2^*	measurable spin-spin dephasing time shortened by field inhomogeneities
V	volume [m³]
w	the common notation of Cartesian coordinates
W	— transition probability
	— signal width [Hz] for unresolved multiplet
Z	saturation factor
Z_0	saturation factor corresponding to maximal value of signal shape function
α, β	basic eigenfunction of spin corresponding $m = + 1/2$ and $m = - 1/2$
γ, γ_N	magnetogyric ratio of N nucleus [$T^{-1}\ s^{-1}$]
$\not{\gamma}$	$\gamma/2\pi$
δN	chemical shift of nucleus N ppm
δ_{ij}	Krönecker-delta
Δ	difference
$\Delta\Delta$	difference of differences
ΔB_0, ΔB_1, etc.	inhomogeneity in fields B_0, B_1, etc.

Δ_c, Δ_d, Δ_p — Fermi-contact, diamagnetic and pseudocontact shifts in SR-technique

ΔE — differences in electronegativity
— average electronic excitation energy

$\Delta G^{\ddagger}$ — free enthalpy of activation [J M^{-1}]

Δ_i, Δ_{Ti} — SR's induced shift in the complex LS

$\Delta\nu$, $\Delta\nu XY$, $\Delta\delta$, $\Delta\delta XY$ — chemical shift difference for X and Y nuclei, $\Delta\nu XY = \nu X - \nu Y$ [Hz], $\Delta\delta XY = \delta X - \delta Y$ [ppm]

$\Delta\nu_{1/2}$ — half band width (band width at the half height of a signal)

$\Delta\nu_i$ — spectral width

$\Delta\sigma$ — anisotropy in σ, $\Delta\sigma = \sigma_{\parallel} - \sigma_{\perp}$

$\Delta\chi$ — anisotropy in χ, $\Delta\chi = \chi_{\parallel} - \chi_{\perp}$

ϵ — sign-factor in CIDNP referring to reaction path

ζ — signal-to-noise ratio ("NMR sensitivity", receptivity)

η — viscosity, P [m^{-1} kg s^{-1}]
— NOE-factor
— assymmetry factor in $e^2qQ/\hbar$

θ, Θ — dihedral or valence bond angle
— polar coordinate
— pulse angle

κ — rate constant of exchange processes

λ — wavelength [m]
— eigenvalues of various operators

μ — magnetic dipole moment, T

μ — magnetic dipole moment in a given direction (e.g., μ_{B_0} or μ_w, etc.)
— sign-factor in CIDNP referring to electronic state of radical pair precursor
— rationalized permeability

μ_0 — permeability of vacuum

μ' — nonrationalized permeability

ν — frequency, Hz [s^{-1}]

ν_0 — measuring frequency (operating frequency of an NMR spectrometer)

ν_1, ν_2, — first, second, etc. excitation frequencies

ν_N — Larmor (precession) frequency of a nucleus N

ν_c — carrier frequency

π — π-electrons

ξ — steric correction substituent factors [ppm]

ρ — electron density [C], [A s]
— substituent constant [ppm]
— L/S ratio in SR measurements

σ — shielding constant
— LS/L ratio in SR measurements
— σ-electrons
— sign-factor in CIDNP, referring to interacting nuclei in a common radical, or in different radicals

σ_N — shielding constant of a nucleus N

$\boldsymbol{\sigma}$ — shielding constant tensor

σ_{vw} — shielding constant tensor components

$\sigma_{\parallel}$, $\sigma_{\perp}$ — shielding constant components parallel or perpendicular to

	the molecular axis
σ^d, σ^p	diamagnetic and paramagnetic contributions to the shielding constant
τ	— time [s]
	— lifetime of individual species in exchange processes, and generally of a spin state
	— time between pulses in pulse sequences
	— chemical shift (thau scale) [ppm]
τ_c	correlation time of molecular tumblings
τ_J	correlation time of molecular rotation at a given ν_{rot} frequency
τ_{SC}	correlation time of a spin state in scalar relaxation
τ_e	correlation time in nuclear-electron relaxation
φ	— product eigen function
	— polar coordinate
	— dihedral or valence bond angle
χ	magnetic susceptibility
ψ	basic function of spin systems, eigenfunction of electronic quantum states
ω	angular velocity [s^{-1}]
	ω_N, ω_0, ω_1, ω_2, etc. see at ν_N,..., etc.

REFERENCES

1. **Abragam, A.,** *The Principles of Nuclear Magnetism,* Oxford University Press, Oxford, 1961; (a) chap. 2, 8; (b) chap. 9.
2. **Abraham, R. J.,** *J. Chem. Soc. B,* 1022, 1969.
3. **Abraham, R. J. and Bernstein, H. J.,** *Can. J. Chem.,* 39, 905, 1961.
4. **Abraham, R. J., Bullock, E., and Mitra, S. S.,** *Can. J. Chem.,* 37, 1859, 1959.
5. **Abraham, R. J., Jackson, A. J., and Kenner, G. W.,** *J. Chem. Soc.,* 3468, 1961.
6. **Abraham, R. J., Lapper, R. D., Smith, K. M., and Unsworth, J. F.,** *J. Chem. Soc. Perkin Trans. 2,* 1004, 1974.
7. **Abraham, R. J., MacDonald, D. B., and Pepper, E. S.,** *J. Am. Chem. Soc.,* 90, 147, 1968.
8. **Abraham, R. J. and Pachler, K. G. R.,** *Mol. Phys.,* 7, 164, 1964.
9. **Abraham, R. J. and Siverns, T. M.,** *Org. Magn. Reson.,* 5, 253, 1973.
10. **Abraham, R. J. and Thomas, W. A.,** *J. Chem. Soc.,* 3739, 1964.
11. **Abraham, R. J. and Thomas, W. A.,** *J. Chem. Soc.,* 335, 1965.
12. **Abriktsen, P.,** *Acta Chem. Scand.,* 27, 3889, 1973.
13. **Abruscato, G. J., Ellis, P. D., and Tidwell, T. T.,** *Chem. Commun.,* 988, 1972.
14. **Adler, G. and Lichter, R. L.,** *J. Org. Chem.,* 39, 3547, 1974.
15. **Ahmad, N., Bhacca, N. S., Selbin, J., and Wander, J. D.,** *J. Am. Chem. Soc.,* 93, 2564, 1971.
16. **Aime, S. and Milone, L.,** Dynamic ^{13}C NMR spectroscopy of metal carbonyls, in *Progress in Nuclear Magnetic Resonance Spectroscopy,* Vol. 11, Emsley, J. W., Feeney, J., and Sutcliffe, L. H., Eds., Pergamon Press, Oxford, 1977, 183.
17. **Aito, Y., Matsuo, T., and Aso, C.,** *Bull. Chem. Soc. Jpn.,* 40, 130, 1967 .
18. **Albert, A. and Sergeant, E. P.,** *Ionization Constants of Acids and Bases,* Methuen, London, 1962.
19. **Alder, F. and Yu, F. C.,** *Phys. Rev.,* 81, 1967, 1951.
20. **Alei, M., Florin, A. E., Litchman, W. M., and O'Brien, J. F.,** *J. Phys. Chem.,* 75, 932, 1971.
21. **Alger, T. D., Grant, D. M., and Harris, R. K.,** *J. Phys. Chem.,* 76, 281, 1972.
22. **Alger, T. D., Grant, D. M., and Lyerla, J. R., Jr.,** *J. Phys. Chem.,* 75, 2539, 1971.
23. **Alger, T. D., Grant, D. M., and Paul, E. G.,** *J. Am. Chem. Soc.,* 88, 5397, 1966.
24. **Al-Iraqi, M. A., Al-Rawi, J. M. A., and Khuthier, A. H.,** *Org. Magn. Reson.,* 14, 161, 1980.
25. **Allerhand, A. and Cochran, D. W.,** *J. Am. Chem. Soc.,* 92, 4482, 1970.
26. **Allerhand, A. and Doddrell, D.,** *J. Am. Chem. Soc.,* 93, 2777, 1971.
27. **Allerhand, A., Doddrell, D., and Komoroski, R.,** *J. Chem. Phys.,* 55, 189, 1971.
28. **Allerhand, A. and Komoroski, R. A.,** *J. Am. Chem. Soc.,* 95, 8228, 1973.
29. **Allerhand, A. and Oldfield, E.,** *Biochemistry,* 12, 3428, 1973.
30. **Allingham, Y., Cookson, R. C., and Crabb, T. A.,** *Tetrahedron,* 24, 1989, 1968.
31. **Allred, A. L. and Rochow, E. G.,** *J. Am. Chem. Soc.,* 79, 5361, 1957.
32. **Aminova, R. M. and Samitov, Yu. Yu.,** *Zh. Strukt. Khim.,* 15, 607, 1974.
33. **Ammon, R. V. and Fischer, R. D.,** *Angew. Chem.,* 84, 737, 1972.
34. **Ammon, R. V. and Fischer, R. D.,** *Angew. Chem. Int., Ed. Engl.,* 11, 675, 1972.
35. **Anderson, C. B. and Sepp, D. T.,** *Chem. Ind. (London),* 2054, 1964.
36. **Anderson, W. A., Freeman, R., and Hill, H. D. W.,** *Pure Appl. Chem.,* 32, 27, 1972.
37. **Andersson, L. O., Mason, J., and van Bronswijk, W.,** *J. Chem. Soc. A,* 296, 1970.
38. **Andrew, E. R., Bradbury, A., and Eades, R. G.,** *Arch. Sci.,* 11, 223, 1958; *Nature,* 182, 1659, 1958.
39. **Anet, F. A. L.,** *J. Am. Chem. Soc.,* 84, 747, 1962.
40. **Anet, F. A. L.,** *J. Am. Chem. Soc.,* 84, 1053, 1962.
41. **Anet, F. A. L.,** *J. Am. Chem. Soc.,* 86, 458, 1964.
42. **Anet, F. A. L., Ahmad, M., and Hall, L. D.,** *Proc. Chem. Soc.,* 145, 1964.
43. **Anet, F. A. L. and Bock, L. A.,** *J. Am. Chem. Soc.,* 90, 7130, 1968.
44. **Anet, F. A. L. and Bourn, A. J. R.,** *J. Am. Chem. Soc.,* 87, 5250, 1965.
45. **Anet, F. A. L., Bradley, C. H., and Buchanan, G. W.,** *J. Am. Chem. Soc.,* 93, 258, 1971.
46. **Anet, F. A. L., Trepka, R. D., and Cram, D. J.,** *J. Am. Chem. Soc.,* 89, 357, 1967.
47. **Anet, F. A. L. and Yavari, I.,** *J. Org. Chem.,* 41, 3589, 1976.
48. **Anet, R.,** *Can. J. Chem.,* 40, 1249, 1962.
49. **Angyal, S. J.,** *Angew. Chem.,* 81, 172, 1969.
50. **Angyal, S. J. and Pickles, V. A.,** *Aust. J. Chem.,* 25, 1695, 1972.
51. *Annual Review of NMR Spectroscopy,* Mooney, E. F., Ed., Academic Press, London, Volumes 1-7, 1968.
52. *Applied Spectroscopy Reviews,* Brame, E. G., Jr., Ed., Marcel Dekker, New York, Volumes 1-7, 1968.
53. **Apsimon, J. W., Beierbeck, H., and Saunders, J. K.,** *Can. J. Chem.,* 53, 338, 1975.
54. **Apsimon, J. W., Craig, W. G., Demarco, P. V., Mathieson, D. W., Saunders, L., and Whalley, W. B.,** *Tetrahedron,* 23, 2339, 1967.

55. **Armarego, W. L. F.,** *J. Chem. Soc. B*, 191, 1966.
56. **Armitage, I. M., Hall, L. D., Marshall, A. G., and Werbelow, L. G.,** Determination of molecular configuration from lanthanide-induced proton NMR chemical shifts, in *Nuclear Magnetic Resonance Shift Reagents,* Sievers, R. F., Ed., Academic Press, New York, 1973, 313.
57. **Arnold J. T.,** *Phys. Rev.,* 102, 136, 1956.
58. **Arnold, J. T., Dharmatti, S. S., and Packard, M. E.,** *J. Chem. Phys.,* 19, 507, 1951.
59. **Arnold, J. T. and Packard, M. E.,** *J. Chem. Phys.,* 19, 1608, 1951.
60. **Asakawa, J., Kasai, R., Yamasaki, K., and Tanaka, O.,** *Tetrahedron,* 33, 1935, 1977.
61. **Aue, W. P., Bartholdi, E., and Ernst, R. R.,** *J. Chem. Phys.,* 64, 2229, 1976.
62. **Aue, W. P., Karhan, J., and Ernst, R. R.,** *J. Chem. Phys.,* 64, 4226, 1976.
63. **Axenrod, T., Mangiaracina, P., and Pregosin, P. S.,** *Helv. Chim. Acta,* 59, 1655, 1976.
64. **Axenrod, T., Pregosin, P. S., and Milne, G. W. A.,** *Chem. Commun.,* 702, 1968.
65. **Axenrod, T., Pregosin, P. S., and Milne, G. W. A.,** *Tetrahedron Lett.,* 5293, 1968.
66. **Axenrod, T., Pregosin, P. S., Wieder, M. J., Becker, E. D., Bradley, R. B., and Milne, G. W. A.,** *J. Am. Chem. Soc.,* 93, 6536, 1971.
67. **Axenrod, T., Pregosin, P. S., Wieder, M. J., and Milne, G. W. A.,** *J. Am. Chem. Soc.,* 91, 3681, 1969.
68. **Axenrod, T. and Wieder, M. J.,** *J. Am. Chem. Soc.,* 93, 3541, 1971.
69. **Axenrod, T. and Wieder, M. J.,** *Org. Magn. Reson.,* 8, 350, 1976.
70. **Axenrod, T., Wieder, M. J., and Milne, G. W. A.,** *Tetrahedron Lett.,* 401, 1969.
71. **Bachmann, P., Aue, W. P., Müller, L., and Ernst, R. R.,** *J. Magn. Reson.,* 28, 29, 1977.
72. **Bacon, M. R. and Maciel, G. E.,** *J. Am. Chem. Soc.,* 95, 2413, 1973.
73. **Baldeschwieler, J. D. and Randall, E. W.,** *Chem. Rev.,* 63, 81, 1963.
74. **Bak, B., Dambmann, C., Nicolaisen, F., Pedersen, E. J., and Bhacca, N. S.,** *J. Mol. Spectrosc.,* 26, 78, 1968.
75. **Baker, E. B.,** *J. Chem. Phys.,* 37, 911, 1962.
76. **Baker, E. B. and Popov, A.I.,** *J. Phys. Chem.,* 76, 2403, 1972.
77. **Banks, R. E., Barlow, M. G., Davies, W. D., Haszeldine, R. N., Mullen, K., and Taylor, D. R.,** *Tetrahedron Lett.,* 3909, 1968.
78. **Banks, R. E., Barlow, M. G., and Mullen, K.,** *J. Chem. Soc. C,* 1331, 1969.
79. **Banks, R. E., Braithwaite, A., Haszeldine, R. N., and Taylor, D. R.,** *J. Chem. Soc. C,* 2593, 1968.
80. **Banwell, C. N. and Sheppard, N.,** *Mol. Phys.,* 3, 351, 1960.
81. **Barbarella, G. and Dembech, P.,** *Org. Magn. Reson.,* 13, 282, 1980.
82. **Barbarella, G., Dembech, P., Garbesi, A., and Fava, A.,** *Org. Magn. Reson.,* 8, 108, 469, 1976.
83. **Barfield, M.,** *J. Chem. Phys.,* 44, 1836, 1966.
84. **Barfield, M. and Gearhart, H. L.,** *Mol. Phys.,* 27, 899, 1974.
85. **Barfield, M. and Grant, D. M.,** *J. Am. Chem. Soc.,* 85, 1899, 1963.
86. **Bargon, J. and Fischer, H.,** *Z. Naturforsch.,* 22a, 1556, 1967.
87. **Barlow, M. G.,** *Chem. Commun.,* 703, 1966.
88. **Bartholdi, E. and Ernst, R. R.,** *J. Magn. Reson.,* 11, 9, 1973.
89. **Bartuska, V. J. and Maciel, G. E.,** *J. Magn. Reson.,* 7, 36, 1972.
90. **Batchelor, J. G.,** *J. Am. Chem. Soc.,* 97, 3410, 1975.
91. **Batchelor, J. G., Prestegard, J. H., Cushley, R. J., and Lipsky, S. R.,** *J. Am. Chem. Soc.,* 95, 6358, 1973.
92. **Battiste, D. R. and Traynham, J. G.,** *J. Org. Chem.,* 40, 1239, 1975.
93. **Bauld, N. L. and Rim, Y. S.,** *J. Org. Chem.,* 33, 1303, 1968.
94. **Beauté, C., Wolkowski, Z. W., and Thoai, N.,** *Chem. Commun.,* 700, 1971.
95. **Becconsall, J. K. and Hampson, P.,** *J. Mol. Phys.,* 10, 21, 1965.
96. **Becconsall, J. K. and Jones, R. A. Y.,** *Tetrahedron Lett.,* 1103, 1962.
97. **Becconsall, J. K., Jones, R. A. Y., and McKenna, J.,** *J. Chem. Soc.,* 1726, 1965.
98. **Beck, W., Becker, W., Nöth, H., and Wrackmeyer, B.,** *Chem. Ber.,* 105, 2883, 1972.
99. **Becker, E. D.,** *High Resolution NMR Theory and Chemical Applications,* Academic Press, New York, London, 1969.
99a. **Becker, E. D.,** *High Resolution NMR,* Appendix A, Academic Press, New York, 1969.
100. **Becker, E. D.,** Nuclei other than hydrogen, in *A Review of Some Nuclear Properties and a Discussion of Their Relaxation Mechanisms,* Axenrod, T. and Webb, G. A., Eds., Wiley-Interscience, New York, 1974, 1.
101. **Becker, E. D. and Bradley, R. B.,** *J. Chem. Phys.,* 31, 1413, 1959.
102. **Becker, E. D., Bradley, R. B., and Axenrod, T.,** *J. Magn. Reson.,* 4, 136, 1971.
103. **Becker, E. D., Bradley, R. B., and Watson, C. J.,** *J. Am. Chem. Soc.,* 83, 3743, 1961.
104. **Becker, E. D. and Farrar, T. C.,** *Science,* 178, 316, 1972.
105. **Becker, E. D., Ferretti, J. A., and Farrar, T. C.,** *J. Am. Chem. Soc.,* 91, 7784, 1969.

106. **Becker, E. D., Liddel, U., and Shoolery, J. N.,** *J. Mol. Spectrosc.,* 2, 1, 1958.
107. **Becker, G., Lutther, W., and Schrumpf, G.,** *Angew. Chem. Int. Ed. Engl.,* 12, 339, 1973.
108. **Beckett, A. H., Taylor, J. F., Casy, A. F., and Hassan, M. M. A.,** *J. Pharm. Pharmacol.,* 20, 754, 1968.
109. **Beierbeck, H., Saunders, J. K., and ApSimon, J. W.,** *Can. J. Chem.,* 55, 2813, 1977.
110. **Bell, C. L. and Danyluk, S. S.,** *J. Am. Chem. Soc.,* 88, 2344, 1966.
111. **Bell, R. A., Chan, C. L., and Sayer, B. G.,** *Chem. Commun.,* 67, 1972.
112. **Bell, R. A. and Saunders, J. K.,** *Can. J. Chem.,* 48, 1114, 1970.
113. **Bentrude, W. G. and Tan, H. W.,** *J. Am. Chem. Soc.,* 98, 1850, 1976.
114. **Benz, F. W., Feeney, J., and Roberts, G. C. K.,** *J. Magn. Reson.,* 8, 114, 1972.
115. **Berger, S. and Rieker, A.,** *Tetrahedron,* 28, 3123, 1972.
116. **Berkeley, P. J., Jr. and Hanna, M. W.,** *J. Phys. Chem.,* 67, 846, 1963.
117. **Berkeley, P. J. and Hanna, M. W.,** *J. Am. Chem. Soc.,* 86, 2990, 1964.
118. **Berlin, K. D. and Rengaraju, S.,** *J. Org. Chem.,* 36, 2912, 1971.
119. **Bernáth, G., Göndös, Gy., Kovács, K., and Sohár, P.,** *Tetrahedron,* 29, 981, 1973.
120. **Bernáth, G., Sohár, P., Láng, K. L., Tornyai, I., and Kovács, Ö. K. J.,** *Acta Chim. Acad. Sci. Hung.,* 64, 81, 1970.
121. **Bernet, W. A.,** *J. Chem. Educ.,* 44, 17, 1967.
122. **Bernheim, R. A. and Lavery, B. J.,** *J. Am. Chem. Soc.,* 89, 1279, 1967.
123. **Bernstein, H. J., Pople, J. A., and Schneider, W. G.,** *Can. J. Chem.,* 35, 65, 1957.
124. **Bernstein, H. J., Schneider, W. G., and Pople, J. A.,** *Proc. R. Soc. London Ser. A,* 236, 515, 1956.
125. **Bernstein, H. J. and Sheppard, N.,** *J. Chem. Phys.,* 37, 3012, 1962.
126. **Bertrand, R. D., Grant, D. M., Allred, E. L., Hinshaw, J. C., and Strong, A. B.,** *J. Am. Chem. Soc.,* 94, 997, 1972.
127. **Besserre, D. and Coffi-Nketsia, S.,** *Org. Magn. Reson.,* 13, 235, 1980.
128. **Beugelmans, R., Shapiro, R. H., Durham, L. J., Williams, D. H., Budzikiewicz, H., and Djerassi, C.,** *J. Am. Chem. Soc.,* 86, 2832, 1964.
129. **Bhacca, N. S., Giannini, D. D., Jankowski, W. S., and Wolff, M. E.,** *J. Am. Chem. Soc.,* 95, 8421, 1973.
130. **Bhacca, N. S., Horton, D., and Paulsen, H.,** *J. Org. Chem.,* 33, 2484, 1968.
131. **Bhacca, N. S. and Williams, D. H.,** *Applications of NMR Spectroscopy in Organic Chemistry. Illustrations from the Steroid Field,* Holden-Day, San Francisco, 1964.
132. **Bidló-Iglóy, M., Méhesfalvi-Vajna, Zs., and Nagy, J.,** *Acta Chim. Acad. Sci. Hung.,* 79, 1, 1973.
133. **Biellmann, J. F. and Callot, H.,** *Bull. Soc. Chim. (France),* 397, 1967.
134. **Binsch, G.,** The study of intramolecular rate processes by dynamic nuclear magnetic resonance, in *Topics in Stereochemistry,* Vol. 3, Eliel, E. L. and Allinger, N. L., Eds., Wiley-Interscience, New York, 1968, 97.
135. **Binsch, G., Lambert, J. B., Roberts, B. W., and Roberts, J. D.,** *J. Am. Chem. Soc.,* 86, 5564, 1964.
136. **Binsch, G. and Roberts, J. D.,** *J. Am. Chem. Soc.,* 87, 5157, 1965.
137. **Binst, G. and Tourwe, D.,** *Heterocycles,* 1, 257, 1973.
138. **Birchall, T. and Jolly, W. L.,** *J. Am. Chem. Soc.,* 87, 3007, 1965.
139. **Bjorgo, J., Boyd, D. R., Watson, C. G., and Jennings, W. B.,** *J. Chem. Soc. Perkin Trans. 2,* 757, 1974.
140. **Black, P. J. and Heffernan, M. L.,** *Aust. J. Chem.,* 16, 1051, 1963.
141. **Black, P. J. and Heffernan, M. L.,** *Aust. J. Chem.,* 18, 353, 1965.
142. **Blackman, R. B. and Tukey, J. W.,** *The Measurement of Power Spectra,* Dover, New York, 1958.
143. **Bleaney, B.,** *J. Magn. Reson.,* 8, 91, 1972.
144. **Bleaney, B., Dobson, C. M., Levine, B. A., Martin, R. B., Williams, R. J. P., and Xavier, A. V.,** *Chem. Commun.,* 791, 1972.
145. **Blizzard, A. C. and Santry, D. P.,** *Chem. Commun.,* 1085, 1970.
146. **Blizzard, A. C. and Santry, D. P.,** *J. Chem. Phys.,* 55, 950, 1971.
147. **Blizzard, A. C. and Santry, D. P.,** *J. Chem. Phys.,* 58, 4714, 1973.
148. **Bloch, F.,** *Phys. Rev.,* 70, 460, 1946.
149. **Bloch, F.,** *Phys. Rev.,* 94, 496, 1954.
150. **Bloch, F., Hansen, W. W., and Packard, M.,** *Phys. Rev.,* 69, 127, 1946.
151. **Bloch, F., Hansen, W. W., and Packard, M.,** *Phys. Rev.,* 70, 474, 1946.
152. **Bloch, F. and Siegert, A.,** *Phys. Rev.,* 57, 522, 1940.
153. **Block, R. E. and Maxwell, G. P.,** *J. Magn. Reson.,* 14, 329, 1974.
154. **Bloembergen, N., Purcell, E. M., and Pound, R. V.,** *Phys. Rev.,* 73, 679, 1948.
155. **Bloom, A. L. and Shoolery, J. N.,** *Phys. Rev.,* 151, 102, 1965.
156. **Blunt, J. W. and Munro, M. H. G.,** *Org. Magn. Reson.,* 13, 26, 1980.
157. **Blunt, J. W. and Stothers, J. B.,** *Org. Magn. Reson.,* 9, 439, 1977.

158. **Boča, R., Pelikan, P., Valko, L., and Miertus, S.,** *J. Chem. Phys.,* 11, 229, 1975.

159. **Bock, K. and Pedersen, C.,** *J. Chem. Soc. Perkin Trans. 2,* 293, 1974.

160. **Bock, K. and Wiebe, L.,** *Acta Chem. Scand.,* 27, 2676, 1973.

161. **Bodenhausen, G., Freeman, R., Morris, G., and Turner, D. L.,** *J. Magn. Reson.,* 31, 75, 1978.

162. **Bodenhausen, G., Freeman, R., Niedermeyer, R., and Turner, D. L.,** *J. Magn. Reson.,* 24, 291, 1976.

163. **Bodenhausen, G., Freeman, R., Niedermeyer, R., and Turner, D. L.,** *J. Magn. Reson.,* 26, 133, 1977.

164. **Bodenhausen, G., Freeman, R., and Turner, D. L.,** *J. Chem. Phys.,* 65, 839, 1976.

165. **Boekelheide, V. and Phillips, J. B.,** *J. Am. Chem. Soc.,* 89, 1695, 1967.

166. **Bohlmann, F. and Zeisberg, R.,** *Chem. Ber.,* 108, 1043, 1975.

167. **Bohman, O. and Allenmark, S.,** *Acta Chem. Scand.,* 22, 2716, 1968.

168. **Booth, H.,** Applications of ¹H NMR spectroscopy to the conformational analysis of cyclic compounds, in *Progress in Nuclear Magnetic Resonance Spectroscopy,* Vol. 5, Emsley, J. W., Feeney, J., and Sutcliffe, L. H., Eds., Pergamon Press, London, 1969, 149.

169. **Booth, H. and Little, J. H.,** *Tetrahedron,* 23, 291, 1967.

170. **Booth, G. E. and Ouelette, R. J.,** *J. Org. Chem.,* 31, 544, 1966.

171. **Bordás, B., Sohár, P., Matolcsy, Gy., and Berencsi, P.,** *J. Org. Chem.,* 37, 1727, 1972.

172. **Borgen, G.,** *Acta Chem. Scand.,* 26, 1740, 1972.

173. **Bothner-By, A. A. and Moser, E.,** *J. Am. Chem. Soc.,* 90, 2347, 1968.

174. **Bothner-By, A. A. and Naar-Colin, C.,** *J. Am. Chem. Soc.,* 84, 743, 1962.

175. **Bottini, A. T. and O'Rell, M. K.,** *Tetrahedron Lett.,* 423, 429, 1967.

176. **Bottini, A. T. and Roberts, J. D.,** *J. Am. Chem. Soc.,* 80, 5203, 1958.

177. **Bourn, A. J . R. and Randall, E. W.,** *Mol. Phys.,* 8, 567, 1964.

178. **Bovey, F. A.,** *NMR Data Tables for Organic Compounds,* Interscience, New York, 1957.

179. **Bovey, F. A.,** *Nuclear Magnetic Resonance Spectroscopy,* Academic Press, New York, 1969.

180. **Bovey, F. A.,** *High Resolution Nucler Magnetic Resonance of Macromolecules,* Academic Press, New York, 1972.

181. **Bovey, F. A., Hood, F. P., III, Anderson, E. W., and Kornegay, R. L.,** *J. Chem. Phys.,* 41, 2041, 1964.

182. **Bovey, F. A. and Tiers, G. V. D.,** *J. Am. Chem. Soc.,* 81, 2870, 1959.

183. **Bowie, J. H., Cameron, D. W., Schütz, P. E., Williams, D. H., and Bhacca, N. S.,** *Tetrahedron,* 22, 1771, 1966.

184. **Bracewell, R.,** *The Fourier Transform and Its Physical Applications,* McGraw-Hill, New York, 1955.

185. **Bradbury, E. M. and Crane-Robinson, C.,** *Nature (London),* 220, 1079, 1968.

186. **Bradley, C. H., Hawkes, G. E., Randall, E. W., and Roberts, J. D.,** *J. Am. Chem. Soc.,* 97, 1958, 1975.

187. **Bramwell, M. R. and Randall, E. W.,** *Chem. Commun.,* 250, 1969.

188. **Braun, S. and Frey, G.,** *Org. Magn. Reson.,* 7, 194, 1975.

189. **Brederode, H. and Huysmans, G. B.,** *Tetrahedron Lett.,* 1695, 1971.

190. **Breitmaier, E., Jung, G., and Voelter, W.,** *Angew. Chem.,* 83, 659, 1971.

191. **Breitmaier, E., Jung, G., and Voelter, W.,** *Angew. Chem. Int. Ed. Engl.,* 10, 73, 1971.

192. **Breitmaier, E., Jung, G., and Voelter, W.,** *Chimia,* 26, 136, 1972.

193. **Breitmaier, E., Jung. G., Voelter, W., and Pohl, L.,** *Tetrahedron,* 29, 2485, 1973.

194. **Breitmaier, E. and Spohn, K. H.,** *Tetrahedron,* 29, 1145, 1973.

195. **Breitmaier, E. and Voelter, W.,** *Tetrahedron,* 29, 227, 1973.

196. **Breitmaier, E. and Voelter, W.,** ¹³*C NMR Spectroscopy,* Verlag Chemie, Weinheim, 1974, (a) pp. 195-202, (b) p. 98, (c) pp. 268-274, (d) p. 83.

197. **Breitmaier, E., Voelter, W., Jung, G., and Tänzer, C.,** *Chem. Ber.,* 104, 1147, 1971.

198. **Breslow, R. and Ryan, G.,** *J. Am. Chem. Soc.,* 89, 3073, 1967.

199. **Brewer, J. P. N., Heaney, H., and Marples, B. A.,** *Chem. Commun.,* 27, 1967.

200. **Brey, W. S. and Ramey, K. C.,** *J. Chem. Phys.,* 39, 844, 1963.

201. **Briggs, J. M., Farnell, L. F., and Randall, E. W.,** *Chem. Commun.,* 680, 1971.

202. **Briggs, J., Frost, G. H., Hart, F. A., Moss, G. P., and Staniforth, M. L.,** *Chem. Commun.,* 749, 1970.

203. **Briggs, J., Hart, F. A., and Moss, G. P.,** *Chem. Commun.,* 1506, 1970.

204. **Briggs, J., Hart, F. A., Moss, G. P., and Randall, E. W.,** *Chem. Commun.,* 364, 1971.

205. **Briggs, J. M., Moss, G. P., Randall, E. W., and Sales, K. D.,** *Chem. Commun.,* 1180, 1972.

206. **Briggs, J. M., Rahkamaa, E., and Randall, E. W.,** *J. Magn. Reson.,* 12, 40, 1973.

207. **Briguet, A., Duplan, J. C., and Delmau, J.,** *Mol. Phys.,* 29, 837, 1975.

208. **Brimacombe, J. S. and Tucker, L. C. N.,** *Carbohydrate Res.,* 5, 36, 1967.

209. **Brois, S. J.,** *J. Org. Chem.,* 27, 3532, 1962.

210. **Brois, S. J.,** *J. Am. Chem. Soc.,* 90, 506, 508, 1968.

211. **Brois, S. J. and Beardsley, G. P.,** *Tetrahedron Lett.,* 5113, 1966.

212. **Brookhart, M., Levy, G. C., and Winstein, S.,** *J. Am. Chem. Soc.,* 89, 1735, 1967.

213. **Brouant, P., Limouzin, Y., and Maire, J. C.,** *Helv. Chim. Acta,* 56, 2057, 1973.

214. **Brown, R. J. C., Gutowsky, H. S., and Shimomura, K.,** *J. Chem. Phys.,* 38, 76, 1963.

215. **Browne, D. T., Kenyon, G. L., Packer, E. L., Sternlicht, H., and Wilson, D. M.,** *J. Am. Chem. Soc.,* 95, 1316, 1973; *Biochem. Biophys. Res. Commun.,* 50, 42, 1973.

216. **Brownstein, S., Dunogues, J., Lindsay, D., and Ingold, K. U.,** *J. Am. Chem. Soc.,* 99, 2073, 1977.

217. **Bruce, J. M. and Knowles, P.,** *Proc. Chem. Soc.,* 294, 1964.

218. (a) Bruker: *High Power Pulsed NMR Applications,* Bruker Analytic GMBH, Karlsruhe, Rheinstetten, West-Germany; (b) Bruker WM Series: *High Field Multinuclear NMR,* Bruker Analytic GMBH, Karlsruhe, Rheinstetten, West-Germany.

219. Bruker: *High Resolution NMR in Solids by Magic Angle Spinning,* Bruker Analytic GMBH, Karlsruhe, Rheinstetten, West Germany.

220. **Bucci, P.,** *J. Am. Chem. Soc.,* 90, 252, 1968.

221. **Buchanan, G. W. and Dawson, B. A.,** *Can. J. Chem.,* 54, 790, 1976.

222. **Buchanan, G. W. and Dawson, B. A.,** *Org. Magn. Reson.,* 13, 293, 1980.

223. **Buchanan, G. W. and Stothers, J. B.,** *Can. J. Chem.,* 47, 3605, 1969.

224. **Buckingham, A. D.,** *Can. J. Chem.,* 38, 300, 1960.

225. **Buckingham, A. D. and McLauchan, K. A.,** *Proc. Chem. Soc.,* 144, 1963.

226. **Buckingham, A. D., Schaefer, T., and Schneider, W. G.,** *J. Chem. Phys.,* 32, 1227, 1960.

227. **Budhram, R. S. and Uff, B. C.,** *Org. Magn. Reson.,* 13, 89, 1980.

228. **Bulman, M. J.,** *Tetrahedron,* 25, 1433, 1969.

229. **Bulusu, S., Autera, J. R., and Axenrod, T.,** *Chem. Commun.,* 602, 1973.

230. **Burke, J. J. and Lauterbur, P. C.,** *J. Am. Chem. Soc.,* 86, 1870, 1964.

231. **Burton, R., Hall, L. D., and Steiner, P. R.,** *Can. J. Chem.,* 49, 588, 1971.

232. **Bystrov, V. F.,** Spin-spin coupling and the conformational states of peptide systems, in *Progress in Nuclear Magnetic Resonance Spectroscopy,* Vol. 10, Emsley, J. W., Feeney, J., and Sutcliffe, L. H., Eds., Pergamon Press, Oxford, 1976, 41.

233. **Bystrov, V. F., Gavrilov, Y. D., and Solkan, V. N.,** *J. Magn. Reson.,* 19, 123, 1975.

234. **Bystrov, V. F., Ivanov, V. T., Portnova, S. L., Balashova, T. A., and Ovchinnikov, Yu. A.,** *Tetrahedron,* 29, 873, 1973.

235. **Bystrov, V. F., Portnova, S. L., Csetlin, V. I., Ivanov, V. T., and Ovcsinnikov, V. A.,** *Tetrahedron,* 25, 493, 1969.

236. **Bystrov, V. F. and Stepanyants, A. U.,** *J. Mol. Spectrosc.,* 21, 241, 1966.

237. **Caddy, B., Martin-Smith, M., Norris, R. K., Reid, S. T., and Sternhell, S.,** *Aust. J. Chem.,* 21, 1853, 1968.

238. **Calder, I. C. and Sondheimer, F.,** *Chem. Commun.,* 904, 1966.

239. **Cameron, D. W., Kingston, D. G. J., Sheppard, N., and Todd, J.,** *J. Chem. Soc.,* 98, 1964.

240. **Campaigne, E., Chamberlain, N. F., and Edwards, B. E.,** *J. Org. Chem.,* 27, 135, 1962.

241. **Campbell, I. D., Dobson, C. M., and Williams, R. J. P.,** *Proc. R. Soc. London Ser. B,* 189, 485, 1975.

242. **Campbell, I. D. and Freeman, R.,** *J. Magn. Reson.,* 11, 143, 1973.

243. **Canet, C., Goulon-Ginet, C., and Marchal, J. P.,** *J. Magn. Reson.,* 22, 539, 1976 (with an erratum in 25, 397, 1977).

244. **Cantacuzène, J., Jantzen, R., Tordeux, M., and Chachaty, C.,** *Org. Magn. Reson.,* 7, 407, 1975.

245. **Caputo, J. F. and Martin, A. R.,** *Tetrahedron Lett.,* 4547, 1971.

246. **Carman, C. J., Tarpley, A. R., and Goldstein, J. H.,** *J. Am. Chem. Soc.,* 93, 2864, 1971.

247. **Carr, H. Y. and Purcell, F. M.,** *Phys. Rev.,* 94, 630, 1954.

248. **Carrington, A. and McLachlan, A. D.,** *Introduction to Magnetic Resonance,* Harper & Row, New York, 1967; (a) p. 187; (b) chap. 11.

249. **Cartledge, F. K. and Riedel, K. H.,** *J. Organomet. Chem.,* 34, 11, 1972.

250. **Caspi, E., Wittstruck, T. A., and Piatak, D. M.,** *J. Org. Chem.,* 27, 3183, 1962.

251. **Castellano, S., Sun, C., and Kostelnik, R. J.,** *J. Chem. Phys.,* 46, 327, 1967.

252. **Casu, B., Reggiani, M., Gallo, G. G., and Vigevani, A.,** *Tetrahedron,* 22, 3061, 1966.

253. **Casu, B., Reggiani, M., Gallo, G. G., and Vigevani, A.,** *Tetrahedron,* 24, 803, 1968.

254. **Casy, A. F.,** *PMR Spectroscopy in Medicinal and Biological Chemistry,* Academic Press, New York, 1971.

255. **Carey, P. R., Kroto, H. W., and Turpin, M. A.,** *Chem. Commun.,* 188, 1969.

256. **Caughey, W. S. and Iber, P. K.,** *J. Org. Chem.,* 28, 269, 1963.

257. **Cavalli, L.,** Fluorine-19 NMR spectroscopy, in *Annual Reports on NMR Spectroscopy,* Vol. 6B, Mooney, E. F., Ed., Academic Press, New York, 1976, 43.

258. **Cavanaugh, J. R.,** *J. Chem. Phys.,* 39, 2378, 1963.

259. **Cavanaugh, J. R. and Dailey, B. P.,** *J. Chem. Phys.,* 34, 1099, 1961.

260. **Celotti, J. C., Reisse, J., and Chiurdoglu, G.,** *Tetrahedron,* 22, 2249, 1966.

261. **Chadwick, D. J. and Williams, D. H.,** *J. Chem. Soc. Perkin Trans. 2,* 1202, 1974.

262. **Chalmers, A. A., Pachler, K. G. R., and Wessels, P. L.,** *J. Magn. Reson.,* 15, 419, 1974.
263. **Chambers, R. D., Hole, M., Musgrave, W. K. R., Storey, R. A., and Iddon, B.,** *J. Chem. Soc. C,* 2331, 1966.
264. **Chambers, R. D. and Palmer, A. J.,** *Tetrahedron Lett.,* 2799, 1968.
265. **Champeney, D. C.,** *Fourier Transforms and Their Physical Applications,* Academic Press, New York, 1973.
266. **Chapman, A. C., Homer, J., Mowthorpe, D. J., and Jones, R. T.,** *Chem. Commun.,* 121, 1965.
267. **Chapman, D. and Magnus, P. D.,** *Introduction to Practical High Resolution Nuclear Magnetic Resonance Spectroscopy,* Academic Press, New York, 1966.
268. **Chapman, O. L. and King, R. W.,** *J. Am. Chem. Soc.,* 86, 1256, 1964.
269. **Cheney, B. V. and Grant, D. M.,** *J. Am. Chem. Soc.,* 89, 5319, 1967.
270. **Cherry, P. C., Cottrell, W. R. T., Meakins, G. D., and Richards, E. E.,** *J. Chem. Soc. C,* 459, 1968.
271. **Chew, K. F., Derbyshire, W., Logan, N., Norbury, A. H., and Sinha, A. I. P.,** *Chem. Commun.,* 1708, 1970.
272. **Chew, K. F., Healy, M. A., Khalil, M. J., Logan, N., and Derbyshire, W.,** *J. Chem. Soc. (Dalton),* 1315, 1975.
273. **Chiang, Y. and Whipple, E. B.,** *J. Am. Chem. Soc.,* 85, 2763, 1963.
274. **Chow, Y. L.,** *Angew. Chem. Int. Ed. Educ.,* 6, 75, 1968.
275. **Christ, H. A.,** *Helv. Phys. Acta,* 33, 572, 1960.
276. **Christ, H. A. and Diehl, P.,** *Helv. Phys. Acta,* 36, 170, 1963.
277. **Christ, H. A., Diehl, P., Schneider, H., and Dahn, H.,** *Helv. Chim. Acta,* 44, 865, 1961.
278. **Christl, M., Reich, H. J., and Roberts, J. D.,** *J. Am. Chem. Soc.,* 93, 3463, 1971.
279. **Christl, M., Warren, J. P., Hawkins, B. L., and Roberts, J. D.,** *J. Am. Chem. Soc.,* 95, 4392, 1973.
280. **Claret, P. A. and Osborne, A. G.,** *Org. Magn. Reson.,* 8, 147, 1976.
281. **Clerc, T. and Pretsch, E.,** *Kernresonanzspektroskopie,* Akademische Verlagsgesellschaft, Frankfurt am Main, 1970.
282. **Closs, G. L.,** *J. Am. Chem. Soc.,* 91, 4552, 1969.
283. **Closs, G. L.,** *23rd International Congress of Pure and Applied Chemistry,* Vol. 4, Boston, Butterworths, London, 1971, 19.
284. **Closs, G. L.,** Chemically induced dynamic nuclear polarization, in *Advances in Magnetic Resonance,* Vol. 7, Waugh, J. S., Ed., Academic Press, New York, 1974, 157.
285. **Closs, G. L. and Closs, L. E.,** *J. Am. Chem. Soc.,* 85, 2022, 1963.
286. **Closs, G. L. and Closs, L. E.,** *J. Am. Chem. Soc.,* 91, 4549, 1969.
287. **Closs, G. L. and Moss, R. A.,** *J. Am. Chem. Soc.,* 86, 4042, 1964.
288. **Closs, G. L. and Trifunac, A. D.,** *J. Am. Chem. Soc.,* 91, 4454, 1969.
289. **Cohen, A. D., Sheppard, N., and Turner, J. J.,** *Proc. Chem. Soc.,* 118, 1958.
290. **Coles, B. A.,** *J. Natl. Cancer Inst.,* 57, 389, 1976.
291. **Colli, H. N., Gold, V., and Pearson, J. E.,** *Chem. Commun.,* 408, 1973.
292. **Connes, J.,** *Rev. Opt.,* 40, 45, 116, 171, 231, 1961.
293. **Connor, T. M. and Reid, C.,** *J. Mol. Spectrosc.,* 7, 32, 1961.
294. **Conti, F., Segre, A., Pini, P., and Porri, L.,** *Polymer,* 15, 5, 1974.
295. **Cooley, J. W. and Tukey, J. W.,** *Math. Comput.,* 19, 297, 1965.
296. **Cooper, R. A., Lichter, R. L., and Roberts, J. D.,** *J. Am. Chem. Soc.,* 95, 3724, 1973.
297. **Cope, F. W.,** *Biophys. J.,* 9, 303, 1969.
298. **Cope, F. W.,** *Physiol. Chem. Phys.,* 10, 535, 541, 547, 1978.
299. **Cope, F. W. and Damadian, R.,** *Physiol. Chem. Phys.,* 11, 143, 1979.
300. **Corio, P. L.,** *Structure of High Resolution NMR Spectra,* Academic Press, New York, 1966.
301. **Corio, P. L.,** *Chem. Rev.,* 60, 363, 1960.
302. **Corio, P. L. and Dailey, B. P.,** *J. Am. Chem. Soc.,* 78, 3043, 1956.
303. **Cornwell, C. D.,** *J. Chem. Phys.,* 44, 874, 1966.
304. **Cottam, G. L., Vasek, A., and Lusted, D.,** *Res. Commun. Chem. Pathol. Pharmacol.,* 4, 495, 1972.
305. **Cowley, A. H. and Schweiger, J. R.,** *J. Am. Chem. Soc.,* 95, 4179, 1973.
306. **Cox, P. F.,** *J. Am. Chem. Soc.,* 85, 380, 1963.
307. **Cox, R. H. and Bothner-By, A. A.,** *J. Phys. Chem.,* 72, 1642, 1968.
308. **Coxon, B.,** *Tetrahedron,* 22, 2281, 1966.
309. **Coxon, B.,** *Ann. N.Y. Acad. Sci.,* 222, 953, 1973.
310. **Crabb, T. A.,** Nuclear magnetic resonance of alkaloids, in *Annual Reports on NMR Spectroscopy,* Vol. 6A, Mooney, E. F., Ed., Academic Press, New York, 1975, 250.
311. **Crabb, T. A.,** Nuclear magnetic resonance of alkaloids, in *Annual Reports on NMR Spectroscopy,* Vol. 8, Webb, G. A., Academic Press, New York, 1978, 2.
312. **Cramer, R. E. and Dubois, R.,** *Chem. Commun.,* 936, 1973.
313. **Cramer, R. E., Dubois, R., and Seff, K.,** *J. Am. Chem. Soc.,* 96, 4125, 1974.

314. **Cramer, R. E. and Seff, K.,** *Chem. Commun.,* 400, 1972.
315. **Creagh, L. T. and Truitt, P.,** *J. Org. Chem.,* 33, 2956, 1968.
316. **Crecely, K. M., Crecely, R. W., and Goldstein, J. H.,** *J. Phys. Chem.,* 74, 2680, 1970.
317. **Crecely, K. M., Crecely, R. W., and Goldstein, J. H.,** *J. Mol. Spectrosc.,* 37, 252, 1971.
318. **Crecely, K. M., Watts, V. S., and Goldstein, J. H.,** *J. Mol. Spectrosc.,* 30, 184, 1969.
319. **Cremer, S. and Srinivasan, R.,** *Tetrahedron Lett.,* 24, 1960.
320. **Crump, D. R., Sanders, J. K. M., and Williams, D. H.,** *Tetrahedron Lett.,* 4949, 1970.
321. **Crutchfield, M. M., Dungan, C. H., Letcher, J. H., Mark, V., and Van Wazer, J. R.,** *Topics in Phosphorus Chemistry,* Vol. 5, Grayson, M. and Griffith, E. J., Eds., Interscience, New York, 1967.
322. **Cudby, M. E. A. and Willis, H. A.,** The nuclear magnetic resonance spectra of polymers, in *Annual Reports on NMR Spectroscopy,* Vol. 4, Mooney, E. F., Ed., Academic Press, New York, 1971, 363.
323. **Cushley, R. J., Codington, J. F., and Fox, J. J.,** *Carbohydr. Res.,* 5, 31, 1967.
324. **Cushley, R. J., Naugler, D., and Ortig, C.,** *Can. J. Chem.,* 53, 3419, 1975.
325. **Cushley, R. J., Watanabe, K. A., and Fox, J. J.,** *J. Am. Chem. Soc.,* 89, 394, 1967.
326. **Dahlquist, K. I. and Forsén, S.,** *J. Phys. Chem.,* 69, 4062, 1965.
327. **Dalling, D. K. and Grant, D. M.,** *J. Am. Chem. Soc.,* 89, 6612, 1967.
328. **Dalling, D. K. and Grant, D. M.,** *J. Am. Chem. Soc.,* 94, 5318, 1972.
329. **Damadian, R.,** *Science,* 171, 1151, 1971.
330. **Damadian, R.,***Biophys. J.,* 11, 739, 773, 1971.
331. **Damadian, R.,** U.S. Patent, 3,789,832, Field 17.03, 1972.
332. **Damadian, R. and Cope, F. W.,** *Physiol. Chem. Phys.,* 6, 309, 1974.
333. **Damadian, R., Goldsmith, M., and Minkoff, L.,** *Physiol. Chem. Phys.,* 9, 97, 1977.
334. **Damadian, R., Minkoff, L., Goldsmith, M., Stanford, M., and Koutcher, J. A.,** *Science,* 194, 1430, 1976.
335. **Damadian, R., Zaner, K. S., Hor, D., Dimaio, T., Minkoff, L., and Goldsmith, M.,** *Ann. N.Y. Acad. Sci.,* 222, 1048, 1973.
336. **Daniel, A. and Pavia, A. A.,** *Tetrahedron Lett.,* 1145, 1967.
337. **Daunis, J., Follet, M., and Marzin, C.,** *Org. Magn. Reson.,* 13, 330, 1980.
338. **Davis, J. B., Jackman, L. M., Siddons, P. T., and Weedon, B. C. L.,** *J. Chem. Soc. C,* 2154, 1966.
339. **Davis, M. and Hassel, O.,** *Acta Chem. Scand.,* 17, 1181, 1963.
340. **Davison, A. and Rakita, P. E.,** *J. Am. Chem. Soc.,* 90, 4479, 1968.
341. **Davoust, D., Massias, M., and Molho, D.,** *Org. Magn. Reson.,* 13, 218, 1980.
342. **Dear, R. E. A. and Gilbert, E. E.,** *J. Org. Chem.,* 33, 819, 1968.
343. **Debye, D.,** *Polar Molecules,* Dover, New York, 1948.
344. **Dehmlow, E. V., Zeisberg, R., and Dehmlow, S. S.,** *Org. Magn. Reson.,* 7, 418, 1975.
345. **De'Kowalewski, D. G. and Kowalewski, V. J.,** *J. Chem. Phys.,* 37, 1009, 1962.
346. **Delbaere, L. T. J., James, M. N. G., and Lemieux, R. U.,** *J. Am. Chem. Soc.,* 95, 7866, 1973.
347. **Della, E. W.,** *Chem. Commun.,* 1558, 1968.
348. **Demarco, P. V., Elzey, T. K., Lewis, R. B., and Wenkert, E.,** *J. Am. Chem. Soc.,* 92, 5734, 1970.
349. **Demarco, P. V., Farkas, E., Doddrell, D., Mylari, B. L., and Wenkert, E.,** *J. Am. Chem. Soc.,* 90, 5480, 1968.
350. **De Mare, G. R. and Martin, S. J.,** *J. Am. Chem. Soc.,* 88, 5033, 1966.
351. **Dennison, D. M.,** *Proc. R. Soc London Ser. A,* 155, 483, 1927.
352. **Deslauriers, R. and Smith, I. C. P.,** Conformation and structure of peptides, in *Topics in Carbon-13 NMR Spectroscopy,* Vol. 2, Levy, G. C., Ed., Wiley-Interscience, New York, 1976.
353. **Deslauriers, R., Walter, R., and Smith, I. C. P.,** *FEBS Lett.,* 37, 27, 1973.
354. **Desreux, J. F., Fox, L. E., and Reilley, C. N.,** *Anal. Chem.,* 44, 2217, 1972.
355. **Deverell, C.,** *Mol. Phys.,* 18, 319, 1970.
356. **Dewey, R. S., Schönewaldt, E. F., Joshua, H., Paleveda, W. J., Schwam, H., Barkemeyer, H., Arison, B. H., Veber, D. F., Denkenwalter, R. G., and Hirschmann, R.,** *J. Am. Chem. Soc.,* 90, 3254, 1968.
357. **Deyrup, J. A. and Greenwald, R. B.,** *J. Am. Chem. Soc.,* 87, 4538, 1965.
358. **Dhami, K. S. and Stothers, J. B.,** *Can. J. Chem.,* 43, 479, 1965.
359. **Dhami, K. S. and Stothers, J. B.,** *Can. J. Chem.,* 43, 498, 1965.
360. **Dhami, K. S. and Stothers, J. B.,** *Can. J. Chem.,* 43, 510, 1965.
361. **Dhami, K. S. and Stothers, J. B.,** *Can. J. Chem.,* 44, 2855, 1966.
362. **Dhami, K. S. and Stothers, J. B.,** *Can. J. Chem.,* 45, 233, 1967.
363. **Dickerman, S. C. and Haase, R. J.,** *J. Am. Chem. Soc.,* 89, 5458, 1967.
364. **Dickinson, W. C.,** *Phys. Rev.,* 77, 736, 1950.
365. **Dickinson, W. C.,** *Phys. Rev.,* 80, 563, 1951.
366. **Diehl, P.,** *Helv. Chim. Acta,* 44, 829, 1961.
367. **Dinesh, Rogers, M. T.,** *J. Magn. Reson.,* 7, 30, 1972.

368. **Ditchfield, R. and Ellis, P. D.,** *Chem. Phys. Lett.,* 17, 342, 1972.
369. **Ditchfield, R. and Ellis, P. D.,** Theory of ^{13}C chemical shifts, in *Topics in Carbon-13 NMR Spectroscopy,* Vol. 1, Levy, G. C., Ed., Wiley-Interscience, New York, 1974, 1.
370. **Doddrell, D. M. and Allerhand, A.,** *J. Am. Chem. Soc.,* 93, 1558, 1971.
371. **Doddrell, D. M., Burfitt, J., Grutzner, J. B., and Barfield, M.,** *J. Am. Chem. Soc.,* 96, 1241, 1974.
372. **Doddrell, D. M., Glushke, V., and Allerhand, A.,** *J. Chem. Phys.,* 56, 3683, 1972.
373. **Doddrell, D. M., Khong, P. W., and Lewis, K. G.,** *Tetrahedron Lett.,* 2381, 1974.
374. **Doddrell, D. M. and Wells, P. R.,** *J. Chem. Soc. Perkin Trans. 2,* 1333, 1973.
375. **Doleschall, G. and Lempert, K.,** *Tetrahedron,* 29, 639, 1973.
376. **Doomes, E. and Cromwell, N. H.,** *J. Org. Chem.,* 34, 310, 1969.
377. **Dorman, D. E., Angyal, S. J., and Roberts, J. D.,** *J. Am. Chem. Soc.,* 92, 1351, 1970.
378. **Dorman, D. E., Bauer, D., and Roberts, J. D.,** *J. Org. Chem.,* 40, 3729, 1975.
379. **Dorman, D. E., Jautelat, M., and Roberts, J. D.,** *J. Org. Chem.,* 36, 2757, 1971.
380. **Dorman, D. E., Jautelat, M., and Roberts, J. D.,** *J. Org. Chem.,* 38, 1026, 1973.
381. **Dorman, D. E. and Roberts, J. D.,** *Proc. Natl. Acad. Sci.,* 65, 19, 1970.
382. **Dorman, D. E. and Roberts, J. D.,** *J. Am. Chem. Soc.,* 92, 1355, 1970.
383. **Dorman, D. E. and Roberts, J. D.,** *J. Am. Chem. Soc.,* 93, 4463, 1971.
384. **Dorman, D. E. and Bovey, F. A.,** *J. Org. Chem.,* 38, 2379, 1973.
385. **Douglas, A. W.,** *J. Chem. Phys.,* 40, 2413, 1964.
386. **Douglas, A. W.,** *J. Chem. Phys.,* 45, 3465, 1966.
387. **Douglas, A. W. and Dietz, D.,** *J. Chem. Phys.,* 46, 1214, 1967.
388. **Downing, A. P., Ollis, W. D., and Sutherland, I. O.,** *Chem. Commun.,* 171, 1967.
389. **Downing, A. P., Ollis, W. D., Sutherland, I. O., and Mason, J.,** *Chem. Commun.,* 329, 1968.
390. **Dörnyei, G., Bárczai-Beke, M., Majoros, B., Sohár, P., and Szántai, Cs.,** *Acta Chim. Acad. Sci. Hung.,* 90, 275, 1976.
391. **Drakenberg, T. and Carter, R. E.,** *Org. Magn. Reson.,* 7, 307, 1975.
392. **Drakenberg, T., Jost, R., and Sommer, J.,** *Chem. Commun.,* 1011, 1974.
393. **Dreeskamp, H.,** *Z. Phys. Chem.,* 59, 321, 1968.
394. **Dreeskamp, H., Hilderbrand, K., and Pfisterer, G.,** *Mol. Phys.,* 17, 429, 1969.
395. **Dreeskamp, H. and Sackman, E.,** *Z. Phys. Chem.,* 34, 273, 1962.
396. **Dreeskamp, H. and Stegmeier, G.,** *Z. Naturforsch.,* 22a, 1458, 1967.
397. **Duch, M. W. and Grant, D. M.,** *Macromolecules,* 3, 165, 1970.
398. **Dudek, E. P. and Dudek, G. O.,** *J. Org. Chem.,* 32, 823, 1967.
399. **Dudek, G. O. and Dudek, E. P.,** *J. Am. Chem. Soc.,* 86, 4283, 1964; 88, 2407, 1966.
400. **Dudek, G. O. and Dudek, E. P.,** *Tetrahedron,* 23, 3245, 1967.
401. **Dudek, G. O. and Holm, R. H.,** *J. Am. Chem. Soc.,* 83, 2099, 1961.
402. **Dungan, C. H. and Van Wazer, J. R.,** *Compilation of Reported F^{19} NMR Chemical Shifts 1951-1967,* Wiley-Interscience, New York, 1970.
403. **Dunham, L., Nichols, S., and Brunschwig, A.,** *Cancer Res.,* 6, 230, 1946.
404. **Durette, P. L. and Horton, D.,** *J. Org. Chem.,* 36, 2658, 1971.
405. **Duus, F., Jakobsen, P., and Lawesson, S. O.,** *Tetrahedron,* 24, 5323, 1968.
406. **Dvornik, D. and Schilling, G.,** *J. Med. Chem.,* 8, 466, 1965.
407. **Dyer, D. S., Cunningham, J. A., Brooks, J. J., Sievers, R. E., and Rondeau, R. F.,** Interactions of nucleophiles with lanthanide shift reagents, in *Nuclear Magnetic Resonance Shift Reagents,* Sievers, R. F., Ed., Academic Press, New York, 1973, 21.
408. **Dyer, J. and Lee, J.,** *Trans. Faraday Soc.,* 62, 257, 1966.
409. *Dynamic Nuclear Magnetic Resonance Spectroscopy,* **Jackman, L. M. and Cotton, F. A.,** Eds., Academic Press, New York, 1975.
410. **Eaton, D. R., Josey, A. D., Phillips, W. D., and Benson, R. E.,** *J. Chem. Phys.,* 39, 3513, 1963.
411. **Ebsworth, E. A. V. and Frankiss, S. G.,** *Trans. Faraday Soc.,* 59, 1518, 1963.
412. **Echols, R. E. and Levy, G. C.,** *J. Org. Chem.,* 39, 1321, 1974.
413. **Edward, J. T.,** *Chem. Ind. (London),* 1102, 1955.
414. **Eggert, H. and Djerassi, C.,** *J. Am. Chem. Soc.,* 95, 3710, 1973.
415. **Ejchart, A.,** *Org. Magn. Reson.,* 9, 351, 1977.
416. **Ejchart, A.,** *Org. Magn. Reson.,* 10, 263, 1977.
417. **Ejchart, A.,** *Org. Magn. Reson.,* 13, 368, 1980.
418. **Elguero, J., Fruchier, A., and Pardo, M. C.,** *Can. J. Chem.,* 54, 1329, 1976.
419. **Elguero, J., Jacquier, R., and Tarrago, G.,** *Tetrahedron Lett.,* 4719, 1965.
420. **Elguero, J., Jacquier, R., and Tarrago, G.,** *Bull. Soc. Chim.,* 2981, 1966.
421. **Elguero, J., Johnson, B. L., Pereillo, J. M., Pouzard, G., Rajzman, M., and Randall, E. W.,** *Org. Magn. Reson.,* 9, 145, 1977.

422. **Eliel, E. L., Allinger, N. L., Angyal, S. J., and Morrison, G. A.,** *Conformational Analysis,* Interscience, New York, 1965.

423. **Eliel, E. L., Bailey, W. F., Kopp, L. D., Willer, R. L., Grant, D. M., Bertrand, R., Christensen, K. A., Dalling, D. K., Duch, M. W., Wenkert, E., Schell, F. M., and Cochran, D. W.,** *J. Am. Chem. Soc.,* 97, 322, 1975.

424. **Eliel, E. L. and Pietrusiewicz, K. M.,** ^{13}C NMR of nonaromatic heterocyclic compounds, in *Topics in Carbon-13 NMR Spectroscopy,* Vol. 3, Levy, G. C., Ed., Wiley-Interscience, New York, 1979, chap. 3.

425. **Eliel, E. L. and Pietrusiewicz, K. M.,** *Org. Magn. Reson.,* 13, 193, 1980.

426. **Eliel, E. L., Rao, V. S., and Pietrusiewicz, K. M.,** *Org. Magn. Reson.,* 12, 461, 1979.

427. **Eliel, E. L., Rao, V. S., and Riddel, F. G.,** *J. Am. Chem. Soc.,* 98, 3583, 1976.

428. **Elleman, D. D. and Manatt, S. L.,** *J. Mol. Spectrosc.,* 9, 477, 1962.

429. **Elleman, D. D., Manatt, S. L., and Pearce, C. D.,** *J. Chem. Phys.,* 42, 650, 1965.

430. **Ellis, G. E. and Jones, R. G.,** *J. Chem. Soc. Perkin Trans. 2,* 437, 1972.

431. **Ellis, G. E., Jones, R. G., and Papadopoulos, M. G.,** *J. Chem. Soc. Perkin Trans. 2,* 1381, 1974.

432. **Ellis, J., Jackson, A. H., Kenner, G. W., and Lee, J.,** *Tetrahedron Lett.,* 23, 1960.

433. **Elvidge, J. A. and Ralph, P. D.,** *J. Chem. Soc. B,* 249, 1966.

434. **Emsley, J. W., Feeney, J., and Sutcliffe, L. H.,** *High Resolution Nuclear Magnetic Resonance Spectroscopy,* Pergamon Press, London, 1965; (a) Vol. 1, p. 31; (b) Vol. 2, pp. 1011-1031.

435. **Emsley, J. W. and Phillips, L.,** Fluorine chemical shifts, in *Progress in Nuclear Magnetic Resonance Spectroscopy,* Vol. 7, Emsley, J. W., Feeney, J., and Sutcliffe, L. H., Eds., Pergamon Press, New York, 1971, 1.

436. **Emsley, J. W., Phillips, L., and Wray, V.,** Fluorine coupling constants, in *Progress in Nuclear Magnetic Resonance Spectroscopy,* Vol. 10, Emsley, J. W., Feeney, J., and Sutcliffe, L. H., Eds., Pergamon Press, New York, 1976, 83.

437. **Engelhardt, G., Radeglia, R., Jancke, H., Lippmaa, E., and Mägi, M.,** *Org. Magn. Reson.,* 5, 561, 1973.

438. **Eremenko, L. T. and Borisenko, A. A.,** *Izv. Akad. Nauk. (SSSR) Ser. Khim.,* 675, 1968.

439. **Ernst, E.,** *Biophysics of Straited Muscle,* Akadémiai Kiadó, Budapest, 1963.

440. **Ernst, E.,** *Acta Biochem. Biophys. Acad. Sci. Hung.,* 10, 95, 1975.

441. **Ernst, E. and Hazlewood, C. F.,** *Inorg. Perspectives Biol. Med.,* 2, 27, 1978.

442. **Ernst, E., Tigyi, J., and Zahorcsek, A.,** *Acta Physiol. Acad. Sci. Hung.,* 1, 5, 1950.

443. **Ernst, L.,** *Chem. Ber.,* 108, 2030, 1975.

444. **Ernst, L.,** *Z. Naturforsch.,* 30b, 788, 1975.

445. **Ernst, L.,** *Z. Naturforsch.,* 30b, 794, 1975.

446. **Ernst, L.,** *J. Magn. Reson.,* 20, 544, 1975.

447. **Ernst, L.,** *J. Magn. Reson.,* 22, 279, 1976.

448. **Ernst, L.,** *Org. Magn. Reson.,* 8, 161, 1976.

449. **Ernst, L. and Mannschreck, A.,** *Tetrahedron Lett.,* 3023, 1971.

450. **Ernst, R. R.,** *J. Chem. Phys.,* 45, 3845, 1966.

451. **Ernst, R. R.,** *Adv. Magn. Reson.,* 2, 1, 1967.

452. **Ernst, R. R.,** *J. Magn. Reson.,* 3, 10, 1970.

453. **Ernst, R. R.,** *Chimia,* 29, 179, 1975.

454. **Ernst, R. R. and Anderson, W. A.,** *Rev. Sci. Instrum.,* 37, 93, 1966.

455. **Espersen, W. G. and Martin, R. B.,** *J. Phys. Chem.,* 80, 741, 1976.

456. **Ettinger, R., Blume, P., Patterson, A., Jr., and Lauterbur, P. C.,** *J. Chem. Phys.,* 33, 1597, 1960.

457. **Evans, D. F.,** *Proc. Chem. Soc.,* 115, 1958.

458. **Evans, D. F., Tucker, J. N., and de Villardi, G. C.,** *Chem. Commun.,* 205, 1975.

459. **Evans, D. F. and Wyatt, M.,** *Chem. Commun.,* 312, 1972.

460. **Evans, H. B., Jr., Tarpley, A. R., and Goldstein, J. H.,** *J. Phys. Chem.,* 72, 2552, 1968.

461. **Ewing, D. F.,** Two-bond coupling between protons and carbon-13, in *Annual Reports on NMR Spectroscopy,* Vol. 6A, Mooney, E. F., Ed., Academic Press, New York, 1975, 389.

462. **Faller, J. W., Adams, M. A., and La Mar, G. N.,** *Tetrahedron Lett.,* 699, 1974.

463. **Farminer, A. R. and Webb, G. A.,** *Org. Magn. Reson.,* 8, 102, 1976.

464. **Farrar, T. C. and Becker, E. D.,** *Pulse and Fourier Transform NMR,* Academic Press, New York, 1971; (a) p. 41; (b) p. 34; (c) p. 54, (d) p. 63; (e) p. 26; (f) p. 61.

465. **Farrar, T. C., Druck, S. J., Shoup, R. R., and Becker, E. D.,** *J. Am. Chem. Soc.,* 94, 699, 1972.

466. **Faure, R., Clinas, J.-R., Vincent, E.-J., and Larice, J.-L.,** *C. R. Acad. Sci. Paris, Ser. C,* 279, 717, 1974.

467. **Faure, R., Galy, J.-P., Vincent, E.-J., and Elguero, J.,** *Can. J. Chem.,* 56, 46, 1978.

468. **Feeney, J. and Partington, P.,** *Chem. Commun.,* 611, 1973.

469. **Feeney, J., Partington, P., and Roberts, G. C. K.,** *J. Magn. Reson.,* 13, 268, 1974.

470. **Feeney, J., Pauwells, P., and Shaw, D.,** *Chem. Commun.,* 554, 1970.
471. **Feeney, J. and Sutcliffe, L. H.,** *J. Chem. Soc.,* 1123, 1962.
472. **Feeney, J. and Sutcliffe, L. H.,** *Spectrochim. Acta,* 24a, 1135, 1968.
473. **Fellgett, P.,** *J. Phys. Radium,* 19, 187, 1958.
474. **Fessenden, R. W. and Waugh, J. S.,** *J. Chem. Phys.,* 31, 996, 1959.
475. **Fields, R.,** Fluorine-19 nuclear magnetic resonance spectroscopy, in *Annual Reports on NMR Spectroscopy,* Vol. 5A, Mooney, E. F., Ed., Academic Press, New York, 1972, 99.
476. **Fields, R.,** Fluorine-19 NMR spectroscopy of fluoroalklyl and fluoroaryl derivatives of transition metals, in *Annual Reports on NMR Spectroscopy,* Vol. 7, Webb, G. A., Ed., Academic Press, New York, 1977, 1.
477. **Figeys, H. P., Geerlings, P., Raeymaekers, P., van Lommen, G., and Defay, N.,** *Tetrahedron,* 31, 1731, 1975.
478. **Filleux-Blanchard, M. L., Durand, H., Bergeon, M. T., Clesse, F., Quiniou, H., and Martin, G. J.,** *J. Mol. Structure,* 3, 351, 1969.
479. **Finch, E. D., Harmon, J. F., and Muller, B. H.,** *Arch. Biochem. Biophys.,* 147, 299, 1971.
480. **Finer, E. G. and Harris, R. K.,** Spin-spin couplings between phosphorus nuclei, in *Progress in Nuclear Magnetic Resonance Spectroscopy,* Vol. 6, Emsley, J. W., Feeney, J., and Sutcliffe, L. H., Pergamon Press, New York, 1971, 61.
481. **Firl, J. R. and Runge, W.,** *Z. Naturforsch.,* 29b, 393, 1974.
482. **Firl, J. R., Runge, W., Hartmann, W., and Utikal, H. P.,** *Chem. Lett.,* 51, 1975.
483. **Fischer, H.,** New Series, Landolt-Börnstein, Group II, in *Magnetic Properties of Free Radicals,* Vol. 1, Hellwege, K. H., Ed., Springer-Verlag, Berlin, 1965.
484. **Fischer, H. and Bargon, J.,** *Acc. Chem. Res.,* 2, 110, 1969.
485. **Fleming, I., Hanson, S. W., and Sanders, J. K. M.,** *Tetrahedron Lett.,* 3733, 1971.
486. **Fleming, I. and Williams, D. H.,** *Tetrahedron,* 23, 2747, 1967.
487. **Fletcher, J. R. and Sutherland, I. O.,** *Chem. Commun.,* 1504, 1969.
488. **Flohé, L., Breitmaier, E., Günzler, W. A., Voelter, W., and Jung, G.,** *Hoppe-Seyler's Z. Physiol. Chem.,* 353, 1159, 1972.
489. **Flygare, W. H.,** *J. Chem. Phys.,* 41, 793, 1964.
490. **Flygare, W. H. and Goodisman, J.,** *J. Chem. Phys.,* 49, 3122, 1968.
491. **Fodor, L., Szabó, J., and Sohár, P.,** *Tetrahedron,* 37, 963, 1981.
492. **Foote, C. S.,** *Tetrahedron Lett.,* 579, 1963.
493. **Foreman, M. I.,** Medium effects. NMR shift reagents, in *Specialist Periodical Reports, Nuclear Magnetic Resonance,* Vol. 1, Harris, R. K., Ed., The Chemical Society, London, 1972, 310.
494. **Foreman, M. I.,** Medium effects. NMR shift reagents, in *Specialist Periodical Reports, Nuclear Magnetic Resonance,* Vol. 2, Harris, R. K., Ed., The Chemical Society, London, 1972, 355.
495. **Foreman, M. I.,** Medium effects. NMR shift reagents, in *Specialist Periodical Reports, Nuclear Magnetic Resonance,* Vol.. 3, Harris, R. K., Ed., The Chemical Society, London 1972, 342.
496. **Foreman, M. I.,** Medium effects. NMR shift reagents, in *Specialist Periodical Reports, Nuclear Magnetic Resonance,* Vol. 4, Harris, R. K., Ed., The Chemical Society, London, 1972, 294.
497. **Foreman, M. I.,** Medium effects on chemical shifts and coupling constants, in *Specialist Periodical Reports, Nuclear Magnetic Resonance,* Vol. 5, Harris, R. K., Ed., The Chemical Society, London, 1973, 292.
498. **Foreman, M. I.,** Solvent effects. NMR shift reagents, in *Specialist Periodical Reports, Nuclear Magnetic Resonance,* Vol. 6, Abraham, R. J., Ed., The Chemical Society, London, 1975, 244.
499. **Formacek, V., Desnayer, L., Kellerhals, H. P., Keller, T., and Clerk, J. T.,** *Bruker Data Bank ^{13}C,* Vol. 1, Bruker Physic Morisch OHG, Karlsruhe, 1976.
500. **Forsén, S.,** *Acta Chem. Scand.,* 13, 1472, 1959.
501. **Forsén, S. and Norin, T.,** *Tetrahedron Lett.,* 2845, 1964.
502. **Foster, A. B., Inch, T. D., Quadir, M. H., and Webber, J. M.,** *Chem. Commun.,* 1086, 1968.
503. **Fourier, J. B. J.,** *Theorie Analytique de la Chaleur,* Paris, 1822.
504. **Fraenkel, G. and Franconi, C.,** *J. Am. Chem. Soc.,* 82, 4478, 1960.
505. **Frankiss, S. G.,** *J. Phys. Chem.,* 67, 752, 1963.
506. **Franklin, N. C. and Feltkamp, H.,** *Tetrahedron,* 22, 2801, 1966.
507. **Frasca, A. R. and Dennler, E. B.,** *Chem. Ind. (London),* 509, 1967.
508. **Fraser, M., McKenzie, S., and Reid, D. H.,** *J. Chem. Soc. B,* 44, 1966.
509. **Fraser, R. R.,** *Can. J. Chem.,* 38, 2226, 1960.
510. **Fraser, R. R., Petit, M. A., and Miskow, M.,** *J. Am. Chem. Soc.,* 94, 3253, 1972.
511. **Freeburger, M. E. and Spialter, L.,** *J. Am. Chem. Soc.,* 93, 1894, 1971.
512. **Freed, I. H. and Pedersen, J. B.,** The theory of chemically induced dynamic spin polarization in *Advances in Magnetic Resonance,* Vol. 8, Waugh, J. S., Ed., Academic Press, New York, 1976, 2.
513. **Freeman, J. P.,** *J. Org. Chem.,* 28, 2508, 1963.

514. **Freeman, R. and Anderson, W. A.**, *J. Chem. Phys.*, 39, 1518, 1963.

515. **Freeman, R. and Hill, H. D. W.**, *J. Chem. Phys.*, 51, 3140, 1969.

516. **Freeman, R. and Hill, H. D. W.**, *J. Chem. Phys.*, 54, 301, 1971.

517. **Freeman, R. and Hill, H. D. W.**, *J. Chem. Phys.*, 54, 3367, 1971.

518. **Freeman, R. and Hill, H. D. W.**, *J. Chem. Phys.*, 55, 1985, 1971.

519. **Freeman, R. and Hill, H. D. W.**, *Molecular Spectroscopy 1971*, Institute of Petroleum, London, 1971.

520. **Freeman, R. and Hill, H. D. W.**, *J. Magn. Reson.*, 4, 366, 1971.

521. **Freeman, R. and Hill, H. D. W.**, *J. Magn.Reson.*, 5, 278, 1971.

522. **Freeman, R., Hill, H. D. W., and Kaptein, R.**, *J. Magn. Reson.*, 7, 82, 1972.

523. **Freeman, R., Hill, H. D. W., and Kaptein, R.**, *J. Magn. Reson.*, 7, 327, 1972.

524. **Freeman, R., Pachler, K. G. R., and La Mar, G. N.**, *J. Chem. Phys.*, 55, 4586, 1971.

525. **Freeman, R. and Whiffen, D. H.**, *Mol. Phys.*, 4, 321, 1961.

526. **Frei, K. and Bernstein, H. J.**, *J. Chem. Phys.*, 38, 1216, 1963.

527. **Frey, H. E., Knispel, R. R., Kruuv, J., Sharp, A. R., Thompson, R. T., and Pintar, M. M.**, *J. Natl. Cancer Inst.*, 49, 903, 1972.

528. **Friedel, R. A. and Retcofsky, H. L.**, *J. Am. Chem. Soc.*, 85, 1300, 1963.

529. **Friedel, R. A. and Retcofsky, H. L.**, *Chem. Ind. (London)*, 455, 1966.

530. **Frith, P. G. and McLauchlan, K. A.**, Chemically induced dynamic nuclear polarization, in *NMR Specialist Periodical Reports*, Vol. 3, Harris, R. K., Ed., The Chemical Society, London, 1974, 378.

531. **Fritz, O. G., Jr. and Swift, T. J.**, *Biophys. J.*, 7, 675, 1967.

532. **Fronza, G., Gamba, A., Mondelli, R., and Pagani, G.**, *J. Magn. Reson.*, 12, 231, 1973.

533. **Fukumi, T., Arata, Y., and Fujiwara, S.**, *J. Mol. Spectrosc.*, 27, 443, 1968.

534. **Fung, B. M.**, *Biochim. Biophys. Acta*, 362, 209, 1974.

535. **Gagnaire, D. and Vincendon, M.**, *Chem. Commun.*, 509, 1977.

536. **Gagnaire, D. and Vottero, P.**, *Bull. Soc. Chim. France*, 2779, 1963.

537. **Gainer, J., Howarth, G. A., and Hoyle, W.**, *Org. Magn. Reson.*, 8, 226, 1976.

538. **Gansow, O. A., Burke, A. R., and La Mar, G. N.**, *Chem. Commun.*, 456, 1972.

539. **Gansow, O. A., Loeffler, P. A., Willcott, M. R., and Lenkinski, E.**, *J. Am. Chem. Soc.*, 95, 3389, 1973.

540. **Gansow, O. A., Willcott, M. R., and Lenkinski, R. E.**, *J. Am. Chem. Soc.*, 93, 4297, 1971.

541. **Ganter, G., Pokras, S. M., and Roberts, J. D.**, *J. Am. Chem. Soc.*, 88, 4235, 1966.

542. **Garbisch, E. W., Jr. and Griffith, M. G.**, *J. Am. Chem. Soc.*, 90, 3590, 1968.

543. **Garbisch, E. W., Jr. and Griffith, M. G.**, *J. Am. Chem. Soc.*, 90, 6543, 1968.

544. **Garnier, R., Faure, R., Babadjamian, A., and Vincent, E.-J.**, *Bull. Soc. Chim. France*, 3, 1040, 1972.

545. **Garrat, P. J.**, *Aromaticity*, McGraw-Hill, London, 1971, 46.

546. **Gash, V. W. and Bauer, D. J.**, *J. Org. Chem.*, 31, 3602, 1966.

547. **Gassmann, P. G. and Heckert, D. C.**, *J. Org. Chem.*, 30, 2859, 1965.

548. **Gault, I., Price, B. J., and Sutherland, I. O.**, *Chem. Commun.*, 540, 1967.

549. **Georgian, V., Kerwin, J. F., Wolff, M. E., and Owings, F. F.**, *J. Am. Chem. Soc.*, 84, 3594, 1962.

550. **Gerlach, H. and Zagalak, B.**, *Chem. Commun.*, 274, 1973.

551. **Gerlach, W. and Stern, O.**, *Ann. Phys. Leipzig*, 74, 673, 1924.

552. **Ghersetti, S., Lunazzi, L., Maccagnani, G., and Mangini, A.**, *Chem. Commun.*, 834, 1969.

553. **Gibbons, W. A., Sogn, J. A., Stern, A., Craig, L. C., and Johnson, L. F.**, *Nature (London)*, 227, 840, 1970.

554. **Gierki, T. D. and Flygare, W. H.**, *J. Am. Chem. Soc.*, 94, 7277, 1972.

555. **Gil, V. M. S. and Geraldes, C. F. G. C.**, *J. Magn. Reson.*, 11, 268, 1973.

556. **Gilles, D. G. and Shaw, D.**, The application of Fourier transformation to high resolution nuclear magnetic resonance spectroscopy, in *Annual Reports on NMR Spectroscopy*, Vol. 5A, Mooney, E. F., Ed., Academic Press, New York, 1972, 557.

557. **Gillespie, R. P. and Birchall, T.**, *Can. J. Chem.*, 41, 148, 1963.

558. **Giovanella, B. C., Morgan, A. C., Stehlin, J. S., and Williams, L. J.**, *Cancer Res.*, 33, 2568, 1973.

559. **Girard, P., Kagan, H., and David, S.**, *Tetrahedron*, 27, 5911, 1971.

560. **Goe, G. L.**, *J. Org. Chem.*, 38, 4285, 1974.

561. **Goering, H. L., Eikenberry, J. N., and Koermer, G. S.**, *J. Am. Chem. Soc.*, 93, 5913, 1971.

562. **Goering, H. L., Eikenberry, J. N., Koermer, G. S., and Lattimer, C. J.**, *J. Am. Chem. Soc.*, 96, 1493, 1974.

563. **Goldman, M.**, *Spin Temperature and Nuclear Magnetic Resonance in Solids*, Oxford University Press, Oxford, 1970.

564. **Goldsmith, M. and Damadian, R.**, *Physiol. Chem. Phys.*, 7, 263, 1975.

565. **Goldsmith, M., Damadian, R., Stanford, M., and Lipkowitz, M.**, *Physiol. Chem. Phys.*, 9, 105, 1977.

566. **Goldsmith, M., Koutcher, J. A., and Damadian, R.**, *Br. J. Cancer*, 36, 235, 1977.

567. **Goldstein, J. H., Watts, V. S., Rattet, L. S.,** ^{13}CH satellite NMR spectra, in *Progress in Nuclear Magnetic Resonance Spectroscopy*, Vol. 8, Emsley, J. W., Feeney, J., and Sutcliffe, L. H., Eds., Pergamon Press, Oxford, 1971, 104.

568. **Goodman, B. A. and Raynor, J. B.,** *Adv. Inorg. Chem. Radiochem.,* 13, 135, 1970.

569. **Gorin, P. A. J.,** *Can. J. Chem.,* 52, 458, 1974.

570. **Gorkom, M. and Hall, G. E.,** *Quart. Rev. Chem. Soc.,* 22, 14, 1968.

571. **Gorodetsky, M., Luz, Z., and Mazur, Y.,** *J. Am. Chem. Soc.,* 89, 1183, 1967.

572. **Gorter, C. J.,** *Physica,* 3, 995, 1936.

573. **Gorter, C. J. and Broer, L. F. J.,** *Physica,* 9, 591, 1942.

574. **Gough, J. L., Guthrie, J. P., and Stothers, J. B.,** *Chem. Commun.,* 979, 1972.

575. **Govill, G.,** *Mol. Phys.,* 21, 953, 1971.

576. **Graham, D. M. and Holloway, C. E.,** *Can. J. Chem.,* 41, 2114, 1963.

577. **Granacher, I.,** *Helv. Phys. Acta,* 34, 272, 1961.

578. **Granger, P. and Maugras, M.,** *Org. Magn. Reson.,* 7, 598, 1975.

579. **Grant, D. M. and Cheney, B. V.,** *J. Am. Chem. Soc.,* 89, 5315, 1967.

580. **Grant, D. M., Hirst, R. C., and Gutowsky, H. S.,** *J. Chem. Phys.,* 38, 470, 1963.

581. **Grant, D. M. and Paul, E. G.,** *J. Am. Chem. Soc.,* 86, 2984, 1964.

582. **Gray, G. A., Cremer, S. E., and Marsi, K. L.,** *J. Am. Chem. Soc.,* 98, 2109, 1976.

583. **Gray, G. A., Ellis, P. D., Traficante, D. D., and Maciel, G. E.,** *J. Magn. Reson.,* 1, 41, 1969.

584. **Gray, G. A., Maciel, G. E., and Ellis, P. D.,** *J. Magn. Reson.,* 1, 407, 1969.

585. **Greene, J. L., Jr. and Shevlin, P. B.,** *Chem. Commun.,* 1092, 1971.

586. **Griffith, D. L. and Roberts, J. D.,** *J. Am. Chem. Soc.,* 87, 4089, 1965.

587. **Grim, S. O. and McFarlane, W.,** *Nature (London),* 208, 995, 1965.

588. **Grinter, R.,** Nuclear spin-spin coupling, in *Specialist Periodical Reports,* Vol. 3, Harris, R. K., Ed., The Chemical Society, London, 1974, 50.

589. **Grinter, R. and Mason, J.,** *J. Chem. Soc. A.,* 2196, 1970.

590. **Grishin, Y. K., Sergeyev, N. M., Subbotin, O. A., and Ustynyuk, Y. A.,** *Mol. Phys.,* 25, 297, 1973.

591. **Grishin, Y. K., Sergeyev, N. M., and Ustynyuk, Y. A.,** *J. Organomet. Chem.,* 22, 361, 1970.

592. **Gronowitz, S., Hörnfeldt, A., Gestblom, B., and Hoffman, R. A.,** *Arkiv. Kemi,* 18, 133, 1962.

593. **Gronowitz, S., Sörlin, G., Gestblom, B., and Hoffman, R. A.,** *Arkiv. Kemi,* 19, 483, 1963.

594. **Grover, S. H., Guthrie, J. P., Stothers, J. B., and Tan, C. T.,** *J. Magn. Reson.,* 10, 227, 1973.

595. **Grover, S. H. and Stothers, J. B.,** *Can. J. Chem.,* 52, 870, 1974.

596. **Grunwald, E., Loewenstein, A., and Meiboom, S.,** *J. Chem. Phys.,* 27, 630, 1957.

597. **Grunwald, E., Loewenstein, A., and Meiboom, S.,** *J. Chem. Phys.,* 27, 641, 1957.

598. **Grutzner, J. B., Jautelat, M., Dence, J. B., Smith, R. A., and Roberts, J. D.,** *J. Am. Chem. Soc.,* 92, 7107, 1970.

599. **Grutzner, J. B. and Santini, R. E.,** *J. Magn. Reson.,* 19, 173, 1975.

600. **Gust, D., Moon, R. B., and Roberts, J. D.,** *Proc. Natl. Acad. Sci. U.S.A.,* 72, 4696, 1975.

601. **Gutowsky, H. S.,** *J. Chem. Phys.,* 37, 2196, 1962.

602. **Gutowsky, H. S., Belford, G. G., and McMahon, P. E.,** *J. Chem. Phys.,* 36, 3353, 1962.

603. **Gutowsky, H. S. and Holm, C. H.,** *J. Chem. Phys.,* 25, 1228, 1956.

604. **Gutowsky, H. S., Holm, C. H., Saika, A., and Williams, G. A.,** *J. Am. Chem. Soc.,* 79, 4596, 1957.

605. **Gutowsky, H. S., Karplus, M., and Grant, D. M.,** *J. Chem. Phys.,* 31, 1278, 1959.

606. **Gutowsky, H. S. and McCall, D. W.,** *J. Chem. Phys.,* 22, 162, 1954.

607. **Gutowsky, H. S., McCall, D. W., and Slichter, C. P.,** *J. Chem. Phys.,* 21, 279, 1953.

608. **Gutowsky, H. S. and Saika, A.,** *J. Chem. Phys.,* 21, 1688, 1953.

609. **Günther, H. and Herrig, W.,** *Chem. Ber.,* 106, 3938, 1973.

610. **Günther, H. and Keller, T.,** *Chem. Ber.,* 103, 3231, 1970.

611. **Günther, H. and Schmickler, H.,** *Angew. Chem. Int. Ed. Engl.,* 12, 243, 1973.

612. **Günther, H. and Ulmen, J.,** *Tetrahedron,* 30, 3781, 1974.

613. **Haake, P. and Miller, W. B.,** *J. Am. Chem. Soc.,* 85, 4044, 1963.

614. **Haake, P., Miller, W. B., and Tyssee, D. A.,** *J. Am. Chem. Soc.,* 86, 3577, 1964.

615. **Haeberlen, U. and Waugh, J. S.,** *Phys. Rev.,* 175, 453, 1960.

616. **Hagen, R. and Roberts, J. D.,** *J. Am. Chem. Soc.,* 91, 4504, 1969.

617. **Hahn, E. L.,** *Phys. Rev.,* 80, 580, 1950.

618. **Hahn, E. L. and Maxwell, D. E.,** *Phys. Rev.,* 88, 1070, 1952.

619. **Hall, L. D.,** *Adv. Carbohydr. Chem.,* 19, 51, 1964.

620. **Hall, L. D. and Manville, J. F.,** *Chem. Ind. (London),* 991, 1965.

621. **Hall, L. D. and Manville, J. F.,** *Chem. Commun.,* 37, 1968.

622. **Halpern, B., Nitecki, D. E., and Weinstein, B.,** *Tetrahedron Lett.,* 3075, 1967.

623. **Hamlow, H. P., Okuda, S., and Nakagawa, N.,** *Tetrahedron Lett.,* 2553, 1964.

624. **Hammel, J. C. and Smith, J. A. S.,** *J. Chem. Soc. A,* 2883, 1969.
625. **Hampson, P. and Mathias, A.,** *Mol. Phys.,* 11, 541, 1966.
626. **Hampson, P. and Mathias, A.,** *Chem. Commun.,* 371, 1967.
627. **Hansen, J. R.,** *Biochim. Biophys. Acta,* 230, 482, 1971.
628. **Hansen, P. E., Poulsen, O. K., and Berg, A.,** *Org. Magn. Reson.,* 7, 23, 1975.
629. **Hansen, P. E., Poulsen, O. K., and Berg, A.,** *Org. Magn. Reson.,* 7, 405, 1975.
630. **Hansen, P. E., Poulsen, O. K., and Berg, A.,** *Org. Magn. Reson.,* 7, 475, 1975.
631. **Hansen, P. E., Poulsen, O. K., and Berg, A.,** *Org. Magn. Reson.,* 9, 649, 1977.
632. **Harris, R. K.,** *Chem. Soc. Rev.,* 5, 1, 1976.
633. **Hartmann, S. R. and Hahn, E. L.,** *Phys. Rev.,* 128, 2042, 1962.
634. **Hatada, K., Nagata, K., and Yuki, H.,** *Bull. Chem. Soc. Jpn.,* 43, 3195, 3267, 1970.
635. **Hatada, K., Takeshita, M., and Yuki, H.,** *Tetrahedron Lett.,* 4621, 1968.
636. **Hausser, K. H. and Stehlik, D.,** Dynamic nuclear polarization in liquids, in *Advances in Magnetic Resonance,* Vol. 3, Waugh, J. S., Ed., Academic Press, New York, 1968, 79.
637. **Hawkes, G. E., Herwig, K., and Roberts, J. D.,** *J. Org. Chem.,* 39, 1021, 1974.
638. **Hawkes, G. E., Randall, E. W., Elguero, J., and Marzin, C. J.,** *J. Chem. Soc. Perkin Trans. 2,* 1024, 1977.
639. **Hawkes, G. E., Randall, E. W., Hull, W. E., Gattegno, D., and Conti, F.,** *Biochemistry,* 17, 3986, 1978.
640. **Hayamizu, K. and Yamamoto, O.,** *Org. Magn. Reson.,* 13, 460, 1980.
641. **Hazlewood, C. F., Chang, D. C., Medina, D., Cleveland, G., and Nichols, B. L.,** *Proc. Natl. Acad. Sci. U.S.A.,* 69, 1478, 1972.
642. **Hazlewood, C. F., Cleveland, G., and Medina, D.,** *J. Natl. Cancer Inst.,* 52, 1849, 1974.
643. **Hazlewood, C. F., Nichols, B. L., Chang, D. C., and Brown, B.,** *Johns Hopkins Med. J.,* 128, 117, 1971.
644. **Heap, N. and Whitham, G. H.,** *J. Chem. Soc. B,* 164, 1966.
645. **Henneike, H. F., Jr. and Drago, R. S.,** *Inorg. Chem.,* 7, 1908, 1968.
646. **Herbison-Evans, D. and Richards, R. E.,** *Mol. Phys.,* 8, 19, 1964.
647. **Hess, R. E., Schaeffer, C. D., and Yoder, C. H.,** *J. Org. Chem.,* 36, 2201, 1971.
648. **Heyd, W. E. and Cupas, C. A.,** *J. Am. Chem. Soc.,* 91, 1559, 1969.
649. *High Resolution NMR Spectra,* SADTLER Research Laboratories, Philadelphia; Heyden and Son, London, 1967-1975.
650. **Hill, E. A. and Roberts, J. D.,** *J. Am. Chem. Soc.,* 89, 2047, 1967.
651. **Hill, R. K. and Chan, T. H.,** *Tetrahedron,* 21, 2015, 1965.
652. **Hinckley, C. C.,** *J. Am. Chem. Soc.,* 91, 5160, 1969.
653. **Hinckley, C. C.,** *J. Org. Chem.,* 35, 2834, 1970.
654. **Hinckley, C. C., Boyd, W. A., and Smith, G. V.,** *Tetrahedron Lett.,* 879, 1972.
655. **Hinckley, C. C., Boyd, W. A., and Smith, G. V.,** Chemistry of lanthanide shift reagents secondary deuterium isotope effects, in *Nuclear Magnetic Resonance Shift Reagents,* Sievers, R. F., Ed., Academic Press, New York, 1973, 1.
656. **Hinshaw, W. S., Bottomley, P. A., and Holland, G. N.,** *Nature (London),* 270, 722, 1977.
657. **Hirayama, M. and Hanyu, Y.,** *Bull. Chem. Soc. Jpn.,* 46, 2687, 1973.
658. **Hirsch, J. A. and Havinga, E.,** *J. Org. Chem.,* 41, 455, 1976.
659. **Hirsike, E.,** *J. Phys. Soc. Jpn.,* 15, 270, 1960.
660. **Hochmann, J. and Kellerhals, H. P.,** *J. Magn. Reson.,* 38, 23, 1980.
661. **Hochmann, J., Rosanske, R. C., and Levy, G. C.,** *J. Magn. Reson.,* 33, 275, 1979.
662. **Hoffman, R. A. and Forsen, S.,** High resolution nuclear magnetic double and multiple resonance, in *Progress in Nuclear Magnetic Resonance Spectroscopy,* Vol. 1, Emsley, J. W., Feeney, J., and Sutcliffe, L. H., Eds., Pergamon Press, Oxford, 1966, 15.
663. **Hoffman, R. A. and Gronowitz, S.,** *Arkiv. Kemi,* 16, 501, 515, 563, 1960.
664. **Hoffmann, T. A. and Ladik, J.,** *Adv. Chem. Phys.,* 7, 94, 1969.
665. **Hofman, W., Stefaniak, L., Urbanski, T., and Witanowski, M.,** *J. Am. Chem. Soc.,* 86, 554, 1964.
666. **Hogeveen, H.,** *Rec. Trav. Chim.,* 85, 1072, 1966.
667. **Holland, C. V., Horton, D., and Jewell, J. S.,** *J. Org. Chem.,* 32, 1818, 1967.
668. **Hollis, D. P., Saryan, L. A., Economou, J. S., Eggleston, J. C., Czeisler, J. L., and Morris, H. P.,** *J. Natl. Cancer Inst.,* 53, 807, 1974.
669. **Hollis, D. P., Saryan, L. A., and Morris, H. P.,** *Johns Hopkins Med. J.,* 131, 441, 1972.
670. **Hollósi, M., Radics, L., and Wieland, T.,** *J. Peptide Protein Res.,* 10, 1286, 1977.
671. **Holly, S. and Sohár, P.,** Theoretical and technical introduction, in *Absorption Spectra in the Infrared Region,* Láng, L. and Prichard, W. H., Eds., Akadémiai Kiadó, Budapest, 1975.
672. **Homer, J.,** *Appl. Spectrosc. Rev.,* 9, 1, 1975.

673. **Homer, J.,** Intramolecular effects. NMR shift reagents, in *Specialist Periodical Reports, Nuclear Magnetic Resonance,* Vol. 7, Abraham, R. J., Ed., The Chemical Society, London, 1976, 330.

674. **Homer, J. and Thomas, L. F.,** *Trans. Faraday Soc.,* 59, 2431, 1963.

675. **Horn, R. R. and Everett, G. W., Jr.,** *J. Am. Chem. Soc.,* 93, 7173, 1971.

676. **Horrocks, W. DeW., Jr.,** *J. Am. Chem. Soc.,* 96, 3022, 1974.

677. **Horrocks, W. DeW., Jr. and Hall, D. DeW.,** *Coord. Chem. Rev.,* 6, 147, 1971.

678. **Horrocks, W. DeW., Jr., Sipe, J. P., III, and Luber, J. R.,** *J. Am Chem. Soc.,* 93, 5258, 1971.

679. **Horsley, W. J. and Sternlicht, H.,** *J. Am. Chem. Soc.,* 90, 3738, 1968.

680. **Horsley, W. J., Sternlicht, H., and Cohen, J. S.,** *Biochem. Biophys. Res. Commun.,* 37, 47, 1969.

681. **Horsley, W. J., Sternlicht, H., and Cohen, J. S.,** *J. Am. Chem. Soc.,* 92, 680, 1970.

682. **Horton, D. and Jewell, J. S.,** *Carbohydr. Res.,* 2, 251, 1966; 3, 255, 1966.

683. **Horton, D. and Turner, W. N.,** *J. Org. Chem.,* 30, 3387, 1965.

684. **Horváth, T., Sohár, P., and Ábrahám, G.,** *Carbohydr. Res.,* 73, 277, 1979.

685. **Hoult, D. I. and Lauterbur, P. C.,** *J. Magn. Reson.,* 34, 425, 1979.

686. **House, H. O., Latham, R. A., and Whitesides, G. M.,** *J. Org. Chem.,* 32, 2481, 1967.

687. **Howard, B. B., Linder, B., and Emerson, M. T.,** *J. Chem. Phys.,* 36, 485, 1962.

688. **Howarth, O. W., and Lynch, R. J.,** *Mol. Phys.,* 15, 431, 1968.

689. **Hubbard, P. S.,** *Phys. Rev.,* 131, 1155, 1963.

690. **Huggins, C. M., Pimentel, G. C., and Shoolery, J. N.,** *J. Phys. Chem.,* 60, 1311, 1956.

691. **Huggins, M. L.,** *J. Am. Chem. Soc.,* 75, 4123, 1953.

692. **Huitric, A. C., Roll, D. B., and Deboer, J. R.,** *J. Org. Chem.,* 32, 1661, 1967.

693. **Ihrig, A. M. and Marshall, J. L.,** *J. Am. Chem. Soc.,* 94, 1756, 1972.

694. **Iijima, N. and Fujii, N.,** *JEOL (Jpn. Electron Opt. Lab.) News,* 9a, 5, 1972.

695. **Imanari, M., Kohno, M., Ohuchi, M., and Ishizu, K.,** *Bull. Chem. Soc. Jpn.,* 47, 708, 1974.

696. **Inamoto, N., Kushida, K., Masuda, S., Ohta, H., Satoh, S., Tamura, Y., Tokumaru, K., Tori, K., and Yoshida, M.,** *Tetrahedron Lett.,* 3617, 1974.

697. **Inch, T. D.,** Nuclear magnetic resonance spectroscopy in the study of carbohydrates and related compounds, in *Annual Reports on NMR Spectroscopy,* Vol. 2, Mooney, E. F., Ed., Academic Press, New York, 1969, 35.

698. **Inch, T. D.,** Nuclear magnetic resonance spectroscopy in the study of carbohydrates and related compounds, in *Annual Reports on NMR Spectroscopy,* Vol. 5A, Mooney, E. F., Ed., Academic Press, New York, 1972, 305.

699. **Inch, T. D., Plimmer, J. R., and Fletcher, H. G., Jr.,** *J. Org. Chem.,* 31, 1825, 1966.

700. **Inch, W. R., McCredie, J. A., Geiger, C., and Boctor, Y.,** *J. Natl. Cancer Inst.,* 53, 689, 1974.

701. **Inch, W. R., McCredie, J. A., Knispel, R. R., Thompson, R. T., and Pintar, M. M.,** *J. Natl. Cancer Inst.,* 52, 353, 1974.

702. **Ingold, C. K.,** *Structure and Mechanism in Organic Chemistry,* G. Bell and Sons, London, 1953.

703. **Inoue, Y., Chûjô, R., and Nishioka, A.,** *Polym. J.,* 4, 244, 1973.

704. **Irving, C. S. and Lapidot, A.,** *Chem. Commun.,* 43, 1976.

705. **Jackman, L. M. and Kelley, D. P.,** *J. Chem. Soc. B,* 102, 1970.

706. **Jackman, L. M., Sondheimer, F., Amiel, Y., Ben-Efraim, D. A., Gaoni, Y., Wolovsky, R., and Bothner-By, A. A.,** *J. Am. Chem. Soc.,* 84, 4307, 1962.

707. **Jackson, J. A., Lemons, J. F., and Taube, H.,** *J. Chem. Phys.,* 32, 553, 1960.

708. **Jaeckle, H., Haeberlen, U., and Schweitzer, D.,** *J. Magn. Reson.,* 4, 198, 1971.

709. **Jakobsen, P. and Treppendahl, S.,** *Org. Magn. Reson.,* 14, 133, 1980.

710. **James, T. L. and Gillen, K. T.,** *Biochim. Biophys. Acta,* 286, 10, 1972.

711. **Jameson, C. J. and Damasco, M. C.,** *Mol. Phys.,* 18, 491, 1970.

712. **Jameson, C. J. and Gutowsky, H. S.,** *J. Chem. Phys.,* 51, 2790, 1969.

713. **Jardine, R. V. and Brown, R. K.,** *Can. J. Chem.,* 41, 2067, 1963.

714. **Jautelat, M., Grutzner, J. B., and Roberts, J. D.,** *Proc. Natl. Acad. Sci.,* 65, 288, 1970.

715. **Jeener, J.,** Ampere Int. Summer School II, *Basko Polje, Yugoslavia, 1971.*

716. **Jennings, W. B., Boyd, D. R., Watson, C. G., Becker, E. D., Bradley, R. B., and Jerina, D. M.,** *J. Am. Chem. Soc.,* 94, 8501, 1972.

717. **Jensen, F. R., Noyce, D. S., Sederholm, C. H., and Berlin, A. J.,** *J. Am. Chem. Soc.,* 84, 386, 1962.

718. **Jensen, F. R. and Smith, L. A.,** *J. Am. Chem. Soc.,* 86, 956, 1964.

719. *JEOL (Jpn. Electron Opt. Lab.) News,* 7C, 16, 1970.

720. *JEOL NMR Applications,* JEOL Co., Ltd., Tokyo, Japan.

721. **Jesson, J. P., Meakin, P., and Kneissel, G.,** *J. Am. Chem. Soc.,* 95, 618, 1973.

722. **Johns, S. R. and Willing, R. J.,** *Aust. J. Chem.,* 29, 1617, 1976.

723. **Johnson, B. F. G., Lewis, J., McArdle, P., and Norton, J. R.,** *Chem. Commun.,* 535, 1972.

724. **Johnson, C. E. and Bovey, F. A.,** *J. Chem. Phys.,* 29, 1012, 1958.

725. Johnson, C. S., Jr., Weiner, M. A., Waugh, J. S., and Seyferth, D., *J. Am. Chem. Soc.*, 83, 1306, 1961.

726. Johnson, F. P., Melera, A., and Sternhell, S., *Aust. J. Chem.*, 19, 1523, 1966.

727. Johnson, L. F., *VARIAN Assoc. Techn. Inform. Bull.*, 6, 1965.

728. Johnson, L. F., Heatley, F., and Bovey, F. A., *Macromolecules*, 3, 175, 1970.

729. Johnson, L. F. and Jankowski, W. C., *C-13 NMR Spectra*, Wiley-Interscience, New York, 1972.

730. Johnston, M. D., Jr., Shapiro, B. L., Shapiro, M. J., Proulx, T. W., Godwin, A. D., and Pearce, H. L., *J. Am. Chem. Soc.*, 97, 542, 1975.

731. Jolley, K. W., Sutcliffe, L. H., and Walker, S. M., *Trans. Faraday Soc.*, 64, 269, 1968.

732. Jonathan, N., Gordon, S., and Dailey, B. P., *J. Chem. Phys.*, 36, 2443, 1962.

733. Jones, A. J., Alger, T. D., Grant, D. M., and Litchman, W. M., *J. Am. Chem. Soc.*, 92, 2386, 1970.

734. Jones, A. J., Beeman, C. P., Hasan, M. U., Casy, A. F., and Hassan, M. M. A., *Can. J. Chem.*, 54, 126, 1976.

735. Jones, A. J., Gardner, P. D., Grant, D. M., Litchman, W. M., and Boekelheide, V., *J. Am. Chem. Soc.*, 92, 2395, 1970.

736. Jones, A. J. and Grant, D. M., *Chem. Commun.*, 1670, 1968.

737. Jones, A. J., Grant, D. M., and Kuhlmann, K., *J. Am. Chem. Soc.*, 91, 5013, 1969.

738. Jones, A. J., Grant, D. M., Winkley, M. W., and Robins, R. K., *J. Am. Chem. Soc.*, 92, 4079, 1970.

739. Jones, K. and Mooney, E. F., Fluorine-19 nuclear magnetic resonance spectroscopy, in *Annual Reports on NMR Spectroscopy*, Vol. 3, Mooney, E. F., Ed., Academic Press, New York, 1970, 261.

740. Jones, K. and Mooney, E. F., Fluorine-19 nuclear magnetic resonance spectroscopy, in *Annual Reports on NMR Spectroscopy*, Vol. 4, Mooney, E. F., Ed., Academic Press, New York, 1971, 391.

741. Joseph-Nathan, P., Herz, J. E., and Rodriguez, V. M., *Can. J. Chem.*, 50, 2788, 1972.

742. Kaiser, R., *J. Chem. Phys.*, 42, 1838, 1965.

743. Kaiser, R., *J. Magn. Reson.*, 3, 28, 1970.

744. Kalinowski, H. and Kessler, H., *Angew. Chem. Int. Ed. Engl.*, 13, 90, 1974.

745. Kametani, T., Fukumoto, K., Ihara, M., Ujiie, A., and Koizumi, H., *J. Org. Chem.*, 40, 3280, 1975.

746. Kametani, T., Ujiie, A., Ihara, M., Fukumoto, K., and Koizumi, H., *Heterocycles*, 3, 371, 1975.

747. Kaptein, R., *Chem. Commun.*, 732, 1971.

748. Kaptein, R. and Oosterhoff, J. L., *Chem. Phys. Lett.*, 4, 195, 214, 1969.

749. Karabatsos, G. J., Graham, J. D., and Vane, F. M., *J. Am. Chem. Soc.*, 84, 37, 1962.

750. Karabatsos, G. J. and Vane, F. M., *J. Am. Chem. Soc.*, 85, 3886, 1963.

751. Karplus, M., *J. Chem. Phys.*, 30, 11, 1959.

752. Karplus, M., *J. Chem. Phys.*, 33, 1842, 1960.

753. Karplus, M., *J. Am. Chem. Soc.*, 84, 2458, 1962.

754. Karplus, M. and Anderson, D. H., *J. Chem. Phys.*, 30, 6, 1959.

755. Karplus, M. and Grant, D. M., *Proc. Natl. Acad. Sci. U.S.A.*, 45, 1269, 1959.

756. Karplus, M. and Pople, J. A., *J. Chem. Phys.*, 38, 2803, 1963.

757. Kato, H. and Yonezawa, T., *Bull. Chem. Soc. Jpn.*, 43, 1921, 1970.

758. Kato, Y. and Saika, A., *J. Chem. Phys.*, 46, 1975, 1967.

759. Katritzky, A. R. and Maine, F. W., *Tetrahedron*, 20, 299, 1964.

760. Kawazoe, Y. and Tsuda, M., *Chem. Pharm. Bull.*, 15, 1405, 1967.

761. Kawazoe, Y., Tsuda, M., and Ohniski, M., *Chem. Pharm. Bull.*, 15, 51, 1967.

762. Keller, C. E. and Petitt, R., *J. Am. Chem. Soc.*, 88, 604, 606, 1966.

763. Kellie, G. M. and Riddell, F. G., *J. Chem. Soc. B*, 1030, 1971.

764. Kessler, H., *Angew. Chem.*, 82, 237, 1970.

765. Khuong-Huu, F., Sangare, M., Chari, V. M., Bekaert, A., Devys, M., Barbier, M., and Lukacs, G., *Tetrahedron Lett.*, 1787, 1975.

766. Kiefer, E. F., Gericke, W., and Amimoto, S. T., *J. Am. Chem. Soc.*, 90, 6246, 1968.

767. King, M. M., Yeh, H. J. C., and Dudek, G. O., *Org. Magn. Reson.*, 8, 208, 1976.

768. Kintzinger, J. P. and Lehn, J. M., *Helv. Chim. Acta*, 58, 905, 1975.

769. Kitching, W., Bullpitt, M., Garsthore, D., Adcock, W., Khor, T. C., Doddrell, D., and Rae, I. D., *J. Org. Chem.*, 42, 2411, 1977.

770. Kleinfelter, D. C., *J. Am. Chem. Soc.*, 89, 1734, 1967.

771. Klesper, E., Johnson, A., Gronski, W., and Wehrli, F. W., *Macromol. Chem.*, 176, 1071, 1975.

772. Klinck, R. E. and Stothers, J. B., *Can. J. Chem.*, 40, 2329, 1962.

773. Knight, S. A., *Org. Magn. Reson.*, 6, 603, 1973.

774. Knight, W. D., *Phys. Res.*, 76, 1259, 1949.

775. Knox, L. H., Velarde, E., and Cross, A. D., *J. Am. Chem. Soc.*, 85, 2533, 1963.

776. Koenig, S. H., Hallenga, K., and Shporer, M., *Proc. Natl. Acad. Sci. U.S.A.*, 72, 2667, 1975.

777. Koer, F. J., de Hoog, A. J., and Altona, C., *Recl. Trav. Chim. Pays-Bas*, 94, 75, 1975.

778. **Komoroski, R. A., Peat, I. R., and Levy, G. C.,** ¹³C NMR studies of biopolymers, in *Topics in Carbon-13 NMR Spectroscopy,* Vol. 2, Levy, G. C., Ed., Wiley-Interscience, New York, 1976, 180.

779. **Korver, P. K., Haak, P. J., Steinberg, H., and Deboer, T. J.,** *Recl. Trav. Chim.,* 84, 129, 1965.

780. **Koutcher, J. A. and Damadian, R.,** *Physiol. Chem. Phys.,* 9, 181, 1977.

781. **Kowalsky, A. and Cohn, M.,** *Annu. Rev. Biochem.,* 33, 481, 1964.

782. **Kraus, W. and Suhr, H.,** *Annalen,* 695, 27, 1966.

783. **Kreishman, G. P., Witkowski, J. T., Robins, R. K., and Schweizer, M. P.,** *J. Am. Chem. Soc.,* 94, 5894, 1972.

784. **Kristiansen, P. and Ledaal, T.,** *Tetrahedron Lett.,* 4457, 1971.

785. **Kroschwitz, J. I., Winokur, M., Reich, H. J., and Roberts, J. D.,** *J. Am. Chem. Soc.,* 91, 5927, 1969.

786. **Krow, G. R. and Ramey, K. C.,** *Tetrahedron Lett.,* 3143, 1971.

787. **Kruczynski, L., Lishingman, L. K. K., and Takats, J.,** *J. Am. Chem. Soc.,* 96, 4006, 1974.

788. **Krueger, G. L., Kaplan, F., Orchin, M., and Faul, W. H.,** *Tetrahedron Lett.,* 3979, 1965.

789. **Kuhlmann, K. F. and Grant, D. M.,** *J. Chem. Phys.,* 55, 2998, 1971.

790. **Kuhlmann, K. F., Grant, D. M., and Harris, R. K.,** *J. Chem. Phys.,* 52, 3439, 1970.

791. **Kulka, M.,** *Can. J. Chem.,* 42, 2791, 1964.

792. **Kumar, A., Aue, W. P., Bachmann, P., Karhan, J., Müller, L., and Ernst, R. R.,** Two-dimensional spin-echo spectroscopy. A means to resolve proton and carbon NMR spectra, in *Magnetic Resonance and Related Phenomena. Proc. XIXth Congr. Ampère, Heidelberg, 1976,* Brunner, H., Hausser, K. H., and Schweitzer, D., Eds., Groupement Ampère, Heidelberg, 1976, 473.

793. **Kumar, A. and Ernst, R. R.,** *Chem. Phys. Lett.,* 37, 162, 1976.

794. **Kumar, A. and Ernst, R. R.,** *J. Magn. Reson.,* 24, 425, 1976.

795. **Kumar, A., Welti, D., and Ernst, R. R.,** *J. Magn. Reson.,* 18, 69, 1975.

796. **Kurland, R. J., Rubin, M. B., and Wyse, W. B.,** *J. Chem. Phys.,* 40, 2426, 1964.

797. **Kuszmann, J. and Sohár, P.,** *Carbohydr. Res.,* 14, 415, 1970.

798. **Kuszmann, J. and Sohár, P.,** *Carbohydr. Res.,* 21, 19, 1972.

799. **Kuszmann, J. and Sohár, P.,** *Carbohydr. Res.,* 27, 157, 1973.

800. **Kuszmann, J. and Sohár, P.,** *Carbohydr. Res.,* 35, 97, 1974.

801. **Kuszmann, J. and Sohár, P.,** *Acta Chim. Acad. Sci. Hung.,* 83, 373, 1974.

802. **Kuszmann, J., Sohár, P., and Horváth, Gy.,** *Tetrahedron,* 27, 5055, 1971.

803. **Kuszmann, J., Sohár, P., Horváth, Gy., and Méhesfalvi-Vajna, Zs.,** *Tetrahedron,* 30, 3905, 1974.

804. **Kuszmann, J. and Vargha, L.,** *Carbohydr. Res.,* 17, 309, 1971.

805. **La Mar, G. N. and Faller, J. W.,** *J. Am. Chem. Soc.,* 95, 3818, 1973.

806. **La Mar, G. N., Horrocks, W. DeW., Jr., and Allen, L. C.,** *J. Chem. Phys.,* 41, 2126, 1964.

807. **Lamb, W. E.,** *Phys. Rev.,* 60, 817, 1941.

808. **Lambert, J. B., Binsch, G., and Roberts, J. D.,** *Proc. Natl. Acad. Sci. U.S.A.,* 51, 735, 1964.

809. **Lambert, J. B. and Keske, R. G.,** *J. Am. Chem. Soc.,* 88, 620, 1966.

810. **Lambert, J. B., Netzel, D. A., Sun, H.-N., and Lilianstrom, K. K.,** *J. Am. Chem. Soc.,* 98, 3778, 1976.

811. **Lambert, J. B., Roberts, B. W., Binsch, G., and Roberts, J. D.,** ¹⁵N Magnetic resonance spectroscopy, in *Nuclear Magnetic Resonance in Chemistry,* Pesce, B., Ed., Academic Press, New York, 1965, 269.

812. **Lambert, J. B. and Roberts, J. D.,** *J. Am. Chem. Soc.,* 85, 3710, 1963.

813. **Lambert, J. B. and Roberts, J. D.,** *J. Am. Chem. Soc.,* 87, 3884, 3891, 1965.

814. **Lambert, J. B. and Roberts, J. D.,** *J. Am. Chem. Soc.,* 87, 4087, 1965.

815. **László, P.,** Solvent effects and NMR, in *Progress in Nuclear Magnetic Resonance Spectroscopy,* Vol. 3, Emsley, J. W., Feeney, J., and Sutcliffe, L. H., Eds., Pergamon Press, New York, 1967, 231.

816. **Lauterbur, P. C.,** *J. Chem. Phys.,* 26, 217, 1957.

817. **Lauterbur, P. C.,** *Ann. N. Y. Acad. Sci.,* 70, 841, 1958.

818. **Lauterbur, P. C.,** *Tetrahedron Lett.,* 274, 1961.

819. **Lauterbur, P. C.,** *J. Am. Chem. Soc.,* 83, 1838, 1961.

820. **Lauterbur, P. C.,** *J. Am. Chem. Soc.,* 83, 1846, 1961.

821. **Lauterbur, P. C.,** Nuclear magnetic resonance spectra of elements other than hydrogen and fluorine, in *Determination of Organic Structures by Physical Methods,* Vol. 2, Nachod, F. C. and Phillips, W. D., Eds., Academic Press, New York, 1962, 465.

822. **Lauterbur, P. C.,** *J. Chem. Phys.,* 38, 1406, 1963.

823. **Lauterbur, P. C.,** *J. Chem. Phys.,* 38, 1415, 1963.

824. **Lauterbur, P. C.,** *J. Chem. Phys.,* 38, 1432, 1963.

825. **Lauterbur, P. C.,** *J. Chem. Phys.,* 43, 360, 1965.

826. **Lauterbur, P. C.,** *Nature (London),* 242, 190, 1973.

827. **Lauterbur, P. C.,** *NMR in Biology,* Dwek, R. A., Campbell, I. D., Richards, R. E., and Williams, R. J. P., Academic Press, London, 1977, 323.

828. **Lauterbur, P. C. and King, R. B.,** *J. Am. Chem. Soc.,* 87, 3266, 1965.

829. **Lawler, R. G.,** *J. Am. Chem. Soc.,* 89, 5519, 1967.
830. **Lawler, R. G.,** *Acc. Chem. Res.,* 5, 25, 1972.
831. **Lawler, R. G.,** Chemically induced dynamic nuclear polarization, in *Progress in NMR Spectroscopy,* Vol. 9, Emsley, J. W., Feeney, J., and Sutcliffe, L. H., Eds., Pergamon Press, Oxford, 1973, 145.
832. **Lee, K. and Anderson, W. A.,** *Handbook of Chemistry and Physics,* 55th ed., Weast, R. C., Ed., Chemical Rubber Co., Cleveland, 1974, E69.
833. **Le Févre, C. G. and Le Févre, R. J. W.,** *J. Chem. Soc.,* 3549, 1956.
834. **Lehn, J. M. and Riehl, J. J.,** *J. Mol. Phys.,* 8, 33, 1964.
835. **Lehn, J. M. and Wagner, J.,** *Chem. Commun.,* 148, 1968.
836. **Lemieux, R. U.,** in *Molecular Rearrangements,* De Mayo, A., Ed., Interscience, New York, 1963.
837. **Lemieux, R. U.,** *Ann. N. Y. Acad. Sci.,* 222, 915, 1973.
838. **Lemieux, R. U., Kullnig, R. K., Bernstein, H. J., and Schneider, W. G.,** *J. Am. Chem. Soc.,* 79, 1005, 1957.
839. **Lemieux, R. U. and Stevens, J. D.,** *Can. J. Chem.,* 44, 249, 1966.
840. **Lempert-Sréter, M. and Sohár, P.,** *Acta Chim. Acad. Sci. Hung.,* 54, 203, 1967.
841. **Lepley, A. R. and Closs, G. L.,** *Chemically Induced Magnetic Polarization,* Wiley-Interscience, New York, 1973.
842. **Levin, R. H. and Roberts, J. D.,** *Tetrahedron Lett.,* 135, 1973.
843. **Levy, G. C.,** *Chem. Commun.,* 352, 1972.
844. **Levy, G. C.,** *J. Magn. Reson.,* 8, 122, 1972.
845. **Levy, G. C.,** *Tetrahedron Lett.,* 3709, 1972.
846. **Levy, G. C.,** *Acc. Chem. Res.,* 6, 161, 1973.
847. **Levy, G. C. and Cargioli, J. D.,** *J. Magn. Reson.,* 6, 143, 1972.
848. **Levy, G. C. and Cargioli, J. D.,** *J. Magn. Reson.,* 10, 231, 1973.
849. **Levy, G. C., Cargioli, J. D., and Anet, F. A. L.,** *J. Am. Chem. Soc.,* 95, 1527, 1973.
850. **Levy, G. C. and Dittmer, D. C.,** *Org. Magn. Reson.,* 4, 107, 1972.
851. **Levy, G. C. and Edlund, U.,** *J. Am. Chem. Soc.,* 97, 5031, 1975.
852. **Levy, G. C. and Komoroski, R. A.,** *J. Am. Chem. Soc.,* 96, 678, 1974.
853. **Levy, G. C., Komoroski, R. A., and Echols, R. E.,** *Org. Magn. Reson.,* 7, 172, 1975.
854. **Levy, G. C., Komoroski, R. A., and Halstead, J. A.,** *J. Am. Chem. Soc.,* 96, 5456, 1974.
855. **Levy, G. C. and Nelson, G. L.,** *Carbon-13 Nuclear Magnetic Resonance for Organic Chemists,* Wiley-Interscience, New York, 1972, (a) p. 110, 121, 123; (b) p. 125; (c) p. 81; (d) pp. 176-198.
856. **Levy, G. C. and Nelson, G. L.,** *J. Am. Chem. Soc.,* 94, 4897, 1972.
857. **Levy, G. C., White, D. M., and Anet, F. A. L.,** *J. Magn. Reson.,* 6, 453, 1972.
858. **Lewis, W. B., Jackson, J. A., Lemons, J. F., and Taube, H.,** *J. Chem. Phys.,* 36, 694, 1962.
859. **Lichtenthaler, F. W.,** *Chem. Ber.,* 96, 845, 2047, 1963.
860. **Lichter, R. L.,** *Determination of Organic Structures by Physical Methods,* Vol. 4, Nachod, F. C. and Zuckerman, J. J., Eds., Academic Press, New York, 1971, 195.
861. **Lichter, R. L., Dorman, D. E., and Wasylishen, R.,** *J. Am. Chem. Soc.,* 96, 930, 1974.
862. **Lichter, R. L. and Roberts, J. D.,** *J. Chem. Phys.,* 74, 912, 1970.
863. **Lichter, R. L. and Roberts, J. D.,** *J. Am. Chem. Soc.,* 93, 5218, 1971.
864. **Lichter, R. L. and Roberts, J. D.,** *J. Am. Chem. Soc.,* 94, 2495, 1972.
865. **Lichter, R. L. and Roberts, J. D.,** *J. Am. Chem. Soc.,* 94, 4904, 1972.
866. **Lillien, I. and Doughty, R. A.,** *J. Am. Chem. Soc.,* 89, 155, 1967.
867. **Lindeman, L. P. and Adams, J. Q.,** *Anal. Chem.,* 43, 1245, 1971.
868. **Lindström, G.,** *Phys. Res.,* 78, 1817, 1950.
869. **Ling, G. N.,** *A Physical Theory of the Living State,* Blaisdel, New York, 1962.
870. **Ling, G. N.,** *Ann. N. Y. Acad. Sci.,* 125, 401, 1965.
871. **Lippert, E. and Prigge, H.,** *Ber. Bunsenges Phys. Chem.,* 67, 415, 1963.
872. **Lippmaa, E., Magi, M., Novikov, S. S., Khmelnitski, L. I., Prihodko, A. S., Lebedev, O. V., and Epishina, L. V.,** *Org. Magn. Reson.,* 4, 153, 1972.
873. **Lippmaa, E., Magi, M., Novikov, S. S., Khmelnitski, L. I., Prihodko, A. S., Lebedev, O. V., and Epishina, L. V.,** *Org. Magn. Reson.,* 4, 197, 1972.
874. **Lippmaa, E. and Pehk, T.,** *Kem. Teollisuus,* 24, 1001, 1967.
875. **Lippmaa, E. and Pehk, T.,** *Eesti NSV Tead. Akad. Toim. Keem. Geol.,* 17, 210, 1968.
876. **Lippmaa, E. and Pehk, T.,** *Eesti NSV Tead. Akad. Toim. Keem. Geol.,* 17, 287, 1968.
877. **Lippmaa, E., Pehk, T., Andersson, K., and Rappe, C.,** *Org. Magn. Reson.,* 2, 109, 1970.
878. **Lippmaa, E., Pehk, T., and Past, J.,** *Eesti NSV Tead. Akad. Toim. Fuus. Mat.,* 16, 345, 1967.
879. **Lippmaa, E., Saluvere, T., and Laisaar, S.,** *Chem. Phys. Lett.,* 11, 120, 1971.
880. **Liska, K. J., Fentiman, A. F., Jr., and Foltz, R. L.,** *Tetrahedron Lett.,* 4657, 1970.
881. **Litchman, W. M. and Grant, D. M.,** *J. Am. Chem. Soc.,* 89, 6775, 1967.

882. **Litchman, W. M. and Grant, D. M.,** *J. Am. Chem. Soc.,* 90, 1400, 1968.

883. **Llinas, J.-R., Vincent, E.-J., and Peiffer, G.,** *Bull. Soc. Chim. France,* 3209, 1973.

884. **Looney, C. E., Phillips, W. D., and Reilly, E. L.,** *J. Am. Chem. Soc.,* 79, 6136, 1957.

884a. **Löw, M., Kisfaludy, L., and Sohár, P.,** *Z. Physiol. Chem.,* 359, 1643, 1978.

885. **Lukacs, G., Khuong-Huu, F., Bennett, C. R., Buckwalter, B. L., and Wenkert, E.,** *Tetrahedron Lett.,* 3515, 1972.

886. **Lunazzi, L., Macciantelli, D., and Boicelli, C. A.,** *Tetrahedron Lett.,* 1205, 1975.

887. **Lunazzi, L., Macciantelli, D., and Taddei, F.,** *Mol. Phys.,* 19, 137, 1970.

888. **Lustig, E., Benson, W. R., and Duy, N.,** *J. Org. Chem.,* 32, 851, 1967.

889. **Lustig, E., Ragelis, E. P., and Duy, N.,** *Spectrochim. Acta,* 23a, 133, 1967.

890. **Lyerla, J. R., Jr., Grant, D. M., and Bertrand, R. D.,** *J. Phys. Chem.,* 75, 3967, 1971.

891. **Lyerla, J. R., Jr., Grant, D. M., and Harris, R. K.,** *J. Phys. Chem.,* 75, 585, 1971.

892. **Lyerla, J. R., Jr. and Levy, G. C.,** Carbon-13 nuclear spin relaxation, in *Topics in Carbon-13 NMR Spectroscopy,* Vol. 1, Levy, G. C., Ed., Wiley-Interscience, New York, 1974, 79; (a) p. 88; (b) p. 96.

893. **Lynch, D. M. and Cole, W.,** *J. Org. Chem.,* 31, 3337, 1966.

894. **Lynden-Bell, R. M. and Sheppard, N.,** *Proc. R. Soc. London Ser. A,* 269, 385, 1962.

895. **Maciel, G. E.,** *J. Phys. Chem.,* 69, 1974, 1965.

896. **Maciel, G. E.,** ^{13}C-^{13}C coupling constants, in *Nuclear Magnetic Resonance Spectroscopy of Nuclei Other Than Protons,* Axenrod, T. and Webb, G. A., Eds., Wiley-Interscience, New York, 1974, 187.

897. **Maciel, G. E. and Beatty, D. A.,** *J. Phys. Chem.,* 69, 3920, 1965.

898. **Maciel, G. E., Dallas, J. L., and Miller, D. P.,** *J. Am. Chem. Soc.,* 98, 5074, 1976.

899. **Maciel, G. E., Ellis, P. D., and Hofer, D. C.,** *J. Phys. Chem.,* 71, 2160, 1967.

900. **Maciel, G. E., Ellis, P. D., Natterstad, J. J., and Savitsky, G. B.,** *J. Magn. Reson.,* 1, 589, 1969.

901. **Maciel, G. E. and James, R. V.,** *J. Am. Chem. Soc.,* 86, 3893, 1964.

902. **Maciel, G. E., McIver, J. W., Jr., Ostlund, N. S., and Pople, J. A.,** *J. Am. Chem. Soc.,* 92, 1, 11, 1970.

903. **Maciel, G. E. and Natterstad, J. J.,** *J. Chem. Phys.,* 42, 2752, 1965.

904. **Maciel, G. E. and Savitsky, G. B.,** *J. Phys. Chem.,* 69, 3925, 1965.

905. **Maciel, G. E. and Traficante, D. D.,** *J. Am. Chem. Soc.,* 88, 220, 1966.

906. **MacNicol, D. D.,** *Tetrahedron Lett.,* 3325, 1975.

907. **Magi, M., Erashko, V. I., Shevelev, S. A., and Fainzil'berg, A. A.,** *Eesti Nsv. Tead. Toim. Keem. Geol.,* 20, 297, 1971.

908. **Maia, H. L., Orrell, K. G., and Rydon, H. N.,** *Chem. Commun.,* 1209, 1971.

909. **Maksić, Z. B., Eckert-Maksić, M., and Randic, M.,** *Theor. Chim. Acta,* 22, 70, 1971.

910. **Malinowski, E. R.,** *J. Am. Chem. Soc.,* 83, 4479, 1961.

911. **Malinowski, E. R., Pollara, L. Z., and Larmann, J. P.,** *J. Am. Chem. Soc.,* 84, 2649, 1962.

912. **Malinowski, E. R., Vladimirov, T., and Tavares, R. F.,** *J. Phys. Chem.,* 70, 2046, 1966.

913. **Manatt, S. L.,** *J. Am. Chem. Soc.,* 88, 1323, 1966.

914. **Manatt, S. L. and Bowers, J.,** *J. Am. Chem. Soc.,* 91, 4381, 1969.

915. **Manatt, S. L., Elleman, D. D., and Brois, S. J.,** *J. Am. Chem. Soc.,* 87, 2220, 1965.

916. **Mandel, F. S., Cox, R. H., and Taylor, R. C.,** *J. Magn. Reson.,* 14, 235, 1974.

917. **Mandel, M.,** *J. Biol. Chem.,* 240, 1586, 1965.

918. **Mann, B. E.,** Dynamic ^{13}C NMR spectroscopy, in *Progress in Nuclear Magnetic Resonance Spectroscopy,* Vol. 11, Emsley, J. W., Feeney, J., and Sutcliffe, L. H., Eds., Pergamon Press, Oxford, 1977, 95.

919. **Mann, B. E.,** The common nuclei. Fluorine ^{19}F, in *NMR and the Periodic Table,* Harris, R. K. and Mann, B. E., Eds., Academic Press, New York, 1978, 98.

920. **Mann, B. E.,** The common nuclei. Phosphorus ^{31}P, in *NMR and the Periodic Table,* Harris, R. K. and Mann, B. E., Eds., Academic Press, New York, 1978, 100.

921. **Mannschreck, A.,** *Tetrahedron Lett.,* 1341, 1965.

922. **Mannschreck, A., Seitz, W., and Staab, H. A.,** *Ber. Bunsenges Phys. Chem.,* 67, 471, 1963.

923. **Mansfield, P. and Maudsley, A. A.,** *Br. J. Radiol.,* 50, 188, 1977.

924. **Marchal, J. P. and Canet, D.,** *J. Am. Chem. Soc.,* 97, 6581, 1975.

925. **Marcus, S. H. and Miller, S. I.,** *J. Am. Chem. Soc.,* 88, 3719, 1966.

926. **Markley, J. L., Horsley, W. J., and Klein, M. P.,** *J. Chem. Phys.,* 55, 3604, 1971.

927. **Markowski, V., Sullivan, G. R., and Roberts, J. D.,** *J. Am. Chem. Soc.,* 99, 714, 1977.

928. **Marr, D. H. and Stothers, J. B.,** *Can. J. Chem.,* 43, 596, 1965.

929. **Marr, D. H. and Stothers, J. B.,** *Can. J. Chem.,* 45, 225, 1967.

930. **Marshall, J. L., Faehl, L. G., Ihrig, A. M., and Barfield, M.,** *J. Am. Chem. Soc.,* 98, 3406, 1976.

931. **Marshall, J. L. and Ihrig, A. M.,** *Org. Magn. Reson.,* 5, 235, 1971.

932. **Marshall, J. L., Ihrig, A. M., and Miiller, D. E.,** *J. Magn. Reson.,* 16, 439, 1974.

933. **Marshall, J. L. and Miiller, D. E.,** *J. Am. Chem. Soc.,* 95, 8305, 1973.

934. **Marshall, J. L., Miiller, D. E., Conn, S. A., Seitwell, R., and Ihrig, A. M.,** *Acc. Chem. Res.,* 7, 333, 1974.

935. **Marshall, T. W. and Pople, J. A.,** *Mol. Phys.,* 1, 199, 1958.

936. **Marshall, T. W. and Pople, J. A.,** *Mol. Phys.,* 3, 339, 1960.

937. **Martin, G. J., Gouesnard, J. P., Dorie, J., Rabilles, C., and Martin, M. L.,** *J. Am. Chem. Soc.,* 99, 1381, 1977.

938. **Martin, J. S. and Dailey, B. P.,** *J. Chem. Phys.,* 39, 1722, 1963.

939. **Martin, L. L., Chang, C. J., Floss, H. G., Mabe, J. A., Hagaman, E. W., and Wenkert, E.,** *J. Am. Chem. Soc.,* 94, 8942, 1972.

940. **Martin, R. H., Defay, N., Figeys, H. P., F.-Barbieux, M., Cosyn, J. P., Gelbcke, M., and Schurter, J. J.,** *Tetrahedron,* 25, 4985, 1969.

941. **Maryott, A. A., Farrar, T. C., and Malmberg, M. S.,** *J. Chem. Phys.,* 54, 64, 1971.

942. **Masamune, S., Kemp-Jones, A. V., Green, J., Rabenstein, D. L., Yasunami, M., Takase, K., and Nozoe, T.,** *Chem. Commun.,* 283, 1973.

943. **Mason, J.,** *J. Chem. Soc. A,* 1038, 1971.

944. **Mason, J. and van Bronswijk, W.,** *J. Chem. Soc. A,* 1763, 1970.

945. **Mason, J. and van Bronswijk, W.,** *J. Chem. Soc. A,* 791, 1971.

946. **Mason, J. and Vinter, J. G.,** *J. Chem. Soc. (Dalton),* 2522, 1975.

947. **Mathias, A.,** *Tetrahedron,* 22, 217, 1966.

948. **Matsuo, T. and Shosenji, H.,** *Chem. Commun.,* 501, 1969.

949. **Matsuura, S. and Goto, T.,** *Tetrahedron Lett.,* 1499, 1963.

950. **Matsuzaki, K., Ito, H., Kawamura, T., and Uryu, T.,** *J. Polym. Sci.,* 11, 971, 1973.

951. **Matsuzaki, K., Kanai, T., Kawamura, T., Matsumoto, S., and Uryu, T.,** *J. Polym. Sci.,* 11, 961, 1973.

952. **Matter, U. E., Pascual, C., Pretsch, E., Pross, A., Simon, W., and Sternhell, S.,** *Tetrahedron,* 25, 691, 1969.

953. **Matusch, R.,** *Angew. Chem.,* 87, 283, 1975.

954. **Mavel, G.,** NMR studies of phosphorus compounds, in *Annual Reports on NMR Spectroscopy,* Vol. 5B, Mooney, E. F., Ed., Academic Press, New York, 1973, 1.

955. **Maxwell, L. R. and Bennett, L. H.,** *Physiol. Chem. Phys.,* 10, 59, 1978.

956. **Mayo, R. E. and Goldstein, J. H.,** *J. Mol. Spectrosc.,* 14, 173, 1964.

957. **McConnel, H. M.,** *J. Chem. Phys.,* 24, 460, 1956.

958. **McConnel, H. M.,** *J. Chem. Phys.,* 27, 226, 1957.

959. **McConnel, H. M.,** *J. Chem. Phys.,* 28, 430, 1958.

960. **McConnel, H. M. and Robertson, R. E.,** *J. Chem. Phys.,* 29, 1361, 1958.

961. **McCreary, M. D., Lewis, D. W., Wernick, D. L., and Whitesides, G. M.,** *J. Am. Chem. Soc.,* 96, 1038, 1974.

962. **McDonald, G. G. and Leigh, J. S., Jr.,** *J. Magn. Reson.,* 9, 358, 1973.

963. **McFarlane, H. C. E. and McFarlane, W.,** *Org. Magn. Reson.,* 4, 161, 1972.

964. **McFarlane, H. C. E., McFarlane, W., and Rycroft, D. S.,** *J. Chem. Soc. (Faraday II),* 68, 1300, 1972.

965. **McFarlane, W.,** *Quart. Rev. Chem. Soc.,* 23, 187, 1969.

966. **McInnes, A. G., Walter, J. A., Wright, J. L. C., and Vining, L. C.,** ^{13}C NMR biosynthetic studies, in *Topics in Carbon-13 NMR Spectroscopy,* Vol. 2, Levy, G. C., Ed., Wiley-Interscience, New York, 1976, 123.

967. **McKeever, L. D., Waack, R., Doran, M. A., and Baker, E. B.,** *J. Am. Chem. Soc.,* 91, 1057, 1969.

968. **McKenna, J., McKenna, J. M., Tulley, A., and White, J.,** *J. Chem. Soc.,* 1726, 1965.

969. **McLachlan, A. D.,** *J. Chem. Phys.,* 32, 1263, 1960.

970. **McLaughlin, A. C., McDonald, G. G., and Leigh, J. S.,** *J. Magn. Reson.,* 11, 107, 1973.

971. **Mechin, B., Richer, J. C., and Odiot, S.,** *Org. Magn. Reson.,* 14, 79, 1980.

972. **Meiboom, S. and Gill, D.,** *Rev. Sci. Instrum.,* 29, 688, 1958.

973. **Meiboom, S. and Snyder, L. C.,** *J. Am. Chem. Soc.,* 89, 1038, 1967.

974. **Meinwald, J. and Meinwald, Y. C.,** *J. Am. Chem. Soc.,* 85, 2541, 1963.

975. **Merrill, J. R.,** *J. Phys. Chem.,* 65, 2023, 1961.

976. **Merrit, R. F.,** *J. Am. Chem. Soc.,* 89, 609, 1967.

977. **Méhesfalvi-Vajna, Zs.,** *Ph.D. thesis,* Budapest, 1971.

978. **Méhesfalvi-Vajna, Zs., Neszmélyi, A., Baicz, E., and Sohár, P.,** *Acta Chim. Acad. Sci. Hung.,* 86, 159, 1975.

979. **Mislow, K. and Raban, M.,** Stereoisomeric relationships of groups in molecules, in *Topics in Stereochemistry,* Vol. 1, Allinger, N. L. and Eliel, E. L., Eds., Interscience, New York, 1967, 1.

980. **Mison, P., Chaabouni, R., Diab, Y., Martino, R., Lopez, A., Lattes, A., Wehrli, F. W., and Wirthlin, T.,** *Org. Magn. Reson.,* 8, 79, 1976.

981. **Mitchell, R. H., Klopfenstein, C. E., and Boekelheide, V.,** *J. Am. Chem. Soc.,* 91, 4931, 1969.

982. **Miyajima, G. and Takahashi, K.,** *J. Phys. Chem.,* 75, 331, 1971.

983. **Miyajima, G. and Takahashi, K.,** *J. Phys. Chem.,* 75, 3766, 1971.

984. **Miyajima, G., Takahashi, K., and Nishimoto, K.,** *Org. Magn. Reson.,* 6, 413, 1974.

985. **Miyajima, G., Utsumi, Y., and Takahashi, K.,** *J. Phys. Chem.,* 73, 1370, 1969.

986. **Mochel, V. D.,** *Rev. Macromol. Chem.,* C8, 289, 1972.

987. **Mollere, P. D., Houk, K. N., Bomse, D. S., and Morton, T. H.,** *J. Am. Chem. Soc.,* 98, 4732, 1976.

988. **Mondelli, R. and Merlini, L.,** *Tetrahedron,* 22, 3253, 1966.

989. **Moniz, W. B. and Gutowsky, H. S.,** *J. Chem. Phys.,* 38, 1155, 1963.

990. **Moniz, W. B., Poranski, C. F., Jr., and Sojka, S. A.,** ^{13}C CIDNP as a mechanistic and kinetic probe, in *Topics in Carbon-13 NMR Spectroscopy,* Vol. 3, Levy, G. C., Ed., Wiley-Interscience, New York, 1979, 362.

991. **Montaudo, G., Librando, V., Caccamese, S., and Maravigna, P.,** *J. Am. Chem. Soc.,* 95, 6365, 1973.

992. **Mooberry, E. S. and Krugh, T. R.,** *J. Magn. Reson.,* 17, 128, 1975.

993. **Mooney, E. F.,** *An Introduction to ^{19}F NMR Spectroscopy,* Heyden, London, 1970.

994. **Mooney, E. F. and Winson, P. H.,** Fluorine-19 nuclear magnetic resonance spectroscopy, in *Annual Reports on NMR Spectroscopy,* Vol. 1, Mooney, E. F., Ed., Academic Press, London, 1968, 243.

995. **Mooney, E. F. and Winson, P. H.,** Nitrogen magnetic resonance spectroscopy, in *Annual Reports on NMR Spectroscopy,* Vol. 2, Mooney, E. F., Ed., Academic Press, London, 1969, 125.

996. **Mooney, E. F. and Winson, P. H.,** Carbon-13 nuclear magnetic resonance spectroscopy, carbon-13 chemical shifts and coupling constants, in *Annual Reports on NMR Spectroscopy,* Vol. 2, Mooney, E. F., Ed., Academic Press, New York, 1969, 153; (a) p. 176.

997. **Moore, G. G. I., Kirk, A. R., and Newmark, R. A.,** *J. Heterocyclic Chem.,* 16, 789, 1979.

998. **Moreland, C. G. and Carroll, F. I.,** *J. Magn. Reson.,* 15, 596, 1974.

999. **Mori, N., Omura, S., Yamamoto, O., Suzuki, T., and Tsuzuki, Y.,** *Bull. Chem. Soc. Jpn.,* 36, 1401, 1963.

1000. **Moriarty, R. M. and Kliegman, J. M.,** *J. Org. Chem.,* 31, 3007, 1966.

1001. **Morishima, I., Mizuno, A., Yonezawa, T., and Goto, K.,** *Chem. Commun.,* 1321, 1970.

1002. **Morishima, I., Yoshikawa, K., Okada, K., Yonezawa, T., and Goto, K.,** *J. Am. Chem. Soc.,* 95, 165, 1973.

1003. **Morrill, T. C., Opitz, R. J., and Mozzer, R.,** *Tetrahedron Lett.,* 3715, 1973.

1004. **Morris, D. G. and Murray, A. M.,** *J. Chem. Soc. Perkin Trans. 2,* 1579, 1976.

1005. **Mortimer, F. S.,** *J. Mol. Spectrosc.,* 5, 199, 1960.

1006. **Moy, D. and Young, A. R.,** *J. Am. Chem. Soc.,* 87, 1889, 1965.

1007. **Neszmélyi, A.,** ^{13}C Data Bank of Central Research Institute of Chemistry of the Hungarian Academy of Sciences, Budapest, Hungary.

1008. **Muller, N.,** *J. Chem. Phys.,* 36, 359, 1962.

1009. **Muller, N. and Pritchard, D. E.,** *J. Chem. Phys.,* 31, 768, 1959.

1010. **Muller, N. and Pritchard, D. E.,** *J. Chem. Phys.,* 31, 1471, 1959.

1011. **Muller, N. and Rose, P. I.,** *J. Am. Chem. Soc.,* 85, 2173, 1963.

1012. **Murrell, J. N.,** *Progr. Nucl. Magn. Reson. Spectrosc.,* 6, 1, 1970.

1013. **Murthy, A. S. N. and Rao, C. N. R.,** Spectroscopic studies of the hydrogen bond, in *Applied Spectroscopy Reviews,* Vol. 2, Brame, E. G., Jr., Ed., Marcel Dekker, New York, 1969, 69.

1014. **Musher, J. I.,** *J. Chem. Phys.,* 34, 594, 1961.

1015. **Musher, J. I. and Corey, E. J.,** *Tetrahedron,* 18, 791, 1962.

1016. **Müller, L., Kumar, A., and Ernst, R. R.,** *J. Chem. Phys.,* 63, 5490, 1975.

1017. **Müller, L., Kumar, A., and Ernst, R. R.,** *J. Magn. Reson.,* 25, 383, 1977.

1018. **Nagal, Y., Ohtsuki, M. A., Nakano, T., and Watanabe, H.,** *J. Organometal. Chem.,* 35, 81, 1972.

1019. **Nagayama, K., Bachmann, P., Wüthrich, K., and Ernst, R. R.,** *J. Magn. Reson.,* 31, 133, 1978.

1020. **Nagayama, K., Wüthrich, K., Bachmann, P., and Ernst, R. R.,** *Biochem. Biophys. Res. Commun.,* 78, 99, 1977.

1021. **Nagayama, K., Wüthrich, K., Bachmann, P., and Ernst, R. R.,** *Naturwissenschaften,* 64, 581, 1977.

1022. **Nakagawa, N. and Saito, S.,** *Tetrahedron Lett.,* 1003, 1967.

1023. **Nakanishi, H. and Yamamoto, O.,** *Tetrahedron Lett.,* 1803, 1974.

1024. **Narasimhan, P. T. and Rogers, M. T.,** *J. Am. Chem. Soc.,* 82, 5983, 1960.

1025. **Narasimhan, P. T. and Rogers, M. T.,** *J. Chem. Phys.,* 34, 1049, 1961.

1026. **Nasfay Scott, K.,** *J. Am. Chem. Soc.,* 94, 8564, 1972.

1027. **Naulet, N., Beljean, M., and Martin, G. J.,** *Tetrahedron Lett.,* 3597, 1976.

1028. **Negrebetskii, V. V., Bogdanov, V. S., and Kessenikh, A. V.,** *Z. Strukt. Khim.,* 12, 716, 1971.

1029. **Neiman, Z. and Bergmann, F.,** *Chem. Commun.,* 1002, 1968.

1030. **Nelson, G. L., Levy, G. C., and Cargioli, J. D.,** *J. Am. Chem. Soc.,* 94, 3089, 1972.

1031. **Neszmélyi, A., Lipták, A., Nánási, P., and Szejtli, J.,** *J. Am. Chem. Soc.,* in press.

1032. **Neudert, W. and Röpke, H.,** *Atlas of Steroid Spectra,* Springer-Verlag, Berlin, 1965.

1033. **Neuss, N., Nash, C. N., Lemke, P. A., and Grutzner, J. B.,** *J. Am. Chem. Soc.,* 93, 2337, 1971.

1034. **Newmark, R. A. and Hill, J. R.,** *Org. Magn. Reson.,* 13, 40, 1980.

1035. **Newmark, R. A. and Sederholm, C. H.,** *J. Chem. Phys.,* 39, 3131, 1963.

1036. **Newmark, R. A. and Sederholm, C. H.,** *J. Chem. Phys.,* 43, 602, 1965.

1037. **Newton, M. D., Schulman, J. M., and Manns, M. M.,** *J. Am. Chem. Soc.,* 96, 17, 1974.

1038. **Ng, S. and Sederholm, C. H.,** *J. Chem. Phys.,* 40, 2090, 1964.

1039. **Nickon, A., Castle, M. A., Harada, R., Berkoff, C. E., and Williams, R. O.,** *J. Am. Chem. Soc.,* 85, 2185, 1963.

1040. **Witanowski, M. and Webb, G. A., Eds.,** *Nitrogen NMR,* Plenum Press, London, 1973.

1041. **Nixon, J. F. and Pidcock, A.,** Phosphorus-31 nuclear magnetic resonance spectra of co-ordination compounds, in *Annual Reports on NMR Spectroscopy,* Vol. 2, Mooney, E. F., Ed., Academic Press, New York, 1969, 346.

1042. *NMR Basic Principles and Progress, Grundlagen und Fortschritte 1-8,* Diehl, P., Fluck, E., and Kosfeld, R., Eds., Springer-Verlag, Berlin, 1969.

1043. *NMR Specialist Periodical Reports,* Harris, R. K. and Abraham, R. J., Eds., Pergamon Press, Oxford; (a) Vol. 3, p. 79; (b) Vol. 1, p. 281.

1044. *NMR Spectra Catalog,* Compiled by Bhacco, N. S., Hollis, D. P., Johnson, L. F., Pier, G. A., and Shoolery, J. N., VARIAN Associates, Palo Alto, Calif., 1962-1963.

1045. **Noggle, J. H. and Schirmer, R. E.,** *The Nuclear Overhauser Effect; Chemical Applications,* Academic Press, New York, 1971.

1046. **Nógrádi, M., Ollis, W. D., and Sutherland, I. O.,** *Chem. Commun.,* 158, 1970.

1047. **Nonhebel, D. C.,** *Tetrahedron,* 24, 1869, 1968.

1048. **Nouls, J. C., Binst, G., and van Martin, R. H.,** *Tetrahedron Lett.,* 4065, 1967.

1049. **Nöth, H. and Wrackmeyer, B.,** *Chem. Ber.,* 107, 3070, 1974.

1050. **Nöth, H. and Wrackmeyer, B.,** *Chem. Ber.,* 107, 3089, 1974.

1051. *Nuclear Magnetic Resonance for Organic Chemists,* Mathieson, D. W., Ed., Academic Press, New York, 1967.

1052. *Nuclear Magnetic Resonance Spectra and Chemical Structure,* Brügel, W., Ed., Academic Press, New York, 1967.

1053. **Axenrod, T. and Webb, G. A., Eds.,** *Nuclear Magnetic Resonance Spectroscopy of Nuclei Other than Protons,* Wiley-Interscience, New York, 1974.

1054. **Ogg, R. A., Jr. and Diehl, P.,** *Helv. Phys. Acta,* 30, 251, 1957.

1055. **Ogg, R. A. and Ray, J. D.,** *J. Chem. Phys.,* 26, 1340, 1957.

1056. **Ogg, R. A. and Ray, J. D.,** *J. Chem. Phys.,* 26, 1515, 1957.

1057. **Ohtsuru, M. and Tori, K.,** *J. Mol. Spectrosc.,* 27, 296, 1968.

1058. **Ohtsuru, M. and Tori, K.,** *Tetrahedron Lett.,* 4043, 1970.

1059. **Okamura, W. H. and Sondheimer, F.,** *J. Am. Chem. Soc.,* 89, 5991, 1967.

1060. **Oki, M., Iwamura, H., and Hayakava, N.,** *Bull. Chem. Soc. Jpn.,* 36, 1542, 1963.

1061. **Oláh, G. A., Bollinger, J. M., and Brinich, J.,** *J. Am. Chem. Soc.,* 90, 2587, 1968.

1062. **Oláh, G. A., DeMember, J. R., and Schlosberg, R. H.,** *J. Am. Chem. Soc.,* 91, 2112, 1969.

1063. **Oláh, G. A., Denis, J.-M., and Westerman, P. W.,** *J. Org. Chem.,* 39, 1206, 1974.

1064. **Oláh, G. A. and Kiovsky, T. E.,** *J. Am. Chem. Soc.,* 90, 4666, 1968.

1065. **Oláh, G. A. and Kreienbühl, P.,** *J. Am. Chem. Soc.,* 89, 4756, 1967.

1066. **Oláh, G. A. and Matescu, G. D.,** *J. Am. Chem. Soc.,* 92, 1430, 1970.

1067. **Oláh, G. A. and Pittman, C. U., Jr.,** *Advances in Physical Organic Chemistry,* Vol. 4, Gold, V., Ed., Academic Press, New York, 1966.

1068. **Oláh, G. A. and White, A. M.,** *J. Am. Chem. Soc.,* 89, 7072, 1967.

1069. **Oláh, G. A. and White, A. M.,** *J. Am. Chem. Soc.,* 91, 5801, 1969.

1070. **Ollis, W. D. and Sutherland, I. O.,** *Chem. Commun.,* 402, 1966.

1071. **O'Reilly, D. E.,** *J. Chem. Phys.,* 36, 274, 1962.

1072. **Oth, J. F. M., Merényi, R., Röttele, H., and Schröder, G.,** *Tetrahedron Lett.,* 3941, 1968.

1073. **Oth, J. F. M., Müllen, K., Gilles, J. M., and Schröder, G.,** *Helv. Chim. Acta,* 57, 1415, 1974.

1074. **Ouelette, R. J.,** *J. Am. Chem. Soc.,* 86, 4378, 1964.

1075. **Overberger, C. G., Kurtz, T., and Yaroslavsky, S.,** *J. Org. Chem.,* 30, 4363, 1965.

1076. **Overhauser, A.,** *Phys. Rev.,* 89, 689, 1953; 92, 411, 1953.

1077. **Ozubko, R. S., Buchanan, G. W., and Smith, I. C. P.,** *Can. J. Chem.,* 52, 2493, 1974.

1078. **Padwa, A., Shefter, E., and Alexander, E.,** *J. Am. Chem. Soc.,* 90, 3717, 1968.

1079. **Page, J. E.,** Nuclear magnetic resonance spectra of steroids, in *Annual Reports on NMR Spectroscopy,* Vol. 3, Mooney, E. F., Ed., Academic Press, New York, 1970, 149.

1080. **Page, T. F., Alger, T., and Grant, D. M.,** *J. Am. Chem. Soc.,* 87, 5333, 1965.

1081. **Paolillo, L. and Becker, E. D.**, *J. Magn. Reson.*, 2, 168, 1970.
1082. **Paolillo, L., Tancredi, T., Temussi, P. A., Trivellone, E., Bradbury, E. M., and Crane-Robinson, C.**, *Chem. Commun.*, 335, 1972.
1083. **Parker, R. G. and Roberts, J. D.**, *J. Am. Chem. Soc.*, 92, 743, 1970.
1084. **Parker, R. G. and Roberts, J. D.**, *J. Org. Chem.*, 35, 996, 1970.
1085. **Parmigiani, A., Perotti, A., and Riganti, V.**, *Gazz. Chim. Ital.*, 91, 1148, 1961.
1086. **Parrish, R., Kurland, D., Janese, W., and Bakay, L.**, *Science*, 183, 438, 1974.
1087. **Pascual, C., Meier, J., and Simon, W.**, *Helv. Chim. Acta*, 49, 164, 1966.
1088. **Patel, D. J., Howden, M. E. H., and Roberts, J.D.**, *J. Am. Chem. Soc.*, 85, 3218, 1963.
1089. **Patrick, T. B. and Patrick, P. H.**, *J. Am. Chem. Soc.*, 95, 6230, 1972.
1090. **Patt, S. L. and Sykes, D. B.**, *J. Chem. Phys.*, 56, 3182, 1972.
1091. **Patterson, A., Jr. and Ettinger, R.**, *Z. Elektrochem.*, 64, 98, 1960.
1092. **Paudler, W. W. and Blewitt, H. L.**, *J. Org. Chem.*, 30, 4081, 1965.
1093. **Paudler, W. W. and Dunham, D. E.**, *J. Heterocyclic Chem.*, 2, 410, 1965.
1094. **Paudler, W. W. and Kress, T. J.**, *Chem. Ind. (London)*, 1557, 1966.
1095. **Paudler, W. W. and Kuder, J. E.**, *J. Heterocyclic Chem.*, 3, 33, 1966.
1096. **Paudler, W. W. and Kuder, J. E.**, *J. Org. Chem.*, 31, 809, 1966.
1097. **Paul, E. G. and Grant, D. M.**, *J. Am. Chem. Soc.*, 85, 1701, 1963.
1098. **Pauli, W.**, *Naturwissenschaften*, 12, 741, 1924.
1099. **Peake, A. and Thomas, L. F.**, *Trans. Faraday Soc.*, 62, 2980, 1966.
1100. **Pehk, T. and Lippmaa, E.**, *Eesti NSV Tead. Akad. Toim. Keem. Geol.*, 17, 291, 1968.
1101. **Pehk, T. and Lippmaa, E.**, *Org. Magn. Reson.*, 3, 679, 1971.
1102. Perkin-Elmer Model R-10 60 MHz Spectrometer, Perkin-Elmer Corp., Norwalk, Conn.
1103. *Perkin-Elmer NMR Q.*, 2, 11, 1971.
1104. *Perkin-Elmer NMR Q.*, 5, 12, 1972.
1105. **Perlin, A. S., Casu, B., and Koch, H. J.**, *Can. J. Chem.*, 48, 2599, 1970.
1106. **Perlin, A. S., Cyr, N., Koch, H. J., and Korsch, B.**, *Ann. N. Y. Acad. Sci.*, 222, 935, 1973.
1107. **Perlin, A. S. and Koch, H. J.**, *Can. J. Chem.*, 48, 2639, 1970.
1108. **Petrakis, L. and Sederholm, C. H.**, *J. Chem. Phys.*, 35, 1174, 1961.
1109. **Pfeffer, H. U. and Klessinger, M.**, *Org. Magn. Reson.*, 9, 121, 1977.
1110. **Philipsborn, W.**, *Angew. Chem.*, 83, 470, 1971.
1111. **Piette, L. H., Ray, J. D., and Ogg, R. A., Jr.**, *J. Mol. Spectrosc.*, 2, 66, 1958.
1112. **Pihlaja, K. and Pasanen, P.**, *Suom. Kemisbil.*, 46, 273, 1973.
1113. **Pimentel, G. C. and McClennan, A. L.**, *The Hydrogen Bond*, W. H. Freeman, San Francisco, 1960.
1114. **Pinhey, J. T. and Sternhell, S.**, *Tetrahedron Lett.*, 275, 1963.
1115. **Pirkle, W. H.**, *J. Am. Chem. Soc.*, 88, 1837, 1966.
1116. **Pirkle, W. H. and Beare, S. D.**, *J. Am. Chem. Soc.*, 89, 5485, 1967.
1117. **Pirkle, W. H. and Beare, S. D.**, *J. Am. Chem. Soc.*, 91, 5150, 1969.
1118. **Pitcher, E., Buckingham, A. D., and Stone, F. G. A.**, *J. Chem. Phys.*, 36, 124, 1962.
1119. **Pitzer, K. S. and Donath, W. E.**, *J. Am. Chem. Soc.*, 81, 3213, 1959.
1120. **Pohland, A. E., Badger, R. C., and Cromwell, N. H.**, *Tetrahedron Lett.*, 4369, 1965.
1121. **Pomerantz, M. and Fink, R.**, *Chem. Commun.*, 430, 1975.
1122. **Pomerantz, M., Fink, R., and Gray, G. A.**, *J. Am. Chem. Soc.*, 98, 291, 1976.
1123. **Pomerantz, M. and Hillenbrand, D. F.**, *J. Am. Chem. Soc.*, 95, 5809, 1973.
1124. **Pomerantz, M. and Hillenbrand, D. F.**, *Tetrahedron*, 31, 217, 1975.
1125. **Pople, J. A.**, *Proc. R. Soc. London Ser. A*, 239, 541, 1957.
1126. **Pople, J. A.**, *Dis. Faraday Soc.*, 34, 7, 1962.
1127. **Pople, J. A.**, *J. Chem. Phys.*, 37, 53, 60, 1962.
1128. **Pople, J. A.**, *Proc. R. Soc. London Ser. A*, 239, 550, 1957.
1129. **Pople, J. A. and Bothner-By, A. A.**, *J. Chem. Phys.*, 42, 1339, 1965.
1130. **Pople, J. A. and Gordon, M. S.**, *J. Am. Chem. Soc.*, 89, 4253, 1967.
1131. **Pople, J. A. and Santry, D. P.**, *Mol. Phys.*, 8, 1, 1964.
1132. **Pople, J. A., Schneider, W. G., and Bernstein, H. J.**, *High-Resolution Nuclear Magnetic Resonance*, McGraw-Hill, New York, 1959.
1133. **Pople, J. A. and Untch, K. G.**, *J. Am. Chem. Soc.*, 88, 4811, 1966.
1134. **Poranski, C. F. and Moniz, W. B.**, *J. Phys. Chem.*, 71, 1142, 1967.
1135. **Porte, A. L., Gutowsky, H. S., and Hunsberger, I. M.**, *J. Am. Chem. Soc.*, 82, 5057, 1960.
1136. **Porter, R., Marks, T. J., and Shriver, D. F.**, *J. Am. Chem. Soc.*, 95, 3548, 1973.
1137. **Portoghese, P. S. and Telang, V. G.**, *Tetrahedron*, 27, 1823, 1971.
1138. **Posner, T. B., Markowski, V., Loftus, P., and Roberts, J. D.**, *Chem. Commun.*, 769, 1975.
1139. **Pouchoulin, G., Llinas, J. R., Buomo, G., and Vincent, E.-J.**, *Org. Magn. Reson.*, 8, 518, 1976.

1140. **Pregosin, P. S. and Randall, E. W.,** *Chem. Commun.,* 399, 1971.

1141. **Pregosin, P. S. and Randall, E. W.,** ^{13}C Nuclear magnetic resonance, in *Determination of Organic Structures by Physical Methods,* Vol. 4, Nachod, F. C. and Zuckerman, J. J., Eds., Academic Press, New York, 1971, chap. 6.

1142. **Pregosin, P. S., Randall, E. W., and White, A. I.,** *J. Chem. Soc. Perkin Trans. 2,* 1, 1972.

1143. **Pretsch, E., Clerc, T., Seibl, J., and Simon, W.,** *Tabellen zur Strukturaufklärung organischer Verbindungen mit spektroskopischen Methoden,* Springer-Verlag, Berlin, 1976; (a) p. C250; (b) p. C10; (c) p. C15; (d) p. C75; (e) p. C50; (f) p. C40; (g) p. C70; (h) p. C100; (i) p. C90; (j) p. C110; (k) p. C170; (l) p. C175, p. C180, p. C185; (m) p. B10; (n) p. C120, p. C125; (o) p. C150; (p) p. C140, p. C145; (r) p. C160; (s) p. C220; (t) p. C230.

1144. **Price, D., Suschitzky, H., and Hollies, J. I.,** *J. Chem. Soc. C,* 1967, 1966.

1145. **Proctor, W. G. and Yu, F. C.,** *Phys. Rev.,* 77, 717, 1950.

1146. **Proctor, W. F. and Yu, F. C.,** *Phys. Rev.,* 81, 20, 1951.

1147. *Progress in Nuclear Magnetic Resonance Spectroscopy,* Emsley, J. W., Feeney, J., and Sutcliffe, L. H., Eds., Pergamon Press, New York, 1966, (a) Vol. 11, 1977, p. 2, p. 3.

1148. **Pugmire, R. J. and Grant, D. M.,** *J. Am. Chem. Soc.,* 90, 697, 1968.

1149. **Pugmire, R. J. and Grant, D. M.,** *J. Am. Chem. Soc.,* 90, 4232, 1968.

1150. **Pugmire, R. J. and Grant, D. M.,** *J. Am. Chem. Soc.,* 93, 1880, 1971.

1151. **Pugmire, R. J., Grant, D. M., Robins, M. J., and Robins, R. K.,** *J. Am. Chem. Soc.,* 91, 6381, 1969.

1152. **Pugmire, R. J., Robins, M. J., Grant, D. M., and Robins, R. K.,** *J. Am. Chem. Soc.,* 93, 1887, 1971.

1153. **Purcell, E. M., Torrey, H. C., and Pound, R. V.,** *Phys. Rev.,* 69, 37, 1946.

1154. **Raban, M. and Mislow, K.,** *Tetrahedron Lett.,* 3961, 1966.

1155. **Rabenstein, D. L.,** *Anal. Chem.,* 43, 1599, 1971.

1156. **Rabenstein, D. L. and Sayer, T. L.,** *J. Magn. Reson.,* 24, 27, 1976.

1157. **Rabi, I. I., Millman, S., Kusch, P., and Zacharias, J. R.,** *Phys. Rev.,* 55, 526, 1939.

1158. **Rácz, P., Tompa, K., and Pocsik, I.,** *Exp. Eye Res.,* 28, 129, 1979; 29, 601, 1979.

1159. **Radeglia, R., Storek, W., Engelhardt, G., Ritschil, F., Lippmaa, E., Pehk, T., Magi, M., and Martin, D.,** *Org. Magn. Reson.,* 5, 419, 1973.

1160. **Rader, C. P.,** *J. Am. Chem. Soc.,* 88, 1713, 1966.

1161. **Ramey, K. C., Lini, D. C., and Krow, G.,** General review of nuclear magnetic resonance, in *Annual Reports on NMR Spectroscopy,* Vol. 6A, Mooney, E. F., Ed., Academic Press, New York, 1975, 147.

1162. **Ramsey, N. F.,** *Phys. Rev.,* 77, 567, 1950.

1163. **Ramsey, N. F.,** *Phys. Rev.,* 78, 699, 1950.

1164. **Ramsey, N. F.,** *Phys. Rev.,* 86, 243, 1952.

1165. **Ramsey, N. F.,** *Phys. Rev.,* 91, 303, 1953.

1166. **Ramsey, N. F. and Purcell, E. M.,** *Phys. Rev.,* 85, 143, 1952.

1167. **Ranade, S. S., Shah, S., Korgaonkar, K. S., Kasturi, S. R., Chaughule, R. S., and Vijayaraghavan, R.,** *Physiol. Chem. Phys.,* 8, 131, 1976.

1168. **Randall, E. W. and Gilles, D. G.,** Nitrogen nuclear magnetic resonance, in *Progress in Nuclear Magnetic Resonance Spectroscopy,* Vol. 6, Emsley, J. W., Feeney, J., and Sutcliffe, L. H., Eds., Pergamon Press, Oxford, 1971, 119.

1169. **Randall, E. W. and Shaw, D.,** *Spectrochim. Acta,* 23a, 1235, 1967.

1170. **Ranft, J.,** *Ann. Physik.,* 10, 399, 1963.

1171. **Rao, C. N. R.,** *Can. J. Chem.,* 40, 963, 1962.

1172. **Rattet, L. S., Williamson, A. D., and Goldstein, J. H.,** *J. Phys. Chem.,* 72, 2954, 1968.

1173. **Read, J. M., Mayo, R. E., and Goldstein, J. H.,** *J. Mol. Spectrosc.,* 22, 419, 1967.

1174. **Reddy, G. S. and Goldstein, J. H.,** *J. Am. Chem. Soc.,* 83, 2045, 1961.

1175. **Reddy, G. S. and Goldstein, J. H.,** *J. Phys. Chem.,* 65, 1539, 1961..

1176. **Reddy, G. S., Hobgood, R. T., Jr., and Goldstein, J. H.,** *J. Am. Chem. Soc.,* 84, 336, 1962.

1177. **Reddy, G. S., Mandell, L., and Goldstein, J. H.,** *J. Am. Chem. Soc.,* 83, 1300, 1961.

1178. **Redfield, A. G.,** *Methods Enzymol.,* 49, 253, 1978.

1179. **Redfield, A. G. and Gupta, R. K.,** *J. Chem. Phys.,* 54, 1418, 1971.

1180. **Redfield, A. G., Kunz, S. D., and Ralph, E. K.,** *J. Magn. Reson.,* 19, 114, 1975.

1181. **Reeves, L. W.,** *Trans. Faraday Soc.,* 55, 1684, 1959.

1182. **Reeves, L. W.,** *Adv. Phys. Org. Chem.,* 3, 187, 1965.

1183. **Reeves, L. W., Riveros, J. M., Spragg, R. A., and Vavin, J. A.,** *Mol. Phys.,* 25, 9, 1973.

1184. **Reeves, L. W. and Schneider, W. G.,** *Can. J. Chem.,* 36, 793, 1958.

1185. **Reeves, L. W. and Schneider, W. G.,** *Trans. Faraday Soc.,* 54, 314, 1958.

1186. **Reich, H. J., Jautelat, M., Messe, M. T., Weigert, F. J., and Roberts, J. D.,** *J. Am. Chem. Soc.,* 91, 7445, 1969.

1187. **Reilly, C. A. and Swalen, J. D.,** *J. Chem. Phys.,* 32, 1378, 1960.

1188. **Reiter, J., Sohár, P., Lipták, J., and Toldy, L.,** *Tetrahedron Lett.,* 1417, 1970.

1189. **Reiter, J., Sohár, P., and Toldy, L.,** *Tetrahedron Lett.,* 1411, 1970.

1190. **Retcofsky, H. L. and Friedel R. A.,** *J. Phys. Chem.,* 71, 3592, 1967; 72, 290, 2619, 1968.

1191. **Retcofsky, H. L. and Friedel, R. A.,** *J. Magn. Reson.,* 8, 398, 1972.

1192. **Retcofsky, H. L. and McDonald, F. R.,** *Tetrahedron Lett.,* 2575, 1968.

1193. **Reuben, J.,** Effects of chemical equilibrium and adduct stoichiometry in shift reagent studies, in *Nuclear Magnetic Resonance Shift Reagents,* Sievers, R. F., Ed., Academic Press, New York, 1973, 341.

1194. **Reuben, J.,** *J. Am. Chem. Soc.,* 95, 3534, 1973.

1195. **Reuben, J.,** *J. Magn. Reson.,* 11, 103, 1973.

1196. **Reuben, J.,** Paramagnetic lanthanide shift reagents in NMR spectroscopy; principles, methodology and applications, in *Progress in Nuclear Magnetic Resonance Spectroscopy,* Vol. 9, Emsley, J. W., Feeney, J. and Sutcliffe, L. H., Eds., Pergamon Press, Oxford, 1973, 1.

1197. **Reuben, J. and Fiat, D.,** *J. Chem. Phys.,* 51, 4909, 1969.

1198. **Reuben, J. and Leight, J. S., Jr.,** *J. Am. Chem. Soc.,* 94, 2789, 1972.

1199. **Reuben, J., Tzalmona, A., and Samuel, D.,** *Proc. Chem. Soc.,* 353, 1962.

1200. **Reutov, O. A., Shatkina, T. N., Lippmaa, E. T., and Pehk, T. I.,** *Dokl. Akad. Nauk. SSSR,* 181, 1400, 1968.

1201. **Reutov, O. A., Shatkina, T. N., Lippmaa, E. T., and Pehk, T. I.,** *Tetrahedron,* 25, 5757, 1969.

1202. **Revel, M., Roussel, J., Navech, J., and Mathis, J.,** *Org. Magn. Reson.,* 8, 399, 1976.

1203. **Ricca, G. S., Danieli, B., Palmisano, G., Duddeck, H., and Elgamal, M. H. A.,** *Org. Magn. Reson.,* 11, 163, 1978.

1204. **Richard, C. and Granger, P.,** Chemically induced dynamic nuclear and electron polarizations — CIDNP and CIDEP, in *NMR Basic Principles and Progress,* Vol. 8, Diehl, P., Fluck, E., and Kosfeld, R., Eds., Springer-Verlag, Berlin, 1974.

1205. **Richards, R. E. and Schaefer, T. P.,** *Trans. Faraday Soc.,* 54, 1280, 1958.

1206. **Richards, R. E. and Thomas, N. A.,** *J. Chem. Soc. Perkin Trans. 2,* 368, 1974.

1207. **Riddell, F. G.,** *J. Chem. Soc. B,* 331, 1970.

1208. **Riddell, F. G. and Lehn, J. M.,** *J. Chem. Soc. B,* 1224, 1968.

1209. **Roberts, B. W., Lambert, J. B., and Roberts, J. D.,** *J. Am. Chem. Soc.,* 87, 5439, 1965.

1210. **Roberts, J. D.,** *J. Am. Chem. Soc.,* 78, 4495, 1956.

1211. **Roberts, J. D.,** *An Introduction to the Analysis of Spin-Spin Splitting in High-Resolution Nuclear Magnetic Resonance Spectra,* Benjamin, New York, 1962.

1212. **Roberts, J. D., Lutz, R. P., and Davis, D. R.,** *J. Am. Chem. Soc.,* 83, 246, 1961.

1213. **Roberts, J. D., Weigert, F. J., Kroschwitz, J. I., and Reich, H. J.,** *J. Am. Chem. Soc.,* 92, 1338, 1970.

1214. **Roberts, R. T. and Chachaty, C.,** *Chem. Phys. Lett.,* 22, 348, 1973.

1215. **Rodger, C., Sheppard, N., McFarlane, C., and McFarlane, W.,** Group VI — oxygen, sulphur, selenium and tellurium, in *NMR and the Periodic Table,* Harris, R. K. and Mann, B. E., Eds., Academic Press, New York, 1978, 383.

1216. **Rogers, E. H.,** *NMR-EPR Workshop Notes,* VARIAN Associates, Palo Alto, Calif.,

1217. **Rogers, M. T. and LaPlanche, L. A.,** *J. Phys. Chem.,* 69, 3648, 1965.

1218. **Ronayne, J. and Williams, D. H.,** *Chem. Commun.,* 712, 1966.

1219. **Ronayne, J. and Williams, D. H.,** *J. Chem. Soc. C,* 2642, 1967.

1220. **Ronayne, J. and Williams, D. H.,** Solvent effects in proton magnetic resonance spectroscopy, in *Annual Reports on NMR Spectroscopy,* Vol. 2, Mooney, E. F., Ed., Academic Press, New York, 1969, 83.

1221. **Rondeau, R. E., Berwick, M. A., Steppel, R. N., and Servé, M. P.,** *J. Am. Chem. Soc.,* 94, 1096, 1972.

1222. **Rondeau, R. E. and Sievers, R. E.,** *J. Am. Chem. Soc.,* 93, 1522, 1971.

1223. **Rosenberg, D., DeHaan, J. W., and Drenth, W.,** *Rec. Trav. Chim. Pays-Bas,* 87, 1387, 1968.

1224. **Rosenberg, D. and Drenth, W.,** *Tetrahedron,* 27, 3893, 1971.

1225. **Rousselot, M. M.,** *C. R. Acad. Sci. France C,* 262, 26, 1966.

1226. **Rowe, J. J. M., Hinton, J., and Rowe, K. L.,** *Chem. Rev.,* 70, 1, 1970.

1227. **Royden, V.,** *Phys. Rev.,* 96, 543, 1954.

1228. **Runsink, J. and Günther, H.,** *Org. Magn. Reson.,* 13, 249, 1980.

1229. **Sackmann, E. and Dreeskamp, H.,** *Spectrochim. Acta,* 21, 2005, 1965.

1230. *SADTLER Nuclear Magnetic Resonance Spectra,* SADTLER Research Laboratories, Philadelphia, 1965.

1231. **Saika, A. and Slichter, C. P.,** *J. Chem. Phys.,* 22, 26, 1954.

1232. **Saito, H. and Smith, I. C. P.,** *Arch. Biochem. Biophys.,* 158, 154, 1973.

1233. **Saito, H., Tanaka, Y., and Nagata, S.,** *J. Am. Chem. Soc.,* 95, 324, 1973.

1234. **Sanders, J. K. M. and Williams, D. H.,** *Chem. Commun.,* 422, 1970.

1235. **Sanders, J. K. M. and Williams, D. H.,** *J. Am. Chem. Soc.,* 93, 641, 1971.

1236. **Sanders, J. K. M. and Williams, D. H.,** *Tetrahedron Lett.,* 2813, 1971.

1237. **Sasaki, Y., Kawaki, H., and Okazaki, Y.,** *Chem. Pharm. Bull. (Jpn.),* 21, 2488, 1973.

1238. **Saunders, M. and Hyne, J. B.,** *J. Chem. Phys.,* 29, 253, 1319, 1958.

1239. **Savitsky, G. B.,** *J. Phys. Chem.,* 67, 2723, 1963.

1240. **Savitsky, G. B., Ellis, P. D., Namikawa, K., and Maciel, G. E.,** *J. Chem. Phys.,* 49, 2395, 1968.

1241. **Savitsky, G. B., Namikawa, K., and Zweifel, G.,** *J. Phys. Chem.,* 69, 3105, 1965.

1242. **Savitsky, G. B., Pearson, R. M., and Namikawa, K.,** *J. Phys. Chem.,* 60, 1425, 1965.

1243. **Schaefer, J.,** *Macromolecules,* 2, 210, 1969.

1244. **Schaefer, J.,** *Macromolecules,* 4, 105, 1971.

1245. **Schaefer, J.,** The carbon-13 NMR analysis of synthetic high polymers, in *Topics in Carbon-13 NMR Spectroscopy,* Vol. 1, Levy, G. C., Ed., Wiley-Interscience, New York, 1974, 149.

1246. **Schaefer, J. and Stejskal, E. O.,** High-resolution ^{13}C NMR of solid polymers, in *Topics in Carbon-13 NMR Spectroscopy,* Vol. 3, Levy, G. C., Ed., Wiley-Interscience, New York, 1979, 284.

1247. **Schaefer, J., Stejskal, E. O., and Buchdahl, R.,** *Macromolecules,* 8, 291, 1979.

1248. **Schaefer, T.,** *J. Chem. Phys.,* 36, 2235, 1962.

1249. **Schaefer, T., Reynolds, W. F., and Yonemoto, T.,** *Can. J. Chem.,* 41, 2969, 1963.

1250. **Schaefer, T. and Schneider, W. G.,** *Can. J. Chem.,* 38, 2066, 1960.

1251. **Schaefer, T. and Schneider, W. G.,** *J. Chem. Phys.,* 32, 1224, 1960.

1252. **Scheiner, P. and Litchman, W. M.,** *Chem. Commun.,* 781, 1972.

1253. **Schiemenz, G. P.,** *Tetrahedron,* 29, 741, 1973.

1254. **Schmidbaur, H.,** *Chem. Ber.,* 97, 1639, 1964.

1255. **Schmidt, C. F. and Chan, S. I.,** *J. Chem. Phys.,* 55, 4670, 1971.

1256. **Schmidt, C. F. and Chan, S. I.,** *J. Magn. Reson.,* 5, 151, 1971.

1257. **Schmutzler, R.,** *J. Chem. Soc.,* 4551, 1964.

1258. **Schneider, H. J., Freitag, W., and Hoppen, V.,** *Org. Magn. Reson.,* 13, 266, 1980.

1259. **Schneider, H. J., Price, R., and Keller, T.,** *Angew. Chem. Int. Ed. Engl.,* 10, 730, 1971.

1260. **Schneider, W. G., Bernstein, H. J., and Pople, J. A.,** *J. Chem. Phys.,* 28, 601, 1958.

1261. **Schraml, J., Duc-Chuy, N., Chvalovsky, V., Mägi, M., and Lippmaa, E.,** *Org. Magn. Reson.,* 7, 379, 1975.

1262. **Schröder, G.,** *Angew. Chem. Int. Ed. Engl.,* 4, 752, 1965.

1263. **Schulman, J. M. and Venanzi, T.,** *J. Am. Chem. Soc.,* 98, 4701, 1976.

1264. **Schultheiss, H. and Fluck, E.,** *Z. Naturforsch.,* 32b, 257, 1977.

1265. **Schwarcz, J. A. and Perlin, A. S.,** *Can. J. Chem.,* 50, 3667, 1972.

1266. **Schwarz, R. M. and Rabjohn, N.,** *Org. Magn. Reson.,* 13, 9, 1980.

1267. **Schwyzer, R. and Ludescher, U.,** *Helv. Chim. Acta,* 52, 2033, 1969.

1268. **Sears, R. E. J.,** *J. Chem. Phys.,* 56, 983, 1971.

1269. **Seel, F., Hartmann, V., and Gombler, W.,** *Z. Naturforsch.,* 27b, 325, 1972.

1270. **Seel, H., Aydin, R., and Günther, H.,** *Z. Naturforsch.,* 33, 353, 1978.

1271. **Segre, A. and Musher, J. I.,** *J. Am. Chem. Soc.,* 89, 706, 1967.

1272. **Sen, B. and Wu, W. C.,** *Anal. Chim. Acta,* 46, 37, 1969.

1273. **Senda, Y., Ishiyama, J.-I., and Imaizumi, S.,** *Bull. Chem. Soc. Jpn.,* 50, 2813, 1977.

1274. **Seo, S., Tomita, Y., and Tori, K.,** *Chem. Commun.,* 270, 1975.

1275. **Sepulchre, A. M., Septe, B., Lukacs, G., Gero, S. D., Voelter, W., and Breitmaier, E.,** *Tetrahedron,* 30, 905, 1974.

1276. **Sergeyev, N. M. and Solkan, V. N.,** *Chem. Commun.,* 12, 1975.

1277. **Servis, K. L. and Fang, K. N.,** *J. Am. Chem. Soc.,* 90, 6712, 1968.

1278. **Sewell, P. R.,** The nuclear magnetic resonance spectra of polymers, in *Annual Reports on NMR Spectroscopy,* Vol. 1, Mooney, E. F., Ed., Academic Press, New York, 1968, 165.

1279. **Shadowitz, A.,** *The Electromagnetic Field,* McGraw-Hill, New York, 1975.

1280. **Shafer, P. R., Davis, D. R., Vogel, M., Nagarajan, K., and Roberts, J. D.,** *Proc. Natl. Acad. Sci. U.S.A.,* 47, 49, 1961.

1281. **Shapiro, B. L., Hlubucek, J. R., Sullivan, G. R., and Johnson, L. F.,** *J. Am. Chem. Soc.,* 93, 3281, 1971.

1282. **Shapiro, B. L. and Johnston, M. D., Jr.,** *J. Am. Chem. Soc.,* 94, 8185, 1972.

1283. **Shapiro, B. L., Johnston, M. D., Jr., and Towns, R. L. R.,** *J. Am. Chem. Soc.,* 94, 4381, 1972.

1284. **Shaw, D.,** Fourier transform NMR, in *NMR Specialist Periodical Reports,* Vol. 3, Harris, R. K., Ed., The Chemical Society, London, 1974, 249.

1285. **Shaw, D.,** Fourier transform NMR, in *NMR Specialist Periodical Reports,* Vol. 5, Harris, R. K., Ed., The Chemical Society, London, 1976, 188.

1286. **Shaw, D.,** *Fourier Transform NMR Spectroscopy,* Elsevier, Amsterdam, 1976, (a) p. 191; (b) pp. 131-134; (c) p. 189; (d) p. 195; (e) p. 184; (f) p. 141; (g) p. 335.

1287. **Sheinblatt, M.,** *J. Am. Chem. Soc.,* 88, 2845, 1966.

1288. **Sheppard, N. and Turner, J. J.,** *Proc. R. Soc. London Ser. A,* 252, 506, 1959.

1289. **Sheppard, N. and Turner, J. J.,** *Mol. Phys.,* 3, 168, 1960.

1290. **Shimanouchi, T.,** *Pure Appl. Chem.,* 12, 287, 1966.

1291. **Shoolery, J. N.,** *VARIAN Tech. Inform. Bull.,* 2, 3, 1959.

1292. **Shoolery, J. N. and Alder, B.,** *J. Chem. Phys.,* 23, 805, 1955.

1293. **Shoolery, J. N. and Rogers, M. T.,** *J. Am. Chem. Soc.,* 80, 5121, 1958.

1294. **Shoppee, C. W., Johnson, F. P., Lack, R. E., Shannon, J. S., and Sternhell, S.,** *Tetrahedron Suppl.,* 8, 421, 1966.

1295. **Shporer, M. and Civan, M. M.,** *Biochim. Biophys. Acta,* 385, 81, 1975.

1296. **Shporer, M., Haas, M., and Civan, M. M.,** *Biophys. J.,* 16, 601, 1976.

1297. **Shouf, R. R. and Van der Hart, D. L.,** *J. Am. Chem. Soc.,* 93, 2053, 1971.

1298. **Shoup, R. R., Becker, E. D., and Farrar, T. C.,** *J. Magn. Reson.,* 8, 290, 1972.

1299. **Siddall, T. H.,** *J. Phys. Chem.,* 70, 2249, 1966.

1300. **Siddall, T. H., III.,** *Chem. Commun.,* 452, 1971.

1301. **Siemion, I. Z. and Sucharda-Sobczyk, A.,** *Tetrahedron,* 26, 191, 1970.

1302. **Sievers, R. E.,** *Nuclear Magnetic Resonance Shift Reagents,* Academic Press, New York, 1973.

1303. **Silver, B. L. and Luz, Z.,** *Q. Rev. Chem. Soc.,* 21, 458, 1967.

1304. **Simmons, H. E. and Park, C. H.,** *J. Am. Chem. Soc.,* 90, 2428, 2429, 2431, 1968.

1305. **Simon, W. and Clerc, T.,** *Strukturaufklärung organischer Verbindungen mit spektroskopischen Methoden,* Vol. 7, Akademische Verlagsgesellschaft, Frankfurt am Main, 1967.

1306. **Simonnin, M. P., Lecourt, M. J., Terrier, F., and Dearing, C. A.,** *Can. J. Chem.,* 50, 3558, 1972.

1307. **Singh, R. D. and Singh, S. N.,** *J. Magn. Reson.,* 16, 110, 1974.

1308. **Sinha, S. P.,** *Europium,* Springer-Verlag, Berlin, 1967.

1309. **Slichter, C. P.,** *Principles of Magnetic Resonance,* Harper, New York, 1963, (a) pp. 16-22.

1310. **Slomp, G. and McKellar, F.,** *J. Am. Chem. Soc.,* 82, 999, 1960.

1311. **Smith, G. V., Boyd, W. A., and Hinckley, C. C.,** *J. Am. Chem. Soc.,* 93, 6319, 1971.

1312. **Smith, G. V. and Kriloff, H.,** *J. Am. Chem. Soc.,* 85, 2016, 1963.

1313. **Smith, I. C. and Schneider, W. G.,** *Can. J. Chem.,* 39, 1158, 1961.

1314. **Smith, W. B.,** Carbon-13 NMR spectroscopy of steroids, in *Annual Reports on NMR Spectroscopy,* Vol. 8, Webb, G. A., Ed., Academic Press, New York, 1978, 199; (a) p. 212, 216; (b) p. 215, pp. 218-220; (c) p. 205; (d) p. 202.

1315. **Snyder, E. I., Altman, L. J., and Roberts, J. D.,** *J. Am. Chem. Soc.,* 84, 2004, 1962.

1316. **Snyder, E. I. and Roberts, J. D.,** *J. Am. Chem. Soc.,* 84, 1582, 1962.

1317. **Sogn, J. A., Gibbons, W. A., and Randall, E. W.,** *Biochemistry,* 12, 2100, 1973.

1318. **Sohár, P.,** *Magy. Kém. Foly.,* 76, 577, 1970 (in Hungarian).

1319. **Sohár, P. and Bernáth, G.,** *Acta Chim. Acad. Sci. Hung.,* 87, 285, 1975.

1320. **Sohár, P.,** *Magy. Kém. Lapja,* 30, 100, 309, 1975 (in Hungarian).

1321. **Sohár, P.,** unpublished results.

1322. **Sohár, P. and Bernáth, G.,** *Org. Magn. Reson.,* 5, 159, 1973.

1323. **Sohár, P., Fehér, G., and Toldy, L.,** *Org. Magn. Reson.,* 11, 9, 1978.

1324. **Sohár, P., Fehér, Ö., and Tihanyi, E.,** *Org. Magn. Reson.,* 12, 205, 1979.

1325. **Sohár, P., Gera, L., and Bernáth, G.,** *Org. Magn. Reson.,* 14, 204, 1980.

1326. **Sohár, P. and Hajós, A.,** unpublished results.

1327. **Sohár, P., Horváth, T., and Ábrahám, G.,** *Acta Chim. (Budapest),* 103, 95, 1980.

1328. **Sohár, P., Kosáry, J., and Kasztreiner, E.,** *Acta Chim. Acad. Sci. Hung.,* 84, 201, 1975.

1329. **Sohár, P. and Kuszmann, J.,** *Org. Magn. Reson.,* 3, 647, 1971.

1330. **Sohár, P. and Kuszmann, J.,** *Org. Magn. Reson.,* 6, 407, 1974.

1331. **Sohár, P. and Kuszmann, J.,** *Acta Chim. Acad. Sci. Hung.,* 86, 285, 1975.

1332. **Sohár, P., Kuszmann, J., Horváth, Gy., and Méhesfalvi-Vajna, Zs.,** *Kém. Közlemények,* 46, 481, 1976 (in Hungarian).

1333. **Sohár, P., Kuszmann, J., Ullrich, E., and Horváth, Gy.,** *Acta Chim. Acad. Sci. Hung.,* 79, 457, 1973.

1334. **Sohár, P. and Lázár, J.,** *Acta Chim. Acad. Sci. Hung.,* 105, 105, 1980.

1335. **Sohár, P., Mányai, Gy., Hideg, K., Hankovszky, H. O., and Lex, L.,** *Org. Magn. Reson.,* 14, 125, 1980.

1336. **Sohár, P., Medgyes, G., and Kuszmann, J.,** *Org. Magn. Reson.,* 11, 357, 1978.

1337. **Sohár, P., Méhesfalvi-Vajna, Zs., and Bernáth, G.,** *Kém. Közlemények,* 46, 487, 1976 (in Hungarian).

1338. **Sohár, P., Nyitrai, J., Zauer, K., and Lempert, K.,** *Acta Chim. Acad. Sci. Hung.,* 65, 189, 1970.

1339. **Sohár, P., Ocskay, Gy., and Vargha, L.,** *Acta Chim. Acad. Sci. Hung.,* 84, 381, 1975.

1340. **Sohár, P., Reiter, J., and Toldy, L.,** *Org. Magn. Reson.,* 3, 689, 1971.

1341. **Sohár, P. and Sipos, Gy.,** *Acta Chim. Acad. Sci. Hung.,* 67, 365, 1971.

1342. Sohár, P., Széll, T., and Dudás, T., *Acta Chim. Acad. Sci. Hung.*, 70, 355, 1971.
1343. Sohár, P., Széll, T., Dudás, T., and Sohár, I., *Tetrahedron Lett.*, 1101, 1972.
1344. Sohár, P. and Toldy, L., *Acta Chim. Acad. Sci. Hung.*, 75, 99, 1973.
1345. Sohár, P. and Varsányi, Gy., *Acta Chim. Acad. Sci. Hung.*, 55, 189, 1968.
1346. Solomon, I., *Phys. Rev.*, 99, 559, 1955.
1347. Solomon, I., *C. R. Acad. Sci. Paris*, 248, 92, 1959.
1348. Solomon, I., *Phys. Rev. Lett.*, 2, 301, 1959.
1349. Sólyom, S., Sohár, P., Toldy, L., Kálmán, A., and Párkányi, L., *Tetrahedron Lett.*, 48, 4245, 1977.
1350. Sorensen, S., Hansen, M., and Jakobsen, H. J., *J. Magn. Reson.*, 12, 340, 1973.
1351. Sorensen, S., Hansen, R. S., and Jakobsen, H. J., *J. Magn. Reson.*, 14, 243, 1974.
1352. *Spectrometry of Fuels*, Friedel, R. A., Ed., Plenum Press, New York, 1970, 90.
1353. Spiesecke, H., *Z. Naturforsch.*, 23a, 467, 1968.
1354. Spiesecke, H. and Schneider, W. G., *J. Chem. Phys.*, 35, 722, 1961.
1355. Spiesecke, H. and Schneider, W. G., *J. Chem. Phys.*, 35, 731, 1961.
1356. Spiesecke, H. and Schneider, W. G., *Tetrahedron Lett.*, 468, 1961.
1357. Spiess, H. W., Schweitzer, D., Haeberlen, U., and Hausser, K. H., *J. Magn. Reson.*, 5, 101, 1971.
1358. Spotswood, T. M. and Tänzer, C. I., *Tetrahedron Lett.*, 911, 1967.
1359. Springer, C. S., Bruder, A. H., Tanny, S. R., Pickering, M., and Rockefeller, H. A., Ln(fod)$_3$ complexes as NMR shift reagents, in *Nuclear Magnetic Resonance Shift Reagents*, Academic Press, New York, 1973, 283.
1360. Srinivasan, P. R. and Lichter, R. L., *Org. Magn. Reson.*, 8, 198, 1976.
1361. Staab, H. A., Brettschneider, H., and Brunner, H., *Chem. Ber.*, 103, 1101, 1970.
1362. Stájer, G., Szabó, E. A., Pintye, J., Klivényi, F., and Sohár, P., *Chem. Ber.*, 107, 299, 1974.
1363. Staley, S. W. and Kingsley, W. G., *J. Am. Chem. Soc.*, 95, 5805, 1973.
1364. Stefaniak, L., *Spectrochim. Acta*, 32a, 345, 1976.
1365. Stefaniak, L., *Tetrahedron*, 32a, 1065, 1976.
1366. Stefaniak, L., *Tetrahedron*, 33, 2571, 1977.
1367. Stefaniak, L. and Grabowska, A., *Bull. Acad. Polon. Sci. Ser. Sci. Chim.*, 22, 267, 1974.
1368. Sternhell, S., *Q. Rev. Chem. Soc.*, 23, 236, 1969.
1369. Steur, R., Van Dongen, J. P. C. M., De Bie, M. J. A., Drenth, W., De Haan, J. W., and Van de Ven, L. J. M., *Tetrahedron Lett.*, 3307, 1971.
1370. Stolow, R. D. and Gallo, A. A., *Tetrahedron Lett.*, 3331, 1968.
1371. Stothers, J. B., *Q. Rev. Chem. Soc.*, 19, 144, 1965.
1372. Stothers, J. B., *Carbon-13 NMR Spectroscopy*, Academic Press, New York, 1972, (a) p. 105; (b) p. 118; (c) p. 134; (d) p. 135; (e) p. 168; (f) p. 153; (g) p. 272; (h) p. 70, p. 71; (i) p. 145, p. 151; (j) p. 151; (k) p. 291; (l) p. 442; (m) pp. 433-439; (n) p. 371; (o) p. 372; (p) pp. 362-370, pp. 375-381; (r) p. 432.
1373. Stothers, J. B., ^{13}C NMR studies of reaction mechanisms and reactive intermediates, in *Topics in Carbon-13 NMR Spectroscopy*, Vol. 1, Levy, G. C., Ed., Wiley-Interscience, New York, 1974, 229; (a) pp. 238-244.
1374. Stothers, J. B. and Lauterbur, P. C., *Can. J. Chem.*, 42, 1563, 1964.
1375. Sterehlow, H., *Magnetische Kernresonanz und chemische Struktur*, Vol. 7, (Fortschritte der physikalischen Chemie), Steinkoff-Verlag, Darmstadt, 1968.
1376. Su, J.-A., Siew, E., Brown, E. V., and Smith, S. L., *Org. Magn. Reson.*, 10, 122, 1977.
1377. Suhr, H., *Chem. Ber.*, 96, 1720, 1963.
1378. Suhr, H., *J. Mol. Phys.*, 6, 153, 1963.
1379. Suhr, H., *Anwendungen der Kernmagnetischen Resonanz in der Organischen Chemie*, Springer-Verlag, Berlin, 1965.
1380. Sullivan, G. R. and Roberts, J. D., *J. Org. Chem.*, 42, 1095, 1977.
1381. Sunners, B., Piette, L. H., and Schneider, W. G., *Can. J. Chem.*, 38, 681, 1960.
1382. Sutherland, I. O., The investigation of the kinetics of conformational changes by nuclear magnetic resonance spectroscopy, in *Annual Reports on NMR Spectroscopy*, Vol. 4, Mooney, E. F., Ed., Academic Press, New York, 1971, 71.
1383. Szántay, Cs., Novák, L., and Sohár, P., *Acta Chim. Acad. Sci. Hung.*, 57, 335, 1968.
1384. Szarek, W. A., Vyas, D. M., Sepulchre, A. M., Gero, S. D., and Lukacs, G., *Can. J. Chem.*, 52, 2041, 1974.
1385. Sztraka, L., *Basic Principles of Fourier-Transform IR Spectrometry*, A kémia ujabb eredményei, Vol. 36, Csákvári, B., Ed., Akadémiai Kiadó, Budapest, 1977 (in Hungarian).
1386. Szymanski, S., Witanowski, M., Gryff-Keller, A., Problems in theory and analysis of dynamic nuclear magnetic spectra, in *Annual Reports on NMR Spectroscopy*, Vol. 8, Webb, G. A., Ed., Academic Press, New York, 1978, 227.
1387. Tadokoro, S., Fujiwara, S., and Ichihara, Y., *Chem. Lett.*, 849, 1973.

1388. **Taillandier, M., Liquier, J., and Taillandier, E.,** *J. Mol. Struct.,* 2, 437, 1968.
1389. **Takahashi, K.,** *Bull. Chem. Soc. Jpn.,* 39, 2782, 1966.
1390. **Takahashi, K., Sone, T., and Fujieda, K.,** *J. Phys. Chem.,* 74, 2765, 1970.
1391. **Takeuchi, Y.,** *Org. Magn. Reson.,* 7, 181, 1975.
1392. **Takeuchi, Y., Chivers, P. J., and Crabb, T. A.,** *Chem. Commun.,* 210, 1974.
1393. **Takeuchi, Y. and Dennis, N.,** *J. Am. Chem. Soc.,* 96, 3657, 1974.
1394. **Tanabe, M. and Detre, G.,** *J. Am. Chem. Soc.,* 88, 4515, 1966.
1395. **Tangerman, A. and Zwannenburg, B.,** *Tetrahedron Lett.,* 5195, 1973.
1396. **Tarpley, A. R. and Goldstein, J. H.,** *J. Mol. Spectrosc.,* 37, 432, 1971.
1397. **Tarpley, A. R. and Goldstein, J. H.,** *J. Mol. Spectrosc.,* 39, 275, 1971.
1398. **Temple, C., Thorpe, M. C., Coburn, W. C., and Montgomery, J. A.,** *J. Org. Chem.,* 31, 935, 1966.
1399. **Terui, Y., Aono, K., and Tori, K.,** *J. Am. Chem. Soc.,* 90, 1069, 1968.
1400. **Thétaz, C., Wehrli, F. W., and Wentrup, C.,** *Helv. Chim. Acta,* 59, 259, 1976.
1401. **Thomas, H. A.,** *Phys. Rev.,* 80, 901, 1950.
1402. **Thomas, W. A.,** NMR spectroscopy in conformational analysis, in *Annual Reports on NMR Spectroscopy 1-5,* Vol. 1, Mooney, E. F., Ed., Academic Press, New York, 1968.
1403. **Thomas, W. A.,** NMR and conformations of amino acids, peptides and proteins, in *Annual Reports on NMR Spectroscopy,* Vol. 6B, Mooney, E. F., Ed., Academic Press, New York, 1976, 1.
1404. **Thomlinson, B. L. and Hill, H. D. W.,** *J. Chem. Phys.,* 59, 1775, 1973.
1405. **Thorpe, M. C., Cobrun, W. C., and Montgomery, J. A.,** *J. Magn. Reson.,* 15, 98, 1974.
1406. **Tiers, G. V. D.,** *J. Phys. Chem.,* 62, 1151, 1958.
1407. **Tiers, G. V. D.,** *Proc. Chem. Soc.,* 1960, 1960.
1408. **Toldy, L., Sohár, P., Faragó, K., Tóth, I., and Bartalits, L.,** *Tetrahedron Lett.,* 2167, 1970.
1409. **Toldy, L., Sohár, P., Faragó, K., Tóth, I., and Bartalits, L.,** *Tetrahedron Lett.,* 2177, 1970.
1410. *Topics in Carbon-13 NMR Spectroscopy,* Levy, G. C., Ed., Wiley-Interscience, New York, Vol. 1 to 3, 1974 to 1979, a) Vol. 2, p. 433.
1411. **Torchia, D. A. and Vanderhart, D. L.,** High-Power double-resonance studies in fibrous proteins, proteoglycans, and model membranes, in *Topics in Carbon-13 NMR Spectroscopy,* Vol. 3, Levy, G. C., Ed., Wiley-Interscience, New York, 1979, 325.
1412. **Tori, K., Kitahonoki, K., Takano, Y., Tanida, H., and Tsuji, T.,** *Tetrahedron Lett.,* 559, 1964.
1413. **Tori, K., Kitahonoki, K., Takano, Y., Tanida, H., and Tsuji, T.,** *Tetrahedron Lett.,* 869, 1965.
1414. **Tori, K., Komeno, T., Sangare, M., Septe, B., Delpech, B., Ahand, A., and Lukacs, G.,** *Tetrahedron Lett.,* 1157, 1974.
1415. **Tori, K. and Kondo, E.,** *Tetrahedron Lett.,* 645, 1963.
1416. **Tori, K., Muneyuki, R., and Tanida, H.,** *Can. J. Chem.,* 41, 3142, 1963.
1417. **Tori, K. and Nakagawa, T.,** *J. Phys. Chem.,* 68, 3163, 1964.
1418. **Tori, K. and Ogata, M.,** *Chem. Pharm. Bull.,* 12, 272, 1964.
1419. **Tori, K. and Ohtsuru, M.,** *Chem. Commun.,* 886, 1966.
1420. **Tori, K., Ohtsuru, M., Aono, K., Kawazoe, Y., and Ohnishi, M.,** *J. Am. Chem. Soc.,* 89, 2765, 1967.
1421. **Tori, K. and Yoshimura, Y.,** *Tetrahedron Lett.,* 3127, 1973.
1422. **Torrey, H. C.,** *Phys. Rev.,* 76, 1059, 1949.
1423. **Towl, A. D. C. and Schaumburg, K.,** *Mol. Phys.,* 22, 49, 1971.
1424. **Traficante, D. D. and Maciel, G. E.,** *J. Phys. Chem.,* 69, 1348, 1965.
1425. **Trager, W. F., Nist, B. J., and Huitric, A. C.,** *Tetrahedron Lett.,* 2931, 1965.
1426. **Tran-Dihn, S., Fermandjian, S., Sala, E., Mermet-Bouvier, R., Cohen, M., and Fromageot, P.,** *J. Am. Chem. Soc.,* 96, 1484, 1974.
1427. **Tran-Dihn, S., Fermandjian, S., Sala, E., Mermet-Bouvier, R., and Fromageot, P.,** *J. Am. Chem. Soc.,* 97, 1267, 1975.
1428. **Triplett, J. W., Digenis, G. A., Layton, W. J., and Smith, S. L.,** *Spectrosc. Lett.,* 10, 141, 1977.
1429. **Tronchet, J. M. J., Barbalat-Rey, F., and Le-Hong, N.,** *Helv. Chim. Acta,* 54, 2615, 1971.
1430. **Troshin, A. S.,** *Problems of Cell Permeability,* Pergamon Press, New York, 1966.
1431. **Truce, W. E. and Brady, D. G.,** *J. Org. Chem.,* 31, 3543, 1966.
1432. **Tulloch, A. P.,** *Can. J. Chem.,* 55, 1135, 1977.
1433. **Turner, A. B., Heine, H. W., Irwing, J., and Bush, J. B., Jr.,** *J. Am. Chem. Soc.,* 87, 1050, 1965.
1434. **Turner, T. E., Fiora, V. C., and Kendrick, W. M.,** *J. Chem. Phys.,* 23, 1966, 1955.
1435. **Uhlenbeck, G. E. and Goudsmit, S.,** *Naturwissenschaften,* 13, 953, 1925; *Nature,* 117, 264, 1926.
1436. **Untch, K. G. and Wysocki, D. C.,** *J. Am. Chem. Soc.,* 88, 2608, 1966.
1437. **Vanasse, G. A. and Sakai, H.,** Fourier spectroscopy, in *Progress in Optics,* Vol. 6, Wolf, E., Ed., North-Holland, Amsterdam, 1967, 261.
1438. **Van Der Veen, J. M.,** *J. Org. Chem.,* 28, 564, 1963.
1439. **Van Vleck, J. H.,** *The Theory of Electric and Magnetic Susceptibilities,* Oxford University Press, Oxford, 1932.

1440. **Van' Wazer, J. R., Callis, C. F., Shoolery, J. N., and Jones, R. C.,** *J. Am. Chem. Soc.,* 78, 5715, 1956.

1441. **Vargha, L., Kuszmann, J., and Sohár, P.,** *Magy. Kém. Lapja,* 356, 1972 (in Hungarian).

1442. **Vargha, L., Kuszmann, J., Sohár, P., and Horváth, Gy.,** *J. Heterocyclic Chem.,* 9, 341, 1972.

1443. *VARIAN Instrum. Appl.,* 2, 4, 1968.

1444. **Vegar, M. R. and Ewlis, R. J.,** *Tetrahedron Lett.,* 2847, 1971.

1445. **Venien, F.,** *C. R. Acad. Sci. France C,* 269, 642, 1969.

1446. **Verchére, C., Rousselle, D., and Viel, C.,** *Org. Magn. Reson.,* 13, 110, 1980.

1447. **Verkade, J. G., McCarley, R. E., Hendricker, D. G., and King, R. W.,** *Inorg. Chem.,* 4, 228, 1965.

1448. **Versmold, H. and Yoon, C.,** *Ber. Bunsenges. Phys. Chem.,* 76, 1164, 1972.

1449. **Vincent, E.-J. and Metzger, J.,** *C. R. Acad. Sci. Paris,* 261, 1964, 1965.

1450. **Vinkler, E., Németh, P., Stájer, G., Sohár, P., and Jerkovich, Gy.,** *Arch. Pharm.,* 309, 265, 1976.

1451. **Voelter, W. and Breitmaier, E.,** *Org. Magn. Reson.,* 5, 311, 1973.

1452. **Voelter, W., Breitmaier, E., and Jung, G.,** *Angew. Chem.,* 83, 1011, 1961; *Angew. Chem. Int. Ed. Engl.,* 10, 935, 1961.

1453. **Voelter, W., Breitmaier, E., Jung, G., Keller, T., and Hiss, D.,** *Angew. Chem. Int. Ed. Engl.,* 9, 803, 1970.

1454. **Voelter, W., Breitmaier, E., Price, R., and Jung, G.,** *Chimia,* 25, 168, 1971.

1455. **Voelter, W., Jung, G., Breitmaier, E., and Bayer, E.,** *Z. Naturforsch.,* 26b, 213, 1971.

1456. **Voelter, W., Jung, G., Breitmaier, E., and Price, R.,** *Hoppe-Seyler's Z. Physiol. Chem.,* 352, 1034, 1971.

1457. **Voelter, W. and Oster, O.,** *Z. Naturforsch.,* 28b, 370, 1973.

1458. **Voelter, W., Zech, K., Grimminger, W., Breitmaier, E., and Jung, G.,** *Chem. Ber.,* 105, 3650, 1972.

1459. **Vo-Kim-Yen, Papoušková, Z., Schraml, J., and Chvalovský, V.,** *Collect. Czech. Chem. Commun.,* 38, 3167, 1973.

1460. **Vold, R. L., Waugh, J. S., Klein, M. P., and Phelps, D. E.,** *J. Chem. Phys.,* 48, 3831, 1968.

1461. **Vögeli, U. and von Philipsborn, W.,** *Org. Magn. Reson.,* 5, 551, 1973.

1462. **Vögeli, U. and von Philipsborn, W.,** *Org. Magn. Reson.,* 7, 617, 1975.

1463. **Vögtle, F., Mannschreck, A., and Staab, H. A.,** *Liebigs Ann. Chem.,* 708, 51, 1967.

1464. **Wallach, D.,** *J. Chem. Phys.,* 47, 5258, 1967.

1465. **Wallach, D.,** *J. Phys. Chem.,* 73, 307, 1969.

1466. **Walter, J. A. and Hope, A. B.,** Nuclear magnetic resonance and the state of water in cells, in *Progress in Biophysics and Molecular Biology,* Vol. 23, Pergamon Press, Oxford, 1971, 1.

1467. **Walter, R., Havran, R. T., Swartz, I. L., and Johnson, L. F.,** *Peptides 1969,* Scoffone, E., Ed., North-Holland, Amsterdam, 1971.

1468. **Walter, R. and Johnson, L. F.,** *Biophys. J.,* 9a, 159, 1969.

1469. **Wang, C. and Kingsbury, C. A.,** *J. Org. Chem.,* 40, 3811, 1975.

1470. **Wang, C. H., Grant, D. M., and Lyerla, J. R., Jr.,** *J. Chem. Phys.,* 55, 4674, 1971.

1471. **Wang, M. C. and Uhlenbeck, G. E.,** *Rev. Mod. Phys.,* 17, 323, 1945.

1472. **Ward, H. R.,** *Acc. Chem. Res.,* 5, 18, 1972.

1473. **Ward, H. R. and Lawler, R. G.,** *J. Am. Chem. Soc.,* 89, 5518, 1967.

1474. **Warren, J. P. and Roberts, J. D.,** *J. Phys. Chem.,* 78, 2507, 1974.

1475. **Wasserman, H. H. and Keehn, P. M.,** *J. Am. Chem. Soc.,* 91, 2374, 1969.

1476. **Wasylishen, R. E.,** Spin-spin coupling between carbon-13 and the first row nuclei, in *Annual Reports on NMR Spectroscopy,* Vol. 7, Webb, G. A., Ed., Academic Press, New York, 1977, 245; (a) pp. 287-291.

1477. **Wasylishen, R. E. and Schaefer, T.,** *Can. J. Chem.,* 50, 2710, 1972.

1478. **Wasylishen, R. E. and Schaefer, T.,** *Can. J. Chem.,* 50, 2989, 1972.

1479. **Watts, V. S. and Goldstein, J. H.,** *J. Phys. Chem.,* 70, 3887, 1966.

1480. **Watts, V. S. and Goldstein, J. H.,** *J. Chem. Phys.,* 46, 4165, 1967.

1481. **Watts, V. S., Loemker, J., and Goldstein, J. H.,** *J. Mol. Spectrosc.,* 17, 348, 1965.

1482. **Waugh, J. S.,** *J. Mol. Spectrosc.,* 35, 298, 1970.

1483. **Waugh, J. S. and Fessenden, R. W.,** *J. Am. Chem. Soc.,* 79, 846, 1957.

1484. **Waugh, J. S., Huber, L. M., and Haeberlen, U.,** *Phys. Rev. Lett.,* 20, 180, 1968.

1485. **Webb, R. G., Haskell, M. W., and Stammer, C. H.,** *J. Org. Chem.,* 34, 576, 1969.

1486. **Webster, D. E.,** *J. Chem. Soc.,* 5132, 1960.

1487. **Wehrli, F. W.,** *Chem. Commun.,* 379, 1973.

1488. **Wehrli, F. W.,** *Solvent Suppression Technique in Pulsed NMR,* VARIAN Application Note NMR, 75, 2, 1975.

1489. **Wehrli, F. W.,** Organic structure assignments using ^{13}C spin-relaxation data, in *Topics in Carbon-13 NMR Spectroscopy,* Vol. 2, Levy, G. C., Ed., Wiley-Interscience, New York, 1976, 343. (a) pp. 362-364; (b) p. 373.

1490. **Wehrli, F. W., Giger, W., and Simon, W.,** *Helv. Chim. Acta,* 54, 229, 1971.
1491. **Wehrli, F. W., Jeremic, D., and Mihailovic, M. L., and Milosavljevic, S.,** *Chem. Commun.,* 302, 1978.
1492. **Wehrli, F. W. and Wirthlin, T.,** *Interpretation of Carbon-13 NMR Spectra,* Heyden, London, 1976; (a) p. 68; (b) p. 248, p. 249; (c) p. 145; (d) p. 37; (e) pp. 129-151.
1493. **Weigert, F. J. and Roberts, J. D.,** *J. Am. Chem. Soc.,* 89, 2967, 1967.
1494. **Weigert, F. J. and Roberts, J. D.,** *J. Am. Chem. Soc.,* 90, 3543, 1968.
1495. **Weigert, F. J. and Roberts, J. D.,** *J. Am. Chem. Soc.,* 90, 3577, 1968.
1496. **Weigert, F. J. and Roberts, J. D.,** *J. Am. Chem. Soc.,* 92, 1347, 1970.
1497. **Weigert, F. J. and Roberts, J. D.,** *J. Am. Chem. Soc.,* 94, 6021, 1972.
1498. **Weigert, F. J., Winokur, M., and Roberts, J. D.,** *J. Am. Chem. Soc.,* 90, 1566, 1968.
1499. **Weiler, L.,** *Can. J. Chem.,* 50, 1975, 1972.
1500. **Weinstein, B.,** *Peptides,* Marcel Dekker, New York, 1970.
1501. **Weisman, I. D., Bennett, L. H., Maxwell, L. R., Woods, M. W., and Burke, D.,** *Science,* 178, 1288, 1972.
1502. **Weissman, S. I.,** *J. Am. Chem. Soc.,* 93, 4928, 1971.
1503. **Weitkamp, H. and Korte, F.,** *Chem. Ber.,* 95, 2896, 1962.
1504. **Weitkamp, H. and Korte, F.,** *Tetrahedron Suppl.,* 75, 1966.
1505. **Wellman, K. M. and Bordwell, F. G.,** *Tetrahedron Lett.,* 173, 1963.
1506. **Wells, E. J. and Abramson, K. H.,** *J. Magn. Reson.,* 1, 378, 1969.
1507. **Wells, P. R., Arnold, D. P., and Doddrell, D.,** *J. Chem. Soc. Perkin Trans. 2,* 1745, 1974.
1508. **Wenkert, E. and Buckwalter, B. L.,** *J. Am. Chem. Soc.,* 94, 4367, 1972.
1509. **Wenkert, E., Buckwalter, B. L., Burfitt, I. R., Gasic, M. J., Gottlieb, H. E., Hagaman, E. W., Schell, F. M., Wovkulich, P. M., and Zheleva, A.,** Carbon-13 NMR Spectroscopy, in *Topics in Carbon-13 NMR Spectroscopy,* Vol. 2, Levy, G. C., Ed., Wiley-Interscience, New York, 1976.
1510. **Wenkert, E., Clouse, A. O., Cochran, D. W., and Doddrell, D.,** *Chem. Commun.,* 1433, 1969.
1511. **Wenkert, E., Cochran, D. W., Hagaman, E. W., Lewis, R. B., and Schell, F. M.,** *J. Am. Chem. Soc.,* 93, 6271, 1971.
1512. **Whipple, E. B., Goldstein, J. H., and Mandell, L.,** *J. Chem. Phys.,* 30, 1109, 1959.
1513. **Whipple, E. B., Goldstein, J. H., and Stewart, W. E.,** *J. Am. Chem. Soc.,* 81, 4761, 1959.
1514. **White, D. M. and Levy, G. C.,** *Macromolecules,* 5, 526, 1972.
1515. **Whitesides, G. M. and Fleming, J. S.,** *J. Am. Chem. Soc.,* 89, 2855, 1967.
1516. **Whitesides, G. M. and Lewis, D. W.,** *J. Am. Chem. Soc.,* 92, 6979, 1970.
1517. **Whitlock, H. W.,** *J. Am. Chem. Soc.,* 84, 3412, 1962.
1518. **Whitman, D. R.,** *J. Mol. Spectrosc.,* 10, 250, 1963.
1519. **Wiberg, K. B. and Nist, B. J.,** *J. Am. Chem. Soc.,* 83, 1226, 1961.
1520. **Wieland, T. and Bende, H.,** *Chem. Ber.,* 98, 504, 1965.
1521. **Williams, D. H. and Bhacca, N. S.,** *Tetrahedron,* 21, 2021, 1965.
1522. **Williams, D. H. and Fleming, I.,** *Spektroskopische Methoden in der Organischen Chemie,* Thieme-Verlag, Stuttgart, 1968.
1523. **Williams, D. H., Ronayne, J., Moore, H. W., and Shelden, H. R.,** *J. Org. Chem.,* 33, 998, 1968.
1524. **Williams, F., Sears, B., Allerhand, A., and Cordes, E. H.,** *J. Am. Chem. Soc.,* 95, 4871, 1973.
1525. **Williamson, K. L., Howell, T., and Spencer, T. A.,** *J. Am. Chem. Soc.,* 88, 325, 1966.
1526. **Williamson, K. L., Lanford, C. A., and Nicholson, C. R.,** *J. Am. Chem. Soc.,* 86, 762, 1964.
1527. **Williamson, K. L., Li Hsu, Y.-F., Hall, F. H., Swager, S., and Coulter, M. S.,** *J. Am. Chem. Soc.,* 90, 6717, 1968.
1528. **Williamson, K. L. and Roberts, J. D.,** *J. Am. Chem. Soc.,* 98, 5082, 1976.
1529. **Williamson, M. P., Kostelnik, R. J., and Castellano, S. M.,** *J. Chem. Phys.,* 49, 2218, 1968.
1530. **Wilson, N. K. and Zehr, R. D.,** *J. Org. Chem.,* 43, 1768, 1978.
1531. **Wing, R. M., Uebel, J. J., and Anderson, K. K.,** *J. Am. Chem. Soc.,* 95, 6046, 1973.
1532. **Witanowski, M.,** *Tetrahedron,* 23, 4299, 1967.
1533. **Witanowski, M.,** *J. Am. Chem. Soc.,* 90, 5683, 1968.
1534. **Witanowski, M.,** *Pure Appl. Chem.,* 37, 225, 1974.
1535. **Witanowski, M. and Januszewski, H.,** *J. Chem. Soc. B,* 1063, 1967.
1536. **Witanowski, M. and Januszewski, H.,** *Can. J. Chem.,* 47, 1321, 1969.
1537. **Witanowski, M. and Stefaniak, L.,** *J. Chem. Soc. B,* 1061, 1967.
1538. **Witanowski, M., Stefaniak, L., Januszewski, H., Bahadur, K., and Webb, G. A.,** *J. Cryst. Mol. Struct.,* 5, 137, 1975.
1539. **Witanowski, M., Stefaniak, L., Januszewski, H., Grabowski, Z., and Webb, G. A.,** *Bull. Acad. Polon. Sci. Ser. Sci. Chim.,* 20, 917, 1972.
1540. **Witanowski, M., Stefaniak, L., Januszewski, H., and Piotrowska, H.,** *Bull. Acad. Polon. Sci. Ser. Sci. Chim.,* 23, 333, 1975.

1541. **Witanowski, M., Stefaniak, L., Januszewski, H., Szymansky, S., and Webb, G. A.,** *Tetrahedron,* 29, 2833, 1973.

1542. **Witanowski, M., Stefaniak, L., Januszewski, H., Voronkov, M. G., and Tandura, S. N.,** *Bull. Acad. Polon. Sci. Ser. Sci. Chim.,* 24, 281, 1976.

1543. **Witanowski, M., Stefaniak, L., Januszewski, H., and Webb, G. A.,** *Bull. Acad. Polon. Sci . Ser. Sci. Chim.,* 21, 71, 1973.

1544. **Witanowski, M., Stefaniak, L., Szymanski, S., Grabowski, Z., and Webb, G. A.,** *J. Magn. Reson.,* 21, 185, 1976.

1545. **Witanowski, M., Stefaniak, L., Szymanski, S., and Januszewski, H.,** *J. Magn. Reson.,* 28, 217, 1977.

1546. **Witanowski, M., Stefaniak, L., Szymanski, S., and Webb, G. A.,** *Tetrahedron,* 32, 2127, 1976.

1547. **Witanowski, M., Stefaniak, L., and Webb, G. A.,** *J. Chem. Soc. B,* 1065, 1967.

1548. **Witanowski, M., Stefaniak, L., and Webb, G. A.,** Nitrogen NMR spectroscopy, in *Annual Reports on NMR Spectroscopy,* Vol. 7, Webb, G. A., Ed., Academic Press, London, 1977, 117.

1549. **Witanowski, M., Urbanski, T., and Stefaniak, L.,** *J. Am. Chem. Soc.,* 86, 2569, 1964.

1550. **Witanowski, M. and Webb, G. A.,** Nitrogen NMR spectroscopy, in *Annual Reports on NMR Spectroscopy,* Vol. 5A, Mooney, E. F., Ed., Academic Press, London, 1972, 395.

1551. **Wittstruck, T. A. and Cronan, J. F.,** *J. Phys. Chem.,* 72, 4243, 1968.

1552. **Wittstruck, T. A., Malhotra, S. K., and Ringold, H. J.,** *J. Am. Chem. Soc.,* 85, 1699, 1963.

1553. **Wokaun, A. and Ernst, R. R.,** *Chem. Phys. Lett.,* 52, 407, 1977.

1554. **Wolkowski, Z. W.,** *Tetrahedron Lett.,* 825, 1971.

1555. **Woods, W. G. and Strong, P. L.,** *J. Am. Chem. Soc.,* 88, 4667, 1966.

1556. **Woodward, R. B. and Skarie, V.,** *J. Am. Chem. Soc.,* 83, 4676, 1961.

1557. **Woolfenden, W. R. and Grant, D. M.,** *J. Am. Chem. Soc.,* 88, 1496, 1966.

1558. **Wright, G. E.,** *Tetrahedron Lett.,* 1097, 1973.

1559. **Wright, G. E. and Tang-Wei, T. Y.,** *J. Pharm. Sci.,* 61, 299, 1972.

1560. **Wright, G. E. and Tang-Wei, T. Y.,** *Tetrahedron,* 29, 3775, 1973.

1561. **Wüthrich, K.,** *FEBS Lett.,* 25, 104, 1972.

1562. **Wüthrich, K., Meiboom, S., and Snyder, L. C.,** *J. Chem. Phys.,* 52, 230, 1970.

1563. **Yamagishi, T., Hayashi, K., Mitsuhashi, H., Imanari, M., and Matsushita, K.,** *Tetrahedron Lett.,* 3527, 1973.

1564. **Yamagishi, T., Hayashi, K., Mitsuhashi, H., Imanari, M., and Matsushita, K.,** *Tetrahedron Lett.,* 3531, 1973.

1565. **Yamaguchi, I.,** *Bull. Chem. Soc. Jpn.,* 34, 353, 1961.

1566. **Yamamoto, O., Watabe, M., and Kikuchi, O.,** *Mol. Phys.,* 17, 249, 1969.

1567. **Yamazaki, M., Usami, T., and Takeuchi, T.,** *Nippon Kagaku Kaishi,* 11, 2135, 1973; *Chem. Abstr.,* 80, 47030b, 1974.

1568. **Yee, K. C. and Bentrude, W. G.,** *Tetrahedron Lett.,* 2775, 1971.

1569. **Yeh, H. J. C., Ziffer, H., Jerina, D. M., and Body, D. R.,** *J. Am. Chem. Soc.,* 95, 2741, 1973.

1570. **Yoder, C. H., Griffith, D. R., and Schaeffer, C. D.,** *J. Inorg. Nucl. Chem.,* 32, 3689, 1970.

1571. **Yoder, C. H., Sheffy, F. K., Howell, R., Hess, R. E., Pacala, L., Schaeffer, C. D., Jr., and Zuckerman, J. J.,** *J. Org. Chem.,* 41, 1511, 1976.

1572. **Yoder, C. H., Tuck, R. H., and Hess, R. E.,** *J. Am. Chem. Soc.,* 91, 539, 1969.

1573. **Yonezawa, T. and Morishima, I.,** *J. Mol. Spectrosc.,* 27, 210, 1968.

1574. **Yonezawa, T., Morishima, I., and Fukuta, K.,** *Bull. Chem. Soc. Jpn.,* 41, 2297, 1968.

1575. **Yonezawa, T., Morishima, I., and Kato, H.,** *Bull. Chem. Soc. Jpn.,* 39, 1398, 1966.

1576. **Young, J. A., Grasselli, J. G., and Ritchey, W. M.,** *J. Magn. Reson.,* 14, 194, 1974.

1577. **Zaner, K. S. and Damadian, R.,** *Physiol. Chem. Phys.,* 7, 437, 1975.

1578. **Zaner, K. S. and Damadian, R.,** *Science,* 189, 729, 1975.

1579. **Zaner, K. S. and Damadian, R.,** *Physiol. Chem. Phys.,* 9, 473, 1977.

1580. **Zimmerman, J. R. and Brittin, W. E.,** *J. Phys. Chem.,* 61, 1328, 1957.

1581. **Zürcher, R. F.,** *Helv. Chim. Acta,* 46, 2054, 1963.

INDEX

A

Absolute configuration, determination of, 114

Absorption signal, 259

Accidental isochrony, detection of, 130—131

Acetaldehyde
'H NMR chemical shifts, of acetyl group, 3
^{17}O NMR chemical shifts, 260

Acetaldehyde, diethyl acetal 2 phenoxy, part of 'H NMR spectrum, 5

Acetaldoxime, 256
(E) and (Z) isomers, 66
'H NMR spectrum, 63

Acetamide, protonated, structure of, 108

Acetamides, NH-, and methyl signal, 74

Acetic acid
anhydride
^{17}O NMR chemical shift, 260
trifluoro (TFA) as NMR solvent, 62, 66, 89
^{19}F NMR chemical shift, 262
like reference material in fluorine resonance, 262
trifluoro, deuterio, as NMR solvent, 97, 109
trifluoro-fluoride
^{19}F NMR chemical shift, 263
vinyl ester
^{17}O NMR chemical shift, 260

Acetic acid and halogenated derivatives, concentration dependence, 104

Acetic acid derivatives, 'H NMR chemical shifts of acetyl signal, 2—3

Aceto-acetic acid diethylamide
determination of keto-enol tautomeric ratio, 102
'H NMR spectrum, 103

Acetols, ^{13}C chemical shift range of, 171

Acetone
hexadeuterio, as NMR solvent, 76, 80
'H NMR chemical shifts, 3
^{17}O NMR chemical shift, 260
perfluoro
^{19}F NMR chemical shift, 262
perfluoro-tio-
^{19}F NMR chemical shift, 262

Acetonitrile, 238, 240
^{13}C NMR data of, 173

Acetophenone, 'H NMR chemical shift of acetyl signal, 3

3β-Acetoxycholest-5-one, ^{13}C NMR spectrum of, 195, 198—199

Acetyl acetone
hydroxy signal of enolized form ('H NMR), 101
^{17}O NMR chemical shifts, 260
tautomeric study with oxygen resonance, 260

Acetyl bromide, ^{17}O NMR chemical shift, 260

Acetyl-chloride, 'H NMR chemical shift of acetyl signal, 3

Acetylene, 46—49
association shift of, 98

^{13}C chemical shifts of, 152
derivatives of coupling constants of, 47
distinction of protons by solvent effect of, 47
'H NMR signal, 46

Acetylene, accidental isochrony of, 47
'H NMR chemical shift of, 46
ranging among spin-spin systems of, 47

Acetylene, p-nitro-phenyl, 'H NMR chemical shift of, 46

Acetylene, perfluoro-, ^{19}F NMR chemical shift, 267

Acetylene, phenyl, 'H NMR chemical shift, 46

Acetylene, vinyl, coupling constants of, 47

Acetylene derivatives, 179—180

Acetyl-methyl signal, 74

Acid-base equilibria, 108

Acid halides, 260

Acidification, 111

Acquisition time, 258

Acridine, 'H NMR chemical shifts of, 88

Acrylic acid
β-chloro and β-bromo, Z and E, carbonyl signals in, 182
coupling constants, 52
'H NMR chemical shift, 51

Acrylnitrite
coupling constants, 52
'H NMR chemical shift, 51

Activating free enthalpy (energy), determination of, by temperature dependence of the NMR spectrum, 17, 21, 24, 26, 39, 56, 59, 80, 92, 94—95, 97, 106

Activation parameters, determination of, 235

Acyclic alkanes, see Saturated acyclic alkanes

Acyloxy derivatives, 167—170

Adamantane, 128, 161
'H NMR chemical shifts, 30

Additivity, 214
of effect of chlorination in ^{13}C NMR, 179
of effects, in ^{13}C NMR, 161
parameters for substituted steroids, 200

Additivity rule, 49, 69—70
nitrogen resonance and, 246—249
substituted pyrroles and substituted thiophenes and, 191

Adenine, 'H NMR chemical shifts, 88

Adenosine derivatives, conformational properties of, 91

Adiabatic nuclear polarization, 134—138

Alanine-s, methyl ester hydrochloride, part of 'H NMR spectrum, 114—115

Alanyl, phenyl, enantiomers, part of 'H NMR spectrum, 115—116

Alanyl-glycine-type dipeptide, 'H NMR spectrum, 108, 110

Alcohol, trityl, association shift of, 99

Alcohols, 167—170
acetoxy derivatives, ^{13}C NMR chemical shifts of, 167—168

association, temperature, concentration, solvent, and pH dependence of OH signal, 99
C_α and C_β signals of, 159
^{1}H NMR chemical shift of hydroxyl signal of, 99
Aldehydes, 61—67, 260
^{1}H NMR signal of, 61
long-range coupling of, 62, 81
protonated, *Z-E* isomery, 62
Aldoses, in aqueous solution, equilibrium of, 171
Aliphatic compounds, 46
Aliphatic nitro compounds, 63
Aliphatic protons, 59
Alkaloids, 201—202
^{13}C NMR spectroscopy of, 201
quinolizidine, 201
Alkanes, 159—161
^{13}C NMR chemical shifts, 160
monosubstituted, ^{13}C NMR chemical shifts of, substituent constants derived from, 153
normal, additivity role in, 159
Alkenes, 175—179
Allene
^{1}H NMR chemical shift, 49
J(H,H) in, 210
shifts of allenic fluorine, 266
Allene, bromo, ^{1}H NMR chemical shifts, 49
Allene derivatives, 46—49, 179—180
coupling constants of, 49
Allyl cyanide, methylene signal, ^{1}H NMR chemical shift of, 62
Amides, 61—67, 244
CH-NH *vicinal* and N-H heteronuclear couplings, 106
^{1}H NMR chemical shift of NH signal, 107
hindered rotation in, 183
site of protonation, 67, 108
temperature, concentration, solvent, and pH dependence of NH signal, 107
Amides, *N,N*-dimethyl, investigation of hindered rotation, 107
Amine, dimethyl, 253
Amine, methyl-, 253
Amine oxides, γ- and β-effects in, 173
Amines, 243
^{13}C chemical shift of, 171
CH-NH *vicinal* and N-H heteronuclear couplings, 104
^{1}H NMR chemical shift of NH signal, 106
macrocyclic tertiary, structure of, 108
protonation of, 108
relationship between pK_a and Δ_i of, 128
Amine salts, 171
Amines and their salts, 171—172
Amino acids, 109, 203—207, 244
^{13}C NMR chemical shifts, 204
pH dependence of ^{1}H NMR spectra, 109
Ammonia, 240, 253
Ammonium, 240
Ammonium, tetramethyl-, 240
Ammonium chloride, 257
Ammonium ion, 257

quaternary, symmetrically tetrasubstituted alkyl-, 257
Analogues containing more nitrogens, 87—92
"Anchor", role of, 231
Androstan, chemical shift of methyl signal (^{1}H NMR), 30
Androstan-3β-acetoxy-4α-bromo, ^{1}H NMR spectrum, 37
Androstane, stereoisomers of ^{1}H NMR chemical shift, 31
Androstane, 2α,3α-epoxy-17β-acetoxy, 200
Androstan-6-hydroxy, ^{1}H NMR spectrum, 34
Androstan-7-one, ^{1}H NMR chemical shifts, 33
Androstan-11-one, ^{1}H NMR spectrum, 30—31
Androstanones, 197
Androsterone, and epi-
^{1}H NMR chemical shift differences between *equatorial* and *axial* ring protons, 28
Angle of torsion, 19
Angular moments of orbitals, vector operators of, 150
Aniline, chloro, *meta*
part of ^{1}H NMR spectrum, 72
spin system of ring protons, 71
Aniline, *p*-dimethyl-amino-, 253
Aniline, 2,4-dinitro-, 253
Aniline, 3,4,5-trimethoxy and 3,4,5-trimethyl, ^{1}H NMR chemical shifts, 71
Anisochronic nuclei, 264
Anisochrony, methyl groups, 67
Anisole, 2,6-dimethyl, ^{1}H NMR chemical shift, 71
Anisotropic, thermal motion of small molecules, 230
Anisotropic dipole interactions, 113
Anisotropic effect, see Anisotropic neighboring group effect
Anisotropic neighboring group effect, 50, 111, 114, 122, 158, 167, 187
acetylene group (C≡C triple bond), 46, 62
of aromatic rings, 13, 18, 37, 61, 67, 73
carbon-carbon single bond, 27, 31
nitrile group, 50
C–O bond, 20
Anisotropic neighboring group effect
carbonyl group, 2, 30, 32, 43, 51, 61, 67, 102
halogen atom, 4, 67
hydroxyl group, 63—64, 101
lone pairs, 89
meta hydrogens, 67
methylene chain, 109
nitrile group, 62
nitro group, 67
nitroso group, 67
olefinic band, 37
tetracyclic ring, 20
tricyclic rings, 13—14, 18, 55
Anisotropy
carbonyl group, 132
ellipsoid, 122
rotation, increase in, 230
triple bond, 46, 179

Annulenes
  aromatic and antiaromatic character of, 60
  C_{18}, hexahydro, ^1H NMR chemical shift, 60
  C_{24}, ^1H NMR chemical shift, 60
Annulenes, dehydro
  aromatic and antiaromatic character of, 60
  isomerization, 61
Anomeric effect, 40
Anthracene spin system, 72
Anthranylic acid, *N*-acetyl, spin system of ring protons, 71
Antiaromatic annulenes, 60
Antiaromatic dehydroannulene, 61
Antiaromaticity, 185
Anti hydroxy group, 217
Apical carbonyl signals, 235
Antraldehyde, 9-, ^1H NMR chemical shift of aldehyde signal, 61
Arabinose, tetraacetate, α-L-, and β-D-
  ^1H NMR chemical shift differences between *equatorial* and *axial* ring protons, 28
Aromatic aldehydes, 61
Aromatic annulenes, 60
Aromatic compounds, 67—75, 185—188
  ^{13}C NMR chemical shifts, 186
Aromaticity, determination of, 207
Aromatic protons, 68
Aromatic rings, 61, 87, 258
Aromatic solvent induced shift (ASIS), see Solvent effect
Arrhenius equation, 94
Aryl-methyl groups, 73
ASIS, see Solvent effect
Assignment problems, 158
  solution of, by the help of shift reagents, 131—132
Associated state, 98
Association, 191
Association (hydrogen bond, hydrogen bridge), 9, 81, 87, 98, 106, 113
  dimerization constant, 100, 104
Association shift, 98—99, 105
  mercaptans, 113
  thioacids and thioenols, 113
Asymmetric isomers, 171
Audio-frequency modulation, 259
Axial acetoxy groups, 40, 170
Axial hydrogens, 25, 29, 34, 37, 39
Axial α-hydrogens, 43
Axial-em hydrogens, 24
Axial hydroxy group, 171, 265
Axial methoxy group, 171
Axial methyl groups, 28—29, 154—155, 165
Axial position, 28, 38, 40, 43
Axial-ring protons, 27—28
Axial substituent, 40, 170
Axial symmetry, 125—126
Azethidines, ring inversion of, 20
Azides, 61—67, 244
Azines, 247—249
Aziride, ^1H NMR chemical shifts of, inversion, 16

Aziride, 1-chloro-2,2-dimethyl, activation free energy of inversion, 16
Aziride, 1-ethyl, nonequivalency of ring protons, chemical shifts of (^1H NMR), 17
Aziride derivatives, ^1H NMR chemical shift-dependence from substituents, 16
Aziride, 1-para tolyl, ^1H NMR chemical shift of ring protons, 17
Aziridine, 16—19
  ^{15}N shift of, 243
Aziridine, 1,2-diphenyl, ^1H NMR chemical shifts, 17—18
Aziridine, 1(*N*)-ethyl-2,3-dimethyl, temperature dependence of ^1H NMR spectrum, 16
Aziridine derivatives
  1-allyl-2,3-disubstituted, ranging to spin system, 18
  2-aryl, coupling constants, 18
  1,2,3-trisubstituted, temperature dependence of ^1H NMR spectra, 17
Azo, 61—67
Azo compounds, ^1H NMR chemical shifts of, 63
Azo-derivatives, 250
Azole derivatives, ^1H NMR chemical shifts of N-methyl signal, 90
Azoles, 245—247
Azulene, ^1H NMR chemical shifts, 60

B

BB excitation, 224
BBDR, 148, 209, 239, 259
  T_1 measurement and, 222
Benzaldazine, ^1H NMR chemical shift of azomethine group, 63
Benzaldehyde, *ortho-nitro*, spin system of ring protons, 71
Benzaldehyde derivatives, long-range couplings of, 62
Benzene
  AB- or *AX*-like multiplet of asymmetrically *p*-disubstituted derivatives, 71—72
  difluoro-chloro-methyl-^{19}F NMR chemical shift, 263
  fluoro-
    ^{19}F NMR chemical shift, 260, 267
    like reference material, 262
  fused compounds
    ^1H NMR chemical shifts, 74
    interactions between protons belonging to different rings, 68, 74
  hexadeuterio causing anisotropic solvent effect, 47
  causing temperature solvent effect, 74
  hexafluoro-, like reference material, 262
  ^1H NMR chemical shift, 67
  2-methyl-4-hydroxy-chloro-, 189—190
  monosubstituted, ^{13}C NMR chemical shift, 187
  nitro-, 240
  O-dichloro, *AA'BB'* spin system and magnetic nonequivalency of ring protons, 71

para-chloro-mercapto, accidental isochrony of ring protons, 71

para-diethoxy, ¹H NMR chemical shift of aromatic signal, 71

para-dinitro, ¹H NMR chemical shift of ring protons, 71

pentafluoro-ethyl-, ¹⁹F NMR chemical shift, 263

substituted, relaxation mechanism of quaternary carbons in, 225

sulfonic acid fluoride-, ¹⁹F NMR chemical shift, 262

trifluoro-methyl, ¹⁹F NMR chemical shift, 262—263

Benzene derivatives (aromatic compounds)
coupling constants, 68
¹H NMR chemical shift, 67—68, 71
monosubstituted, ¹H NMR chemical shifts and spin systems, 67

Benzene ring, ¹H NMR signals, 130

Benzenesulfonic amide, *meta*-dinitro, spin system of ring protons, 71

Benzofuran, ¹H NMR chemical shifts, 88

Benzofuran oxime 2-aceto, ¹H NMR spectrum, 64—65

Benzoic acid, fluoride-, ¹⁹F NMR chemical shift, 262

Benzonitrile, relaxation times in, 228

Benzophenone, 2,2′,4,4′-oxime, ¹H NMR spectrum of, 64

Benzothiophene, ¹H NMR chemical shifts, 88

Benzoxazole, ¹H NMR chemical shifts, 88

Benzpyrazole, ¹H NMR chemical shifts, 88

Benzthiazole, ¹H NMR chemical shifts, 88

Benzyl alcohol
association shift of, 99
in the presence of shift reagent, 133

Benzyl alcohol, α-trifluoromethyl, as chiralic NMR solvent, 114

Bicycloheptadiene, isochronous, 57

Bicycloheptan-1,4-diene, ¹H NMR spectrum, 58

Bicyclo[2.2.0]hexane, ¹H NMR chemical shift, 30

Binomial coefficients, 140

Biosyntheses
investigation of, 232—234
with use of ¹³C NMR, 232

Bipyridil, α,α′, ¹H NMR chemical shifts, 90

Bloch equations, 94

Boat conformation, 24, 91

Boltzmann distribution, 136

Boltzmann factor, 94

π-Bond order, linear correlation of, to ²*J*(C,C) and ³*J*(C,C) couplings, 220

Bond parameters, for the sp² carbons, 176

Borane adducts, 247

BPh₄, as anion in shift reagent, 127

Broad absorption, 30, 32

2-Bromocholestan-3-one, 234

Bromoform, heavy atom effect in, 156

1-Bromopropene-1, distinction of *Z–E* isomers by the help of ¹³C NMR, 177—178

Buckingham equation, 156

Bulky groups, 99

Bulvalene, valence isomerism in, studied by ¹³C DNMR, 236

Bulvalene and derivatives, valence isomerization, 57

Butadiene, 1,1,4,4-tetraphenyl, ¹H NMR chemical shift, 54

Butadiene derivatives, long-range coupling, 54

Butane
heptafluoro-1,1,1,trichloro–, ¹⁹F NMR chemical shift, 263
perfluoro-, ¹⁹F NMR chemical shift, 263
spectral parameters of rotamers of 2,2,3,3-tetrachloro-hexafluoro-, 264
threo and *erithro* isomers of 2-fluoro-3-halogen-*N*-butanes distinguished by ground of their fluorine resonance spectra, 264

Butane derivatives, 3,3-symmetrically disubstituted, conformational study, 8

Butene, F-F coupling constants of 2H-perfluoro-butene-2, 263

Buten-3-ine-1,(2)-1-methoxy ¹H NMR spectrum, 47—48

N,*N*-di-*n*-Butylformamide, 232

Butyronitrile, 4-phenoxy, ¹³C NMR spectrum, 121

Butyrophenone, 2,4,5-trihydroxy, ¹H NMR chemical shift of OH signal, 102

C

¹³C, Δ_c value of, relative to ¹H, 121

¹³C chemical shifts, see Chemical shifts or specific compounds

¹³C enriched samples, 219
of ethane, ethylene, acetylene and benzene, *J*(H,H) determination in, 210, 212

¹³C NMR
dimer of triphenylmethyl, structure identification of, carried out by, 234
formolysis of tosyloxymethylcyclopentane, investigation by, 233
incorporation of tryptophan into antibiotic, studying by, 233
investigation of reaction mechanisms and biosynthesis with the use of, 232
pK determined by, 191

¹³C NMR chemical shifts, see also Chemical shifts, 149, 158

¹³C NMR signals, their sensitivity of hydrogen bond formation, 183

¹³C NMR spectrum
3β-acetoxycolest-5-en
on different field strengths, 195, 198—199
in solid phase, 195, 198—199
cyclodecapeptide, 205
cyclopropane cation, proton coupled, 208
different temperatures, 237
maleic acid-ethylene oxide copolymer, 207
2-methyl-4-hydroxy-chlorobenzene, 188
phenols, 221
4-phenoxybutyronitrile, 121

polymethylmethacrylate, 207
proton-coupled or off-resonance, 209
range of, 148
terpenes, 201
testosterone, 195—196
^{13}C NMR technique, 127
^{13}C spectra, 2
^{13}C side-band technique, 68
"Cage", cyclodextrine or crown ether, 232
Calibration in nitrogen resonance spectroscopy, 239—240
Camphor, assignments in, by use of shift reagent, 131—132
Camphor monoxime, ^{1}H NMR chemical shift of OH signal, 104
Capture reactions, 137—138, 140
Carbohydrates, 38—43, 170—171
    aldohexoses, equilibrium of α- and β-pyranose and furanose anomers, 40—41
    assignment of ^{13}C signals in, by the help of deuteration, 159
    configurational and conformational analysis, 23, 38—39, 41
    coupling constants, 27, 38—39, 41
    empirical rules for structure identification, 38, 40
    ^{1}H NMR chemical shifts of *axial* and *equatorial* protons, 27, 39, 41
    ^{1}H NMR investigation of, 38—39, 41
    multicomponent system, computer method, 41
    open-chain derivatives, conformational investigation of, 41
    polyfunctional, as monodentate ligand, 128
    pyranose rings and, 171
    relative stability of conformers, temperature-dependent investigation, 40
    tetraacetyl-1-fluoro-hexapyranose, determination of conformation by ground of F-H coupling constants, 265—266
    virtual coupling, 41
Carbon
    atom, order of, 152
    hybridization state of, 152
    ionic, shifts for, 207
    quaternary, identification of lines on the basis of intensities, 228
    quaternary and protonated, 227
Carbon chemical shifts
    proportionality between the proton and, 191
    quinoline, as a function of pH, 191
    theoretical interpretation of, 148—159
Carbon dioxide, ^{13}C chemical shift of, 173
Carbon disulfide
    as external reference, 148
    relaxation mechanisms in, 226
Carbonium cations, 207—208
Carbonium ions, 208
Carbon monoxide, ^{13}C chemical shift of, 173
Carbon resonance intensities, 220—221
Carbon resonance spectroscopy, 146—237
    advantages of, 146, 148
    chemical shifts, theoretical interpretation of, 148—159

chemical shifts of various compounds in, 159—208
CIDNP effect studied in, 143
coupling constants, ^{13}C-^{1}H and ^{13}C-^{13}C, 208—220
dynamic, 234—237
investigation of reaction mechanisms and biosyntheses, 232—234
shift reagents, 234
spin-lattice relaxation in, 220—232
spin-spin interaction, 208—220
Carbon signals
    acetyl-acetone, metal complexes, 208
    polyacrylonitrile, 207
    quaternary, weak or not observable, intensification of, 227
    quaternary carbons, intensity of, 221
Carbon tetrachloride as NMR solvent, 67, 74, 107, 113
Carbonyl compounds, 180—185
    ^{13}C NMR chemical shifts, 182
    ^{17}O NMR chemical shifts, 260
Carbonyl derivatives, ^{1}H NMR chemical shifts, 2—3, 61
Carbonyl signal
    acetone, measured in different solvents, 158
    carbonate and hydrocarbonate anions, 185
    metal carbonyls, 185
    split
        in methyl-ethyl ketoxime, 183
        in *N*-methylformamide, 183
    thiocarbonyl derivatives, 185
Carboxylic acids
    determination of equilibrium constant of the monomer-dimer equilibrium, 104
    ^{1}H NMR chemical shift of OH signal, 104
    ^{1}H NMR investigation of association, 104
    recognizing of acid protons, using proton acceptor solvents, 104
Carboxylic amide, 243—244
Carboxylic amide, *N*-isopropyl-*N*-benzylmesityl, hindered rotation of, 107
CAT technique, 257
CH-OH splittings, 101
Chair conformation, 24, 26, 30, 38, 91
Charges, partial, alternating, 172
Chemical shift difference, 90, 93
    between α- and β-ring protons of heteroaromatic compounds, 76, 79—80, 83, 86, 89
    between *axial* and *equatorial* pairs of nuclei, in case of six-membered saturated cyclic compounds, 24—28, 31—32, 39—41, 43, 45
    dependence on the solvent, 74, 87
    geminal pairs of nuclei, 34, 75
Chemical shifts, see also specific compounds
    acetaldehyde, 3
    acetone, 3
    acetyl signal, 2—3
    acetylene, 152, 179—180
    acetylenic protons, 46
    acrylnitrite, 51
    adamantane, 30

alcohols, acetoxy derivatives, 167—168
aldehyde signal, 61
alkanes, 160
allene, 49
    bromo, 49
    monosubstituted derivatives, 179
    tetramethyl and tetraphenyl, 179
amines, 171
amino acids, 204
aminoboranes, silyl-, 243
androstan-7-one, 33
aniline, 3,4,5-trimethoxy and 3,4,5-trimethyl, 71
anisole, 2,6-dimethyl, 71
anisotropy, its frequency dependence, 225
aromatic and olefinic carbons, linear relationship
    between, 177
aromatic compounds, 186
azide ion, 244
aziridine, 243
    1,2-diphenyl, 17—18
azo compounds, 63
azoles, 245
azomethine group, 63
azulene, 60
benzaldazine, 63
benzene, 67
    fused compounds, 74
    monosubstituted, 187
    para-diethoxy, 71
    para-dinitro, 71
benzene derivatives, 67—68, 71
    monosubstituted, 67
bicyclo[2.2.0]hexane, 30
bipyridil, α,α', 90
^{13}C, calculated, for *racemic* and *meso* 2,4,6,8-te-
    tramethylnonane, 159, 161—162
^{13}C NMR, see also specific compounds under
    Chemical shifts, 149
^{13}C resonance, 148
calculation by empirical formulas, 5, 49, 78—80,
    86
carbohydrates, 27
carbon dioxide, 173
carbon monoxide, 173
carbon, theoretical interpretation of, 148—159
carbons, ionic, of Me_2C^+H, Me_3C^+, PH_2C^+H,
    and $^2Ph_3C^+$, 207
carbonyl carbon
    of di-*t*-butyl ketone, 181
    of *t*-butyl-*i*-propyl ketone, 181
of cyclic ketones, 181
carbonyl compounds, 182
carbonyl derivatives, 2—3, 61
changing of, effected by electronegativity, induc-
    tive and mesomeric effect of substituents, 1,
    3, 5, 13—14, 16, 21, 27, 30—31, 38, 40—
    41, 45, 49, 52, 63, 66—69, 74, 76, 79—80,
    86, 107—108, 113, 261—262, 266—268
cholestane, 6-halogenated derivatives, 33
citidine, uridine, anhydro-uridine, adenosine, and
    inosine, 202

compound types in N NMR spectra, 242
cyclic compounds, three-membered, 152
cycloalkanes, 161
    bi-, 164
cyclobutane, 20
cyclobutene, 56
cycloheptatriene, 56, 176
cyclohexane, 25, 27
    monosubstituted, 166
cyclohexane derivatives, 63
cyclohexanol, *cis* and *trans* 2- and 3-methyl, 169
cycloocta-1,5-diene, 56
cyclooctatetraene, 67
cycloolefins, 55, 57, 175
cyclopropane derivatives, 13, 37
    ring protons of, 13
cyclopropene, 14, 56
cyclopropene-3-one, 56
dependence from the measuring frequency, 95
diacetyl, 259
diazo derivatives, 63
diazoketones, 63
diethylamine, 16
dimethyl azide, 63
dioxane, 1,3-, 45
π-electron density and, relationship between, 191
enols and chelates, 101
ethane, 152
    monosubstituted derivatives (including ethyl
    halides), 74
ether, diisopropyl, 7
ethyl derivatives, 3—5
ethylene, 49—50, 152
ethylene derivatives, 49
    monosubstituted (vinyl derivatives), 49—50
^{19}F NMR, see specific compounds under Chemi-
    cal shifts
fluorine nuclei, 261—262
fluorine resonance, 262—263
fluorobenzene and fluorine substituted six-mem-
    bered nitrogen heterocycles, 267
formyl signal, 61
furan, 75—76
    2-acetyl-5-methyl, 79
germanium derivatives, 113
α-glucose, 43
^{1}H NMR, see specific compounds under Chemi-
    cal shifts
helicenes, 74
heteroaromatic compounds
    five-membered, 190
    fused, 195
    six-membered, 192
heterocycles, saturated, 174
n-hexane, 27
hydrazines, 243
hydrides, metal, 113
hydroxy groups, 99—105
11α-hydroxyprogesterone and 3-acetoxy-estrone,
    201
hydroxy signals, 100—101

imidazole derivatives, 245
indole, methyl derivatives, 81
internal rotation barrier of the C-N bond, linear
relationship between, 244
isocyanides, ethyl and cyclohexyl derivatives of,
sp carbon in, 173
isoprene, 54
isopropyl groups, 7—8
isoquinolines, 191
linear dependence on the size of the substituent,
177
malonester, acetamide and formamide derivative,
74
metal-organic compounds, 113
β-maltose, 43
methane, trinitro, 7
methane derivatives, halogenated, 167
methine signal, see Methine signal
methyl carbon atoms, 166
methyl-ketones, 181
methyl signal, see Methyl signal
methylene signal, see Methylene signal
mobile (acidic) protons, 46, 101, 105, 107, 109
neopentene methyl signal, 208
nitrocyclohexene, substituted carbon of, 173
nitrogen, see specific compounds under Chemical
shifts
nitromethane, 173, 240
NMR, see specific compounds under Chemical
shifts
^{17}O NMR, see specific compounds under Chemical shifts
olefins, 47, 49—50, 52, 175
oxethane, 20
oxirane, 14
oxirane derivatives, 14—15
oxygen nuclei, 259—261
oxygen resonance, 260
parts per million, 60
pH dependence, 81, 108—109
phenylglioxal aldoxime, 63
phosphorus nuclei, 270
phosphorus resonance, 268
picric acid, 71
piperidine, 43
N-acetyl and N-thioacetyl, 43
N-t-butyl, 43
piperidine derivatives, 43
piperidinium iodide, 1,1,2-trimethyl, 44
porphyrins, 91
propargyl alcohol, 47
purine, azido, 89
pyrane, tetrahydro, 43
pyrazole derivatives, 245
pyridine, 82—83
2-amino-5-bromo, 83
2-amino-4,6-dimethyl, 87
3,4-dicarboxyethyl-5-methoxy, 87
2-ethoxy-3,5-dinitro, 86
N-oxide chlorohydrate, 82
pyridine derivatives, 43

pyrrole
1,5-dimethyl-2-formyl-, 82
1-*para*-tolyl, 78, 80
1,2,5-trimethyl, 76
pyrrolidine, 22
reversed order, 61
ring carbons, 166
ring protons, 87
in five- and six-membered heterocycles and derivatives, 88
scales, 61
shift reagent, effect of, 127—130
silicon compounds, 113
simple carbonyl derivatives, 180
simple molecules, 241
some Me$_n$ M-type metal-organic compounds, 208
steroids, 27
α-halogenated keto derivatives, 33
stilbene sulfide, *cis*, 16
styrene oxide, 14—15
sulfide, diethyl, 16
talopyranose, β,–L–1,4-anhydro-6-desoxy-2,3-O-
isopropylidine, 42—43
tetrahydrofuran, 22
tetrahydropyrane, and 2-substituted derivatives,
38
tetraiodomethane, 156
thiazole and benzothiazole, 245
thiophene, 75—76
2,3-dibromo, 80
thiophene derivatives, 78—80
thirane
derivatives and, 16
phenyl, 16
1,2,4-triazole, 88
1,1,2-trichloroethylene and tetracloroethylene,
comparison of, 179
2,4,6-trinitrotoluene, 71
twistane, 30
variation of, with solvent, concentration and pH,
in nitrogen resonance, 239
vinyl acetate and vinyl-methyl ether, 177
vinyl protons, 51
water, 261
xylene, *ortho* and *meta*, 71
Chemically induced dynamic nuclear polarization
(CIDNP), 134—143
adiabatic nuclear polarization, 134—138
application of, 141—143
dynamic nuclear polarization, 134—138
effect, in ^{19}F NMR, 143
effect, of ^{31}P nuclei, 143
and the Kaptein rules, 140
multiplet effect in, 138—141
nuclear polarization, 237
observed in ^{13}C spectra, 143
organic chemistry, application in, 141—143
relative intensities, 138—141
sign of, 138—141
theoretical interpretations, 134—138
Chiral solvents, 132

Chloroform (CDCl₃) as NMR solvent, 81, 87, 89, 240
  ¹³C signal, solvent effect on, 158
  deuterio, as NMR solvent, 44, 47, 74, 79—80, 100, 102, 104, 107—108, 113
Cholestane, 6-halogenated derivatives
  ¹H NMR chemical shifts of 6*a* and 6*e* protons, 33
Cholestan-7-one
  6α- and 6β-chloro- and 6α- and 6β-fluoro-, chemical shift differences between *equatorial* and *axial* ring protons, 28
Cholestane
  3β–acetoxy–7α–hydrocy–5α–, 200
  methyl signals, 38
Cholest-5-one, 200
CIDNP, see Chemically induced dynamic nuclear polarization
Cinnoline, ¹H NMR chemical shifts, 88
Cis isomer, 12, 17—18, 20, 22—23, 26—28, 43—44, 165
Cis olefinic protons, 57, 68
Cis position, 39, 53
Cis protons, 14, 16—17
Cis substituents, 55
Cis vicinal coupling, 49
Cis vicinal hydrogen, 51
Cis vicinal proton, 51
Cisoid coupling, 53—54, 220
Coalescence method (line shape analysis), 97
Coalescence point, 94
Coalescence temperature, 81, 94—95, 97, 235
Coincident signals, 94
Collisional complex, 47
Complex concentration, correlation with induced shifts, 115—120
Complexation of impurities, 119
Complexes, structure of, 115—120
Computer analysis, 131
Computer methods in NMR spectroscopy, 41
Concentration dependence, 104
Configuration
  absolute, determination of, 133
  determination of, by NMR, 11, 14, 16, 21, 29, 33, 38, 44, 55, 114
Conformation, 230
  peptides, determination of, 218
Conformational analysis
  basis of ¹H NMR spectrum, 8, 11, 23—27, 29, 38—39, 41, 43, 54, 59, 72, 81, 90, 111, 114—115
  cyclobutane derivatives, 19
  cyclohexane derivatives, 23, 25, 30, 38
  cyclopentane derivatives, 21
  cyclophanes, 91
  DNMR, 235
  group of ¹⁹F NMR spectra, 264—265
  macrocyclic aromatic compounds (propeller and helical conformers), 73
  rotamers: constellation, see also s-*cis* and s-*trans* isomers, 53
Conformational equilibrium, 56, 170

cyclophanes, 91
furanophane, 91
Conformational investigation of thienophane, 91
Conformational motions, activation parameters of, determination of, by DNMR, 235
Conformers
  for the cyclohexapeptide (Pro-Pro-Gly)₂, 203, 205
  ratio of, 169
Conjugated dienes and polyenes, 54—55
Conjugation, 175, 181
Constants
  coupling, see Coupling constants
  values for different *a* and *b*, 159—160
Continuously conjugated cyclic polyenes, 59, 61
Contract term, 234
Coordination sites, competition of, 128
Copolymers, 9
Coronene spin system, 72
Correlation time, T꜀, 222, 225, 257
  isotropic, effective, 222
Coupling
  constant, see Constant couplings
  diaxial, 131
  electron spin orbitals, g, 137
  higher-order, CIDNP effect in the case of, 140
  ¹J(¹⁵N,¹³C), 256
  F-H, 261
  ¹⁹F-¹⁴N, 258
  first-order, 2
  ²J*ᵍᵉᵐ*-type, 130
  ¹⁴N isotope and, splitting due to, 251
  ¹⁵N-¹³C, sensitive to *syn-anti* isomerism, 255
  ¹⁴N-¹H, 258
    varied with the dihedral angle, 258
  ¹⁵N-¹⁵N, 256
  solvent dependent, 256
  "W arrangement", 264
Coupling constants, 116, 129, 131, 238—239, 267, 269
  acetylene, vinyl, 47
  allene derivatives, 49
  benzene derivatives (aromatic compounds), 68
  butadiene derivatives, 54
  ¹³C-¹³C, 208—210
  ¹³C-¹H, 208—209
  ¹³C resonance, 148
    determination of the relative sign, 210
  ¹³C-H and ¹³C-D, proportionality of, 216
  carbohydrates, 27, 38—39, 41
  cyclobutane derivatives, nonplanar, 19
  cyclobutene, 56
  cyclohexane derivatives, 27
  cyclopentane derivatives, 22
  cyclopropane derivatives, 14
  cyclopropene, 56
  cyclopropene-3-one, 56
  dependence from electronegativity, inductive and mesomeric effects of the substituent, 15, 19, 23, 34, 45, 52, 54, 67—68, 80
  dependence on conformation (temperature dependence), 53, 57, 101

different nuclei
  H-metal, 113
  H-N, 44, 81, 105
dioxane, 1, 3-, 45
dioxolane, 2,2,4,5-tetramethyl, 23
effect of vicinity of heteroatoms on, 216
ethane derivatives, 4—5
ether, vinyl-methyl, 52
ethylene derivatives, monosubstituted (vinyl de-
    rivatives), 50, 52
F-F, 264
F-H, 264, 267
furan, 81
furan derivatives, 81
H-P, 270
Hammett constants, linear relationship between,
    253
hydrides, metal, 113
hyperfine, 121, 143
hyperfine electron-nucleus, 140
hyperfine splitting, 137
indole, 2,3-dihydro, 92
inductive and mesomer effects of the substituents,
    15
sign, 15, 19
$^1J(C,C)$
  acetate ion, acetic acid and ethyl acetate, 220
  benzene derivatives, 220
  bicyclobutane, 219
  C_1-C_2, of cyclobutanone, -pentanone, -hexan-
    one, 220
  cyclopropane, methinyl, 219
  dicyclopropyl ketone, 219
  diphenylacetylene, 220
  iodo-, bromo- and chlorocyclopropane, 219
  $^1J(C,H)$, linear correlation between, 219
  pH dependence of, in amino acids, 220
  proportionality to the s-character, 219
$^nJ(C,C)$, 220
$^1J(C,H)$, 213—216
  in aldehydes, 216
  for aromatic and heteroaromatic compounds,
    217
  bicyclobutene, 213
  correlation between δC and, for acetylene and
    hydrocarbon derivatives, 214
  cyclopropenone, 214
  dependence of, on solvent, temperature and
    phase, 216
  determined from the proton resonance spectra,
    209
  "direct", 213
  ethane, ethylene and acetylene, 214
  for furan, pyrrole, and thiophene, 216
  Hammett or Taft constants and, linear relation-
    ship between, 214
  HCN, effect of protonation on, 216
  HCN and acetylene, 214
  methyl amine and ethane, 214
  order of the carbon atom and, 214
  relation of $^1J(C,H_a)$ and $^1J(C,H_e)$, 216

relation of $^1J(C,H)^{cis}$ and $^1J(C,H)^{trans}$ for cyclo-
    propanes, 216
relation of $^1J(C,H^o)$, $^1J(C,H^m)$ and $^1J(C,H^p)$, in
    benzene derivatives, 216
relation of $^1J(C,H)_Z$ and $^1J(C,H)_E$, for fluoro-
    ethylene, 216
relation to the s-character of carbon atom, 213
$^1J(^{13}C$-$^1H)$, 209
$^1J(C,X)$, 220
$^1J(N,H)$
  amide groups of dipeptides, 253
  as proof of tautomery, 254
  carboxylic amides and thioamides, 253
  conformational dependence of, in the case of
    N-acetyl-phenylhydrazine, 254
  extremely high, for the N-oxide of trimethyl-
    benzonitrile, 252—253
  for NH_4^+, $C_5H_5NH^+$ and $HCNH^+$ cations, 252
  protonated acetonitrile, aniline, isoquinoline,
    and benzonitrile, 252
  for pyrrole, 253
  for uracil, 253
$^1J(^{14}N,H)$, of ammonia and ammonium nitrate,
    258
$^2J(C,C)$, extremely high, for propionitrile, ace-
    tone and diphenylacetylene, 220
$^2J(C,H)$, 216—218
  for acetaldehyde and its mono-, di- and tri-
    chloro-derivatives, 217
  five-membered heteroaromatic rings, 217
  for iodo- bromo-, chloro-, and fluoroacetylene,
    217
  relationship of
  with the −I effect of substituents, 217
  with the order of the carbon, 217
  thiethane dioxide, 217
$^2J(N,H)$
  in naphthyl-aziridine, 255
  for protonated HCN, formamide and pyridine,
    254
$^3J(C,C)$, its dependence on dihedral angle, 220
$^3J(C,H)$, 218—220
  dependence on dihedral angle, 210
  −I effect of halogens, linear correlation be-
    tween, 218
  $^3J(H,H)$ of olefins, linear relationship between,
    218
  relation to $^2J(C,H)$, distinguishing of *ortho* and
    meta carbons of monsubstituted benzene de-
    rivatives, on the basis of, 218
$^3J(N,H)$
  characteristically different in *Z*– *E* isomers, 255
  dependence on dihedral angle, 255
  effect of quaternization and N-oxidation on,
    255—256
$^3J(^{14}N,H)$, in quaternary N-ethyl, N-isopropyl,
    and N-t-butyl derivatives, 258
$^3J(^{15}M,H)$, 254
$^4J(C,H)$, 219
  for dimethylacetylene, 219
lead, tetraethyl, 114

metal-organic compounds, 113
^{15}N isotope, negative sign of, 251
^{15}N-H, measured in ^{1}H NMR spectra, 259
^{15}N-X and ^{14}N-X, proportionality between, 251
nitrogen atoms, 251—259
olefinic hydrogens, 107
olefinic protons, 52
olefins, 50, 52
oxethane, 2,2-dideutero, 20
oxirane derivatives, 14
P-H, 269
piperidine derivatives, 43
propargyl alcohol, 47
pyridine, 83
pyridine derivatives, 87
pyrrole, 81
pyrrole derivatives, 81
reduced, 213, 251
relationships with dihedral angles, 100
scalar hyperfine, 121
shift reagent, effect of, 127—130
steroids, 27, 37
temperature dependence, 52
thiirane, and derivatives, 16
thiophene and derivatives, 81
vinyl formate, 53
vinyl protons, 52
Cr(ACAC)$_3$, 226
Cross-correlation, graphical, of the ^{1}H and ^{13}C
    NMR spectra, 209
Crotonic acid, fluoride, perfluoro-, ^{19}F NMR chemi-
    cal shifts and F-F coupling constants, 267
Crystallographic investigations, 126
CW
    instruments, 238
        determining ^{15}N-X couplings by. 251
    spectrometers, 145, 147
    spectrum, 259
        phenol, 146
    technique, 220, 261. 268
Cyclic compounds, see Saturated cyclic compounds
Cycloalkanes, 161—166
    ^{13}C NMR chemical shifts, 161
Cycloalkanes, bi-, ^{13}C NMR chemical shifts, 164
Cycloalkenones, increasing of coupling constant
    with the size of the ring, 57
Cyclobutanes, 19—21
    ^{1}H NMR chemical shift, and ring inversion, 20
Cyclobutane derivatives, nonplanar, ^{1}H NMR cou-
    pling constants, 19
Cyclobutane derivatives, 1,3-R-1', 3'-R' and 1,3'-
    R-1'-3-R-tetrasubstituted isomers spin sys-
    tems, 20
Cyclobutanol, (Z) and (E)-3-isopropyl, temperature
    dependence of ^{1}H NMR spectrum, 19
Cyclobutanone, carbonyl signals, 181
Cyclobutene, ^{1}H NMR chemical shifts and coupling
    constant, 56
Cycloheptanone, carbonyl signals, 181
Cycloheptatriene
    ^{13}C chemical shifts of, 176

^{1}H NMR chemical shifts, 56
Cycloheptatriene, dibenzo, inversion barrier of, 56
Cyclohexa-1,4-diene, long-range coupling, 59
Cyclohexane
    acetoxy derivatives, ^{1}H NMR chemical shift of
        methyl signal, 28
    t-butyl derivatives, conformational analysis, 29
    (4-t-butyl) halogenated, temperature dependent
        ^{13}C chemical shifts of, 167
    t-butyl-substituted derivatives, 165
    chemical shifts and coupling constants of cis and
        trans 4H-perfluoromethyl-, 265
    cis and trans 4-t-butyl-trifluoromethyl-, ^{19}F NMR
        chemical shift, 265
    cis 1,2-dimethyl
        ^{13}C DNMR investigation of, 235—236
        ^{13}C NMR investigation of, 165
    configurational and conformational analysis, 25,
        27
    dimethyl, investigation of, 28
    ^{1}H NMR chemical shift, 25, 27
    monohalogenides, ^{1}H NMR chemical shift differ-
        ence between equatorial and axial ring pro-
        tons, 28
    monosubstituted, ^{13}C NMR chemical shift, 166
    perfluoro
        configurational and conformational study of,
        265
        effect of substituting upon the shifts of fluorine
        in, 265
        temperature dependence of ^{19}F NMR spectra
        of, 265
    potential barrier of the ring inversion, 25
    temperature-dependent studies, 25
    1,1,3-trimethyl-, ^{13}C NMR investigation of, 165
Cyclohexane derivatives
    conformational analysis, 38
    coupling constants, 27
    dimethyl-, 235
    ^{1}H NMR chemical shifts, chemical shift differ-
        ences between equatorial and axial ring pro-
        tons, 27—29, 63
    monosubstituted conformation, 27, 29
        ^{1}H NMR chemical shifts of the geminal ring
        protons, 29
    partially deuterated compounds, spin systems of,
        27
Cyclohexane derivatives and their heteroanalogues,
    23—45
    carbohydrates, 38—43
    conformational analysis by ^{1}H NMR spectros-
        copy, 23—26
    cyclohexane derivatives, 27—30
    saturated six-membered heterocyclic compounds,
        43—45
    steroids, 30—38
Cyclohexanol
    acetoxy derivatives, δC- shift of, 170
    cis 4-t-butyl-^{1}H NMR spectrum, 117
        in the presence of Eu(DPM)$_3$, 117—118
    cis and trans-t-butyl, ^{1}H NMR chemical shift of
        OH signal, 100

cis and trans 2- and 3-methyl-, ^{13}C chemical shifts of, 169

dependence of the ^{19}F NMR chemical shifts upon the orientations of the trifluoroacetyl groups in such derivatives, 265

^{1}H NMR chemical shift difference between *equatorial* and *axial* ring protons, 28

^{1}H NMR chemical shifts of the *geminal* ring protons, 29

1-methyl-, acetylation of, 170

Cyclohexanol derivatives

^{13}C shifts of, 169

dependence of CH-OH coupling constants on conformation, 100

^{1}H NMR chemical shifts, chemical shift difference between *equatorial* and *axial* ring protons, 28, 63

^{1}H NMR chemical shifts of the *geminal* ring protons, 29

OH shift dependence on configuration and conformation, 100

Cyclohexanone, carbonyl signals, 181

Cyclohexanone-derivatives, ^{1}H NMR chemical shift difference between *equatorial* and *axial* ring protons, 28

Cyclohexil-halogenides, ^{1}H NMR chemical shift difference between *equatorial* and *axial* ring protons, 28

Cycloocta-1,5-diene, conformational equilibrium and ^{1}H NMR chemical shifts, 56

Cyclooctatetraene, ^{1}H NMR chemical shift, 67

Cyclooctatetraenes, 1,2-disubstituted, valence isomery, 57

Cyclooctatrienone, valence isomery, 57

Cycloolefins, 55—59

bi- and polycyclic, spin systems, ^{1}H NMR chemical shifts, 57

^{13}C NMR chemical shifts, 175

containing exocyclic double bonds, investigation of, 59

^{1}H NMR chemical shifts, 55, 57

olefin and long-range couplings, 57

racemization, 59

symmetric (C_2) *trans* isomers, 59

Cyclopentadiene and methyl trimethyl silyl cyclopentadiene, valence isomerization of π-complexes, 75

Cyclopentane

conformers, 21

perfluoro-, ^{19}F NMR chemical shift, 262

Cyclopentane derivatives

coupling constants, 22

pseudorotation, 21

Cyclopentane derivatives and their heteroanalogues, 21—23

Cyclopentane, iodo-, ^{1}H NMR chemical shift of *geminal* proton, 21

Cyclopentane, methyl, ^{1}H NMR chemical shift of *geminal* proton, 21

Cyclopentanols

cis and trans 2- and 3-methyl-, ^{13}C data of, 169—170

1-methyl, in comparison with its acetoxy derivative, 170

Cyclopentanone, carbonyl signals, 181

Cyclophanes, 91

Cyclopropane

^{1}H NMR chemical shift ring, 13

perfluoro-, ^{19}F NMR chemical shift, 267

Cyclopropane derivatives, 12—14

^{1}H NMR chemical shifts, 14, 37

^{1}H NMR chemical shifts of ring protons, 13

spin systems and coupling constants, 14

Cyclopropane ring, anisotropic effect, 18, 37

Cyclopropene

coupling constant, 56

^{1}H NMR chemical shifts, 14, 56

Cyclopropene derivatives, ^{1}H NMR chemical shifts ring, geometric structure, 14

Cyclopropene-3-one, ^{1}H NMR chemical shift and coupling constant, 56

Cytidine derivatives, conformational study, 91

D

δMe (19), 31

δ-scale, 148

Δ_c contribution, 121

Δ_c/Δ_p ratio, 122

$\Delta G\ddagger$, 235, 237

Δ_i, dependence of, on molecular geometry, 123

in *N*-disubstituted amids, 128

for oximes, *syn* and *anti* isomers, 128

Deshielding, 4, 14, 17, 46, 59, 61—62, 66—67, 75—76, 82, 87, 90, 98, 106, 108, 151, 156, 158, 170, 173, 197, 243—244

ring protons, 72

Deuterated compounds, selectively, model, 209

Deuteration, partial, 133, 216

Deuterium labeling, to distinguish carbon signals by double resonance, 233

Deuterium oxide, ^{17}O NMR chemical shift, 260

Diacetyl, ^{17}O NMR chemical shift, 259

Diamagnetic component of the shift, 121

Diamagnetic contributions, 150

of the shift, 121

Diamagnetic shielding, 2, 12, 14, 16, 20, 67, 73, 98, 262

carbonium ions, 152

contribution in, 239—240, 245

Diamagnetic shift, 18, 22, 47, 50—51, 73—74, 87, 124, 171, 192, 200, 202

Diamagnetic shift reagents, 117

α,ω-Diamines, additivity of substituent parameters in, 171

Diastereomeric pairs, *meso-racemic*, 133

Diastereomers (*meso-racem* and *erythro-threo* isomer pairs, anomers), 8, 30, 32—33, 114, 115, 131

Diastereotopic, 159, 161

Diastereotopic pairs of nuclei, 7, 14, 74, 114, 264

accidentally isochronous, 130—131

methylene signal of, 5
Diaxial couplings, 33—34, 37, 131
Diaxial interaction, 43
1,3-*Diaxial* interaction, 170, 231
Diazirine, 250
Diazo-derivatives, 250
  ^{1}H NMR chemical shifts, 63
Diazoketones, ^{1}H NMR chemical shifts, 63
d,d-Dicamphoryl-methane, 132
1,1-Dichloroethylene, ^{13}C chemical shift differences in, 179
1,2-Dichloroethylene, *Z* and *E*, *J*(H,H) in, 210
Dienes, conjugated, 54—55
Diequatorial substituents, 28
Diethylamine
  ^{1}H NMR chemical shifts, 16
  N-nitroso, ^{17}O NMR chemical shift, 260
  ^{17}O NMR chemical shift, 260
  perfluoro-, ^{19}F NMR chemical shift, 262
Diethyl malonate, T_{1C} of, 221
Difluorodisulfide, ^{19}F NMR chemical shift, 262
Dihedral angles, 8, 12, 14, 19, 21, 24, 38, 57, 92, 100, 124, 218, 220, 255, 258, 264
Dimerization, 119—120, 127
Dimerization constants, 100, 104
Dimethylacetamide, 2
2,3-Dimethylanisole, 188
Dimethyl azide, ^{1}H NMR chemical shifts, 63
Dimethyl formamide, perfluoro-, ^{19}F NMR chemical shift, 263
Dimethylnitrosamine, temperature dependence of ^{1}H NMR spectrum, 93—94
2,6-Dimethyloctane, as model compound to assign ^{13}C shifts for cholesterine side chain, 197
Dimethyl sulfoxide (DMSO), 133
  as polar solvent, 99
  ^{17}O NMR chemical shift, 260
Dimethyl sulfoxide (DMSO-d$_6$), hexadeuterio, as NMR solvent, 86, 89, 99—101, 107, 109, 111
Dinitrogen difluoride, ^{19}F NMR chemical shift, 262
Dioxane
  cis and *trans* 4,6-dimethyl-1,3, ^{13}C data of, 170
  2-methyl-1,3, 170
  cis and *trans* 2,4,4,6-tetramethyl-1,3-, 170
Dioxane, 1,3-, ^{1}H NMR chemical shifts and coupling constants, 45
β-Dioxo compounds, capable of complex coordination, 127
Dioxolane ring, ^{1}H NMR chemical shift of methylene group, 23
Dioxolane, 2,2,4,5-tetramethyl, spin system and coupling constant, 23
Diphenylketimine, ^{1}J(N,H) coupling constant of, 252
Dipolar interactions, 122
Dipolar shift, (Δ_p), 122—123
  R-dependence of, 132
Dipolar spin-lattice relaxation times, application in structure analysis, 227—230
Dipole-dipole interaction, 222, 225

between nuclear and electron spins, 213
between nuclear magnetic moment and electronic orbitals, 213
  ^{13}C resonance and, in peptides, 203
  ^{13}C-^{1}H, 227
  mechanism, 225
  T_{1C} and, 222
Dipole-dipole mechanism, 229
  ^{13}C-^{1}H, 227
  relaxation, 222—224
  dipole relaxation, 222—224, 226
Dipole-dipole term, 229, 259
  determination from partial NOE, 229
Dipole moments, 50, 67, 70
Dispersion curve, 146
Dispersion forces, intramolecular, 158—159
Dispersion signal, 259
Dissociation constant, 120, 128—129, 133
  complexes, 120
  different, for coordination centers, 127
  enantiomers, 133
  shift reagent complexes, 120
Distortion, extent of, 19
Dithianes, 1,3-, irregular behavior of *geminal* α-protons, 44
DMF, as reference, in nitrogen resonance, 240
DMFA
  ^{1}H NMR chemical shift of formyl signal, 61
  O-protonation, 67
DNMR spectroscopy, 234—235
DNP, 134—138
Double resonance (DR), 55, 111, 117, 210, 238, 251
  experiments, 117
  heteronuclear, 78, 83
  spin decoupling, 9
Doublet
  carbohydrates, 42—43
  merger into uncharacteristic maximum, 9
  meta position, 218
  methine signal, 11
  methyl group, 9
  2-methyl group, 44
  methyl group signal, 1
  methyl signal, 7, 109
  methylene signal, 11, 109
  olefinic protons, 49
  olefins, 53
  pyridine, 2-methoxy-6-bromo, 86
Downfield shifts, 78, 98, 101, 106, 111, 114, 117, 169—171, 177, 185, 201, 262
DPM, 127, 133
  complexes, 127
DSS, 148
Dy, Δ_c/Δ_r ratio of, in ^{13}C NMR, 122
Dynamic carbon resonance spectroscopy, 234—237
Dynamic NMR (DNMR), 234—235
Dynamic nuclear polarization (DNP), 134—138

E

α-Effect, 153
   in methyl-substituted norbornane derivatives, 164
β-Effect, 154, 161, 181, 242—244, 246—247,
      249—250
   in alkenes, 175
   in methyl-substituted norbornane derivatives, 164
   "transmitting" by oxygen atom, 167
   "transmitting" by σ–electrons, 180
γ-Effect, 154, 161
   in cyclobutanes, 165
   "transmitting" by oxygen atom, 167
Electron-buffering behavior of nitrogen atom, 79
Electron density, 67, 75, 82, 98, 102, 130, 150—
      152, 154, 158, 172, 177, 180, 245—246,
      252, 262, 266
   carbonium ions, 152
   linear relationship between shielding and, 150
   relative, 159
π-Electron density 52, 57, 68, 81, 151, 187, 191,
      203, 242
   chemical shifts and, relationship between, 191
   relationship between chemical shift and, for ionic
      aromatic systems, 185
s-Electron density, 213
Electron distribution, symmetry of (delocalization of
      electrons), 67, 75, 82, 87
Electronegativity, 152
   of the substituents, 243
      correlation between δC and the, 152
Electronic configuration, isotropy of, 122
Electronic states
   average excitation energy of, 151
   eigenfunctions of, 150
   mixing of ground and excited, 151
Electron-nucleus interaction, 129
Electron-repelling substituents, 12, 51, 78, 86
Electrons
   delocalization, 151, 161, 179, 187
   lone pair, 255
      replacing of, by a covalent bond, effect of, 249
   neighboring, 240
   spin-spin relaxation of, 121, 136
   unpaired, 120—121, 143
σ–Electrons, 68
   delocalization, 151
Electron-withdrawing carboxyl group, 111
Electron-withdrawing substituents, 12, 17—18, 46,
      49, 62, 67, 78, 86, 130, 216, 243, 255, 258
Electrostatic repulsion, 40
Empirical formulas, 252
Empirical rules, see Rules
Enamines, equilibrium between the N- and C-pro-
      tonated forms, 108
Enantiomers (optical isomers), 59, 114
   distinction of, 132—134
      by shift reagent, 132
Endo-axial hydrogens, 155
Endo isomer, 164

Energy levels, rotational-vibrational, 239
Enols and chelates, hydroxyl signal of
   ¹H NMR chemical shift of, 101—102
   ¹H NMR spectra, insensitivity to the variation of
      solvent, concentration and temperature, 101
Enrichment of ¹³C, 146
Envelope conformation, 21
Equatorial acetoxy group, 39, 170
Equatorial carbonyl signals, 235—236
Equatorial hydrogens, 24, 34
Equatorial α-hydrogens, 43
Equatorial hydroxyl group, 39, 169, 265
Equatorial methyl groups, 28—29, 154, 165
Equatorial (endo) 2-methyl group, 164
Equatorial position, 28, 38, 43
Equatorial protons, 25, 29
Equatorial ring hydrogens, 39
Equatorial ring protons, 27—28
Equatorial secondary hydroxyl groups, 40
Equatorial skeletal protons, 39
Equatorial substituents, 37
Equilibrium
   between the nonionic and zwitterionic form, 244
   tautomeric, 245
Equilibrium constant, 24, 26, 119
Equivalence, equivalent nuclei chemical, chemically
      (isochrony, isochronic), 16, 59, 266
   magnetic, magnetically (isogamy, isogamic), 59
Er, Δ_c/Δ_p ratio of, in ¹³C NMR, 122
Escape reactions, 137—138, 140—141
ESR, 140
   hyperfine interaction, 137
Ethane
   association shift of, 98
   bromo-1,1-difluoro-, ¹⁹F NMR chemical shift,
      263
   bromo-2,2,2-trifluoro-, ¹⁹F NMR chemical shift,
      263
   bromo-2,2-trifluoro-1-chloro-, ¹⁹F NMR chemical
      shift, 263
   bromo-2,2,2-trifluoro-1-iodo-, ¹⁹F NMR chemical
      shift, 263
   ¹³C chemical shifts of, 152
   1,1-difluoro-, ¹⁹F NMR chemical shift, 263
   1,1-difluoro-1-chloro-, ¹⁹F NMR chemical shift,
      263
   1,2-difluoro-tetrachloro-, ¹⁹F NMR chemical shift,
      262—263
      like reference substance, 262
   1,2-difluoro-tetrachloro-2-chloro-, ¹⁹F NMR
      chemical shift, 263
   1,2-diphenyl-tetrafluoro-, ¹⁹F NMR chemical
      shift, 263
   1,2-disubstituted, symmetrically, J(H,H) in, 210
   fluoro-, ¹⁹F NMR chemical shift, 263
   monosubstituted derivatives (including the ethyl
      halides), ¹H NMR chemical shift of, 3—4,
      74
   nitro-, ¹⁷O NMR chemical shift, 260
   pentafluoro-, ¹⁹F NMR chemical shift, 263
   pentafluoro-iodo-, ¹⁹F NMR chemical shift, 263
   1,1,1,2-tetrafluoro-, ¹⁹F NMR chemical shift, 263

1,1,1-trifluoro-, ^{19}F NMR chemical shift, 263
2,2,2-trifluoro-diazo-, ^{19}F NMR chemical shift,
   263
Ethane derivatives, ^{1}H NMR coupling constant of,
   4—5
Ethanol
  anhydrous, ^{1}H NMR spectrum, 95—97
  ^{1}H NMR spectrum, 96
    concentration dependence of the OH signal, 99
    dependence on the rate of exchange process,
     95, 97
  ^{17}O NMR chemical shift, 260
  perfluoro-, ^{19}F NMR chemical shift, 263
Ethers, 167—170
  ^{1}H NMR chemical shift of methylene signal, 5,
   14, 20
  diisopropyl, ^{1}H NMR chemical shift of methine
   signal, 7
  dimethyl-perfluoro-, ^{19}F NMR chemical shift, 263
  ethyl-vinyl-, ^{17}O NMR chemical shift, 260
  vinyl-methyl
   ^{1}H NMR chemical shift, 51
   ^{1}H NMR coupling constant, 52
Ethyl acetate, ^{1}H NMR chemical shifts of methylene
  signal, 5
Ethylamine, perfluoro-, ^{19}F NMR chemical shift,
  263
Ethylene
  association shift, 98
  ^{13}C chemical shifts of, 152
  α-carbonyl, the —M effect of carbonyl group, 52
  fluoro-, ^{19}F NMR chemical shift, 267
  ^{1}H NMR chemical shift, 49—50
  nitro-
   vinyl protons, 51
   ^{1}H NMR coupling constant mix of vinyl pro-
    tons, 52
Ethylene derivatives, see also Olefins
  ^{1}H NMR chemical shifts, 49
  monosubstituted (vinyl derivatives)
   calculation of ^{1}H NMR chemical shifts by addi-
    tivity rule, 49
   ^{1}H NMR chemical shifts, 49—51
   ^{1}H NMR coupling constants, 50, 52
   ranging to spin systems, 49
   temperature dependence of coupling constants,
    52
Ethylene-lithium, ^{1}H NMR coupling constants of
  olefinic protons, 51
Ethyl groups, 3—5
Ethyl signals, of acetamides, 74
Eu (DPM)$_3$, see Europium
Eu(FOD)$_3$, see Europium
Eu(FOD)$_3$ · (H$_2$O)$_n$, 134
Eu(NO$_3$)$_3$·6H$_2$O, as shift reagent, 127
Eu(TFC)$_3$, 133
Europium, 126, 129
  chelates, 121
  complexes, 126
  tris-dipivaloyl-methane (Eu(DPM)$_3$), 116—117,
   120, 122—124, 126, 129, 131

cis 3-(α-naphthyl)-cyclohexanol, in the presence of, 124
diacetone glucose with, 126
2,6-diisopropyl-acetanilide, hindered rotation
  of, examined in the presence of, 124
tris-1,1,1,2,2,3,3-heptafluoro-7,7-dimethyloctane-
  4,6-dione (Eu(FOD)$_3$), 117, 120—121, 128,
  133—135
  ^{13}C NMR spectrum of 4-phenoxybutyronitrile
   in the presence of, 121
tris- {[3-(1-hydroxy-2,2-dimethyl)-propylidene]-*d*-
  camphor}-, 132
tris-2,2,6,6-tetramethylheptane-3,5-dione
  (Eu(TMHD)$_3$), 117
Exchange, 193
  H → D, isotope effect of, in ^{13}C NMR, 159
  ligand, of metal-organic compounds, studied in
   carbon resonance, 235
  phenomena, 234, 238
   activation parameters of, 134
   in shift reagent complexes, 127
Exchange process, 78, 92—143, 234
  intermolecular, of mobile (acidic) protons, 43,
   81, 97, 101, 113
  rate of, 44, 81, 92, 97, 101, 105, 134
  temperature-dependent study, 92—93, 97
  time-dependent study, 95
  transition states of, 93
Excitation energies, 240
  average, 252
Excitation field, 145
  relationship to signal-to-noise ratio, 145
Eyring equation, 94

F

Fermi-contact
  interaction, or direct nuclear spin-electron spin interaction, 213
  shift, 121
  term, 121, 252, 256
Field effects, 161, 164—166, 179, 187—188, 197,
  200
  additive, 154
  in aldoses, 171
  in androstane, 17β-acetoxy-, 2α,3α- and 2β,3β-
   epoxy isomers, 200
  in 4-*t*-butyl-chlorocyclohexane, 155
  in 4-*t*-butyl-cyclohexanol, 155
  in *cis* decaline, 155
   in contrast to the *trans* isomer, 164
  as 1,3-*diaxial* interaction, 169, 195
  different shielding in anomers of glycosides, owing to, 171
  in *cis* 1,2- and 1,1-dimethylcyclohexane, 155
  in *trans*-1,2-dimethylcyclohexane, 154
  in dioxanes, 170
  electric, 156
  γ-effect cited as, 154
  in *cis* and *trans* 2-methylcyclohexanol, 167

in methyl substituted norbornane derivatives, 164

in *ortho*-dichlorobenzene, 188

in proline peptides, 203

steric compression shift, 154

in steroids, 200

in *cis-cis* 1,2,3-trimethylcyclohexane, 155

Field gradient, 257

Five-membered aromatic heterocycles and their derivatives, 76—82

Flexible systems, 155

^{1}F NMR chemical shifts, 261—262

^{19}F NMR

chemical shifts, see Chemical shifts

Δ_c value of, relative to ^{1}H, 121

parameters

aromatic and heteroaromatic compounds, 267—268

saturated cyclic compounds, 265—266

saturated open chain compounds, 262—265

unsaturated compounds, 266—267

resonance, CIDNP effect studied, in, 143

Fluorine

^{19}F NMR chemical shift, 262

Fluorine molecule, ^{19}F NMR chemical shift, 262

Fluorine resonance, 261—268

chemical shifts, 262—263

Fluorine resonance spectroscopy, 261—268

FOD, 127

complexes, 127

Formaldehyde, dimethyl acetal, ^{17}O NMR chemical shift, 260

Formic acid

halogenated derivatives, concentration dependence of OH signal, in ^{1}H NMR spectrum, 104

investigation of association, 104

methyl ester, ^{17}O NMR chemical shift, 260

^{17}O NMR chemical shift, 260

Four-membered saturated heterocycles, 19—21

Free energy, 40—41

Frequency

dependence of T_{1C}, 224

measuring, 113

consequences of the further increasing of, 224

increase of, as a tool for structural determination, 67, 108, 111

FT

instruments, 208, 238

method, 145

routine measuring of T_1, 221

spectrometers, 146, 209, 238, 258

RF pulses of, 220

spectrum, of phenol, 146

technique, 38, 113, 251, 258—259

NMR-spectrometers, 259

Fumarate, diethyl, ^{1}H NMR chemical shift of olefinic protons, 52

Furan

2-acetyl-5-methyl, ^{1}H NMR chemical shifts, 79

^{1}H NMR chemical shifts, 75—76

^{1}H NMR coupling constants, 81

^{1}H NMR spectrum, 76—77

^{1}H NMR spin system of ring protons, 76

^{17}O NMR chemical shift, 260

Furan derivatives

calculation of ^{1}H NMR chemical shifts by empirical formulas, 78

chemical shift ranges of ring protons, 79

coupling constants, 81

2,3-dihydro-, ^{1}H NMR chemical shifts, 75

Furancarboxylic acid, 5-nitro-2-, varying of the chemical shift difference with the solvent, 80

Furanophane, conformational equilibrium, 91

Furanose anomers, 40—41

Furfurol

^{1}H NMR spectrum, 81—82

long-range interaction, 81

Fused heterocyclic systems, 87—91

G

Gd, 127

Gd(DPM)$_3$, 227

Gd^{3+} complexes, 122

Gauche position, 254, 264

γ-*Gauche* conformation, 155

S-*Gauche* conformer, 53

Geminal fluorines, 264

Geminal hydrogens, 15, 34, 50—51, 63, 228—229

Geminal methyl group, 170

Geminal olefinic protons, 49

Geminal protons, 12, 16, 25, 29, 32, 49—50, 229

Geminal α-protons, 44

Geminal ring hydrogen, 27—28

Geminal ring protons, 12, 14, 25, 27, 38

Geminal vicinal coupling, 49

Germanium derivatives, ^{1}H NMR chemical shifts, 113

α-Glucose, ^{1}H NMR chemical shifts, 43

Glucose pentaacetate, α- and β-D, ^{1}H NMR chemical shift difference between *axial* and *equatorial* protons, 28

Glycine shifts, 109

Glycosides, anomers of, 171

Glycylalanine, pH dependence of ^{1}H NMR spectra, 111

Gradient ∂**E**/∂**r**, 156

Gramicidine-S, ^{13}C NMR spectrum, 206

g-tensor, 122

Guanine, ^{1}H NMR chemical shifts, 88

Gyromagnetic

constant, negative, 251

factor, of nitrogen isotopes, ratio of, determined by atom beam methods, 238

ratios,

negative, of ^{15}N, 259

nitrogen isotopes, with respect to hydrogen, 238

H

Half-band width, 257—258

Half-chair conformation, 24
Halogen derivatives, 166—167
Hammett constant, 67, 253, 267
Hammett's δ, 158, 182, 187, 214
Heavy atom effect, 156, 158, 167, 176, 179
    interpreted as anisotropic effect, 156
Heavy water (D₂O), as NMR solvent, 40, 44, 109
Helicenes, ¹H NMR chemical shifts and solvent effect, 74
Heteroaromatic compounds, 75—92
    analogues containing more nitrogens, 87—92
    five-membered aromatic heterocycles and their derivatives, 76—82
    fused, ¹³C NMR chemical shifts of, 195
    fused heterocyclic systems, 87—91
    long-range couplings, 92
    pyridine and its derivatives, 82—87
    quasiaromatic heterocyclic systems, 91—92
    six-membered, ¹³C NMR chemical shifts of, 192
Heteroaromatic ring, 43, 87
Heteroaromatic systems, 188—194
Heteroatoms, 44—45, 50, 53, 75—76, 87, 92, 121, 126, 128, 156, 176, 188
Heterocycles, 44—45
Heterocycles, saturated, 173
    ¹³C NMR chemical shifts, 174
Heteronuclear coupling, 76, 78
    amides and imines, 106
    amines and oximes, 104
    pyridine, 82—83
    pyrrole, 81
Hexamethylphosphoramide (HMPA), as NMR solvent, 107
n-Hexane, ¹H NMR chemical shift of methylene signal, 27
Hexane, 2,2,4-trimethyl, ¹H NMR spectrum, 9—10
Higher-order spectra, simplification into first order, 131
Hindered rotation, see Rotation
HMFA, 89
Ho, Δᵢ/Δᵣ ratio of, in ¹³C NMR, 122
Homonuclear couplings, 221
Homotropylium cation, chemical shift difference between methylene protons, 74
Hückel rule, 185
Hybridization state, 213
    carbon, 152, 159
    nitrogen, 252—253
Hydrazines, 243
Hydrides, metal, ¹H NMR chemical shifts and coupling constants, 113
¹H, Δᵢ value of, relative to other nuclei, 121
¹H spectra, 2
Hydrogen bonds, 97—99, 232, 238
    effect of, in ¹³C NMR, 158
Hydrogen chloride, association shift, 98
Hydrogen fluoride
    association shift, 98
    ¹⁹F NMR chemical shift, 262
Hydrogen iodide, association shift, 98
¹H NMR chemical shifts, see Chemical shifts

¹H NMR coupling constants, see Coupling constants
¹H NMR spectra, see also specific compounds, 156
Hydrogen peroxide, ¹⁷O NMR chemical shift, 260
Hydrogens attached to other atoms than
    carbon, 92—114
    exchange processes, 92—97
    hydrogen bonds, 97—99
    hydroxy groups, 99—105
    NH groups, 105—113
    protons attached to atoms other than C, O, N, or S and adjacent to magnetic nuclei, 113—114
    SH groups, 105—113
Hydroxylamine, perfluoro-5-methyl-tio-, ¹⁹F NMR chemical shift, 262
Hydroxy groups, 99—105
Hyperconjugation, 47, 51, 113—114, 155—156
Hypoxanthine, ¹H NMR chemical shifts, 88

I

I effect, 67—68
−*I* effect, 43, 46, 49—50, 52, 61, 67, 88, 108, 130, 153—154, 166, 171, 173, 177, 180—181, 191, 216, 243, 250, 253
    bromine, 158
    carbonyl compounds, 180—181
    hydroxy group, 169
    substituents, 152, 214, 244—245, 262, 269
Idose, D-, anomers, ¹H NMR investigation, 40
Imidaso [1,2a]-pyridine, ¹H NMR chemical shifts, 88
Imidaso [1,5a]-pyridine, ¹H NMR chemical shifts, 88
Imidaso [1,2a]-pyrinidine, ¹H NMR chemical shifts, 88
Imidazole
    diethyl, 246
    1,3-dimethyl-, 246
    effect of protonation in, 246
    ¹H NMR chemical shifts of ring protons, 88
    N-methyl, 245
    potassium salt of, 246
Imines, 61—67
    CH-NH *vicinal* and N-H heteronuclear couplings, 106
    geometric isomery of, 66
    N-oxides, 249
    sidnone, 250
Indole
    ¹H NMR chemical shifts, 88
    2,3-dihydro, ¹H NMR coupling constants, 92
    methyl derivatives, ¹H NMR chemical shifts, 81
Indolisine and 3-substituted derivatives, protonation, 90
Induced shifts
    components of, 120—126
    correlation with complex concentration, 115—120
Inductive effects, 156, 171, 187
Intensities, relative, of first-order multiplets, in CIDNP, 140

Intensity, ^{13}C resonance, 148

Interaction
C-H dipolar, 226
dipolar, 122
hyperfine, 137
nuclear-electron, 226
orbital and dipole, 213
through-space, 121, 264, 266

Internal molecular (conformational) motions
effect on the ^{1}H NMR chemical shifts (on the chemical equivalency), 16, 18, 56—57, 60
effect on the magnitude of coupling constants, 54
equilibrium constant, 24—26
internal rotation
free, 24, 54
hindered, 8, 21, 43, 54, 66, 90, 107
rotamers, 24
inversion
in compounds owing nitrogen and other atoms, having nonplanar valence bonds, 17—18, 43, 66, 107
ring, 18, 20, 24—26, 40, 43, 45, 56, 59, 91
rate constant, 94—95
rotation about the C-C single bonds, 59
temperature dependent, NMR investigations of, 24, 26, 92, 94
time dependent, NMR investigation of, 95

Internal ring currents, 59

Internal rotation, 257

Intramolecular (aromatic, internal, diamagnetic) ring currents in aromatic compounds, 61, 67, 73

Intramolecular paramagnetic ring current in antiaromatic systems, 61
in planar continually conjugated polyenes, 61

Intramolecular ring currents
in fused systems, 74
in heteroaromatic compounds, 75, 87
in quasi-aromatic systems, 74, 91

Inversion
in decalines, 155, 164
ring, see Ring inversion

Iodobenzene, ^1J(C,C) coupling in, 220

Ionic aromatic systems, 185

Ionization, potential, 158

IR spectroscopy, 49

IR spectrum, 46, 156, 179, 181

Isoborneol, 234

Isochrony, accidental, detection of, by shift reagent, 130—131

Isochrony, isochronic nuclei, 57, 59
accidental, accidentally, 42, 47, 49, 54, 71, 130

Isocyanate (MeNCO), ^{13}C NMR data of, 173

Isocyanides
^{13}C chemical shift of sp carbon in, 173

Isomerism, see Isomers, isomery

Isomers, isomery
anti and syn, 18, 63—64, 104, 107, 183
cis and trans
in salts and quaternary derivatives of ring systems containing nitrogen atom, 17, 22, 43—44, 107

in saturated compounds, 12, 17, 19—20, 23, 26, 30, 44
in unsaturated compounds, 52, 63, 66, 73, 107—108
S-cis and S-trans, 52
of ketimines, 183
methyl ester, 182
in unsaturated compounds, 266
valence, amide-tetrazole, 244
Z-E, 62, 66, 173, 177
^3J(N,H) characteristically different in, 255
ratio of, 207

Isonicotinate, ethyl, ^{1}H NMR spectrum, 83

Isonitrile (MeNC), ^{13}C NMR data of, 173

Isoprene, ^{1}H NMR chemical shift, 54

Isopropyl group
^{1}H NMR chemical shift, 7—8
multiplicity of ^{1}H NMR signal, 109

Isoquinoline, ^{1}H NMR chemical shifts, 88

Isothiocyanate (MeNCS), ^{13}C NMR data of, 173

Isotope effect, 129, 159, 193, 210, 230, 238
in ^{13}C NMR, 159

Isotope labeling
in 3-hydroxy-2-picoline amide, 192
in the study of reaction mechanisms and biosynthesis, 159

Isotopes
natural abundance of, 113—114, 256, 259, 261, 268
nitrogen, the complementary nature of, 238

Isotropic electron configuration, 122

Isotropic susceptibility, 122

Isotropy, 122

Isoxazole, 245

J

$J_{a,a}$, 27, 43
$J_{a,e}$, 27, 43
$\mathbf{J}^c$, 14, 16, 18, 22, 92
$^5J^c$, 57
J^{cis}, 52, 56, 62
$J_{e,e}$, 27, 43
J^{gem}, 14, 16, 22, 37, 52
J^{gem}_{AB}, 37
$J^{gem}_{2,2}$, 43
$^2J^{gem}$-type couplings, 130
J^t, 14, 16, 18, 22, 92
$^5J^t$, 57
J^{trans}, 52, 62
^4J allylic coupling, 108
^4J coupling, 113
^5J coupling, 57
^3J coupling constants, 44, 57, 61
^1J(C,C) coupling constant, see Coupling constants
^1J(C,H) coupling constant, see also Coupling constants, 213—217
$^1J(^{13}C-^1H)$ coupling constants, 209
^2J(C,H) coupling constant, see also Coupling constants, 216—218

^3J(C,H) coupling constant, 218—220
^1J(N,H) coupling constants, see also Coupling constants, 252—254
^nJ(N,H) coupling constants, see also Coupling constants, 254—256

K

Kaptein rules, 138, 140
Karplus equation, see Karplus relation
Karplus relation, 14—15, 18—19, 22, 25, 38, 92, 218, 220, 254
   modified, 218
Keto-enol equilibria, study of, 260
Ketones, 62
Ketones, α- and β-hydroxy, association of, 100
Kronecker delta, 150

L

Lactams, 244
Lactams, N-substituted and heteroanalogues, ^{1}H NMR investigation of, 59
Lamb formula, 150
Lanost-8-en-3-one, 2α- and 2β-bromo, ^{1}H NMR chemical shift difference between *equatorial* and *axial* ring protons, 28
Lanthanides, 116—117
La, 127
La^{3+} ion, 121
Lanthanum, shift measured for, 121
Lead, tetraethyl (J(Pb,H)), ^{1}H NMR coupling constants, 114
Lifetime
   mean, see Mean lifetime
   of radical pairs, 138
Ligands, polydentate, 128
Linear combinations of atomic orbitals-molecular orbitals, (LCAO-MO), 151
Line broadening, 116, 129, 145, 258, 261
   because of allyl coupling, 54
   because of dipole interactions, 113
   because of exchange processes, 93, 95, 97, 106
   because of heteronuclear spin-spin interaction, 18, 76, 78, 83, 105
   because of more viscous solutions, 113
   in case of quadrupole nuclei, 97, 106, 238
   interactions of aromatic protons with alpha hydrogens, 68
   viscous solutions, 113
Line narrowing condition, extreme, 222, 224, 226
Line width, 235, 251
   connection with height of signal, 93
Local field, 116
   induced by lanthanides, 132
Lone electron pair, 247, 255
   replaced by covalent bond, 249
   of sp^2 nitrogen, 242
Long paraffin chain compounds, 9—11

Long-range couplings, 1—2, 15, 46—47, 235, 261
   aldehyde, 62, 81
   aldehyde protons, 61
   allylic and homoallylic, 53
   benzaldehyde derivatives, 62
   butadiene derivatives, 54
   cyclohexa-1,4-diene, 59
   cycloolefins, 57
   heteroaromatic compounds, 92
   methylene signal of diastereotopic pairs, 5
   olefinic protons, 51
   oxirane, methyl, 14
   ''W arrangement'', 53
Long-range interaction, furfurol, 81
Low-temperature measurements, 165

M

Macromolecules, importance of carbon resonance spectroscopy, 194
Macropeptides, configuration of, analysis of, by ^{13}C NMR, 205
Magic angle spinning, 195, 198—199, 207
Magnetically equivalent hydrogens, 57
Magnetic anisotropy, carbonyl groups, 3
Magnetic field
   further increasing of, 224
   sensitivity enhancement effect of, 195
Magnetization, 223
Magnetic moments, 145, 261
   ^{13}C isotope, 146
   low, of ^{14}N isotope, 238
Magnetic nuclei, 145
Magnetic resonance, ^{1}H NMR, see Protons
Magnetic susceptibility
   components of, 122
   isotropy of, 122
Magnetogyric factor, 145
Magnets, superconducting, 145
Maleate, diethyl, ^{1}H NMR chemical shifts of olefinic protons, 52
Malonates, acyl, tautomeric equilibrium, 102
Malonester, acetamide and formamide derivative, ^{1}H NMR chemical shifts, 74
β-Maltose, ^{1}H NMR chemical shifts, 43
Mannopyranose derivatives, virtual coupling of 1,6-anhydro-2,3-O-isopropyliden-4-keto-α-D, 43
Mass spectrometry, 232
Matrix, charge density-bond order, 151
McConnel equation, 150
McConnel-Robertson equation, 122
MeCOR derivatives, 2
Me(18), 31
Mean lifetime, 118, 125, 135
   S state of a radical pair, 138
Medium effects, 92—143
M effect, 67—68
−M effect, 52, 67
+M effect, 180

Mercaptan derivatives, ¹H NMR chemical shift of
  SH signal, 113
Meso compounds, 8
Meso isomer, 131, 133
Meso structure, 131
Mesomeric effects, 158, 179, 268
  in carbonyl compounds, 180
Mesomeric structures, 173
  azide ion, 244
Mesomeric system, 250
Mesomery, 63, 253
  amide-, 248
Meta carbon atoms, 230
Meta hydrogens, 67—68
Meta position, 158, 182, 218, 221
Meta ring proton, 62
Meta splitting, 86
Metal atom, 122—123
  paramagneticity of, 121
Metal ion, paramagnetic, 120, 129
Metal-organic compounds, 207—208
  ¹H NMR chemical shifts and coupling constants,
    113
Methane di-*p*-bromobenzoyl, ¹H NMR chemical
    shift of OH signal, 101
  bromo-trifluoro-, ¹⁹F NMR chemical shift, 262
  chloro and fluoro derivatives, heavy atom effect
    in, 156
  dibromo-difluoro-, ¹⁹F NMR chemical shift, 262
  difluoro-, ¹⁹F NMR 4 chemical shift, 262
  difluoro-dichloro-, ¹⁹F NMR chemical shift, 262
  fluoro, ¹⁹F NMR chemical shift, 262
  fluoro-trichloro-
    ¹⁹F NMR chemical shift, 262
    like ¹⁹F NMR reference material, 262
  diphenyl-difluoro-, ¹⁹F NMR chemical shift, 263
  tetra-fluoro-
    ¹⁹F NMR chemical shift, 262—263
    like ¹⁹F NMR reference material, 262
  tetranitro-, 240
  tribromo-fluoro-, ¹⁹F NMR chemical shift, 262
  trifluoro-, ¹⁹F NMR chemical shift, 262—263
  trifluoro-chloro-, ¹⁹F NMR chemical shift, 262
Methane
  trifluoro-iodo-, ¹⁹F NMR chemical shift, 262
  trinitro–, ¹H NMR chemical shifts, 7
Methane derivatives, halogenated, ¹³C NMR chemi-
    cal shifts, 167
Methanol
  ¹⁷O NMR chemical shift, 260
  perfluoro-, ¹⁹F NMR chemical shift, 263
Methine groups, 5—9
Methine quartet, 63
Methine signal
  as uncharacteristic absorption maxima, 9, 30
  ¹H NMR chemical shifts, 1, 5—6
    calculation by additivity rule, 7
    of isopropyl and cyclohexyl halides in *geminal*
      position to the substituent, 4
  relative intensity (¹H NMR), 6
  splitting of, in polymers (¹H NMR), 9, 12

Methoxy, *para-*, 250
Methylamine, 240
  perfluoro-, ¹⁹F NMR chemical shift, 263
3-Methylbutane, 1-chloro- and 1-bromo, ¹³C NMR
    spectrum of, 156—157
Methylcyclohexane, 165
Methylcyclopentane, 165
Methyl doublet, 63
Methylene groups, 5
  ¹H NMR chemical shift of dioxolane ring, 23
Methylene protons, 4
Methylene signal
  of diastereotopic pairs, 5, 14
  of diastereotopic protons, 74
  ¹H NMR chemical shifts, 1, 5, 12, 14, 74
    allyl cyanide, 62
    calculation by Schoolery rule, 5
    ether, diethyl, 5, 14, 20
    ethyl acetate, 5
  splitting of
    due to H-M interaction, 114
    in polymers, 9, 12
  as uncharacteristic absorption maxima, 30, 114
Methyl formate, ¹H NMR chemical shift of formyl
    signal, 61
Methyl groups, 1—3
N-Methyl groups, 3
O-Methyl groups, 3
S-Methyl groups, 3
Methyl iodide
  as an example for CIDNP, 141
  in cyclohexane, solvent effect on its ¹³C signal,
    158
Methyloxethane, γ-effect in, 165
Methyl signal
  broadening of, due to allylic coupling, 54
  of *t*-butyl groups, 2, 9
  cholestane, 38
  in dimethylcyclohexanes, 28
  ¹H NMR chemical shifts, 1—2
    acetone acetaldehyde, acetate anion, acetic
      acid, acetyl chloride and methyl acetate, 181
    acetophenone, comparison with that of acetone,
      181
    in acyl and acetoxy groups, 2, 37, 74
    androstan, 30
    cyclohexane, acetoxy derivatives, 28
    in groups attached to sp² carbon atom, 1, 90
    in metal-organics, 114
    in *N*-methyl groups, 3, 90
    in *O*-methyl groups, 3, 22, 94
    in *S*-methyl groups, 3, 90
    in steroids, 33
  multiplicity of, 1, 114
Methyl signal
  as uncharacteristic absorption maxima, 9, 30, 114
O-Methyl signals, 66
S-Methyl signals, 66
Mobile (acidic) protons, 102
  ¹H NMR chemical shifts, 46, 101, 105, 107, 109

shape and chemical shifts, depending on experimental parameters, 43, 95, 105, 109

signal's
  collapsing with the solvent's signal, 101, 105, 107
  depending on conformation, 101
  disappearing, 75, 95, 107, 109
  selection of, by ASIS, addition of acid or D_2O, 97, 101
  selection of, by means of their temperature, concentration and pH dependence, 97, 101, 105
  splitting of, 75, 95, 97, 101, 105, 107, 109

Mobile hydrogens, 92

Molecular conformational motions
  freezing in of rotamers, 264
  internal rotation, rotamers, 264

Molecular motions
  application of shift reagents for studying, 134
  investigation of, 230—232
  studying by the help of shift reagents, 134

Mole fractions, 25, 94—95, 119

Molten state, 98

Monomers, 9

Multicomponent system, subtracting computer method, 41

Multiplet effect in CIDNP, 138—141

Multiplets, 2
  aziridine, 16, 18
  azulene, 60
  collapsing of, 129
  cyclohexane derivatives, 26
  ethyl groups, 114
  higher-order, separated into first order, by shift reagent, 131
  α-hydrogens, 38
  merger into uncharacteristic maximum, 9
  methyl group signal, 2
  olefinic protons, 56
  pyrrole, 2-formyl-, 82
  signal width, 34
  symmetrical, 116

Mutual steric position, 2, 52

N

Naphthalene
  perfluoro-, direct (*peri*) F-F coupling constant, 268
  spin system, 72

Narrowing condition, extreme, 259

Natural abundance
  ^{13}C isotope, 146
  isotopes, 113—114, 256, 259, 261, 268
  ^{14}N isotope, 238
  ^{15}N, 251
  sensitivity and, 145

Neighbor effect, 239
  atoms, effect of, 150
  electrons, 240

groups, anisotropy of the magnetic susceptibility of, 150

Nd, λ Δ_c/Δ_p ratio of, in ^{13}C NMR, 122

Neopentane, perfluoro-, ^{19}F NMR chemical shift, 262

Neutron diffraction, 24

Nicotinate, ethyl, ^1H NMR spectrum, 83

Nitrate ion, as internal reference, in aqueous solutions, in nitrogen resonance, 240

Nitric acid, 240

Nitriles, 61—67

Nitrobenzene, $^1J(C,C)$ coupling in, 220

Nitro compounds, 61—67

6-Nitrocoumarin, 130

Nitrocyclohexene, ^{13}C chemical shift of substituted carbon, 173

Nitro derivatives, 173, 249—250

NH groups, 105—113

$NH_4^{\oplus}$ ion, 238

Nitrogen, 238
  couplings with atoms other than protons, 256
  difluoride-chloride-, ^{19}F NMR chemical shift, 262
  fluoride-dichloride-, ^{19}F NMR chemical shift, 262
  isotopes, chemical shifts and coupling constants, linearity in, for, 238
  shifts
    comparison, of nitrobenzene, 1-nitropropene and 3-nitropropene, 249
    of nitromethane, 1,1-di- and 1,1,1-tri-, 249
    proportionality to the electronegativities of the substituents, 249
    tetranitromethane, 249
  trifluoride-, ^{19}F NMR chemical shift, 262
  unsaturated, 173—175

^{14}N, Δ_c value of, relative to ^1H, 121

^{14}N isotope, natural abundance of, 238

^{15}N, resonance, CIDNP effect studied in, 143

Nitrogen inversion, 43

N NMR chemical shifts, see Chemical shifts

Nitrogen resonance
  frequency of, the $\nu_{as}NO_2$ IR frequency, proportionality between, 250
  references in, 239—240

Nitrogen resonance spectroscopy, 238—259
  basic problems, 238—239
  calibration, 239—240
  chemical shifts, 239
    molecular structure and, relationships between, 240—251
  coupling constants, 251—259
  spin-spin interactions, 251—259

Nitromethane
  ^{13}C chemical shift of, 173
  as external reference in nitrogen resonance, 239

Nitrosamines, 250
  N,N-dimethyl- and, *N,N*-diethyl-, *Z-E* isomerism in, 173
  mesomery, hindered rotation of, 66

Nitrosobenzenes, 250

Nitroso compounds, 61—67

N-Nitroso-compounds, high frequency signals of, 250

Nitroso derivatives, 250—251
  nitrogen NMR spectra of, 250

NMR
  fluorine resonance, 264
  hydrogen resonance in, 266

NMR inactive (nonmagnetic) nuclei, 113
  chlorine, 97

NMR measurements, optical purity, 114—115

NMR spectra, temperature dependence, variable measurement, 16, 19—20, 24, 26, 40, 45, 55, 57, 60, 75, 90, 92—95, 101, 107

NMR spectrometers
  continuous T wave (CW) techniques, 261, 268
  FT technique, 259

NMR spectroscopy
  ¹³C spectra, 2
  ¹³C NMR spectra, 38, 49
  computer methods, 41, 97
  ¹⁹F NMR spectrum, 114
  FT techniques, 38
  ¹H spectra, 2, 9, 20, 23, 38, 46—47, 49, 57, 61, 99, 104—105, 109
  investigation of protons attached to atoms other than C, O, N, or S and adjacent to magnetic nuclei, 114
  spectrum, 8—9, 12, 18, 23, 26, 30, 38, 66, 88, 109, 111, 114—115

NMR time scale, 93, 118, 134, 254

NOE, 209, 220, 224, 230, 232, 239, 259
  as a consequence of BBDR, 221, 223
    in nitrogen resonance, 239
  dipole-dipole interaction and, 228
  elimination of, 227
  factor, 259
    measured with and without paramagnetic substance, 226
  frequency dependence, 224
  maximal, 224
  measurement, 221, 227
  negative, 227
  paramagnetic concentration and, 226

Nonequivalence, nonequivalent nuclei
  chemical (anisochrony, anisochronic nuclei), 11, 16, 109
  chemical temperature dependent, 16, 73
  magnetic (anisogamy, anisogamic nuclei), 8

Nonplanar conformations, 181

Norbornanes, 201

Nuclear-electron relaxation, 226—227

Nuclear magnetic resonance, see specific elements (nitrogen, oxygen, etc.)

Nuclear polarization, 134—138, 142—143
  ¹³C, 143
  dynamic, DNP, 136
  irregular, 136—137
  net, 138, 140

Nuclear relaxation, rate of, 140

Nuclear spin states, 137

Nucleosides, 202—203

¹H NMR investigation of, 43

Nucleotides, 202—203
  ¹H NMR investigation of, 43

O

Off-resonance spectrum, 179, 181, 209, 216
  1,3-butanediol, 209

Olefinic protons, 14, 91

Olefins, see also Ethylene and vinyl derivatives, 49—54
  ¹³C NMR chemical shifts, 175
  coupling constant, 50, 52
  fluoro- and perfluoro-, ¹⁹F NMR chemical shifts and H-F and F-F coupling constants, 267
  ¹H NMR chemical shifts, 47, 49—50, 52
    calculation of, by additivity rule, 49

Optical isomers, see also Enantiomers
  determination of, 132—134
    by shift reagent, 132
  NMR investigation of (NMR determination of optical purity), 59, 114—115, 261

Orbital expansion, 154

Organic chemistry, application of CIDNP in, 141—143

Organic complexing agents, 126—127

Ortho carbon, 187

Ortho carbon atoms, 230

Ortho couplings, 68, 86

Ortho hydrogens, 68, 87

Ortho position, 61, 67, 70, 86, 181, 218, 221

Ortho protons, 67

Ortho substituents, 188

Ortho substitution, 181—182

Overhauser effect, 136

Overlapping signals, 3, 5, 7, 30, 33—34, 87, 97, 113, 126—127, 228
  separation of, 130
  shift reagents to separate, 116, 130

Oxamine, 1,3-, 3(*N*)-methyl-, 255

1,4-Oxathianes, ¹H NMR chemical shifts of axial and equatorial protons, 44

Oxethane, ¹H NMR chemical shifts, 20

Oxethane, 2,2-dideutero, ¹H NMR coupling constants, 20

N-Oxides, 61—67

Oximes, 61—67, 249—250
  CH-NH *vicinal* and N-H heteronuclear couplings, 104
  ¹H NMR chemical shift of hydroxyl signal, 104
  syn-anti isomery, 63, 102

Oxirane, 14—15
  ¹H NMR chemical shift, 14
  methyl, long-range coupling, 14

Oxirane derivatives
  ¹H NMR chemical shifts, 14—15
    coupling constant and spin systems, 14, 17
  Δ_c value of, relative to ¹H, 121
  resonance, Δ_p/Δ_c values of, 122

¹⁷O NMR chemical shifts, see Chemical shifts

Oxygen, difluoride-, ^{19}F NMR chemical shift, 262
Oxygen resonance, 259—261
Oxygen resonance signal, 260
Oxygen resonance spectroscopy, 259—261
Oxytocine, 220 MHz ^{1}H NMR investigation and spectrum, 111, 112

P

π-p interactions, 143
Para carbon atom, 187
Para coupling, 68
[1,2]-Paracyclophane, 158
Para-dioxo[2,36] para dioxine, hexahydro, ^{1}H NMR chemical shifts of methine protons, 7—8
Paraffins, ^{1}H NMR investigation of, 9, 12
Para hydrogens, 68
Paramagnetic additives, 227
  concentration of, 226
Paramagnetic contributions, 150, 152
  shielding, 158
Paramagneticity of shift reagent, 129
Paramagnetic metal ion, 120, 129
Paramagnetic metal ion induced shift, 115—120
Paramagnetic molecular oxygen in solvents, 227
Paramagnetic ring currents, 61
Paramagnetic shielding, 246, 261—262
  contribution, 239—240
  p-electrons, 243
Paramagnetic shift, 7, 14, 18, 21, 27, 29—30, 32—33, 39, 47, 50, 63, 65, 101, 124, 126
Paramagnetic shift reagents, 117
Paramagnetic species, magnetic moment of, 226
Paramagnetic substance, 226—227
Paramagnetic term, σ^d, 150
Parameters, determination of thermodynamic parameters by temperature dependence of NMR spectra, 39, 41, 60, 93, 95, 108
Para position, 61, 182, 218, 258
Para protons, 67, 86
Partial deuteration, 42, 55, 108
p-character of C-C bonds, its effect on $^1J(C,C)$ couplings, 219
Pentade, 207
n-Pentadecane, ^{1}H NMR spectrum, 9—10
Pentane 3-methyl, ^{1}H NMR spectrum, 9—10
Pentine, 3-hydroxy-4-methyl, ^{1}H NMR sectrum, 7
Peptides, 203—207, 243—244
  -CONH- group in, distinguished by deuteration in, ^{13}C NMR, 159
  determination of optical purity, racemization, 114
  diastereomeric, distinguished by ^{13}C NMR, 205
  ^{1}H NMR investigation, 108, 111
  oligo, ^{1}H NMR investigation of, 109
  pH dependence of the spectra, 111
  poly, ^{1}H NMR investigation of, 111
Perilene, spin system, 72
Perkin-Elmer, NMR spectrum made by PE spectrometer, 269

Perturbation theory, of spin-spin interactions, developed by Ramsey, 213
pH dependence, 106, 108—109, 111, 220
  [13]C spectra, of amino acids and peptides, 203
  nitrogen spectra of peptides, 244
  quinoline, 193
Phenanthrene
  assignments in, proved by deuteration, 230
  direct (*peri*) F-F coupling constant of the 4,5-disubstituted-1,8-difluoro-, 268
  ^{1}H NMR spin system, 72
Phenol
  C_α and C_β signals of, 159
  ^{1}H NMR chemical shift of OH signal in various solvents, 99—100
Phenol, 2-amino-3-nitro, ^{1}H NMR spectra, 70—71
Phenol, o-chloro, intramolecular association, 100
2-Phenyl-butan-2-ol, 133
Phenylglioxal aldoxime, ^{1}H NMR chemical shift of azomethine group, 63
Phlalazine, ^{1}H NMR chemical shifts, 88
Phosphine, monoethyl-, dimethyl-, trimethyl-, ^{31}P NMR chemical shift, 268
  tri-(ethylmercapto-)-, ^{31}P NMR chemical shift, 268
  tri-methylmercapto-, ^{31}P NMR chemical shift, 268
  triphenyl
    like ^{31}P resonance reference material, 268
    ^{31}P NMR chemical shift, 268
  triphenylmercapto-, ^{31}P NMR chemical shift, 268
Phosphinite, ethyl-di-(trifluoromethyl)-, ^{31}P NMR spectrum, 269
Phosphinoxide
  trichloro-tio-, ^{31}P NMR chemical shift, 268
  triethyl, ^{31}P NMR chemical shift, 268
  tri-(ethylmercapto)-, ^{31}P NMR chemical shift, 268
  tri-(ethylmercapto)-tio-, ^{31}P NMR chemical shift, 268
  trimethyl-, ^{31}P NMR chemical shift, 268
  tri-(phenylmercapto)-tio-, ^{31}P NMR chemical shift, 268
  triphenyl-tio-, ^{31}P NMR chemical shift, 268
Phosphite, cyclic 2-(hydroxy-methyl)-1,3 propane diol-, H-P coupling constant, 268
[31]P
  Δ_c value of, relative to ^{1}H, 121
  resonance, 121
    CIDNP effect studied in, 143
    shift of, in the presence of shift reagents, 121
Phosphorus
  difluoride-chloride-, ^{31}P NMR chemical shift, 268
  fluoride-dichloride-, ^{31}P NMR chemical shift, 268
  tetra-, ^{31}P NMR chemical shift, 268
  tribromide-
    like ^{31}P NMR reference material, 268
    ^{31}P NMR chemical shift, 268
  trichloride-, ^{31}P NMR chemical shift, 268
  trifluoride-, ^{31}P NMR chemical shift, 268
  triiodide-, ^{31}P NMR chemical shift, 268
Phosphorus acid
  like ^{31}P NMR reference material, 268

tio-
^{31}P NMR chemical shift, 268
triphenyl ester-, ^{31}P NMR chemical shift, 268
trimethyl ester-, ^{31}P NMR chemical shift, 268
Phosphorus resonance, 268—270
chemical shifts, 268
Phosphorus resonance spectroscopy, 268—270
Picric acid, ^{1}H NMR chemical shift, 71
Piperidine
^{1}H NMR chemical shifts, 43
protonated forms and sales, geometric isomers
and inverses, dependent on the rate of proton
exchange, 44
ring and nitrogen inversion, 43
Piperidine, N-acetyl and N-thioacetyl, ^{1}H NMR
chemical shifts, 43
Piperidine, N-nitroso, ^{1}H NMR chemical shift difference between *equatorial* and *axial* pro-
tons, 66
Piperidine, N-t-butyl, ^{1}H NMR chemical shifts, 43
Piperidine derivatives, ^{1}H NMR chemical shifts and
coupling constants, 43
Piperidinium iodide, 1,1,2-trimethyl, ^{1}H NMR
chemical shift, 44
pK, 108
determined by ^{13}C NMR, 191
Planar conformations, 181
Planar structure, 57, 59
Planar transition state, 56
Planck constant, 94
Polarizability, 158
atom-atom, 213
Polarization, 180
alternating, 153—154
carbonyl bond, 158
negative, 260
Polar solvents, 87, 99, 104, 106, 158
Polybutadiene, 1,4- and 1,2- ratio of, 207
Polyene chain, 56
Polyenes
conjugated, 54—55
determination of configuration, 54
Polymerization, 9
Polymers, 9—11, 207
configurations and conformation of the chain, 9
^{1}H NMR investigation, 9, 12
Polyvinylchloride, usual ^{1}H NMR and DR spec-
trum, 11
Population, rotamer, 130
Porphyrins, ^{1}H NMR chemical shifts, 91
ppm (parts per million) unit, 60
Pr, Δ_c/Δ_p ratio of, in ^{13}C NMR, 122
Pr(DPM)$_3$, 121—123
Pr(FOD)$_3$, 121
Pr(HFC)$_3$, 133
Praseodymium, 126
chelates, 121
complexes, 126
Precession, 137
Product, captured, 138

Progesterone, 11α- and 11β-hydroxy-, ^{1}H NMR
chemical shift differences between *equatorial*
and *axial* ring protons, 28
Proline, cycloglycyl, ^{1}H NMR shift of NH signal,
111
Propane
2-bromo-perfluoro-, ^{19}F NMR chemical shift, 260
cyclic phosphite, 1,3-dihydroxy-2-(hydroxy-
methyl)-, ^{31}P NMR spectrum, 268
1,2-dibromo-2-phenyl, long-range coupling, 1
2-fluoro-2-methyl-, ^{19}F NMR chemical shift, 263
heptafluoro-1-iodo, heptafluoro-2-iodo-, hepta-
fluoro-1-chloro-, ^{19}F NMR chemical shift,
263
1,1,1,2,2,3-hexafluoro-, 1,1,1,3,3,3-hexafluoro-,
1,1,1,2,2,3,3-heptafluoro-, 1,1,1,2,3,3,3-
heptafluoro-, perfluoro-, ^{19}F NMR chemical
shift, 263
^{1}H NMR chemical shift of methylene signal, 12
2-phenyl-perfluoro-, ^{19}F NMR chemical shift, 263
tiol-2, perfluoro-, ^{19}F NMR chemical shift, 262
Propargyl alcohol, ^{1}H NMR chemical shifts and
coupling constants, 47
Propene
2,3-dibromo, ^{1}H NMR spectrum, 53
effect of hyperconjugation, 51
(Z),(E)-1-phenyl-1-fluoro-, F-H coupling constants, 267
2-Propenyl acetate
^{1}H NMR spectrum, and broadened olefinic signal,
because of allylic coupling, 54—55
Propine-1,3-cyano, ^{1}H NMR chemical shift of acetylene signal, 46
Propionic acid
fluoride, ^{17}O NMR chemical shift, 260
α,α,β,β-pentafluoro-, ^{19}F NMR chemical shift,
263
Proportionality
constants, 252
inverse, between T_{1c}^{DD} and τ_c, 227
Propyl nitrite, ^{17}O NMR chemical shift, 260
Protic solvents, 158
Protoberberine derivatives, *cis⇌ trans* isomerization
in, 201
Protonated isomers, tautomers, 44, 81, 87
Protonation, 43—44, 81, 87, 172, 191, 216, 243
determination of salt-base ratio, 108
effect of
in ^{13}C NMR, 158
in imidazole and pyrrole, 191
in nitrogen resonance, 246
for pyrazole, 191
protonated isomers, tautomers, 17, 22, 62, 67,
107—108
of purine, determined by ^{13}C NMR, 191
Proton coupled spectrum, 179, 181, 209, 216, 232
Proton decoupling, 221
broad-bend (BB), 209
Proton exchange, 43—44
Proton magnetic resonance (^{1}H NMR) spectroscopy,
see Protons

Proton resonance, 122, 127, 130
Protons, see also specific topics, 1—143
  acidic, exchanging to deuterium, 229
  aldehydes, 61—67
  amides, 61—67
  aromatic compounds, 67—75
  attached to atoms other than C, O, N, or S and
    adjacent to magnetic nuclei, 113—114
  azides, 61—67
  exchange processes, 92—143
  heteroaromatic compounds, 75—92
  imines, 61—67
  medium effects, 92—143
  nitriles, 61—67
  nitro compounds, 61—67
  nitroso compounds, 61—67
  N-oxides, 61—67
  oximes, 61—67
  saturated acyclic alkanes, 1—11
  saturated cyclic compounds, 11—45
  unsaturated compounds, 46—61
Pseudo-contact dipolar shift, 122
  interactions, 234
  shift, 121—122
Pseudorotation, 21, 169
Pulse angle, 227—228, 259
Pulse sequences, 259
Purine
  ^{1}H NMR chemical shifts of ring protons, 87—88
Purine, azido, chain-ring tautomery, ^{1}H NMR
  chemical shifts, 89
Purine derivative, rotation barrier of hindered rota-
  tion, 91
Pyrane, tetrahydro, ^{1}H NMR chemical shifts, 43
Pyranose anomers, 40—41
  ^{1}H NMR investigation, 40
Pyranoses, pentaacetyl, conformational equilibrium,
  38, 40
Pyrazine, ^{1}H NMR chemical shifts of ring protons,
  88
Pyrazole
  ^{1}H NMR chemical shifts of ring protons 87—88
  lithium salt of, 246
  N-methyl-, 246
Pyrazolo[1,5a]pyridine, ^{1}H NMR chemical shifts of
  ring protons, 87—88
Pyrene, spin system, 72
Pyridazine, ^{1}H NMR chemical shifts of ring pro-
  tons, 88—89
Pyridine, 82—87, 116
  AA'XX' multiplet of γ-substituted derivatives, us-
    ing AX-approximation, 83
  2-amino-5-bromo, ^{1}H NMR chemical shifts, 83
  2-amino-4,6-dimethyl, ^{1}H NMR chemical shifts,
    87
  2-chloro-5-nitro-, ^{1}H NMR spectrum, spin sys-
    tem, 86
  coupling constants, 83
  2,6-diamino, ^{1}H NMR spectrum, spin system,
    85—86

3,4-dicarboxyethyl-5-methoxy, ^{1}H NMR chemical
    shifts, 87
  3,5-dichloro, ^{1}H NMR spectrum, spin system,
    85—86
  differently substituted, ranging to spin system, 83
  2-ethoxy-3-bromo, ^{1}H NMR spectrum, spin sys-
    tem, 84—86
  2-ethoxy-3,5-dinitro, ^{1}H NMR chemical shifts, 86
  heteronuclear coupling, 82—83
  2-methoxy-6-bromo, ^{1}H NMR spectrum, spin sys-
    tem, 85—86
  N-oxide chlorohydrate, ^{1}H NMR chemical shifts,
    82
  solvent effect of, 86
    for recognizing the signals of acid protons, 105
  spin system and ^{1}H NMR chemical shifts, 82—83
Pyridine derivatives, 82—87
  coupling constant, 87
  empirical formula for estimation of ^{1}H NMR
    chemical shifts, 83
  ^{1}H NMR chemical shifts, 43
  substituent constants $ρ_i$ for the estimation of ^{1}H
    NMR chemical shifts, 83, 86
Pyridinium, iodide, N-ethyl, 258
Pyrido[3,2β]pyridine, long-range coupling, "W"
  type, 92
Pyrimidine
  ^{1}H NMR chemical shifts of ring protons, 87—88
  nitrogen shifts of, calculated, 248
Pyrrole, 245
  N-acyl, rotation barrier of hindered rotation, 90
  coupling constants, 81
  1,5-dimethyl-2-formyl-, spin-spin and ^{1}H NMR
    chemical shifts, 82
  2-formyl, interaction between the aldehyde and
    ring protons, 82
  heteronuclear coupling, 75—76, 81
  ^{1}H NMR spectrum, dependence on the solvent,
    75—76, 79
  lithium salt of, 246
  1-para-tolyl, ^{1}H NMR spectrum and chemical
    shifts, 78, 80
  1,2,5-trimethyl, ^{1}H NMR chemical shift, 76
Pyrrole derivatives
  coupling constants, 81
  empirical formula for estimation of ^{1}H NMR
    chemical shifts of ring protons, 79
  pH dependence, 81
  solvent effect, 80—81
Pyrrolidine, ^{1}H NMR chemical shifts, 22
Pyrrolidine derivatives, coupling constants and pH
  dependence of the ^{1}H NMR spectrum, 23
Pyrrolidines, symmetrically 2,5-disubstituted-N-ben-
  zyl, cis-trans isomery, 22
Pyrrolidium salts, cis-trans isomery, 22

Q

Quadrupole broadening, 239
Quadrupole moment, 145, 251, 256—257, 259, 268
  effect on spin-lattice relaxation, 97, 106
  of ^{14}N isotope, 238

Quadrupole relaxation, 238, 257
  of ^{14}N isotope, 256—259
Quantitative analysis, 40, 87, 102, 108, 114
  ratio of conformers, 264
Quantum chemical calculations, *ab initio* and sem-
    iempirical, 239
Quantum chemical methods, 151
  quantitative interpretation of nuclear shielding by,
    148
Quantum mechanics, 94
  application for determination of direction of mag-
    netic dipole, induced by aromatic and anti-
    aromatic ring currents, 61
Quantum state (energy and spin state) energy level,
    term, 97
  magnetic (spin state), averaging of, 97
  mean lifetime, 94—95, 97
Quartets
  acetone, 76
  aziridines, 18
  ethoxy groups, 4—5
  methylene, 109
Quasi-aromatic systems, 74
  heterocyclic, 91—92
Quasi-axial position, 74
Quaternization of *N*-oxidation, 255
Quinazoline, ^{1}H NMR chemical shifts, 88
Quinoline
  ^{1}H NMR chemical shifts, 88
  2- and 8-hydroxy–, nitrogen shifts of, 248
  pH dependence, 193
Quinoxalines
  CH $\rightleftharpoons$ NH tautomery, 89
  ^{1}H NMR chemical shifts, 88
Quintets
  methine, 11, 109
  methylene signal, 11

R

Racemic compounds, 8, 133
Racemic isomer, 161
Racemic solvent, 114
Racemic structure, 131
Racemization, 115, 205
  of peptides, 114
  by rotation about the C-C single bond, 59
Radical
  electron spins, adiabatic interaction of, 137
  pairs
    adiabatic interaction of, in solution, 137
    S and *T*, of benzyl and diphenyl-methyl, 142
Rare earth metal atoms, 126, 127
R-dependence, 132
  in borneol, with Pr(DPM)$_3$, 123
  of dipolar shift, 132
Reaction mechanisms
  investigation of, 232—234
    with use of ^{13}C NMR, 232

Reference, external, in nitrogen resonance, 239—
    240
Reference compounds
  external standard, 262, 268
  in ^{19}F NMR, 262
  internal standard, 261
  in ^{17}O NMR, 260
  in ^{31}P NMR, 268
Reichstein's steroid, structure identification (^{1}H
    NMR) of a derivative, obtained by micro-
    biological hydroxylation, 32
Relaxation
  chemical shift anisotropy contribution in, 225
  1,4-diphenyl-butadiene, SA contribution in, 225
  dipole-dipole, 222—224, 226
  electrons, 137
  electron spin, 136
  mechanisms, 259
    dipolar, 228
    scalar, 226
  nuclear, rates of, 140
  nuclear-electron, 226—227
  processes, fast, in ^{14}N nuclei, 238
  quadrupole, see Quadrupole relaxation
  quaternary carbons, contribution of SR mecha-
    nism to, 225
  rate, 129, 222
  reagents, 227
  slow, of methyl groups, in steroids, 31
  spin-lattice, see Spin-lattice relaxation
  spin-rotational mechanism of, 224
  spin-spin, of electrons, 121
  time, 129, 145, 251
    in *n*-decanol, 232
    difference of $T_{1C}^{DD}(H)$ and $T_{1C}^{DD}(D)$, 230
    dipolar, 231
    for fluorine nuclei, 261
    increasing with temperature, 258
    of ^{15}N nuclei, 258
    for oxygen nuclei, 260
    for phosphorus nuclei, 268
    short, 238, 256
    T_{1C}, of 2,2,4-trimethylpentane, 226
    T_{1C}^{DD}, its proportionality to the order of carbon
      atom, 222
    T_1, relation, for adamantane, 227
Resonance absorption, 93
Resonance signals, assignment of, 97
Retention, 43—44
RF, 221
  pulses, of FT spectrometers, 220
Rigid, fused-ring systems, 20
Rigid systems, 155
Ring anisotropy, 106
Ring currents, 150, 161
  aromatic, 158
Ring inversion, 19—20, 24—26, 43—44, 56, 97,
    155, 235
  cyclobutane, 20
  rate, 237

Ring protons, 12—18, 20, 23, 27, 29, 69, 71, 75,
    78, 81, 86, 91
  chemical shifts, 87
  deshielding, 72
Ring size, 175, 181, 213
Roof structure, 2
Rotamers, 41
  of amides, having different nitrogen-proton cou-
      pling constants, 255
  ^{13}C NMR detection of, of proline peptides, 203
Rotation
  hindered, 43, 53, 66, 97, 107, 124, 203
    in amines, 183
    internal, in benzaldehyde and *p*-nitroso-*N*,*N*-di-
        methyl-aniline, studied by ^{13}C DNMR, 236
    partial, 126
    presence of shift reagent, 128
  substituent, 53
Rotational barrier, 125
  C-N bond, 244
Rubber products, ^{13}C NMR investigation, 207
Rudimentary spectrum, 20
Rules, empirical, to calculate ^{13}C chemical shifts,
    161

S

Salt-base ratio, 108
Salt formation, 105
Salycylic aldehyde, ^{1}H NMR chemical shift of
    phenol group, 101
Sample
  holder, spherical, 239
  isotopically enriched, 251
Satellite lines, 210
Satellites, 210
Satellite signals, 232
Satellite spectroscopy, 209—210
  as a tool of determination of couplings between
      chemically equivalent protons, 210
Satellite spectrum, 68, 210, 259
  of acetaldehyde, partial decoupling of, 210—211
  subtraction of, 251
Satellite technique, difference, 259
Saturated acyclic alkanes, 1—11
  ethyl groups, 3—5
  long paraffin chain compounds, 9—11
  methine groups, 5—9
  methyl groups, 1—3
  methylene groups, 5
  polymers, 9—11
Saturated cyclic compounds, 11—45
  aziridines, 16—19
  cyclobutanes, 19—21
  cyclohexane derivatives and their heteroana-
      logues, 23—45
  cyclopentane derivatives and their heteroana-
      logues, 21—23
  cyclopropane derivatives, 12—14
  four-membered saturated heterocycles, 19—21

oxiranes, 14—15
thiiranes, 15—16
three-membered, ^{13}C chemical shifts of, 152
Saturated six-membered heterocyclic compounds,
    43—45
Saturation, 145, 221, 224, 238, 258—259
  partial, 220
Saturation technique, differential, 259
Scalar (SC) mechanism, 226
s-character, 213—214, 253
  couplings and, relationship between, 252
  ^1J(C,H) and, linear relationship between, 213
  ^2J(C,H), proportionality to, 217
  magnitude of ^1J(N,C), direct proportionality be-
      tween, 256
  nitrogen atom, 252
Schoolery rule, 5—6
Segmental motions, 222, 230—232
Self-association, 106
Sensitivity, 145, 146, 224
Septet, methine signal, 7
SH groups, 105—113
Shielding, see also Diamagnetic shielding; Paramag-
    netic shielding, 2, 62, 67, 74, 180, 187,
    261, 266—268
  anomalous, 60
  aromatic π-electron sextet, 37
  carbon, σ-contribution to, 191
  constant δ, 148
  decomposed into components, 150
  diamagnetic and paramagnetic contributions of, in
      nitrogen resonance, 239—240, 243, 245
  extreme, of carbon atoms in cyclopropane, 161
  factor, paramagnetic component of, 240
  increase in, 154
  linear relationship between electron density and,
      150
  local diamagnetic, 150
  lone pair of an sp^2 nitrogen, 242
  net, 158
  nuclear, quantitative interpretation of, 148
  olefinic protons, 52
  paramagnetic contribution, 158
  ring protons, 17
  tensor, 148
  three-membered rings, 56
Shielding anisotropy, 74
Shielding cone, diamagnetic, of carbonyl group, 33
Shielding constants, 243
  azoles, 246
  nitrogen isotopes, 238
  paramagnetic contribution of, 245
Shielding effect
  bridged structures, 37
  three-membered ring, 14, 16
Shifting compounds, converted by chemical meth-
    ods, 128
  of thioethers, 128
Shift reagents, 115—134
  accidental isochrony, detection of, 130—131

alcohols and their α-deuterated derivatives in the presence of, 129
aqueous solutions, 127
carbon resonance and, 234
chemical shifts, effect on, 127—130
chiral, 133
complexes
  exchange phenomena in, 127
  structure of, 115—120
components of induced shifts, 120—126
concentration, 126
correlation between complex concentration and induced shifts, 115—120
coupling constants of substrate, effect on, 127—130
decomposition, 127
diamagnetic, 117
dimerization of, the use of Eu(FOD)₃ and Eu(DPM)₃, 120
effect of its increasing amount, 118
enantiomers, distinction of, 132—134
enriched in ¹²C, 127
experiments, 120
fully deuterated analogues, 127
low amounts of, 120
magnetic susceptibility of, anisotropy of, 122
molecular motions, application for studying, 134
optically active, 127, 132—133
optical purity, determination of, 132—134
organic complexing agents, 126—127
overlapping signals, separation of, 130
paramagnetic, 117
paramagnetic metal ion induced shift, 115—120
rare earth metal atoms, 126—127
separation of overlapping NMR signals, 116
simplification of higher-order spectra into first order, 131
solution of assignment problems, 131—132
stereoisomers, distinction of, 132
structure elucidation, application of technique in, 130—134
substrate complexes, 119
substrate ratio, 119
technique, 115—116, 120, 124, 134, 234
  combined with DNMR method, 134
  first order analysis by, 115
  main advantage of, 124
ytterbium-based, 234
Shifts, see also Chemical shifts; Downfield shifts; Upfield shifts
anisotropy(SA) mechanism, 225—226
¹³C, of CN⁻, NCO⁻, NCS⁻ ions, 173
¹³C, of cyanides and isothiocyanates, compared to carbon dioxide, 173
carbons, ionic, 207
carbonyl and thiocarbonyl, linear relationship between, 185
diamagnetic, see Diamagnetic shifts
diamagnetic component of, 121
dipolar (Δ_p), 122—123
  R-dependence, 132

increasing with the order of carbon atom, 161
measured for the lanthanum, 121
measured for normal alcohols, in the presence of Eu(DPM)₃, 123
paramagnetic, see Paramagnetic shifts
pseudo-contact, 122
in pyridine-N-oxide, 192
shift-reagent induced, temperature dependence of, 129
Side-bands, 259
Signal broadening, 259
  caused by short relaxation times, due to quadrupole moment, 256
Signal doubling, 107
Signals, overlapping, see Overlapping signals
Signal shape analysis, 135
Signal-to-noise ratio, 143, 145, 239, 259
  of FT spectra, in the presence of T_1 reagents, 227
  in nitrogen resonance, 238
Silicon compounds, ¹H NMR chemical shifts, 113
Singlets
  acetoxy group, 37
  allene, 49
  annulenes, 60
  aromatic compounds, 72
  aziridines, 18
  benzyl group, 43
  bullvalene, 59
  t-butyl groups, 2—3, 9
  carbohydrates, 42
  characteristic type of signal, 3
  cyclobutane, 20
  cyclopentadiene, 75
  electron spin state, 137
  ethylene, 49
  ethylene oxide, 14
  hydroxy signal, 46
  3-methine groups, 64
  methine signal, 6—7
  methyl group signal, 1, 3, 30
  methylene protons, 131
  methylene signals, 16, 109
  olefin, 54
  quasiaromatic porphyrins, 91
Solution of assignment problems, 131—132
Solvation, 119
Solvent cage, 230
Solvent effect, 116, 158—159, 239—240, 242, 247, 249
  aldehyde protons, 62
  application to structure determination
    comparison of spectra made by using different solvents, 47, 63, 71, 75, 80, 86, 98—99, 104, 262
    lifting of apparent equivalency (isochrony), 47, 86
    selection of acidic protons' signals (adding D₂O and/or acid), 47, 97, 106
  in ¹³C NMR, 158—159
  on carbonyl signal of acetone, 183

coupling constant $^1J(N,H)$, in aniline derivatives, 253

helicenes, 74

induced by aromatic solvent, 50, 66, 74, 76, 87

for interpretation of complicated multiplets, 76

pyridine, 101

pyrrole derivatives, 80—81

reduced by aromatic solvents, 101

thiophene, 80

Solvents, see also specific types

    chiral, 59, 132, 261

    dependence, 210

    optically active, 59, 114, 132, 261

    optically inactive, 114

    racemic, 114

    shifts, in nitrogen resonance, 239

sp^2 character of cyclopropane carbons, 213

Specificity, 116

Spectra, simplification of higher-order, into first-order, by shift reagents, 131

Spin

    angular moment, 137

    electrons, 136

    quantum number, 145, 261

    states, singlet electron, 137

    transitions, combined electron-nuclear, 136

Spin-lattice relaxation, 97, 106, 226

    ^{13}C, times, 207

    carbon nuclei, 221—222

    carbon resonance, 220—232

    time, T_1, 220—221, 259

        carbon, 230

        dipolar, application in structure analysis, 227—230

Spin rotation (SR) interaction, 224—225

Spin rotation mechanism, 226

    free rotation, 231

    of relaxation, 225

Spin rotation term, 229

Spin-spin coupling, 6, 91, 95, 207—208, 234, 251

    OH protons, 100

Spin-spin interactions, 210—213

    first-order, 2, 116

    heteronuclear, 18, 44, 76, 78, 81, 83, 105—106, 113

    higher order, 116, 140

        simplified to first-order, by using shift reagents, 116

    nitrogen atoms, 251—259

    olefinic protons, 55

    proton-proton

        allyl, 54, 108

        axial, equatorial, diaxial, diequatorial, 28—29, 34, 38—39, 43

        cisoid and *transoid*, 53

        geminal, 21, 54

        long-range, 5

        through π-electron (unsaturated and conjugated bonds), 45, 47, 51, 59, 68, 80

        through more bonds, 59, 92

        through single electron pairs (heteroatoms), 54, 92

        transferred by d-electrons, 113

        vicinal, 2, 5, 7, 14, 18, 20—21, 23, 39—41, 45, 49, 51—52, 59, 62, 68, 108

        ''W'' type, 2, 15, 22, 54, 62, 81, 92

    transmitted by the electrons, 213

Spin-spin interaction (coupling constant), see also Coupling constant, 261, 270

    first-order, 83

    proton-proton

        geminal, 14, 18—19, 37, 43, 45, 49, 52

Spin-spin interaction Z and E (coupling constants)

    proton-proton

        axial, equatorial, diaxial, diequatorial, 27

        cis and *trans*, 15, 18, 21, 23, 27, 49, 52—53, 55, 59, 63, 92

        ''W'' type, 54

Spin-spin relaxation of electrons, 121

Spin-spin relaxation time, magnitude, 94

Spin-spin splitting, see also Spin-spin interaction

    lacking of

        due to fast exchange process, 95, 97, 105, 107

        in case of quadrupole nuclei, 97

    multiplicity of, first order [(n + 1) rule], 97

Spin systems

    A_2, 18, 43, 133

    A_4, 14

    $AA'BB'$, 14, 20, 71—72, 76, 83

    $A_2A_2'BB'$, 20

    $AA'BB'B''B'''$, 56

    $AA'BB'X$, 27

        simplified by shift reagent, 131

    $AA'BXX'$, 82—83

    $AA'XX'$, 14, 20, 22, 71, 76, 83

    $AA'X_3X'_3$, 8, 23

    AB, 5, 18, 22, 34, 40, 43, 56, 72, 78, 80, 130—131, 133

    AB-like, 72

    AB_2, 72

    AB_3, 47

    A_2B_2, 9

    A_2B_3, 114

    $ABA'XXEAA'BXX'$, 67

    $ABB'CC'$, 67

    $ABB'XX \equiv AA'BXX'$, 82—83

    ABC, 14, 49, 78

    ABC_2, 14

    $AB_2C \equiv ABC_2$, 14

    $ABCD$, 14, 71—72

    $ABCX$, 75

    $ABMNX$, 34

    $ABMX$, 14, 34

    ABX, 14, 27, 34, 49, 72, 78, 83, 210

    ABX_2, 14

    ABX_3, 72, 74

    $AMPX$, 71, 83

    AMX, 14, 34, 47, 49, 78, 83

    AMX_2, 53

    $A_2M_2X_4$, 57

    A_3MX, 210

AX, 71, 78, 82, 223
  NOE factor in, determination of, 223
  represented by C and H atoms, 223
AX-like, 72, 83
AX$_2$, 83
AX$_3$, 2
A$_2$X$_3$, 2
  higher-order coupling, 94
Spin temperature, infinite, 136
Splitting, 1, 5, 33—34, 41—42, 46, 49, 51, 53—
    54, 57, 62, 69, 81, 92, 95, 101, 108, 113—
    114, 129, 131, 159, 193, 197, 216, 218,
    235, 251, 255—256, 258
  ^{13}C-^{2}H, 209
  hyperfine, coupling constant of, 137
  ^{14}N-^{1}H coupling and, in ammonium ions, 257
Standard, external, 239
States, excited electronic, 179
Stereoisomers, distinction of, 132
Steric compression shift, see Field effect
Steric correction factors, 153
Steric effect, 177
Steric factors, 151
Steric hindrance, 126, 159, 231
Steric interactions, 152
Steric positions, 217
Steroids, 23, 30—38, 194—201, 228, 232
  anisotropic effect of carbonyl group, 32
  chemical shift difference of *axial* and *equatorial*
    protons, 27—28, 31—32
  coupling constants, 27, 37
  ^{1}H NMR chemical shifts in methyl signal, 33
  ^{1}H NMR chemical shifts of the ring proton *gem-*
    inal to the substituent, its use in structure
    elucidation, 32, 34
  NMR investigation, 38
  structure elucidation
    by solvent effect, 86
    by Zurcher's data, 31
Steroids, α-halogenated keto derivatives, ^{1}H NMR
  chemical shift of α-hydrogens, 33
E-Z Stilbene, 73
Stilbene sulfide, *cis*, ^{1}H NMR chemical shift, 16
Stoichiometry of the complexes, 120, 125
Structure elucidation, application of shift reagent
  technique in, 130—134
Strychnine, complete ^{13}C assignment of, 202
Styrene oxide, ^{1}H NMR chemical shifts and part of
  spectrum, 14—15
Substituent
  parameters
    for monosubstituted pyridine derivatives, 192
    for monosubstituted thiophenes, 191
Substituent constants
  pyridines, monosubstituted, 194
  thiophenes, monosubstituted, 190
Substituent effects, 86, 152—153, 161, 173, 177,
    179, 187, 200—201, 214, 218, 220
  additivity
    for benzene derivatives, 188
    ^{13}C resonance, 153

in α,ω-diols, ω-halo-alkanols and α,ω-dihalo-
  derivatives, 167
on ^1J(N,H), 253
Sulane, vinychloro, accidental isochrony of olefinic
  protons, 49
Sulfide, diethyl, ^{1}H NMR chemical shift, 16
Sulfur tetrafluoride, ^{19}F NMR chemical shift, 262
Supraconducting magnets, 224—225
Susceptibility, volume, 239
Susceptibility ellipsoid, anisotropic, of the shift re-
  agent, 125
Sweep, fast adiabatic, 146
Sylale, fluoro-, difluoro-, trifluoro-, perfluoro-, ^{19}F
  NMR chemical shift, 262
Symmetric isomers, 171
Symmetry, spherical, 258
Syn-anti isomers, 255—256
  oximes, 104
Syn-axial heteroatoms, 156
Syn-axial hydrogen atoms, 154
Syn-axial hydrogens, 171
Syn-axial interaction, 200
Syn hydroxy group, 217
Syn isomer, 185
Syn position, 164—165

T

T$_1$
  change on deuteration, 229
  benzonitrile, 228
  cholest-5-ene, 3β-chloro-, 228
  codeine, using to assign the carbon signals, 229
  in ferrocene derivatives, 231
  frequency dependent, 222, 225
  salicylic acid methyl ester, 231
  toluene and nitrobenzene, 230
  reagents, 259
  relation, of methyl groups, in 1,2,3-trimethylben-
    zene and dimethylformamide, 231
  values of aniline, 230
  values in *n*-butanol, 227
  values and protonation, 230
T$_1$ reagents, acetyl-acetone (CH$_2$AC$_2$; ACAC), metal
  complexes of, 208
T$_{1c}^{DD}$, determination of, 227
Taft constant, 49, 67, 214, 267
Talo-derivative, 42
Talopyranose, β-L-1,4-anhydro-6-desoxy-2,3-*O*-iso-
  propylidene, ^{1}H NMR chemical shifts and
  spectrum, 42—43
Tautomeric systems, 185
Tautomerism, between the nitroso and oxime
  groups, 250
Tautomers, relative amounts of, determined by nitrogen resonance, 248, 250
Tautomers, tautomery determination of equilibrium
  by ^{1}H NMR spectroscopy, 89, 102
investigation of, by NMR, 89, 102, 109
Tautomery

identification of, by nitrogen resonance, 242

of nitrogen-containing groups, investigation of, 248

of thiourea derivatives, ^{15}N NMR investigation of, 249

of tropolone, studied by ^{13}C DNMR, 237

Tb, Δ_c/Δ_p ratio of, in ^{13}C NMR, 122

Temperature dependence, 17, 19—20, 28, 41, 52, 56—57, 60, 75, 92, 101, 129, 167

of $\Delta G\ddagger$, 235

NMR spectra, measurement at different temperatures, 261, 264—265

Temperature-dependent NMR measurements, 234

Temperature gradient, 239

Testosterone, ^{13}C NMR spectrum of, 195—196

Tetrabromo-methane, heavy atom effect in, 156

Tetrahydrofuran, ^1H NMR chemical shifts, 22

Tetrahydrofuran, 2,5-dimethoxy-3,4-dibromo, configurational analysis, 22

Tetrahydrofuran derivatives, 2,5-dimethoxy, *cis* and *trans* isomers, 22

Tetrahydropyrane, and 2-substituted derivatives, ^1H NMR chemical shifts, 38

Tetraiodomethane, ^{13}C chemical shift of, 156

Tetrazine, nitrogen shifts of, calculated, 248

Tetrazole, trimethylene-, 246

Tetrazole, 3,5-dimethyl, ^1H NMR chemical shift of *N*-methyl signal, 90

Tetrazole, 1-ethyl- and 2-ethyl, ^1H NMR chemical shifts of ring protons, 88

TFM, 261

Thermal equilibrium, 136

Thiazole, 245

^1H NMR chemical shifts of ring protons, 88

Thienophane, conformational investigation of, 91

Thiiranes, 15—16

Thiirane, and derivatives, ^1H NMR chemical shifts and coupling constants, 16

Thiirane, phenyl, ^1H NMR chemical shifts, 16

Thioacetic acid, ^1H NMR chemical shift of SH signal, 113

Thioacids and thioenols, association shift, 113

Thioamides, 243—244

Thiocarbonyl signals

of thioacetamide and thioacetic acid, 185

of γ-thiolactams and five-membered cyclic thioureas, 185

β-Thioketo-thioesters, keto-enol equilibria of, 113

Thiophene

^1H NMR chemical shifts, 75—76

solvent effect, 80

spin system and ^1H NMR spectrum, 76—77

Thiophene, 2,3-dibromo, ^1H NMR chemical shifts, 80

Thiophene, 2,3-dihydro, ^1H NMR chemical shifts, 88

Thiophene and derivatives

coupling constants, 81

empirical formulas to calculate ^1H NMR chemical shifts, 78

^1H NMR chemical shifts, 78—80

spin systems of ring protons, 78

Time scale, see NMR time scale

Tin, tetramethyl, ^1H NMR investigation of, 113

TMS, 126, 148, 208

Toluene, first complete analysis of aromatic multiplet of, 67

Trans configuration, 55

Trans conformations, 8

S-trans conformer, 53

Trans-diaxial interaction, 43

Trans hydrogen, 51

Trans isomers, 12, 17—18, 20, 22—23, 26—29, 43—44, 59, 131, 164, 170

Trans methyl group, 13

Trans olefinic protons, 61

Trans position, 2, 55, 254, 264

Trans protons, 12, 14, 16—17, 61

Trans ring proton, 15

Trans vicinal coupling, 49

Transitions

$S \leftrightarrow T_0$, 137—138

$S \leftrightarrow T_{+1}$, 140

$\pi \rightarrow \pi^*$, 151

$\sigma \rightarrow \sigma^*$, 151

Transoid coupling, 53—54, 220

Triads, 11, 207

iso, syndio, and *heterotactic* and *meso* and *racemic*, 11—12

in polyacrylonitrile, 207

of poly-vinyl-chloride, 207

Triazole, 188

Triazole, 1-methyl-1,2,4-, 246

1,2,4-Triazole, ^1H NMR chemical shifts and tautomery, 88

1,1-Trichloroethane, as an example for CIDNP, 141

Trimethyl-amine, perfluoro-, ^{19}F NMR chemical shift, 263

Trimethyl carbamate, 134

2,4,6-Trinitrotoluene, ^1H NMR chemical shift, 71

Triphenylene, spin system, 72

1,1,2-Triphenylethane, 142

Triplets

aldehyde hydrogen, 62

carbohydrates, 42

electron spin states, three, 137

ethoxy groups, 4—5

furan, 76

merger into uncharacteristic maximum, 9

methine signal, 11

methyl group signal, 1, 9, 109

methylene group, 56

methylene signal, 11

ortho position, 218

para position, 218

pyridine, 2-methoxy-6-bromo, 86

splitting, 37

Twistane, ^1H NMR chemical shifts, 30

Twistboat conformation, 24, 30, 38

Twist conformation, 21—22, 24, 200

U

Universal gas constant, 94
Unpaired electron, 120—121, 143
Unsaturated compounds, 46—61
    acetylene, 46—49
    allene derivatives, 46—49
    conjugated dienes and polyenes, 54—55
    continuously conjugated cyclic polyenes, 59—61
    cycloolefins, 55—59
    olefins, 49—54
Upfield shifts, 78, 99, 107, 111, 114, 170—171,
    185, 191, 201, 260
    olefinic carbon, 179
    terminal carbon atoms, 179
Urea, N,O-diprotonated, structure of, 254
UV, 156, 181

V

Valence bond angle, 268
Valence isomerization, valence isomers, effect on
    chemical shift, 57, 60, 97
Valence isomery, valence isomers, 37, 57, 75
δ-Value, for nitrogen isotopes, direct proportionality
    between, 240
van der Waals' interactions, 158—159
Vicinal aliphatic protons, 68
Vicinal axial ring protons, 28
Vicinal coupling, 2, 7, 108
Vicinal coupling constants, 14, 20, 23, 38, 40, 52,
    62
Vicinal equatorial hydrogens, 29
Vicinal hydrogens, 3, 12, 21, 38, 50—51
Vicinal methyl groups, 55
Vicinal olefinic hydrogens, 48
Vicinal proton-proton interactions, 218
Vicinal protons, 14, 16, 50—51
Vicinal ring hydrogens, 14, 21
Vicinal ring protons, 12, 21
Vicinal spin-spin couplings, 45
Vinyl acetate
    ^{1}H NMR spectrum of, 49—50
    ^{1}H NMR spin system, 49
Vinyl bromide

    ^{1}H NMR chemical shift of vinyl protons, 51
    ^{1}H NMR coupling constant of vinyl protons, 52
    ^{1}H NMR spin system, 49
Vinyl chloride, ^{1}H NMR chemical shift of vinyl
    protons, 51—52
Vinylcyclopropane, ^{1}H NMR conformational study,
    53
Vinylfluoride, ^{1}H NMR chemical shift of vinyl pro-
    tons, 51—52
Vinyl formate, ^{1}H NMR coupling constants, 53
Virtual coupling, 41
Vitamin B_1, 229

W

"W" arrangement, see Spin-spin interaction, pro-
    ton-proton, "W" type
Water
    H-O coupling constant, 261
    ^{17}O NMR chemical shift, 261
Wavefunctions, 240

X

X-ray diffraction, 19, 24, 125
Xylene, *ortho* and *meta*, ^{1}H NMR chemical shifts,
    71

Y

Yb
    complexes of, 127
    Δ_c/Δ_p ratio of, in ^{13}C NMR, 122
Yb(TFN)$_3$, 234

Z

Z-E isomery, 106, 173—177
Zeeman effect, 129
Z–HC≡C–CH=CHOMe molecule, 47
Zurcher's additivity, 31—32, 34
Zwitterion structure, 111